Biologics, Biosimilars, and Biobetters

Biologics, Biosimilars, and Biobetters

An Introduction for Pharmacists, Physicians, and Other Health Practitioners

Edited by

Iqbal Ramzan
Sydney Pharmacy School, Faculty of Medicine and Health
The University of Sydney
Sydney, New South Wales, Australia

Registered Office
John Wiley & Sons, Inc., 111 River Street, Hoboken, NJ 07030, USA

Editorial Office
111 River Street, Hoboken, NJ 07030, USA

For details of our global editorial offices, customer services, and more information about Wiley products visit us at www.wiley.com.

Wiley also publishes its books in a variety of electronic formats and by print-on-demand. Some content that appears in standard print versions of this book may not be available in other formats.

Limit of Liability/Disclaimer of Warranty

Library of Congress Cataloging-in-Publication Data Applied for:

ISBN: 9781119564652

Cover Design: Wiley
Cover Image: Courtesy of David Hibbs. Image based on Protein Database ID (PDBID): 1IGY – Structure of Immunoglobulin, organism Mus musculus (Harris, *et al.* (1998); Crystallographic structure of an intact IgG1 monoclonal antibody at *J. Mol. Biol.* 275: 861–872), created with the Maestro program (Schrödinger LLC). Natural outdoors bokeh background in green and blue tones with sun rays © guvendemir/Getty Images

Set in 9.5/12.5pt STIXTwoText by SPi Global, Pondicherry, India

SKY10023597_011321

Contents

List of Contributors

Jeffry Adiwidjaja
Sydney Pharmacy School
Faculty of Medicine and Health
The University of Sydney
Sydney
New South Wales
Australia

Esteban Cruz
Sydney Pharmacy School
Faculty of Medicine and Health
The University of Sydney
Sydney
New South Wales
Australia

Hans C. Ebbers
Scientific Affairs Biosimilars
Biogen International GmbH
Baar
Switzerland

Paul W. Groundwater
Sydney Pharmacy School
Faculty of Medicine and Health
The University of Sydney
Sydney
New South Wales
Australia

Jane R. Hanrahan
Sydney Pharmacy School
Faculty of Medicine and Health
The University of Sydney
Sydney
New South Wales
Australia

Bryson A. Hawkins
Sydney Pharmacy School
Faculty of Medicine and Health
The University of Sydney
Sydney
New South Wales
Australia

David E. Hibbs
Sydney Pharmacy School
Faculty of Medicine and Health
The University of Sydney
Sydney
New South Wales
Australia

Kevin Huang
Department of Medicine
Harvard Medical School
Boston
MA
USA

Prateek Jain
Independent consultant
Toronto, Canada

Reza Kahlaee
Sydney Pharmacy School
Faculty of Medicine and Health
The University of Sydney
Sydney
New South Wales
Australia

Veysel Kayser
Sydney Pharmacy School
Faculty of Medicine and Health
The University of Sydney
Sydney
New South Wales
Australia

Felcia Lai
Sydney Pharmacy School
Faculty of Medicine and Health
The University of Sydney
Sydney
New South Wales
Australia

Christine Y. Lu
Department of Population Medicine
Harvard Medical School and Harvard
Pilgrim Health Care Institute
Boston
MA
USA

Hemant Malhotra
Department of Medical Oncology
Sri Ram Cancer Center
Mahatma Gandhi Medical College Hospital
Jaipur
India

Andrew J. McLachlan
Sydney Pharmacy School
Faculty of Medicine and Health
The University of Sydney
Sydney
New South Wales
Australia

Cody Midlam
Willis Towers Watson
Cincinnati OH
USA

Sanja Mirkov
School of Pharmacy
The University of Auckland
Auckland
New Zealand
Ramsay Pharmacy Services
Melbourne
Victoria
Australia

Catherine A. Panozzo
Department of Population Medicine
Harvard Medical School and Harvard Pilgrim
Health Care Institute
Boston MA
USA

Ankur Punia
Department of Medical Oncology
Sri Ram Cancer Center
Mahatma Gandhi Medical College Hospital
Jaipur
India

Iqbal Ramzan
Sydney Pharmacy School
Faculty of Medicine and Health
The University of Sydney
Sydney
New South Wales
Australia

Gregory Reardon
School of Pharmacy and Health Sciences
Keck Graduate Institute
Claremont CA
USA

Mouhamad Reslan
Sydney Pharmacy School
Faculty of Medicine and Health
The University of Sydney
Sydney
New South Wales
Australia

Johan Rosman
Medical School
Curtin University Perth
Western Australia
Australia

Mehmet Sen
Department of Biology and Biochemistry
University of Houston
Houston TX
USA

Michael Ward
Clinical and Health Sciences
University of South Australia
Adelaide South Australia
Australia

Lynn Weekes
Health Strategy and Sciences
Sydney New South Wales Australia
School of Pharmacy University of Queensland
St. Lucia Queensland
Australia

Foreword

Tracing back through history, one observes that the treatment of human disease, while always multimodal, has been strongly influenced, and even dominated, by select therapeutic strategies for discrete periods of time. Examples include the use of herbs to treat disease in prehistoric times (herbalism), bloodletting and humorism (starting around 500 years before the Christian era), germ theory, and chemotherapy (in the eighteenth and nineteenth centuries). Although the use of chemicals for medicine may be traced back to Paracelsus in the sixteenth century, pharmacotherapy with small molecule drugs (SMDs) did not dominate medicine until the twentieth century. At present, early in the twenty-first century, thousands of SMDs are in use as treatments for virtually all human diseases and conditions, including infectious disease, cardiovascular disease, mental health, pain, diabetes, and cancer.

The use of antibodies to treat disease may be traced to the 1890s, with the application of antisera to treat and prevent toxicity relating to diphtheria. Exogenous insulin was first used to treat diabetes in 1922, and human, recombinant insulin became available for therapeutic use in 1978. Building on these successes, and through advancements in the fields of protein chemistry, immunology, and molecular biology, we may now be entering a new phase where biological drugs, including peptides, proteins (e.g. antibodies), nucleic acid therapeutics (siRNA, antisense oligonucleotides, etc.), and cell therapies (T-cells, viruses, bacteriophages, etc.) emerge as dominant treatments for human disease.

At the time of writing this text in 2020, biologics account for more than 50% of new therapeutic entities under development at many major pharmaceutical companies, and monoclonal antibodies (mAbs) may be considered as the largest drug class (with ~75 mAbs approved for therapeutic use). Five of the current top 10 selling drugs are mAbs, including the top selling drug (adalimumab).

Relative to SMD, biologic drugs are often more selective in their actions, which translates to an improved ratio of beneficial effects relative to unwanted toxicity. However, biologics are much larger, and much more complex, than typical SMDs. An average mAb is associated with a molecular weight of ~150 000 Da, more than 30-times the average molecular weight of SMDs. Additionally, most biological drugs are not chemically synthesized, but are produced by biological systems (e.g. cells grown in bioreactors) that are subject to biological variability. Consequently, biological drugs may be most appropriately considered as complex distributions of molecular entities, rather than as unique chemical compositions. Variability exists within and between preparations of a biologic with regard to post-translational modifications (e.g. the extent and nature of glycosylation and sialylation), chemical modifications (e.g. deamidation and oxidation of labile functional groups), presence of aggregates, and the presence of host cell proteins (i.e. proteins relating to the cells used for production of the biologic). These and other product variables have been associated with significant effects on the pharmacokinetics, pharmacodynamics, and safety of the biologic product. As such, pharmacists, physicians, and other healthcare professionals have been faced with uncertainties regarding the safety and utility of preparations of biologics that are marketed as being "biosimilar" to an innovator biologic, or preparations that are developed as being superior to an innovator product (i.e. "biobetter").

This text is extremely timely in that it addresses many fundamental scientific, clinical, and regulatory issues relating to innovator biologics, biosimilars, and biobetters, through a thoughtful and detailed collection of 16 chapters. The text, which has been expertly compiled and edited by Dr. Iqbal Ramzan, provides discussion of the major classes of biological drugs, clear presentation of the terminology and nomenclature of the field, review of approved biosimilar and biobetter drugs, biophysical concepts and key biophysical analytical tests, pharmacokinetics, pharmacogenomics, pharmacovigilance, and pharmacoeconomics. The work provides a practical and clinical perspective to the use of biologics and biobetters, including consideration of controversial topics such as the interchangeability of innovator and biosimilar products. This book will serve as an excellent primer for all pharmacists and clinicians as we move forward into what may become a new era of medicine, an era dominated by the use of biological drugs.

Joseph P. Balthasar
University at Buffalo, Buffalo NY USA

Preface

This is a comprehensive primer, study guide, and primary reference text for pharmacists, doctors, and other health practitioners that presents the relevant science, clinical, policy, and regulatory frameworks for biologic medicines. The contents are pitched at a level that is easily understandable and can be immediately applied in everyday practice.

Innovator biologics, their interchangeable equivalents, biosimilars and their more efficacious, successors, biobetters are taking up a larger share of the therapeutics drug market compared to small molecule drugs. They are potent, highly complex in their therapeutic and clinical utility and far more expensive. Pharmacists are the primary healthcare professionals who will be expected to provide advice on these drugs as governments and other third-party payers attempt to contain their costs by introducing interchangeable biologic medicine products.

This book explores the current and emerging scientific and clinical practices. It compares different policy and regulatory approaches across countries. There is a focus on what pharmacists need to discuss with doctors and patients about the regulatory approval principles of biosimilars and evidence for interchangeability. Pharmacists and other clinicians require an understanding of the suite of biophysical tests needed to establish similarity, the likely efficacy, safety, and clinical risk(s) of switching not only from an innovator biologic to a biosimilar or a biobetter but also from any biologic medicine to another. Sound clinical and policy decisions will require health professionals to assimilate new types of information to ensure patients achieve optimal outcomes. This book will help them navigate this complex territory.

The book also provides recommendations for pharmacy educators and accreditors of pharmacy degree programs on the knowledge areas and competency standards to be met by pharmacy students and pharmacists on the entire burgeoning area of biologic medicines. Pragmatic regulatory approaches to dealing with these drugs in the context of rapidly evolving scientific and clinical data and evidence are also provided. A checklist is provided for pharmacists to facilitate conversations with doctors and patients to ensure quality use of medicine for biologic medicines to deliver patient-centered health outcomes.

Like many current health professionals, I had limited or no exposure to biologic medicines when I trained as a pharmacist. However, while serving as Dean of Pharmacy, at University of Sydney, for over 12 years, I had a bird's-eye view of the profession and of many future directions in healthcare. It was clear that pharmacists would be expected to take on a greater educative role with biologic medicines and I did not necessarily believe that they were sufficiently confident or knew enough about all aspects of biologic medicines. I therefore approached Jonathan Rose at Wiley and put forward a book proposal on biologic medicines. With his support, the proposal was approved after several iterations and I managed to assemble a very talented group of scientists and health professionals who were willing to share this journey with me.

Whether you are a pharmacist, a pharmacy student looking forward to entering professional practice, or a family doctor or specialist prescriber, I hope this book will empower you to understand the complexities of biologic medicines so that you can have an evidence-based and objective conversation with your patients. There is much hype and many anecdotes, and it is critical to separate these from the facts and data that support use of these important new medicines.

Editing this book (and writing two chapters) has been a very challenging task, probably because I underestimated the enormity of the challenge. The sheer breath of the scientific and clinical literature on biologic medicines is breathtaking. In addition, the literature and the evidence base are evolving so rapidly. If I had correctly gauged how much effort it would have taken me, I probably would not have embarked on this assignment. I am very pleased with the outcome largely due to the very able group of chapter contributors who have worked tirelessly with me to get the book pitched at the right level for pharmacists, doctors, and patients.

I want to thank all the contributing authors for their dedication to this book and to working with me to translate all aspects of the complex science to a level that is easily understood by busy time-poor pharmacists and doctors. My sincere thanks also go to the team at Wiley led by Jonathan Rose who has been very supportive from the beginning and Aruna Pragasam for assisting on the book submission.

I would also like to thank my wife, Dr. Lynn Weekes, who has been tremendously encouraging and supportive through this challenging project even though she herself wrote her own book during much of this time. Kimberlee and Justen, your encouragement to finish the project is also appreciated.

I dedicate this book to my late mum (Amma) who gave me such a strong work ethic and taught me perseverance.

Professor Iqbal Ramzan
Sydney Pharmacy School,
Faculty of Medicine and Health,
The University of Sydney,
Sydney, New South Wales, Australia
26 June 2020

1

Innovator Biologics, Biosimilars, and Biobetters

Terminology, Nomenclature, and Definitions

Iqbal Ramzan

Sydney Pharmacy School, Faculty of Medicine and Health, The University of Sydney, Sydney, New South Wales, Australia

---

KEY POINTS

- Many different terms are used for innovator biologics, biosimilars, and biobetters internationally.
- Greater harmonization of terminology, definitions, and nomenclature across different regulatory jurisdictions and countries would assist health practitioners and patients in understanding the complex issues of biologic medicines.
- The salient language of innovator biologics, biosimilars, and biobetters are introduced in this chapter to set the context for the rest of the book, which deals with specific issues in greater detail pitched at pharmacists, doctors, and patients.

Abbreviations

Abbreviation	Full name
ABPI	Association of British Pharmaceutical Industry
AfPA	Alliance for Patient Access
BAP	Biosimilars Action Plan
BLA	Biologics License Application
BPC	Biologics Prescribers Collaborative
BPCI Act	Biologics Price Competition and Innovation Act
CAR-T	Chimeric Antigen Receptor Therapy
CHMP	Committee for Medicinal Products for Human Use (EMA)
CIOMS	Council of International Organizations of Medical Sciences
CQAs	Critical Quality Attributes
CVMP	Committee for Medicinal Products for Veterinary Use
Da	Dalton
DNA	Deoxyribonucleic Acid
EMA	European Medicines Agency
EPAR	European Public Assessment Report
EU	European Union
FDA	Food and Drug Administration
GABi/GaBI	Generics and Biosimilars Initiative

Abbreviation	Full name
LDN	Limited Distribution Network
mAbs	monoclonal Antibodies
NHS	National Health Service
NICE	National Institute for Clinical Excellence
NMS	Non-Medical Switching
NOBs	Non-Original Biologics
PBS	Pharmaceutical Benefits Scheme
P&T	Pharmacy & Therapeutics (Committee)
PTMs	Post-Translational Modifications
QbD	Quality by Design
QUM	Quality Use of Medicine
RMP	Risk Management Plan
RPS	Reference Product Sponsor
RWE	Real-World Evidence
SEBs	Subsequent-Entry Biologics
SMD	Small Molecule Drug
UMC	Uppsala Monitoring Centre
UNESCO	United Nations Educational, Scientific and Cultural Organization
USA	United States of America
USD	US Dollar
WHO	World Health Organisation

1.1 Place of Biologics in Modern Therapeutics

Biologic therapies have entirely revolutionized the treatment of many debilitating, life-changing chronic autoimmune diseases like rheumatic arthritis and plaque psoriasis as well as life-threatening cancers for which no viable treatment option has existed previously. They also play a critical therapeutic role in many endocrine disorders and neurodegenerative conditions. Biologics are the fastest growing sector of the drug market[1] and are also the most expensive therapies. As a result, "highly similar" versions of innovator biologics, biosimilars have been introduced to provide cost-effective biologic treatments.

The first innovator biologic was introduced ~40 years ago, and the first biosimilar was introduced in the European Union (EU) and United States (USA) in 2006 and 2015, respectively. Currently, there are over 300 biologics registered worldwide and the EU has over 60 approved biosimilars. In the United States, biosimilars are an emerging market, with 19 approved biosimilars. Biosimilar market access comparison between the United States and EU has shown that market access in the United States is less favorable. This is due to many factors including lack of incentives to prescribe biosimilars in the United States and small price discounts of biosimilars compared to innovator biologics.[2]

In many countries, including Australia and emerging pharmaceutical markets like Brazil, biosimilar use is actively encouraged as governments attempt to contain the costs of expensive innovator biologics. Assuming discounts on off-patent innovator biologics and biosimilars of ~50%, it is predicted that by 2020 there will be annual savings of over €8–10 billion in the EU.[3]

Biologics are complex proteins or protein-like molecules produced using biotechnology techniques in living cells. Their structural, functional, and manufacturing processes lead to clinical concerns/controversy about their efficacy and safety, including the potential for treatment failure and severe immunogenicity reactions. Pharmacists, doctors, and other health professionals, therefore, need to be fully conversant with all aspects of their clinical utility.

1.2 Background to Terminology, Nomenclature, and Definitions

In the biologics field, there are international differences in the various terms, definitions, and abbreviations that are used. This arises due to country/continent differences, different regulatory and policy frameworks, and the specific requirements of the various regulatory agencies.

Definitions, nomenclature, and terminology on biologics will now be reviewed in detail so there is a common understanding among readers.

1.3 Innovator Biologics, Biosimilars, and Biobetters

1.3.1 What Is a Biologic Medicine?

Biologic medicines are active substances made by or derived from a biological source, rather than a chemical source, or synthesized chemically. Biologic medicines are also known as biopharmaceuticals or biotherapies and they are comprised of proteins such as vaccines, hormones, enzymes, blood products, allergenic extracts, monoclonal antibodies (mAbs), human cells and tissues, and gene therapies (Table 1.1). Typically, biologics are proteins or protein-containing fragments. The first biologic (recombinant human insulin) was approved in 1982.[4]

When a biological medicine is administered to a patient, the expectation is that it will function as the

Table 1.1 Broad categories of biologic medicines.

Biologic	Description
Hormone	A substance (peptide or steroid) produced by a tissue or organ to elicit a physiologic action
Vaccine	An agent containing an antigen (live, killed, or attenuated pathogenic agent) to stimulate the immune system
Interferons	Proteins produced by cells in response to bacterial or viral infections
Growth factors	A substance that promotes growth, especially cellular growth
Polypeptides	Peptides containing from 10 to 50 amino acids
Proteins	Naturally occurring or synthetic polypeptides generally of 10 kDa in size
Monoclonal antibodies (MAbs)	A single synthetic immunoglobulin produced by recombinant techniques directed against a single antigen or endogenous molecule
Interleukins	Group of cytokine proteins
Cellular and tissue biotherapies	Like CAR-T
Emerging biotherapies	Like antibody–drug conjugates or bispecific antibodies

natural endogenous protein, resolving clinical symptoms and either preventing or slowing the progression of the disease process. The mechanism(s) by which biologic medicines produce their clinical effects varies from product to product and across different clinical indications and diseases. Biologics may be tailor-made to target the desired receptor or cells in the body.

Terms like "de novo biologic drugs" or "bio-originators" have also been used in the biologics literature. The first (initial) biologic medicine belonging to a specific class or category to be approved and registered (and/or marketed) is known as an innovator biologic or the biologic reference product. The term originator biologic is also used.

Biotechnology techniques are increasingly associated with the production of most biologic medicines. Biotechnology is the application of bioengineering techniques to manipulate living organisms such as bacteria or yeasts or living cells, of bacterial, animal, or human origin to produce biologic compounds for medicinal or other purposes. Genetic engineering is used to produce the required molecules or proteins of interest. The cells have their genes altered or modified, using recombinant DNA techniques so that they produce a specific substance or perform a specific function, that is, the genes for a particular protein are introduced into the genes of a host cell, which then produces the specific protein of interest.

Each innovator biologic manufacturer has its unique cell line and manufacturing process. The production processes are precisely controlled to guarantee the quality and consistency of the final product. The production of biologic medicines is complex and requires a very high level of technical expertise and numerous (hundreds or thousands) of in-process tests during product development and manufacture.

1.3.2 What Is a Biosimilar?

Unlike small molecule drug (SMD) generics, which are identical to their innovator drug, it is not possible to produce an identical copy of an innovator biologic due to their large size, complex structure, and manufacture in a unique living cell line. Instead, a biologic deemed to be "highly similar" to the innovator biologic is known as a biosimilar.

A biosimilar is a non-innovator biologic or biotherapy that is "highly similar" to and has no clinically meaningful differences from an approved innovator (licensed or reference) product. Thus, a biosimilar is deemed as having no clinically meaningful differences to the innovator product in terms of purity, potency, and safety. Manufacturers are generally required to provide sup-

porting evidence that standards for biosimilarity recommended by the World Health Organization (WHO) are met[5] when seeking marketing approval from the USFDA, EMA, or other regulatory agencies.

Regulatory approval of a biosimilar involves a similarity exercise. This may include head-to-head comparability studies, and it is based on a totality of evidence concept[6] that generates a hierarchy of data and evidence to support similarity with the originator product. To support biosimilarity, the product must be deemed "highly similar" to the reference product; demonstrated by extensively characterizing chemical identity (structure), purity, and bioactivity (function) of both the reference and the proposed biosimilar. Minor differences between the reference product and the proposed biosimilar product in clinically *inactive* components are acceptable. There are similarities at molecular and structural levels or critical quality attributes (CQAs) between the reference and biosimilar products are similar.[7]

CQAs are divided into four separate categories: content-related attributes such as protein content; structural attributes; isoform profile; and, impurities and biological activity.[8]

Development of biosimilars is therefore methodologically complex, and it has been remarked that biosimilar development is an "imitation game."[9] A biosimilarity index (based on reproducibility probability) has been proposed to assess biosimilarity.[10] In summary, biosimilars are approved by showing "near fingerprint identity," but the term "near" is not absolute. Highly similar implies (but does not prove) therapeutic equivalency.

The targeted quality profile of biosimilars is strictly defined by the originator's product characteristics.[11] Typically, over time, since the original biologic was introduced, many product enhancements and efficiencies would have been achieved. Moreover, cell lines change over time. An important question, therefore, has been posed: "Is a biologic produced 15 years ago, a biosimilar of itself today?"[12]. In fact, it is hard to envisage that an originator biologic manufactured today would be able to demonstrate similarity to its product manufactured 15 years previously. Therefore, a global reference comparator for each biologic for biosimilar development and testing has been proposed.[13] Mandatory deposit of the original biologic's cell line with the regulator at the time of its approval has been suggested as a remedy.[14] The paradox of sharing the same therapeutic (and adverse) action without full (absolute) chemical identity has also been raised and is the subject of lively debate.[15]

Other terms used in the literature for biosimilars include follow-on biologics, similar biotherapeutic products (SBPs) or subsequent-entry biologics (SEBs) as used by the regulatory authority in Canada[16] or non-original biologics (NOBs). Bio-mimics is also used, but the production of exact molecular copies of biologics (bio-copies) is almost impossible and biosimilars are not replicas of innovator biologics and cannot be regarded as (or be confused with) SMD generics and are not bio-generics. Biosimilar definitions in major regulatory jurisdictions are summarized in Table 1.2.

Questions posed in the biosimilar literature include: are biosimilars identical twins or just siblings; when are biosimilars similar enough[17]; are biosimilars "bio-same or bio-different[18] how dissimilarly similar are biosimilars[19]; biosimilars – how similar or dissimilar are they[20,21]; how far does similarity go[22–25]; should the term semi-similars be used instead of biosimilars[26] or are they overpriced me-toos?"[27]. It has also been asked, "if biosimilars are patentable?"[28].

Some emotive language has also crept into the literature; some fear the adoption of biosimilars while others see it as an opportunity.[29]

The above discussion highlights that internationally accepted terminology is important for biosimilars. An excellent resource on the language of biosimilars is available for historical and contemporary context.[30]

1.3.3 What Is a Biobetter?

Biobetters are related to existing biologics by the target of action but have been intentionally improved in manufacturing attributes, disposition/pharmacokinetics, efficacy, safety, or enhanced stability.[31] Biobetters build on the success of an approved innovator biologic or biosimilar but possess a lower commercial risk for biotechnology companies than a novel class of biologic. Biobetters are also known as second-generation biologics.

Biobetters improve on the relevant property of biologics. Many innovator biologics or biosimilars have less than optimal pharmacokinetic properties (e.g. high clearances or short half-lives). Besides, almost all these proteins are dosed parenterally by injection rather than orally. Thus, modifications to improve their pharmacokinetic behavior have led to biobetters. Examples include pegylated longer half-life version of filgrastim or a more extended half-life version of epoetin α, using fusion proteins.

While biosimilars are comparable to the originator product in terms of quality, safety, and efficacy, biobetters incorporate intentional modifications to the innovator's molecular profile. This distinction between biosimilars and biobetters has essential implications from a regulatory perspective. Biosimilars follow class-specific regulatory guidance whereas biobetters are considered as new molecular entities and have registration requirements of a new drug.

Biobetters may have advantages due to their pharmacologic comparability to innovator biologics, which may accelerate their development. For instance, choice of their dose and biomarkers in both nonclinical and clinical studies is simpler, and prior knowledge from the innovator biologic may reduce the scale/duration of clinical trials and safety monitoring focused on known side effects of the target pathway.

Table 1.2 Definitions of biosimilars by major regulatory agencies.

Regulatory agency/ Country	Definition
European Medicines Agency, EMA	Biologic product is similar to another biologic already authorized for use
World Health Organization, WHO	Biotherapeutic product that is similar (quality, safety, and efficacy) to the licensed reference product
Food and Drug Administration, FDA (USA)	A biologic product that is highly similar to the reference product in safety, purity, and potency; minor differences in clinically inactive components are acceptable
Biologics and Genetic Therapies Directorate, BGTD (Canada)	A biologic entering the market after a version previously authorized; demonstrated similarity to reference
Pharmaceuticals and Medical Devices Agency, PMDA (Japan)	A biotechnological drug developed by a different company, comparable to an approved biotechnology product
Therapeutic Goods Administration, TGA (Australia)	A version of already registered biologic with demonstrated similarity in physicochemical, biologic, and immunologic characteristics (efficacy, safety) based on comparability exercise

1.4 Differences Between Biosimilars and Generic Medicines

Significant differences exist between biologic medicines including innovator biologics, biosimilars, and biobetters compared with chemically synthesized or isolated small molecule drugs (SMDs) and their generics. Biologics and biosimilars are in a different league to their chemical pre-predecessors in terms of molecular complexity and natural variability.

Table 1.3 Pivotal differences between biologics and small molecule drugs.

Biologics	Small molecule drugs (SMDs)
Large/complex molecules or mixtures of these molecules	Well-defined chemical structures
Product is the process: >1000 process steps	Manufactured by chemical synthesis: specific agents are used in an ordered/sequential manner
Living processes that are very sensitive to minor changes in manufacturing: may alter the product and its function (efficacy, safety)	Well-defined chemical synthesis or isolation: subject to lower batch-to-batch variability
Product quality, purity, and function are ensured by "stable" or "consistent" manufacturing	Each individual component of the finished drug product is identified and quantified
Unwanted immune reactions are common	Unwanted immune reactions are rare

First and foremost, biologics are produced in living cells or organisms and are not chemically synthesized. Many of the other factors that hinder the full acceptance of biosimilars stem from these critical differences in the properties of biologics and SMDs. A summary of the pivotal differences between biologics and SMDs is provided in Table 1.3. Pharmacists and doctors need to keep these key differences in mind when having conversations about biosimilar medicines.

1.5 Interchangeability, Switchability, and Substitution

1.5.1 Interchangeability

Interchangeability is defined as the medical practice of changing one medicine for another to achieve the same clinical effect in a given clinical setting and patient, on the initiative of, or with the prescriber's agreement. An interchangeable product is a biosimilar that produces the same clinical outcome in any given patient. Demonstration of interchangeability presents many challenges.[32]

In the United States, registration of a biosimilar does not imply interchangeability and another class of biosimilars, "interchangeable biosimilars" have been introduced into the regulatory framework. To meet this interchangeability designation, a sponsor must demonstrate that the biosimilar produces the same clinical result as the reference product in any given patient and, for a biological product that is administered more than once, that the risk of switching between the biosimilar and reference product is not greater than the risk of maintaining the patient on the reference product. No biosimilar has been granted interchangeability status so far.

1.5.2 Switchability

Switching, on the other hand, is a decision by the physician to exchange one medicine for another with the same therapeutic intent in a given patient. Alternation also refers to switching.[33] Another term in this context is non-medical switching (NMS) referring to when a patient whose current therapy is effective and well-tolerated is switched between therapies, such as an innovator biologic to its biosimilar for an economical, formulary, or other nonmedical reasons, i.e. for reasons other than the patient's health and safety.[34] Generally, NMS may be initiated by a hospital pharmacist, based on the local formulary, or the insurance company providing health insurance in consultation with the patient and the physician. The Biologics Prescribers Collaborative (BPC), representing specialist/general physician prescribers, and Alliance for Patient Access (AfPA) have developed NMS principles and guidelines.[35]

Shared decision-making between physicians, pharmacists, and patients is crucial for successful switching.[36] Patients' attitudes and level of satisfaction with switching to a biosimilar is related to being provided with necessary information about their health.[37] A comprehensive review concluded that evidence gaps around efficacy and safety of switching still exist.[38]

1.5.3 Substitution

Substitution refers to dispensing one medicine instead of another equivalent/interchangeable medicine by the pharmacist without consulting the prescriber.

In some jurisdictions/countries, interchangeability and switching are only permitted or recommended in some patients/conditions and at different treatment periods (for example, initiating therapy versus continuation of therapy).[39] Switch comes with challenges, so there needs to be clear local and national biosimilar substitution and switching policies and switch management strategies are important.[40] Pharmacists should play a pivotal role in patient empowerment as well as raising awareness of biosimilars among physicians and patients and reducing scepticism about the safety of biosimilars.

Key challenges for the integration of biosimilars into routine biologic therapy include questions around interchangeability, switching, and automatic substitution. Additional switch studies and drug registries may enhance our understanding of the safety and effectiveness of switching and a key hurdle to broader adoption of biosimilars is lack of interchangeability with reference biologics.[41]

Chapter 7 deals with interchangeability principles and evidence.

1.6 Other Clinical Considerations with Biosimilars

1.6.1 Indication Extrapolation

For innovator biologics, efficacy and safety must be demonstrated separately for each clinical indication. In contrast, biosimilar clinical trials are not required for all indications approved for the innovator biologic. Indication extrapolation is defined as approval of biosimilars for all indications of the innovator product even though the biosimilar may not have been studied in all indications.[42] The molecular similarity is the key guiding principle for extrapolation to multiple indications; it is an important concept in biosimilar development and is permitted by regulatory agencies, provided it is scientifically justified.[43]

1.6.2 Nocebo Effect

The nocebo effect is defined as a negative treatment effect that is induced by a patient's expectations that are unrelated to the pharmacologic actions of a medicine.[44] In any switching study, the subsequent biologic prescribed (like a biosimilar) is perceived to exert a lower therapeutic benefit due to this nocebo effect. The attitudes of doctors, patients, and payers are therefore crucial for the full acceptance of biosimilars because of the nocebo effect.[45]

1.6.3 Immunogenicity Reactions

An important consideration with all biologics is unwanted immunogenicity as biologics are often manufactured in living cells of nonhuman origin. Unwanted immunogenicity may lead to a reduction or loss of efficacy, altered pharmacokinetics, general immune and hypersensitivity reactions, and neutralization of the natural endogenous counterpart.[46] Immunogenicity of biosimilars would be expected to mirror those of the innovator biologic based on the similarity principle.[47]

1.6.4 Definition of Frequency of Adverse Effects

Pharmacists and doctors need to understand the accepted definitions of the frequency of adverse drug reactions. The Council for International Organizations of Medical Sciences (CIOMS), an international nongovernment organization established jointly by WHO and UNESCO in 1949, and its Uppsala Monitoring Centre, UMC, define[48] the frequency of adverse reactions as:

Very Common (≥1 in 10); **Common/Frequent** (≥1 in 100 and <1 in 10); **Uncommon/Infrequent** (≥1 in 1 000 and <1 in 100); Rare (≥1 in 10 000 & <1 in 1 000); and **Very Rare** (<1 in 10 000).

1.6.5 Pharmacovigilance of Biologics

Any drug may produce adverse reactions, with varying levels of severity and frequency. Not all adverse reactions are, however, identified before the approval of a new drug, some only being observed during post-marketing use when the drug is prescribed more widely to patients, as opposed to only clinical trial participants.

As part of the marketing authorization for biologics, the sponsor must submit a pharmacovigilance plan as part of a risk management plan (RMP) to the relevant authorities in accordance with EU regulations.[49] Applicants seeking biosimilar approval also need to submit an RMP, as required for innovator biologics. The purpose of an RMP is to document the risk management system necessary to identify, characterize, and minimize a drug's significant risks. The plan should incorporate identified and potential risks outlining a plan for pharmacovigilance activities, to characterize and quantify clinically relevant risks, and to identify new adverse reactions and outline risk minimization measures.[50]

1.7 Manufacture, Delivery, and Naming Considerations

1.7.1 Post-Translational Modifications (PTMs)

Biosimilars, like innovator biologics, raise challenges compared to SMDs, due to manufacturing complexity, presence of minor natural variations in the molecular structure (collectively known as microheterogeneity), and post-manufacturing (post-translational) modifications.[51,52] The production of innovator biologics and biosimilars comprises numerous steps and minuscule differences in the product may result in different clinical outcomes. Consistent drug discovery and manufacturing paradigm are likely to minimize product variations. Besides, drift (unnoticed and unplanned deviations) and evolution (planned changes) may lead to divergence, which can also lead to product variability and different product attributes. Divergence means that the biosimilar and the currently marketed innovator differ from the originator product that was first approved and marketed and the innovator product that was used in the comparability exercise.[53] Biotechnology process and manufacturing innovations, needed for regulatory reasons, production scale-up, change in a facility or raw materials, and improving quality or consistency or optimizing production efficiency,[54] may lead to a higher quality biologic.

Identifying and controlling PTMs and demonstrating biosimilarity require specific and sensitive analytical techniques. Pharmacists need to be familiar with such techniques and issues; these are discussed in Chapter 6.

1.7.2 Quality by Design Paradigm

Quality by design (QbD) is an approach that aims to ensure the quality of medicines by employing statistical, analytical, and risk-management methodology in the design, development, and manufacturing of medicines. It focuses on the use of multivariate analysis, often in combination with the modern process and analytical chemistry methods, and knowledge-management tools to enhance the identification and understanding of critical attributes of materials and critical parameters of the manufacturing process. This enhanced understanding of product and process is used to build quality into manufacturing and provide the basis for continuous improvement of products and processes.[55]

One of the goals of QbD is to ensure that all sources of variability affecting a process are identified, explained, and managed by appropriate measures. This enables the finished medicine to meet its predefined characteristics consistently.

The concepts behind QbD were introduced into international pharmaceutical guidelines between 2009 and 2012. EMA accepts applications that include QbD concepts.

1.7.3 Delivery Devices for Biologics

Converting a promising innovator biologic or biosimilar molecule into a pharmaceutical product presents numerous new challenges. For example, biologic medicines are highly viscous and formulated at high concentrations, which makes them more prone to aggregation. In addition, they need to be handled, packaged, stored, and transported carefully.[56] These requirements are driving innovation in packaging and delivery device development as, increasingly, drug companies demand technologies that can protect and administer these high-value medicines safely and conveniently.[57]

Historically, all biologic drugs were freeze-dried and packaged in glass vials and administered, after reconstitution, using glass syringes. While most biologics are still packaged and delivered using glass, a growing number of biologics (including biosimilars) are packaged in plastic vials and administered using plastic syringes.

Devices for biosimilar administration are essential in quality use of medicine (QUM) considerations for biosimilars as for all biologics[58]; they may also have critical practical implications for patients. These devices, either prefilled syringes, pens, or pumps, are important for dosing accuracy and reproducibility as well as long-term patient compliance and adherence. From a patient perspective, one would envisage that the device via which a biosimilar is administered must at least be able to match the innovator biologic's device for convenience and comfort. Inferior usability may also reduce treatment adherence and product uptake by patients. The design and user experience of the delivery device may also serve as a critical market differentiator between the innovator biologic and the biosimilar.

1.7.4 Naming and Labeling of Biosimilars

A critical question that is still eliciting much debate internationally is the naming convention for biosimilars; in other words, what should be their nonproprietary (noncommercial) name?

SMD generic medicines have the same nonproprietary names as their innovator medicines as the active ingredients in generics are identical to that in the innovator drugs. In contrast, biosimilars are not identical to innovator biologics. Giving all biologics, including biosimilars, different (distinguishable) nonproprietary names are consistent with the concept that no two versions of a biologic including a biosimilar are identical.[59]

Views on the naming of biosimilars fall broadly in two groups. The first is that since a biosimilar is highly similar to its innovator biologic, it should have the same name as the innovator. The other view is that for safety reasons, it is critical to have a unique name for each biologic, including a biosimilar so each biologic can be identified individually.[59]

Under the FDA's naming system, each biologic, innovator/reference product, and biosimilar receives a unique nonproprietary name; a "core" name followed by a unique (but meaningless) four-letter suffix. Thus, each biologic has a unique, distinguishable name in the United States.

Europe, Australia, and Canada have adopted a different naming approach that incorporates distinguishable suffixes. These countries allow biologics including biosimilars to share nonproprietary names but have strengthened adverse event monitoring by either mandating inclusion of brand names or nonproprietary names as well as brand names in adverse event or pharmacovigilance reporting. In Australia, for example, the product's trade name, as well as the nonproprietary name, is a mandatory field when reporting an adverse event.[60]

The naming of biosimilars has implications far beyond the marketing and commercial sphere; it may directly affect patients' confidence in switching to biosimilars and traceability of each biosimilar product with respect to its efficacy and safety monitoring once on the market.

1.8 Listing of Approved Biologics

1.8.1 Purple Book in the United States

The *Purple Book* is a compendium of FDA-approved biological products and their biosimilar and interchangeable products. It resembles the *Orange Book*, which is a list of approved SMD generics. Information on each product listed in the *Purple Book* includes its BLA tracking number, product name, product proprietary name, date of licensure, date of first licensure, reference product exclusivity expiration date, indication as to whether the product is interchangeable (I) or biosimilar (B), and whether the product was withdrawn from the market.[61] Other countries have similar lists of approved innovator biologics and biosimilars.

1.8.2 European Generic Medicines Association (EGA) Biosimilars Handbook

This handbook provides information on the current state of biosimilar medicines in the EU. It describes the science and technology behind biosimilar medicines, how they are produced and regulated, and provides answers to many specific questions. These include the terminology used, the meaning of "quality, efficacy and safety" and "comparability," the purposes and methodologies of nonclinical and clinical tests and trials, the role of pharmacovigilance and risk management, and the significance of immunogenicity. Access to medicines, including substitution, interchangeability, and the importance of identification is also included.[62]

1.9 Biosimilar Initiatives and Organizations

Many initiatives and organizations with interest in broader adoption of biosimilars have come into being, driven by governments and/or private organizations and agencies including patient advocacy groups. A summary of these is provided as these initiatives and organizations affect the information flow and influence the uptake of innovator biologics and biosimilars. The intent here is not to discuss comprehensively every national or international initiative on the adoption of biosimilars but to present some prominent exemplars so that the reader is able to map to similar national and local initiative(s) in their own country. If such initiatives are not currently available in a country, then the reader may also be able to facilitate the creation of such an initiative tailored to the specific needs of their country.

1.9.1 Generics and Biosimilars Initiative (GaBi/GaBI)

Generics and Biosimilars Initiative (GaBi/GaBI) was founded in 2008. The mission of GaBI is to foster the efficient use of high quality and safe medicines at an affordable price, thus advancing and supporting the idea of accessible, affordable, and sustainable health care internationally.

GaBI aims to raise the scientific status of SMD generics and biosimilar medicines and to provide comprehensive high-quality, scientifically sound, reliable, well-documented, and up-to-date information about generics and biosimilar medicines in both print and electronic media in an open-access format. To this end, GaBI provides a service for healthcare providers to support them in making cost-effective choices when it comes to treatment option decisions. Physicians and pharmacists are the primary target of GaBI, followed by healthcare policymakers and drug regulators, third-party insurers, and pharmaceutical/biotech industry.[63] This initiative has GaBI Online and GaBI Journal as its principal resources.

1.9.2 Biologics Price Competition and Innovation Act in the United States

The US Congress passed the Biologics Price Competition and Innovation (BPCI) Act in 2009, authorizing the FDA to oversee an "abbreviated pathway" for approval of biologics that are "biosimilar" to already approved biologic products.[64] The BPCI Act (also known as the Affordable Care Act) aligns with the FDA's longstanding policy of permitting appropriate reliance on what is already known about a drug, thereby saving time and resources and avoiding unnecessary testing.

Under the BPCI Act, a sponsor may seek approval of a "biosimilar" product. A biological product may be demonstrated to be "biosimilar" if data show that the product is "highly similar" to the reference product notwithstanding minor differences in clinically inactive components and there are no clinically meaningful differences between the biological product and the reference product in terms of safety, purity, and potency.

In order to meet the higher standard of interchangeability, a sponsor must demonstrate that the biosimilar product can be expected to produce the same clinical result as the reference product in any given patient and, for a biological product that is administered more than once, that the risk of alternating or switching between use of the biosimilar product and the reference product is not greater than the risk of maintaining the patient on the reference product. Interchangeable products may be substituted for the reference product by a pharmacist without the intervention of the prescribing health-care provider.

The BPCI Act intended to facilitate timely approval of and access to biosimilars to US citizens. Recent evidence appears to suggest that this goal has not been achieved and other essential steps are required.[2,65]

1.9.3 Biosimilars Action Plan (USFDA)

In July 2018, the USFDA published its Biosimilars Action Plan, BAP.[66] Key elements of the BAP include (i) improving the efficiency of the biosimilar and interchangeable product development and approval process; (ii) maximizing scientific and regulatory clarity for the biosimilar product development community; (iii) developing effective communication to improve understanding of biosimilars among patients, clinicians, and payers; and (iv) supporting market competition by reducing gaming of FDA requirements or other attempts to unfairly delay biosimilar competition.

1.9.4 NHS England Commissioning Framework for Biological Medicines

NHS England released this framework in late 2017. This document supports NHS commissioners to act promptly to make the most of the opportunity presented by increased competition among biological medicines, including biosimilars. In particular, this framework sets out the importance of taking a collaborative approach to the commissioning of innovator biologics and biosimilars.[67] A companion commentary on preparing for the biologic switch is also available.[68]

1.9.5 PrescQIPP

PrescQIPP is a UK NHS funded not-for-profit organization that supports quality optimized prescribing for patients. It helps NHS organizations to improve medicines-related care to patients, through the provision of accessible and evidence-based resources.[69] PrescQIPP also provides a platform to share innovation, learning, and good practice. It operates for the benefit of NHS patients, commissioners, and organizations. PrescQIPP provides many resources on biologics and other high-cost medicines (for example, on biosimilars of infliximab, insulin, and etanercept, respectively).

1.9.6 The Association of the British Pharmaceutical Industry

The UK government recognizes the Association of the British Pharmaceutical Industry (ABPI) as the industry body negotiating on behalf of the branded pharmaceutical industry for statutory consultation requirements including pricing schemes for medicines in the United Kingdom. The ABPI has partnerships with UK NHS on medicine-related projects including innovator biologics and biosimilars.[70]

1.9.7 NHS Scotland

Healthcare Improvement Scotland, as part of NHS Scotland, has led a biosimilar medicines national prescribing framework to support the safe, effective, and consistent use of biosimilar medicines in Scotland.[71]

1.9.8 National Institute for Health and Care Excellence

National Institute for Health and Care Excellence (NICE) in the United Kingdom supports the managed introduction of biosimilar medicines as part of its key therapeutic topics and initiatives for medicines optimization. NICE has released a biosimilars position statement[72] and also has a position statement on the assessment of biosimilars.[73] NICE provides many other resources.

1.9.9 Australian Biosimilar Awareness Initiative

This initiative was announced in 2015 as part of the Australian Pharmaceutical Benefits Scheme (PBS) Access and Sustainability Package. The aim is to support awareness of, and confidence in, the use of biosimilar medicines for healthcare professionals and consumers.[74]

Several research activities like up-to-date literature review[75] form part of the initiative; these aim to separate the evidence from commentary better; gather data on current awareness and attitudes toward innovator biologics and biosimilars in Australia; and identify critical issues, like barriers to uptake and use of biosimilars.

1.9.10 NPS MedicineWise (Australia)

Several important initiatives are in place in Australia to ensure the health community embraces the full potential of innovator biologics and biosimilars. Within this context, and under the stewardship of NPS MedicineWise, The Biologic and Biosimilar Medicines 2020 Forum was held in 2016, to maximize the opportunities these medicines present to the Australian healthcare system.[76] The Australian National Medicines Policy provided a framework for the Forum to discuss the opportunities and challenges presented by the availability of both innovator biologic and biosimilar medicines. A broad range of perspectives from research, industry, government, medical, pharmacy, and consumer perspectives were considered. The expanding settings in which innovator biologics and biosimilars may be used was also taken into consideration including hospitals, specialist medical centers, primary care, community pharmacy, and nonclinical environments such as the home. The themes that emerged from the forum included: improving the evidence base; optimizing data capture; pharmacovigilance and naming conventions; and building stakeholder confidence and shared decision-making through high-quality information.

1.10 Common Terms Used in the Biologics Literature

This section intends to provide the reader with significant terms that are used in biologic medicine literature internationally.

1.10.1 Real-World Evidence

Real-world evidence (RWE) refers to data on the use of a drug product obtained outside of clinical trials.[77] In other words, the efficacy and safety data collected from medical records, pharmacovigilance records, personal devices, or electronic health applications after the medicine has been marketed, i.e. data and evidence about the drug product that is gathered during its widespread clinical use.

Depending on their design, RWE studies may follow patients for several years, or study treatments in patients not included in clinical trials (e.g. in children, elderly patients, or patients with concomitant diseases) or in clinical indications not studied during clinical trials. RWE studies may enhance the broader adoption of biosimilars.[78] Importantly, RWE studies must be carefully designed to yield credible, reproducible results using sound pharmacoepidemiological principles and practices.

1.10.2 Patent Dance

As mentioned, the BPCI Act in the United States provides for an elaborate process of information exchange, known as the "patent dance"[79] between a biosimilar applicant and an innovator/reference product sponsor (RPS) intended to resolve potential patent disputes in an orderly and expeditious fashion. This procedure (patent dance) has strict timing and sequencing requirements and involves several rounds of information exchange between the innovator/RPS and the biosimilar applicant.

1.10.3 Evergreening

Evergreening refers to the use of various strategies for patent extension, of innovator biologics as also occurs for SMD innovator drugs to delay the introduction of their SMD generics. Among other outcomes, evergreening may limit timely availability of biosimilars and affect their price.[80]

1.10.4 Limited Distribution Network

The limited distribution network, LDN, which restricts the distribution channel for a drug to one or a very small number of distributors, can stifle competition for biosimilars and affect their price.[81]

1.10.5 Drug Tendering

The goal of pharmaceutical procurement is to purchase high-quality products with reliable supplier service and the lowest possible price. One method to contain spending is tendering, a formal procedure using competitive bidding for a particular contract; tendering is used when equivalents for a specific medicine are available, and is defined as "any formal and competitive procurement procedure through which offers are requested, received and evaluated for the procurement of goods, works or services, and as a consequence of which an award is made to the tenderer whose tender/offer is the most advantageous"[82]. Drug tendering may influence biosimilar uptake and price.[82]

1.10.6 Pharmacy and Therapeutics Committees

Pharmacy and Therapeutics (P&T) committees exist in most hospitals and pharmacists are key members of such committees offering objective, unbiased information and advice on all aspects of drug use. Considerations of quality, cost (reimbursement), access, and procurement and interchangeability of biosimilars with innovator biologics[83] will be even more important to P&T committees as new emerging and even more expensive biotherapies enter the market and hospitals and insurers and governments attempt to improve clinical care within enormous budgetary constraints.

1.10.7 Quality Use of Medicine

QUM involves improving medicine use, including prescription, non-prescription, and complementary medicines, and medical devices by health professionals and decision-makers as well as by consumers and the pharmaceutical industry.[84] QUM is also known as rational drug use, responsible drug use, or appropriate use of medicines and includes:

Selecting management options wisely; choosing suitable medicines if a medicine is considered necessary and ensuring that patients and carers have the knowledge and skills to use medicines safely and effectively. QUM concepts apply equally to decisions about medicine use by individuals as well as decisions that affect the health of the population.

QUM concepts and principles as they apply to innovator biologics and biosimilars is the subject of a detailed discussion in Chapter 14.

1.10.8 European Public Assessment Report

EMA publishes detailed information on the medicines assessed by the Committee for Medicinal Products for Human Use (CHMP) and Committee for Medicinal Products for Veterinary Use (CVMP) which are granted (or refused) central marketing authorization by the European Commission. The main vehicle for this information is known as a European Public Assessment Report (EPAR), which is a full scientific assessment report of medicines authorized in the EU.

An essential role of the EPAR is to reflect the scientific conclusions of the relevant EMA committee at the end of the assessment process, providing the grounds for the expert opinion on whether to approve an application.

EPARs are updated periodically to reflect the latest regulatory information on medicines. If the original terms and conditions of a marketing authorization are varied, the EPAR is updated to reflect such changes with an appropriate level of detail.

EPARs are a valuable source of information about innovator biologics and biosimilars.[85]

1.11 Abbreviations Associated with Biologic Medicines

Many abbreviations relating to innovator biologics, biosimilars, and biobetters are used in the literature by many and varied stakeholders. A summary of these frequently used abbreviations is presented in Table 1.4 to familiarize readers with these terms and abbreviations.

1.12 Concluding Remarks

The science behind innovator biologics, biosimilars, and biobetters are complex and the literature is changing rapidly. The scientific and clinical data are evolving at a much faster rate than the ability of pharmacists, doctors, other health practitioners and patients to keep pace with new information. Regulators, as well as policymakers, also find it challenging to keep pace with this change and evolution and to embed regulatory and policy frameworks in a timely and responsible manner. There will also continue to be greater

Table 1.4 Abbreviations used in biologic medicine literature.

Abbreviation	Full name
ADCs	Antibody–Drug Conjugates
ADR	Adverse Drug Reaction
ADE	Adverse Drug Event
AE	Adverse Event
ANDA	Abbreviated New Drug Application
ARTG	Australian Register of Therapeutic Goods
ATMP	Advanced Therapy Medicinal Products
BDMARDs	Biologic Disease-Modifying Anti-Rheumatic Drugs
BIA	Budget Impact Analysis
BLA	Biologics License Application
BPCI Act	Biologics Price Competition and Innovation Act
CAPs	Centrally Authorised Products (EU)
CAR-T	Chimeric Antigen Receptor Therapy
CDMO	Contract Development and Manufacturing Organisation
CEOR	ClinicoEconomics and Outcomes Research
CHMP	Committee for Medicinal Products for Human Use (EMA)
CIOMS	Council for International Organizations of Medical Sciences
CE	Comparability Exercise
CMA	Critical Material Attribute
CPP	Critical Process Parameter
CQA	Critical Quality Attribute
CTD	Common Technical Document
DCP	Decentralised Procedure
DDD	Defined Daily Dose
DDR	Dose-Dense Regimens
EC	European Commission
EMA (EMEA)	European Medicines Agency
EPARs	European Public Assessment Reports
EPO	Erythropoietin (epoetin)
EU	European Union
Eudra	European Drug Regulatory Authorities
FDA	Food and Drug Administration
FD&C	Food, Drug, and Cosmetic
FTC	Federal Trade Commission (in the United States)
GCP	Good Clinical Practice
GH	Growth Hormone
GMP	Good Manufacturing Practice

(Continued)

Table 1.4 (Continued)

Abbreviation	Full name
GVP	Good Pharmacovigilance Practice
HCPCS	Healthcare Common Procedure Coding System
HPLC	High-Performance Liquid Chromatography
IBD	Inflammatory Bowel Disease
ICH	International Conference on Harmonisation of Technical Requirements for Registration of Pharmaceuticals for Human Use
INN	International Non-proprietary Name
IP	Intellectual Property
LDN	Limited Distribution Network
LMWH	Low Molecular Weight Heparins
MA	Marketing Authorisation
MAA	Marketing Authorisation Application
MCOs	Managed Care Organisations
MR	Mutual Recognition
NHS	National Health Service
NDA	New Drug Application
NMS	Non-Medical Switching
NICE	National Institute for Clinical Excellence (UK)
mAb	Monoclonal Antibody
PBAC	Pharmaceutical Benefits Advisory Committee
PBS	Pharmaceutical Benefits Scheme
PFS	Pre-Filled Syringe
PHS	Public Health Service
PMS	Post-Marketing Surveillance
PPRS	Pharmaceutical Price Regulation Scheme (UK)
QbD	Quality by Design
RCT	Randomised Clinical Trial
REMS	Risk Evaluation and Mitigation Strategies
RMOCs	Regional Medicines Optimisation Committees (UK)
RMPs	Risk Management Plans
RMR	Reaction Monitoring Report
RPG	Reference Price Group
RWD	Real-World Data
SB	Synthetic Biology
SEBs	Subsequent-Entry Biologics
SBPs	Similar Biotherapeutic products
SMDs	Small molecule Drugs

(Continued)

Table 1.4 (Continued)

Abbreviation	Full name
SmPAR	Summary of Pharmacovigilance Assessment Report
SmPc	Summary of Product Characteristics
SPBs	Similar Protein Biotherapies
TGA	Therapeutic Goods Administration
TPP	Target Product Profile
TPQP	Target Product Quality Profile
WHO-UMC	World Health Organization-Uppsala Monitoring Centre

economic pressures from governments and payers of all political persuasions for affordable biologics like biosimilars to realize the full benefits of innovator biologic medicines. It is therefore imperative that pharmacists keep abreast of such rapid changes in the information as they will be expected to lead discussions with doctors and patients on these important therapeutic agents.

The subsequent chapters of this book are pitched explicitly to pharmacists and doctors and deal in greater detail with the various scientific, clinical, economic, QUM, and pharmacovigilance aspects of innovator biologics, biosimilars, and biobetters. The material provided in this and subsequent chapters should facilitate discussions by pharmacists with doctors and patients on these expensive and highly effective medicines so that the full therapeutic potential of all biologic medicines is realized in a timely manner.

Acknowledgement

The author thanks Dr Reza Kahlaee for expert assistance with collation of the cited references.

References

1 Davies, N. (2018). The future of biologics. *Pharm Lett* www.thepharmaletter.com/article/the-future-of-biologics (accessed 20 June 2020).

2 Sarpatwari, A., Barenie, R., Curfman, G. et al. (2019). The US biosimilar market: stunted growth and possible reforms. *Clin Pharmacol Ther* 105 (1): 92–100.

3 Dutta, B., Huys, I., Vulto, A.G., and Simoens, S. (2019). Identifying key benefits in European off-patent biologics and biosimilar markets: it is not only about price! *BioDrugs* 34 (Suppl. 3): 159–170. https://doi.org/10.1007/s40259-019-00395-w.

4 Johnson, I. (1983). Human insulin from recombinant DNA technology. *Science* 219 (4585): 632–637.

5 WHO (2009). Health Organization expert committee on biological standardization. Guidelines on evaluation of similar biotherapeutic products (SBPs). https://www.who.int/biologicals/publications/trs/areas/biological_therapeutics/TRS_977_Annex_2.pdf?ua=1 (accessed 17 February 2020).

6 Declerck, P. and Farouk, R.M. (2017). The road from development to approval: evaluating the body of evidence to confirm biosimilarity. *Rheumatology (Oxford)* 56 (suppl 4): iv4–iv13.

7 Beyer, B., Walch, N., Jungbauer, A., and Lingg, N. (2019). How similar is biosimilar? A comparison of infliximab therapeutics in regard to charge variant profile and antigen binding affinity. *Biotechnol J* 14 (4): e1800340.

8 Nick, C. (2015). Quality attributes in the making and breaking of biosimilarity. *Regul Rapporteur* 12 (9): 10–13.

9 Dahodwala, H. and Sharfstein, S.T. (2017). Biosimilars: imitation games. *ACS Med Chem Lett* 8 (7): 690–693.

10 Zhang, A., Tzeng, J.-Y., and Chow, S.-C. (2013). Statistical considerations in biosimilar assessment using biosimilarity index. *J Bioequivalence Bioavailab* 5 (5): 209–214.

11 Beck, A. and Reichert, J.M. (2013). Approval of the first biosimilar antibodies in Europe: a major landmark for the biopharmaceutical industry. *MAbs* 5 (5): 621–623.

12 Mehr, S.R. and Zimmerman, M.P. (2016). Is a biologic produced 15 years ago a biosimilar of itself today? *Am Health Drug Benefits* 9 (9): 515–518.

13 Webster, C.J. and Woollett, G.R. (2017). A 'global reference' comparator for biosimilar development. *BioDrugs* 31 (4): 279–286.

14 Diependaele, L., Cockbain, J., and Sterckx, S. (2018). Similar or the same? Why biosimilars are not the solution. *J Law Med Ethics* 46 (3): 776–790.

15 Pani, L., Montilla, S., Pimpinella, G., and Bertini, M.R. (2013). Biosimilars: the paradox of sharing the same pharmacological action without full chemical identity. *Expert Opin Biol Ther* 13 (10): 1343–1346.

16 Blanchard, A., D'Iorio, H., and Ford, R. (2010). What you need to know to succeed: key trends in Canada's biotech industry. *Insight* 1 (Spring).

17 Mechcatie, E. (2010). FDA asks: when are biosimilars similar enough? *Commun Oncol* 8 (10): 479–480.

18 Murphy, C., Sugrue, K., Mohamad, G. et al. (2015). P505: biosimilar but not the same. *Paper presented at 10th Congress of ECCO2015*, Barcelona, Spain (January 2015).

19 Vijayalakshmi, R., Sabitha, K., and Arvind, K. (2012). How dissimilarly similar are biosimilars? *Int Res J Pharm* 3 (5): 12–16.

20 Ronco, C. (2005). Biosimilars: how similar are they? *Int J Artif Organs* 28 (6): 552–553.

21 Jeffrey, K. and Ferner, R. (2016). How similar are biosimilars? *BMJ Online* 353: i2721.

22 Gupta, P. (2017). Biosimilars: similar but not the same. *JK Sci* 19 (2): 67–69.

23 Dingermann, T. and Zündorf, I. (2007). Biosimilars: similar, but not the same. *Dtsch Apoth Ztg* 147 (38): 68–74.

24 Dingermann, T. and Zündorf, I. (2015). Same, similar or different? How biosimilars can differ from originals. *Dtsch Apoth Ztg* 155 (7).

25 Schellekens, H. (2004). How similar do 'biosimilars' need to be? *Nat Biotechnol* 22 (11): 1357–1359.

26 Schellekens, H. and Moors, E. (2015). Biosimilars or semi-similars? *Nat Biotechnol* 33 (1): 19–20.

27 Rudolf, T., Brinkmann, C., and Bleiziffer, A. (2015). Biosimilars-necessary alternative or overpriced me-toos? *Pharm Ind* 77 (2): 160–165.

28 Rolf, D., Parker, J., and Morgan, M. (2016). Are biosimilars patentable? *Expert Opin Ther Pat* 26 (8): 871–875.

29 Guillon-Munos, A., Daguet, A., and Watier, H. (2014). Antibody biosimilars: fears or opportunities? First LabEx MabImprove Workshop. *MAbs* 6 (4): 805–809.

30 Declerck, P., Danesi, R., Petersel, D., and Jacobs, I. (2017). The language of biosimilars: clarification, definitions, and regulatory aspects. *Drugs* 77 (6): 671–677.

31 Courtois, F., Schneider, C.P., Agrawal, N.J., and Trout, B.L. (2015). Rational design of biobetters with enhanced stability. *J Pharm Sci* 104 (8): 2433–2440.

32 Ebbers, H.C., Crow, S.A., Vulto, A.G., and Schellekens, H. (2012). Interchangeability, immunogenicity and biosimilars. *Nat Biotechnol* 30 (12): 1186–1190.

33 Esplugues, J.V., Flamion, B., and Puig, L. (2016). Putting the "bio" in "biotherapeutics"/checkpoints for biosimilars/application of biosimilars. *New Horiz Transl Med* 3 (3–4): 161.

34 Moots, R., Azevedo, V., Coindreau, J.L. et al. (2017). Switching between reference biologics and biosimilars for the treatment of rheumatology, gastroenterology, and dermatology inflammatory conditions: considerations for the clinician. *Curr Rheumatol Rep* 19 (6): 37.

35 Biologics Prescribers Collaborative (BPC) (2018). *Non-Medical Switching Principles and Guidelines* (16 March). www.biologicsprescribers.org/policy-issues/non-medical-switching (accessed 20 June 2020).

36 French, T. (2018). Shared decision making in switching to biosimilars. *Ann Rheum Dis* 77 (Supplement 2): 7.

37 Cvancarova S.M., Brandvold, M., and Andenaes, R. (2018). Is patients' satisfaction with being switched to a biosimilar medication associated with their level of health literacy? Results from a Norwegian user survey. *Ann Rheum Dis* 77 (Supplement 2): 86.

38 McKinnon, R.A., Cook, M., Liauw, W. et al. (2018). Biosimilarity and interchangeability: principles and evidence. A systematic review. *BioDrugs* 32 (1): 27–52.

39 GaBi Online (2018). International policies for interchangeability, switching and substitution of biosimilars. http://gabionline.net/Reports/International-policies-for-interchangeability-switching-and-substitution-of-biosimilars (accessed 18 February 2020).

40 Milmo, S. (2016). Biosimilars: making the switch comes with challenges. More efforts are needed to raise awareness of biosimilars among physicians and patients in Europe and address scepticisms about the quality and safety of biosimilars. *BioPharm Int* 29 (6): 10–14.

41 Uhlig, T. and Goll, G.L. (2017). Reviewing the evidence for biosimilars: key insights, lessons learned and future horizons. *Rheumatology (Oxford)* 56 (Suppl 4): iv49–iv62.

42 Gerrard, T.L., Johnston, G., and Gaugh, D.R. (2015). Biosimilars: extrapolation of clinical use to other indications. *GaBI J* 4 (3): 118–124.

43 McCamish, M. and Woollett, G. (2017). Molecular "sameness" is the key guiding principle for extrapolation to multiple indications. *Clin Pharmacol Ther* 101 (5): 603–605.

44 Pouillon, L., Socha, M., Demore, B. et al. (2018). The nocebo effect: a clinical challenge in the era of biosimilars. *Expert Rev Clin Immunol* 14 (9): 739–749.

45 Povsic, M., Lavelle, P., Enstone, A., and Rousseau, B. (2018). Patient influence on biosimilar uptake: the nocebo effect. *Value Health* 21 (Supplement 3): S181.

46 Buttel, I.C., Chamberlain, P., Chowers, Y. et al. (2011). Taking immunogenicity assessment of therapeutic proteins to the next level. *Biologicals* 39 (2): 100–109.

47 Strand, V., Goncalves, J., Hickling, T.P. et al. (2018). Immunogenicity of biosimilars for rheumatic diseases: an updated review from regulatory documents and confirmatory clinical trials. *Ann Rheum Dis* 77 (Supplement 2): 1398.

48 World Health Organization, CIOMS (2018). Definitions. https://www.who.int/medicines/areas/quality_safety/safety_efficacy/trainingcourses/definitions.pdf (accessed 20 February 2020).

49 European Medicines Agency (2014). Guideline on similar biological medicinal products containing biotechnology-derived proteins as active substance: non-clinical and clinical issues. https://www.ema.europa.eu/en/documents/scientific-guideline/guideline-similar-biological-medicinal-products-containing-biotechnology-derived-proteins-active_en-2.pdf (accessed 20 February 2020).

50 Calvo, B., Martinez-Gorostiaga, J., and Echevarria, E. (2018). The surge in biosimilars: considerations for effective pharmacovigilance and EU regulation. *Ther Adv Drug Saf* 9 (10): 601–608.

51 Wechsler, J. (2018). Biosimilars raise manufacturing and regulatory challenges. *Pharm Technol* 42 (7): 14–15.

52 Gamez-Belmonte, R., Hernandez-Chirlaque, C., Arredondo-Amador, M. et al. (2018). Biosimilars: concepts and controversies. *Pharmacol Res* 133: 251–264.

53 Ramanan, S. and Grampp, G. (2014). Drift, evolution, and divergence in biologics and biosimilars manufacturing. *BioDrugs* 28 (4): 363–372.

54 Declerck, P., Farouk-Rezk, M., and Rudd, P.M. (2016). Biosimilarity versus manufacturing change: two distinct concepts. *Pharm Res* 33 (2): 261–268.

55 Rathore, A. (2009). Roadmap for implementation of quality by design (QbD) for biotechnology products. *Trends Biotechnol* 27 (9): 546–553.

56 Mitragotri, S., Burke, P., and Langer, R. (2014). Overcoming the challenges in administering biopharmaceuticals: formulation and delivery strategies. *Nat Rev Drug Discov* 13 (9): 655–672.

57 Hooven, M.D. (2017). Opportunities and challenges in biologic drug delivery. American Pharmaceutical Review. https://www.americanpharmaceuticalreview.com/Featured-Articles/345540-Opportunities-and-Challenges-in-Biologic-Drug-Delivery (accessed 20 February 2020).

58 Kuhlmann, M. and Schmidt, A. (2014). Production and manufacturing of biosimilar insulins: implications for patients, physicians, and health care systems. *Biosimilars* 4: 45–58. https://doi.org/10.2147/BS.S36043.

59 Jordan, B. (2019). U.S. biologics and biosimilars need distinguishable names. *Stat News*. https://www.statnews.com/2019/04/23/biologics-biosimilars-distinguishable-names (accessed 20 February 2020).

60 DiGrande, S. (2018). Australian Government announces decision on biosimilar naming conventions. *The Center for Biosimilars*. https://www.centerforbiosimilars.com/news/australian-government-announces-decision-on-biosimilar-naming-conventions (accessed 20 February 2020).

61 U.S. Food and Drug Administration (2020). *Purple Book: Lists of Licensed Biological Products with Reference Product Exclusivity and Biosimilarity or Interchangeability Evaluations*. https://www.fda.gov/drugs/therapeutic-biologics-applications-bla/purple-book-lists-licensed-biological-products-reference-product-exclusivity-and-biosimilarity-or (accessed 18 June 2020).

62 EGA (2010). Biosimilars handbook. https://www.medicinesforeurope.com/wp-content/uploads/2016/03/EGA_BIOSIMILARS_handbook_en.pdf (accessed 20 February 2020).

63 GaBi (2020). About GaBi journal. http://gabi-journal.net/gabi-journal/about-gabi-journal (accessed 18 June 2020).

64 FDA (2016). Implementation of the Biologics Price Competition and Innovation Act of 2009. https://www.fda.gov/drugs/guidance-compliance-regulatory-information/implementation-biologics-price-competition-and-innovation-act-2009 (accessed 20 February 2020).

65 Kabir, E.R., Moreino, S.S., and Sharif Siam, M.K. (2019). The breakthrough of biosimilars: a twist in the narrative of biological therapy. *Biomolecules* 9 (9): 410. http://dx.doi.org/10.3390/biom9090410.

66 FDA (2018). FDA's Biosimilar Action Plan (BAP): balancing innovation and competition. https://www.fda.gov/media/114574/download (accessed 20 February 2020).

67 NHS England (2017). Commissioning framework for biological medicines (including biosimilar medicines). https://www.england.nhs.uk/wp-content/uploads/2017/09/biosimilar-medicines-commissioning-framework.pdf (accessed 20 February 2020).

68 Robinson, J. (2018). Preparing for the big biologic switch. *Pharm J* 301 (7916) https://doi.org/10.1211/PJ.2018.20205278.

69 PrescQIPP (2019). We are an NHS funded not-for-profit organisation that supports quality, optimised

prescribing for patients. https://www.prescqipp.info (accessed 20 February 2020).

70 ABPI (2015). ABPI position on biologic medicines, including biosimilar medicines. https://www.abpi.org.uk/media/4568/abpi-position-on-biosimilar-medicines.pdf (accessed 20 February 2020).

71 Healthcare Improvement Scotland (2018). Biosimilar medicines: a national prescribing framework. http://www.healthcareimprovementscotland.org/our_work/technologies_and_medicines/programme_resources/biosimilar_medicines_framework.aspx (accessed 20 February 2020).

72 National Institute for Health and Care Excellence (2016). NICE's biosimilars position statement. https://www.nice.org.uk/Media/Default/About/what-we-do/NICE-guidance/NICE-technology-appraisals/Biosimilar-medicines-postition-statement-aug-16.pdf (accessed 20 February 2020).

73 National Institute for Health and Care Excellence (2016). Biosimilar medicines. https://www.nice.org.uk/advice/ktt15/resources/biosimilar-medicines-58757954414533 (accessed 20 February 2020).

74 The Department of Health (2015). Biosimilar awareness initiative. Australian Government. https://www1.health.gov.au/internet/main/publishing.nsf/Content/biosimilar-awareness-initiative (accessed 20 February 2020).

75 Ward, M., Lange, S., and Staff, K. (2016). Biosimilars awareness inititiative literature review. Australian Government. http://www.health.gov.au/internet/main/publishing.nsf/Content/biosimilar-literature-review (accessed 20 February 2020).

76 chf.org.au (2016). Biologic and biosimilar medicines 2020: making the most of the opportunities. https://chf.org.au/sites/default/files/biologic_and_biosimilar_medicines_2020_making_the_most_of_the_opportunit.pdf (accessed 16 June 2020).

77 Sherman, R.E., Anderson, S.A., Dal Pan, G.J. et al. (2016). Real-world evidence: what is it and what can it tell us? *N Engl J Med* 375 (23): 2293–2297.

78 Ronnebaum, S. and Atzinger, C. (2018). Enhancing biosimilar adoption with real-world evidence. Value & Outcomes Spotlight July/August: 26–28.

79 Zheng, L. (2018). Shall we (patent) dance? Key considerations for biosimilar applicants. https://www.biosimilardevelopment.com/doc/shall-we-patent-dance-key-considerations-for-biosimilar-applicants-0001 (accessed 18 February 2020).

80 Ohly, C. and Patel, S.K. (2011). Evergreening biologics. *J Gene Med* 8 (3): 132–139.

81 Karas, L., Shermock, K.M., Proctor, C. et al. (2018). Limited distribution networks stifle competition in the generic and biosimilar drug industries. *Am J Manag Care* 24 (4): e122–e127.

82 Dranitsaris, G., Jacobs, I., Kirchhoff, C. et al. (2017). Drug tendering: drug supply and shortage implications for the uptake of biosimilars. *Clinicoecon Outcomes Res* 9: 573–584.

83 Vogenberg, F. and Gomes, J. (2014). The changing roles of P&T Committees: a look back at the last decade and a look forward to 2020. *P&T* 39 (11): 760–772.

84 The Department of Health (2019). Quality use of medicines (QUM). https://www1.health.gov.au/internet/main/publishing.nsf/Content/nmp-quality.htm-copy2 (accessed 20 February 2020).

85 Papathanasiou, P., Brassart, L., Blake, P. et al. (2016). Transparency in drug regulation: public assessment reports in Europe and Australia. *Drug Discov Today* 21 (11): 1806–1813.

2

Approved Biologic Medicines and Biosimilars in Major Regulatory Jurisdictions

Prateek Jain

Independent Consultant, Toronto, Canada

KEY POINTS
• Ensure consistent quality of biologics by controlling the manufacturing process. • Rigorous clinical studies are required in patients to demonstrate safety and efficacy of biologics. • Post-approval monitoring of safety and efficacy via pharmacovigilance plans need to be in place for of all prescribed biologics in patients to generate "real-world" clinical data.

Abbreviations

Abbreviation	Full name	Abbreviation	Full name
ADA	Adenosine Deaminase Deficiency	GMP	Good Manufacturing Practice
ADC	Antibody–Drug Conjugate	HER-2	Human Epidermal Growth Factor Receptor-2
ANDA	Abbreviated New Drug Application	IDL	Import Drug License (China)
BLA	Biologics License Application	IL-5	Interleukin-5
BPCIA	Biologics Price Competition and Innovation Act	IMS	Intercontinental Medical Statistics
CBER	Center for Biologics Evaluation and Research	IV	Intravenous
		JAAME	Japanese Association for the Advancement of Medical Equipment
CDE	Center for Drug Evaluation	KIKO	Another Name for OPSR (Japan)
CDER	Center for Drug Evaluation and Research	LLP	Lipoprotein Lipase Deficiency
CFDA	China Food and Drug Administration	mABs	Monoclonal Antibodies
COPD	Chronic Obstructive Pulmonary Disease	MHLW	Ministry of Health, Labour and Welfare (Japan)
CTA	Clinical Trials Application	MIDAS	Market Information on Global Sales Activities (from IQVIA)
DO	Drug Organization (Japan)		
DRG	Diagnosis-Related Groups	NASDAQ (Index)	National Association of Securities Dealers Automated Quotations (Index)
EGRF	Endothelial Growth Factor Receptor		
EMA	European Medicines Agency	NOBs	Non-Original Biologics
EU	European Union	OPSR	Organization for Pharmaceutical Safety and Research
FDA	Food and Drug Administration		
GCP	Good Clinical Practice	PCSK-9	Proprotein Convertase Subtilisin/Kexin type 9
GLP	Good Laboratory Practice		

Biologics, Biosimilars, and Biobetters: An Introduction for Pharmacists, Physicians, and Other Health Practitioners,
First Edition. Edited by Iqbal Ramzan.
© 2021 John Wiley & Sons, Inc. Published 2021 by John Wiley & Sons, Inc.

Abbreviation	Full name
PMDA	Pharmaceuticals and Medical Devices Agency (Japan)
PMDEC	The Pharmaceuticals and Medical Devices Evaluation Center
R&D	Research and Development
SC	Subcutaneous
SCID	Severe-Combined Immuno-Deficiency
TGA	Therapeutic Goods Administration
TNF	Tumor Necrosis Factor
WHO	World Health Organization

2.1 Regulatory Frameworks

Generic drug manufacturers apply for marketing approval of generic drugs under the Abbreviated New Drug Application (ANDA) pathway established by FDA. Moreover, generic drug applications are termed "abbreviated" because they are generally not required to include preclinical and clinical data to establish safety and effectiveness. The generic manufacturer needs to demonstrate only pharmaceutical equivalence and bioequivalence between the generic and innovator products, in order to gain approval for their generic product. This approach cannot be extrapolated to biosimilars, however, because the active substance of a biopharmaceutical is a collection of large protein isoforms and not a single molecular entity, as is generally true for conventional small molecule drugs. Thus, the active substances in two products are highly unlikely to be identical and, therefore, unlike generics, biosimilars are only highly similar and not identical to the innovator products. These differences imply that biosimilars should not be approved and regulated in the same way as conventional generic drugs. The regulatory pathway for approval of biosimilars is more complex than for the generic innovator product because the design of a scientifically valid study to demonstrate the similarity of a highly process-dependent product is not easy. Further, the analytical tests currently available are not sophisticated enough to detect the slight but important structural differences between innovator and biosimilar products. Modest differences may have clinical implications and pose a significant risk to patient safety. Therefore, it is considered necessary that biosimilars must be assessed for clinical efficacy and safety by valid preclinical and clinical studies before marketing approval.[1]

To maintain a sustainable launch of biosimilars, organizations must plan the market entry by considering the local administrative and regulatory conditions and aligning their launch strategy accordingly. Most developing markets that have such an administrative system have modeled it on those of the World Health Organization (WHO) or the European Medicines Agency (EMA). With the introduction of biosimilars, it was considered that the rules set up for generic small molecules were not reasonable to replicate for the approval of biosimilars. Starting with the EMA in 2005, regulatory jurisdictions around the globe started to create and execute explicit regulatory frameworks for biosimilars to guarantee the generation of safe and effective medications.

> The approach established for generic medicines is not suitable for development, evaluation and licensing of similar biotherapeutic products (SBPs) since biotherapeutics consist of relatively large and complex proteins that are difficult to characterize.
>
> – *The World Health Organization*

There are presently 36 nations, spread across most geographic regions of the world, that have biosimilar guidelines set up that regulate the approval and registration of these medications. Such guidelines provide biosimilar manufacturers/sponsors with guidance and direction on the most suitable method to demonstrate biosimilarity with the innovator/reference biologic including guidance on nomenclature/naming and demonstrating comparability for biosimilar drugs.[2]

2.2 Major Regulatory Jurisdictions

2.2.1 Food and Drug Administration

The Food and Drug Administration (FDA) regulates innovator biologics and biosimilar products through Center for Drug Evaluation and Research (CDER) and Center for Biologics Evaluation and Research (CBER). To obtain licensure for a new biologic, the sponsor (generally the manufacturer of the product) submits to the agency a biologics license application (BLA) with data demonstrating that the biologic, and the facility in which it is manufactured, processed, packed, or held, meet standards to assure that the product is safe, pure, and potent. The Biologics Price Competition and

Innovation Act (BPCIA), enacted as Title VII of the Patient Protection and Affordable Care Act (ACA, P.L. 111–148), established an abbreviated licensure pathway for biosimilar biological products or biosimilars. To obtain licensure for a biosimilar, the sponsor submits to FDA a BLA that provides information demonstrating, among other things, biosimilarity based on data from analytical studies (structural and functional tests), animal studies (toxicity tests), and/or a clinical study or studies (tests in human patients).[3]

Since enactment of the BPCIA in 2009, as of 29 May 2019, 19 biosimilars, for nine reference products, have been licensed in the United States. However, many of these licensed biosimilars are not yet available to patients, primarily due to ongoing litigation, although various other factors may impact the uptake of biosimilars.[4]

2.2.2 European Medicines Agency

The EMA is the European Union (EU) body responsible for coordinating the existing scientific resources put at its disposal by Member States for the evaluation, supervision, and pharmacovigilance of medicinal products. The mission of the EMA is to foster scientific excellence in the evaluation and supervision of medicines, for the benefit of public and animal health. The agency provides the Member States and the institutions of the EU the best-possible scientific advice on any question relating to the evaluation of the quality, safety, and efficacy of medicinal products for human or veterinary use referred to it in accordance with the provisions of EU legislation relating to medicinal products.[5]

2.2.3 Pharmaceuticals and Medical Devices Agency (Japan)

Until recently, reviews and related operations for pharmaceutical medications were handled by two organizations. The Pharmaceuticals and Medical Devices Evaluation Center (PMDEC) was the main product review body comprising of 8–10 specialists while the Organisation for Pharmaceutical Safety and Research (OPSR or KIKO or DO) was an independent consultation body. The PMDEC and KIKO worked closely together and both made use of "experts," which would often comprise the same people. The Ministry of Health, Labour and Welfare (MHLW) supervised PMDEC and KIKO and granted approval. As part of the reforms that have been ongoing recently, the PMDEC and KIKO, along with the Japanese Association for the

Advancement of Medical Equipment (JAAME), were merged in April 2004 to form the Pharmaceuticals and Medical Devices Agency (PMDA). The PMDA now handles the whole process from clinical study stage, providing "face to face" advice, through the approval phase and is also responsible for post marketing safety measures. The PMDA comprises 25 work sites, 6 groupings (or sections), and the Kansai and Hokuriku branches. The PMDA is currently attempting to accomplish objectives under the Third Medium Range Plan (2014–2018). The role of the PMDA is to provide consultations concerning the clinical trials of new drugs and medical devices, and to conduct approval reviews and surveys of the reliability of application data.[6]

2.2.4 Therapeutic Goods Administration (Australia)

The Therapeutic Goods Administration (TGA) is Australia's drug regulatory agency for therapeutic goods. It carries out appraisal and checking to guarantee therapeutic goods accessible in Australia are of an appropriate standard with the aim of guaranteeing that Australians have timely access to safe and efficacious therapeutic advances and innovative treatments.

The regulatory framework for biologicals medicines is based on the authorization premise of guidelines on human tissue and cell-derived products and live animal cells, tissues, or organs that are provided and available in Australia or exported from Australia.

The biologicals enactment was initiated on 31 May 2011, after a suggestion from Ward, State, and Domain wellbeing experts to improve the guidelines on human tissues and cell-based treatments. All items within the purview of the system need to consent to the necessary rules that cover the enactment.

The regulatory framework applies various levels of guidelines to the particular items and the risks associated with their utilization. The regulatory framework and the subsequent guidelines and processes for approval of biologic medicines including biosimilars is intended to be adaptable and agile to suit developing emerging therapeutic innovations.[7]

2.2.5 Centre of Drug Evaluation, CDE (China)

China's hard-to-navigate drug approval system, slow processing, and unfavorable intellectual property protection environment have been major concerns for foreign Biopharma companies, but highly ranked central government officials have pledged to tackle these issues to encourage more advanced foreign drugs into the Chinese market and the Chinese health system. However, while China offers many opportunities, it still poses numerous challenges for biologic drug developers seeking to optimize biologics development and reduce product registration timelines. Global pharmaceutical companies are conducting more clinical trials in China as part of their international multicenter drug development programs.

Success in China depends on the regulatory know-how in navigating the country's evolving drug regulatory landscape. Drug developers need to balance the great advantages of China's large (often chemo-naive) patient population base and efficient study participant enrollment with obstacles posed by registration pathways for new and existing small molecule and biologic drugs.

The China Food and Drug Administration (CFDA), formerly the State Food and Drug Administration, considers drugs that are approved and marketed in other countries as new drugs in China. The CFDA designates previously approved therapies as category III "import drugs" and requires clinical data from trials conducted in China to support an application for an import drug license (IDL). This is currently the requirement for all drugs already marketed in another country.

However, for drugs that have not been approved in another international jurisdiction as yet, drug developers might choose to conduct a full clinical development program in China and submit a new drug application to gain market approval that may be achieved a several years earlier when compared with the category III pathway (China-manufactured generic drug that is only approved outside China).

In China, multinational companies operate in a climate of rapid change and increasing harmonization with research standards and processes of mature drug markets. They also encounter some dramatic differences compared with the regulatory processes of the U.S. Food and Drug Administration (FDA), the EMA, and other major drug regulatory authorities. The CFDA has developed its own standards for good clinical practices (GCP), good laboratory practices (GLP), and good manufacturing practices (GMP). CFDA technical evaluation and administrative review takes from 7 to 12 months, compared to 30 days for the FDA's IND review and 60 days for an EMA CTA review.[8]

2.3 Maturation of the Biologic Market

Small molecule drugs have had a 110-year history of scientific advancement and regulatory and industry evolution. In contrast, the modern biologic industry is relatively nascent. The earliest marketed example of a biologic medicine was only 35 years ago with the approval of the first recombinant therapeutic protein, human insulin. Biologics have enormous potential; yet much of this potential is still largely untapped, in terms of therapeutic spread, medical efficacy, and population access. This potential will gradually be fully realized as biologic technologies are translated into routine treatments, occasionally transformational lifesaving treatments. Within the next 5–10 years, the biologics drug market will undergo a transformational period of rapid evolution and maturation compared with the current biologic drug discovery paradigm and business/financial model:

1) **Biologics entering nontraditional biologic disease areas:** Biologics are entering therapeutic areas where they have not been present historically, such as asthma, dyslipidaemia, and allergy. They will expand treatment options for patients with these disease indications, many of which are underserved currently. Collectively these are important areas for future biologic drug growth but will also present challenges for market creation and maturation.
2) **Disruptive drugs and technologies:** The number of novel biologic molecules approved by the EMA and FDA has surged in the past three years. In 2016, 50% of FDA new chemical entity approvals were for biologics. This period of high biologic innovation output will bring biologic drugs that will compete with and expand the current innovator biologic and biosimilars market. New technologies also have the potential to be game changing, both in terms of efficacy and novel technological platforms.
3) **Biologic asset revaluation:** The biologic drug model, both in pre-commercialization and commercialization stages is now well understood and proven to be effective. Confidence in the growing role that biologics are playing in the pharmaceutical market is impacting on company acquisition trends.
4) **Biosimilars bring value:** We are entering a transformative period where the largest selling biologics will soon face biosimilar competition in all major international markets. Legal opinions and regulatory guidelines adopted during this initial phase of patent loss will have lasting impact beyond 2020.

5) **Competition and market environment:** While previously many new biologics were first-in-class to go to market, now many biologics are entering the market competing with other biologics with the same mechanisms of action, increasing the ferocity of market competition. As third-party payers find they have increasing choice in many therapeutic areas, such as autoimmune diseases, competitive dynamics for biologics increasingly resemble those of mature small molecule drug areas and payers place pressure on prices and discounts.

These five market trends will transform the biologic space in the next five years. Market players with interest in biologics face both challenges and opportunities in this new era; what is clear is that the biologic market will be far more complex and evolve far more rapidly than is the case currently.[9]

2.4 Player Archetypes in a Maturing Market

These biologic maturation events will have differing impact depending on individual player's strategic position.

2.4.1 Established Biologic Players

The largest biologic players are not just large within the biologic drug space, but the scale of their biologic drug success has made them global pharmaceutical leaders. Examples are Roche, Sanofi, and Amgen. These players face pressure to remain key leaders in their therapeutic areas of focus:

- **Revenue erosion.** The greatest challenge is the threat of biosimilars eroding revenues. Follow-on biologics such as PEGylated filgrastim and modern insulins have been successful in capturing and protecting franchise revenue in the past. However, recent follow-on launches such as those in the insulin space have not performed as well as expected or predicted, leaving many large biologics vulnerable to biosimilar erosion. The current payer environment is not as open to innovation on the franchise model. Follow-on innovation remains important for improving patient outcomes, but to achieve wider adoption, they must also be designed with the payer's perspective. Roche's subcutaneous formulation of Herceptin® in European markets is an example of when follow-on can succeed. The new formulation reduces treatment time

from 30–90 minutes to 5 minutes, saving time for the patient and bed/staff resources for the clinic. It has taken 28% of human epidermal growth factor receptor 2 (HER2) franchise sales and their share is growing. HER2-positive breast cancers tend to be more aggressive than other types of breast cancer. They are less likely to be sensitive to hormone therapy, though many people with HER2-positive breast cancer can still benefit from hormone therapy. Along with the other HER2 follow-on Roche products, only 43% of the franchise in Europe is currently vulnerable to biosimilar competition.

- **Greater competition.** Biologic drug classes such as growth factors, insulins, and anti-TNFs have several comparable products available and are therefore highly competitive areas. It took many years for these competitive environments to develop and for many indications particularly within oncology, the market remains relatively uncompetitive. However, this lack of competition is not an intrinsic property of the biologic market; it is a consequence of their relative novelty. A larger pipeline of biologics has meant that many players are developing treatments for the same indication, sometimes with the same mechanism of action. Previously validated disease/treatment pathways also reduce clinical trial risk, facilitating "fast follower" strategies. The result is that the window of opportunity for a first-to-market biologic will be shorter, with less market differentiation. The more competitive future biologic market will impact the return on investment for manufacturers. This can already be seen in many of the recent biologic launches such as immuno-oncologic, respiratory biologics, and PCSK-9 inhibitors.

- **Maintaining leadership.** In times of high innovation output, established players are frequently challenged by new competitors entering their field. These companies can often be more dynamic and agile, making and acting on decisions quickly. Bristol-Myers Squibb (BMS) is an example of a company that broke into a leadership position through partnering early and investing heavily. BMS moved up from rank 11 in oncology to rank 3 between 2011 and 2016 (DRG, Company & Drugs, April 2016), establishing itself as a long-term leader in immuno-oncology and a partner of choice for biotechs. Similarly, Alexion was founded as a small biotech in 1992, but has now become a top 30 biologic player, thanks to its focus on rare diseases. Large established players need to stay on the cutting edge of biologics R&D or risk losing leadership within their space. Business development will remain an important source of this innovation, but the challenge will be to keep it cost-effective given the greater future competition in the market.[10]

2.4.2 Niche Biologic Innovators

Small biotechnology companies are the lifeblood of the biologic drug industry. The positive market environment for biologic products has placed these companies in a position of strength with respect to access to capital. This has given some the ability to push through development while retaining autonomy. However, the gains from deals have never been greater. Licensing leading products while keeping earlier pipeline and the scientific talent is a popular compromise. Our understanding of the science behind biological technologies is improving, and investor confidence has increased. However, the fact remains that many novel technologies pursued by biotechs will be high-risk areas of research.[11]

2.4.3 Players Looking to Enter the Biologic Space

These are companies that predominantly invested in small molecule research and did not previously consider biologics as key to their strategies. Large pharmaceutical companies such as AstraZeneca and GlaxoSmithKline have not historically embraced the biologic wave. Many midsized innovative companies also fall under this category since they have specific disease area focus, often in therapy areas with little biologic use. Biologic therapies are becoming relevant in a greater number of disease areas. These companies will be looking to extend disease area leadership by following opportunities for investment in biologic products and biotech capabilities. For example, AstraZeneca, one of the leading companies in the respiratory space with its small molecule, Symbicort® franchise, is now poised to enter the respiratory biologics space with benralizumab.

The challenge for these incoming biologic players will be to secure deals for the most promising pipeline candidates. They will be competing with other big pharma for increasingly sought after and expensive assets, with disadvantages in areas such as experience, biologics manufacturing infrastructure, and capital in the case of midsized companies. However, these midsized companies do have a greater capability for focus, particularly in niche therapeutic areas and technologies.[12]

2.4.4 Biosimilar Players

Biosimilar players are presented with an opportunity to take sales from 15 of the top 20 biologics in most developed countries by 2020, a market value greater than US$80 billion (DRG, Company & Drugs, April 2016). Investment barriers have meant that these players will face fewer competitors relative to the small molecule generic market. However, the competition that is present has formidable resources to draw from. Leading players have been gaining experience taking biosimilars through regulatory approval. Their legal teams/partners have been setting precedents while clearing patents to prevent at-risk launch. This experience will be invaluable when preparing launches for the many biosimilar targets that will be presently moving forward. Some players such as Novartis/Sandoz, Merck & Co, and Amgen will be playing in both the originator/innovator and the biosimilar space. These hybrid players can leverage their expertise in biologic development and manufacturing to generate synergies. They also have the financial capacity to invest heavily in this space. However, they will face a certain degree of conflict of interest.

Biosimilar players can also bring bio-betters to market. Today's off-patent originator molecules were engineered over 15 years ago. Since then, scientific advancements have enabled biosimilar developers to improve the molecule significantly. Novel screening methods have assisted in the detection and replacement of immunogenic parts of the protein; iterative binding assays have improved specificity and binding strength; better understanding of structure/solution stability and stress tests have improved the temperature stability and shelf life. However, these improvements with respect to biosimilars are limited by the regulatory requirement to keep the molecule like the originator, this is to enable simple switching and to avoid dosage confusion.

The challenge is that there is currently no dedicated FDA or EMA regulatory guidance for bio-betters. Approval through a novel medicine's pathway would require the developer to invest in clinical trials at a scale like creating a new biologic product. If the improvements on the molecule are not transformative, and payers do not see a profound therapeutic value, the return will not be high enough to justify investment. In the longer term, the development of an abbreviated bio-better regulatory pathway remains a possibility. As the biosimilar market becomes more competitive, players looking for product differentiation may explore this bio-better route.[13]

2.5 Outlook: Landscape of the Biologic Market

Biologics markets are concentrated and crowded. Currently, the top 10 biologic therapies account for 36% of all biologic drug spending. The three largest biologic therapy areas (autoimmune, diabetes, and oncology) are worth US$110 billion, over half of all biologic drug revenue. They are represented in 9 of the top 10 biologics and are increasingly relevant due to their contribution to 70% of biologic market growth since 2010.[14]

Over half of the biologic pipeline is in therapy areas with few or no biologic treatments on the market. Their large presence in the pipeline is a sign of biologics broadening their therapeutic focus, bringing new therapeutic areas for growth.[15]

2.5.1 Biologics in Nontraditional Biologic Disease Areas

2015 was the year that two high-profile classes of biologics had their first launches, the anti-PCSK9 mAbs for hypercholesterolemia (Repatha® and Praluent®) and an antiIL-5 mAb for severe asthma (Nucala®). These launches were particularly important because the indications they were approved for have seen either no biologics (hypercholesterolemia) or a single biologic (asthma-Xolair®). Both diseases are highly prevalent and mainly treated by primary care physicians using widely available generics. Healthcare systems have not been accustomed to restructuring drug administration for these patients with biologics, let alone paying for them.[16]

However, it is important to consider that biologics entering nontraditional biologic disease areas may take longer to optimally position within the patient pathway. This is because primary care physicians and patients are not accustomed to prescribing and using biologics, respectively, so it may take longer to benefit from such innovation. Historic examples of slow initial uptake for drugs in this category can be seen in the case of Xolair® and Prolia®; however, both drugs have now surpassed US$1 billion in sales (DRG Company & Drugs, April 2016).[17]

2.6 Technology and Science Innovation in the Long Term?

2.6.1 The Potential of Innovative Technologies

Currently, mAbs hold the lion's share of the biologic market sales and remain the largest protein technology class within the biologic pipeline. However, the mAb dominance we see today could be outperformed by novel biologic technologies currently in the pipeline. In the next 10 years, therapies using nonestablished technologies will have been launched into the market. Although only a handful of launches will occur before 2020, these first few will show us the potential of these therapeutic strategies to change the way we treat disease in the long term. There are four technology classes with significant pipeline scale that will be entering a pivotal stage during their first few launches by 2020.

1) **Antibody–drug conjugates:** A drug (e.g. a cytotoxin) is coupled to an antibody that specifically targets a specific biological marker (e.g. cell surface tumor antigen). The function of the antibody is to act as a vector, enabling targeted delivery of the toxic drug to the antibody target. When compared with standard drug treatment, it allows orders-of-magnitude lower dosage, reducing the undesirable systemic side effects caused by the toxic drug. This means that a drug or certain high drug dosages that may have previously been too toxic for use in treatment can be utilized safely. There are currently two antibody–drug conjugates (ADCs) on the market, Kadcyla® marketed by Roche/Genentech and Adcetris® marketed by Seattle Genetics/Takeda. There are an additional 17 ADCs from Phase II through registration looking to enter the market in the near future. Depending on clinical success and market acceptance, we may see ADCs becoming a more popular pipeline choice.

2) **Antisense/RNAi:** These are two similar naturally occurring biological processes in which RNA molecules modulate the level of gene expression. They have been manipulated for therapeutic benefit in order to prevent the expression of disease-causing proteins with great specificity. These are relatively new technologies, RNAi was only utilized as a scientific technique in 1998, but they are showing great promise in a range of therapy areas from oncology to hyperlipidemia. Improvements in delivery systems have been key to enabling the use of these unstable RNA treatments. Two pioneering antisense RNA treatments were approved by the FDA in 2016, Spinraza® and Exondys®. Spinraza® is the only available treatment for spinal muscular atrophy, an orphan disease with a low life expectancy. Market analysis consensus revenue for Spinraza® is over US$1 billion by 2021 (IMS MIDAS 2016), very substantial considering it is a novel technology. With 44 antisense/RNAi candidates in Phase II and later phases, this could be an important segment for biologic market growth.

3) **Gene therapy:** Gene therapies are treatments in which genetic material is incorporated into the cells of a patient with an intended therapeutic benefit. Much of the gene therapy pipeline candidates function by attempting to correct or replace a genetic defect that underlines the root cause of the disease. The only examples of approved gene therapies are Glybera®, used to treat lipoprotein lipase deficiency, and Strimvelis®, for treating adenosine deaminase deficiency (ADA)-severe combined immuno deficiency, ADA-SCID. The potential for gene therapies is that they aim to be curative. There are also gene therapies going beyond genetic correction and toward non-corrective, with more sophisticated mechanisms of action. Examples are pipeline candidates aiming to stimulate nerve cell growth in patients with Parkinson's disease and stimulating blood vessel growth for heart disease.

4) **Cell therapy:** Cell therapies are treatments in which intact, living, human cells are injected into a patient for therapeutic benefit. Sixty percent of the cell therapies in development are autologous (fully personalized treatments where the cells themselves originate from the patient), the rest are allogeneic (off the shelf). In 2010, the FDA approved the first ever autologous cell therapy vaccine, Provenge®. Although this product was not a commercial success, this area remains very dynamic particularly due to the high-profile CAR T-cell, and T-cell treatments that have been valued so highly during recent company acquisitions.

Collectively these drug technologies make up 18% of the Phase II+ biologic drug pipeline. The performance of each technology class is somewhat dependent on the first few launches. The challenges associated with the cost of groundbreaking curative treatments in the pipeline must be tackled proactively. Innovative approaches to funding will be a necessary prerequisite of success when commercializing such valuable treatments.[18]

2.6.2 Drug Delivery: Calls for Change

There are currently two biologic delivery methods that are used for the great majority of biologic products, intravenous (IV) and subcutaneous (SC). Today, many biologics have subcutaneous formulations available. This has the advantage of enabling patient self-administration and often cutting down on the delivery time. This solves many of the challenges of IV delivery; however, there is still room for improvement and innovation. Much of the work for innovative biologic delivery has been in the diabetes space. This is because diabetes is a primary care area with an extremely large and growing patient population that could see significant benefit and increased compliance with insulin treatment should administration be made easier and less onerous. Once established, these technologies could spread to other disease areas. This will be particularly important in diseases with large patient populations like asthma, chronic obstructive pulmonary disease (COPD), and hypercholesterolemia.

Collectively these drug technologies make up 18% of the Phase II+ biologic drug pipeline. It is still not clear which of these platforms will enter the mainstream market. The performance of each particular technology class is somewhat dependent on the first few launches. If they fail to deliver clinically and commercially, these launches serve as warnings to investors for the platform as a whole. Existing marketed examples of cell and gene therapies have faced multiple challenges in commercialization, particularly in the funding of treatment. The western world's first gene therapy, Glybera®, was priced at €1.1 million in Germany. Glybera® is used to treat an ultra-orphan indication, lipoprotein lipase deficiency, but much of the pipeline similarly aims to cure disease and will likely be priced highly. The challenges associated with the cost of groundbreaking curative treatments in the pipeline must be tackled proactively. Innovative approaches to funding will be a necessary prerequisite to success when commercializing such valuable treatments. Instilling payer confidence in a technology's curative promise is challenging given the inability for clinical trials to model a lifelong cure. Schemes such as the UK's Early Access to Medicines and the Accelerated Access Review can enable early collection of real-world data before approval and enable longer periods of evaluation.[19]

2.6.3 Biologic Asset Deal Frenzy

As biologic pathway targets are validated, competition for the mode of action intensifies. In the 2012–2016 period, the upfront value of a biologic product deal rose from ~US$20 to 60 million (IMS MIDAS, 2016), tripling in four years. These valuations increased due to three factors:

- Innovation output from biotechs has increased in scale and quality. This is possible due to greater scientific understanding of disease, advancement in scientific techniques, and their wider availability.
- Investment strategies are increasingly incorporating biologics into the pipeline with large pharma driving the trend. This competition, particularly between companies with deep pockets, is driving up deal values.

- There has been prolonged availability of capital at low interest rates, promoting deal making across all sectors.

The Valeant, Turing, and Mylan pricing scandals attracted heavy criticism in late 2015 and 2016. The resulting attention from policy makers in the United States has concerned investors, reducing expectations of future pharmaceutical market potential and growth. This has contributed toward the fall of the NASDAQ biotech index by 21% since September 2015. President Trump's pronouncements after his election have served to increase the uncertainty of an already nervous and volatile sector. The number of biologic product deals signed has also risen. Between 2008 and 2012, these numbers were relatively stable at ~250 deals per annum. However, in 2015, the number of deals announced reached 400. A relevant question to be posed is: Where are these biologic assets being sourced from?

Historically, most biologic product deals have been executed early in the drug development cycle. This trend is becoming more pronounced. Between the 2006–2010 and 2011–2015 periods, deals for biologic drugs in development to Phase II increased by over 60%. If we look at deal growth in absolute terms, the bulk of biologic deal increase is coming from very early stage, discovery/preclinical (392 more deals, 71% of deal growth) (DRG Deal Database, 2017).

There are not many high-potential late-stage biologics left to acquire. Demand for pipeline biologic therapies has increased but it will take several years before reactive supply will progress to the late stage.

- High valuation of biologic products is pushing players who are unwilling to invest heavily to look earlier in the development of promising candidates.
- Players in the industry now have many years of experience developing biologics. They have taken them from scientific concept through to market blockbusters. As a result of this experience, more players have comfort in conducting early stage deals.
- The greater risk of early deal making has been balanced with the increased usage of contracted milestones within deals.[20]

2.7 The Arrival of Major Biosimilars

2.7.1 Biosimilar Immediacy

When small molecule drugs lose patent protection, generics enter the market, resulting in lower drug cost burden for payers. These savings are channeled into the funding of new innovative drugs and expanding

access to older ones. The same innovation cycle for biologics is reaching maturity. Many biologic blockbuster products now have biosimilars lined up to take market share. Those biologic makers facing loss of exclusivity on a current marketed product can be partially comforted by the prospect of funding availability for future launches.

- **A jump in biosimilar availability and usage:**
  - The first biosimilar mAb, infliximab, has launched for all originator indications and has taken majority share in several European markets. There are now three competing infliximab biosimilar brands in Europe: Remsima® marketed by MundiPharma, Inflectra® by Pfizer/Hospira, and Flixabi® by Biogen.
  - The list of biosimilar molecules that have gained FDA approval now includes filgrastim, infliximab, adalimumab, and etanercept, with many more entrants expected before the end of the decade.
  - A rich pipeline with over 240 biosimilars in development (including only those that are announced publically) will mean that launches will be coming with increasing frequency and there will greater competition within each molecule.
- **Stakeholders will have biosimilars high in their priorities. They will gain a lot of experience in the space of a few years:**
  - Regulators will be clarifying guidance for biosimilar manufacturers. Many regulatory bodies are aligning guidelines to those of the EMA.
  - Country medicines agencies will be assessing the clinical evidence over time. Important decisions on stance for switching patients to biosimilars will be adopted as a result.
  - Payers will be grappling with barriers to biosimilar uptake in order to find savings and increase leverage.
  - Physician and patient groups will express their views. These will form the backbone of public opinion on biosimilars and have the potential to influence regulatory agency guidance.
  - The biopharma industry, innovative and biosimilar players, will develop new strategies for competition. The level of discounting that a biosimilar business model can sustainably provide and absorb will be more fully understood.

The decisions and opinions developed during this transition period will set precedence moving forward. As a result, keeping up to date with this rapidly changing space will be important for strategic decision-making in the short and long term.

Many of the top 20 biologics are already exposed to biosimilar competition. An estimated 6/20 have lost exclusivity in the United States and 7/20 in Europe. By 2020, these figures will increase to 15/20 and 14/20, respectively (DRG Company & Drugs, 2016).[21]

2.7.2 Regulatory Hurdles for Biosimilar Launch

The regulatory evolution of biosimilars is still relatively immature. The EMA published the world's first biosimilar guidelines in 2005, with the FDA publishing its guidelines in 2012. The convergence between these and other regulatory guidelines has been slow, preventing single cost-effective biosimilar development. Biosimilar legislation is only in its infancy. The Hatch-Waxman Act in the United States in 1984 did not lead to an immediate mature small molecule drug generic market, and neither will this be the case for biosimilars. As regulatory agencies and biosimilar manufacturers gain experience in bringing biosimilars to market, regulatory difficulties and pathways will have a smaller impact on biosimilar uptake.

Players have gained experience developing and taking biosimilars through regulatory procedures. This combined with a greater understanding of the biosimilar business model has meant that biosimilars are now being developed earlier and with greater competition than was the case previously. Moving forward, the lag time between biologic loss of patent protection and biosimilar launch will decrease.[22]

2.7.3 Interchangeability and Substitution

Another factor that may affect uptake of biosimilars is that a biosimilar generally cannot be automatically substituted for the reference product (i.e. brand-name or innovator biologic) at the pharmacy level unless it is determined to be interchangeable with the reference product. An interchangeable product "can be expected to produce the same clinical result as the reference product in any given patient and, for a biological product that is administered more than once, that the risk of alternating or switching between use of the biosimilar product and the reference product is not greater than the risk of maintaining the patient on the reference product. Interchangeable products may be substituted for the reference product by a pharmacist without the intervention of the prescribing health care provider." In January 2017, FDA released draft guidance on interchangeability, and final guidance on 10 May 2019. FDA has not yet approved an interchangeable biologic product.[23]

2.8 Biosimilars in Emerging Markets

The emerging markets typically have relatively low access to biologic medicines when compared with developed markets. Patients in these markets stand to gain the greatest increase in access as a result of biosimilar competition. This has caused emerging market health authorities to put significant effort into encouraging use of non-original biologics (NOBs). NOBs are copy-biologics that have not gone through a biosimilar pathway with strict regulatory scrutiny such as the EMA, FDA, or WHO biosimilar guidelines. They have been preferred in the emerging markets due to their early access and lower price relative to true biosimilars. NOB uptake has been significant, the market was worth US$2.1 billion in 2015 relative to US$1.1 billion globally for true biosimilars (IMS MIDAS, 2016). They equate to 18% of all biologic sales in Pharmerging markets (countries having low positions in the pharmaceutical market but having rapid pace of growth) and are growing at almost twice the speed. However, the biosimilar regulatory environment in the emerging markets is changing rapidly. There has been a marked push for increasing quality of copy biologic medicines but increasing access and affordability will continue be the top priorities for policy makers.[24]

2.9 Top 10 Biologic Drugs in the United States

The 10 drugs on the list all have exceeded the monetary definition of a "blockbuster," which designates a drug as having generated more than US$1 billion in annual sales.[25] Figures are based on 2018 sales data reported by the respective manufacturers for revenues generated in 2017.

2.9.1 Humira®

The anti-inflammatory drug Humira® is not only the best-selling biologic; it is one of the best-selling drugs worldwide, regardless of class.

Indications: Rheumatoid arthritis, plaque psoriasis, Crohn's disease, ulcerative colitis, ankylosing spondylitis, psoriatic arthritis, polyarticular juvenile idiopathic arthritis
Manufacturer: AbbVie, an Abbott Laboratories spinoff
Global Sales in 2017: $18.4 billion
Generic name: Adalimumab
Launch date: 2002

2.9.2 Rituxan®

Rituxan® (rituximab) was developed by IDEC Pharmaceuticals under the name IDEC-C2B8. Rituxan® is currently co-marketed in the United States by Biogen Idec and Roche subsidiary Genentech.

Indication: Non-Hodgkin's lymphoma, chronic lymphocytic leukemia, rheumatoid arthritis
Manufacturer: Roche
Global Sales in 2017: $9.2 billion
Generic name: Rituximab
Launch date: 1997

2.9.3 Enbrel®

Enbrel® was developed by researchers at Immunex. Today the drug is co-marketed in North America by Amgen and Pfizer, by Takeda Pharmaceuticals in Japan, and by Wyeth in the rest of the world.

Indication: Rheumatoid arthritis, plaque psoriasis, psoriatic arthritis
Manufacturer: Pfizer/Amgen
Global Sales in 2017: $7.9 billion
Generic name: Etanercept
Launch date: 1998

2.9.4 Herceptin®

Herceptin® was developed by Genentech, now a Roche subsidiary, and UCLA's Jonsson Comprehensive Cancer Center.

Indication: HER2+ breast cancer
Manufacturer: Roche
Global Sales in 2017: $7.4 billion
Generic name: Trastuzumab
Launch date: 1998

2.9.5 Avastin®

When launched in 2004, Genentech's Avastin® was one of the most expensive drugs on the market, with a US$4400 monthly price tag.

Indication: Breast, colorectal, kidney, non-small-cell lung, glioblastoma, ovarian cancers
Manufacturer: Roche
Global Sales in 2017: US$7.1 billion
Generic name: Bevacizumab
Launch date: 2004

2.9.6 Remicade®

Remicade® was originally developed by Centocor Ortho Biotech, which is now Janssen Biotech, a Johnson & Johnson subsidiary.

Indications: Rheumatoid arthritis, Crohn's disease, ankylosing spondylitis, psoriatic arthritis, plaque psoriasis, ulcerative colitis
Manufacturer: Johnson & Johnson/Merck & Co.
Global Sales in 2017: US$7.1 billion
Generic name: Infliximab
Launch date: 1998

2.9.7 Lantus®

Lantus® was developed at Sanofi-Aventis's biotechnology research center in Frankfurt-Höchst, Germany.

Indication: Diabetes
Manufacturer: Sanofi
Global Sales in 2017: US$5.7 billion
Generic name: Insulin glargine [rDNA origin] injection
Launch date: 2000

2.9.8 Neulasta®

Neulasta® is pegylated version of Neupogen® launched by Amgen in 2002, with extended half-life, which allows for administration as a single dose per chemotherapy cycle compared with daily dosing with Amgen's Neupogen®, for reducing the incidence of infection, as manifested by febrile neutropenia in patients with nonmyeloid malignancies who are receiving myelosuppressive drugs.

Indication: Neutropenia related to cancer chemotherapy
Manufacturer: Amgen
Global Sales in 2017: US$4.7 billion
Generic name: Pegfilgrastim
Launch date: 2002

2.9.9 Avonex®

Avonex® is marketed in the United States by Biogen Idec and by Merck under the brand name Rebif®. Gemany's Fraunhofer Institute for Interfacial Engineering and Biotechnology IGB and CinnaGen Company cloned Interferon beta-1a and since 2006 the drug has been sold as CinnoVex®, a biosimilar, in Iran.

Indication: Multiple sclerosis (MS)
Manufacturer: Biogen Idec
Global Sales in 2017: US$2.1 billion
Generic name: Interferon beta-1a
Launch date: 1996

2.9.10 Lucentis®

Developed by Genentech, ranibizumab, an injectable, is marketed in the United States by Genentech and by Novartis outside the United States.

Indication: Age-related macular degeneration
Manufacturer: Roche, Novartis
Global Sales in 2017: US$1.5 billion
Generic name: Ranibizumab
Launch date: 2006

2.10 Top 10 Biologic Drugs in the EU

The top ranked biologics in the EU are predominantly the same as in the United States.

2.11 Conclusions

The biologic market is large and rapidly expanding. It accounts for over a quarter of pharmaceutical spending, giving it increasing payer attention. The pipeline contains a growing share of biologic drugs preparing to launch into therapy areas that have seen very little biologic use historically, such as Alzheimer's, asthma, and cardiovascular diseases. Extremely large patient populations in these areas will accelerate biologic drug budget growth. Additionally, novel therapeutic technologies in the pipeline such as gene therapies and autologous cell therapies will also be launched with greater frequency. The high cost per patient for some of these potentially curative products will place immense pressure on drug budgets further.

When launching a biologic into this payer environment, pharmaceutical companies should look seriously at alternative funding mechanisms. Pay for performance schemes have successfully been implemented in the United States. However, other novel mechanisms should be explored, with different mechanisms varying in effectiveness depending on the specific treatment and regulatory jurisdiction. Examples include: differential pricing, which can be based on which indication a drug is used for or the severity of the patient condition;

payment in installments, which spreads an acute one-off budget impact into manageable portions, e.g. in indications with patient warehousing:

- Larger patient populations and budget constraints favor a strategy based on volume growth. The growth of biologic manufacturing capability in the Far East has enabled the production of cheaper biologics. Companies should look at opportunities to reduce manufacturing costs with the primary ambition of expanding access, particularly in less developed markets where drug cost is more likely to limit access.
- Launching a biologic into a nontraditional biologic indication creates unique challenges that must be actively overcome. For example, the majority of patients for these biologics will historically have been treated by primary care physicians, who may have had little exposure to biologics. Without appropriate education, these gatekeepers may not efficiently refer patients through to the appropriate specialists who can carry out treatment. A coordinated, multichannel approach can supplement physician education while maintaining commercial cost effectiveness.

Older products in traditional biologic therapy areas are being joined by competitors, both original and biosimilar, fragmenting the at-present concentrated market and applying downward pricing pressure. The scale of the biosimilar pipeline will ensure that in the future, off-patent competition will come rapidly after key patent expiry, giving originators little hope of maintaining unprotected and guaranteed biologic revenue.

Originators planning to protect a franchise from biosimilars must ensure that follow-on biologics are a strong value proposition for payers. Players should direct the development of follow-ons to improve efficacy, reduce side effects or effectiveness in patients not clinically satisfied with current biologics. Easier administration or patient support services are nice to have but may not make the difference between a low-cost biosimilar and a costlier newer agent.

The on-patent innovative biologic market will also be under greater competitive pressure. Recent launches of biologics have more quickly been followed by other originator competitors, often with the same mechanism of action. The pipeline follows this trend with more launches forecast for autoimmune indications, diabetes combinations, oncology checkpoint inhibitors, EGFR inhibitors, and respiratory biologics. In this more competitive landscape, first to market advantage will be short lived. Rapidly maximizing market share is more important than ever, and promotional strategy should reflect this. However, expectations for biologic launches should still be adjusted accordingly and pragmatically applied during business development.

- Players should carefully plan the timing of a product acquisition. The stabilization of deal price in this political environment may present favorable opportunities.
- Companies should look earlier in development for acquisition targets. Strong competition and relatively high prices have left fewer valuable late-stage assets.
- It will be increasingly important for companies to strengthen the competencies required to nurture a biologic pipeline candidate, particularly if they have historically been small molecule focused.

Ranging from incumbent biologic drugs to areas with no currently available biologics, the market is undergoing an unprecedented period of change. Market leadership will be at stake. As more companies bring biologics into the mainstream of their portfolio and biologics become mainstream across more and more therapies, the companies that thrive will be those that are bold in their investments, effective in their product differentiation, and innovative in their commercial model and acumen.

References

1 https://www.researchgate.net/publication/228730641_Biosimilars_An_overview (accessed 20 September 2019).

2 https://researchadvocacy.org/sites/default/files/resources/Biosimilar%20Medicines6_4Final.pdf (accessed 20 September 2019); World Health Organization. *Biological Qualifier: An INN Proposal.*

3 https://www.fda.gov/drugs/therapeutic-biologics-applications-bla/biosimilars (accessed 20 September 2019).

4 Programme on International Nonproprietary Names (INN). Revised draft July 2014. http://www.who.int/medicines/services/inn/bq_innproposal201407.pdf (accessed 20 September 2019).

5 European Medicines Agency. *Product-Information Requirements.* http://www.ema.europa.eu/ema/index.jsp?curl=pages/regulation/general/general_content_000199.jsp; https://www.ema.europa.eu/en/documents/leaflet/biological-medicinal-products_en.pdf (accessed 20 September 2019).

6 https://www.pmda.go.jp/files/000152369.pdf; https://www.ema.europa.eu/en/human-regulatory/research-development/scientific-guidelines/biological-guidelines (accessed 20 September 2019).

7 https://www.tga.gov.au/publication/australian-regulatory-guidelines-biologicals-argb; https://www.tga.gov.au/regulatory-framework-biologicals (accessed 20 September 2019).

8 https://www.ppdi.com/drug-registration-china/home_fierce.htm; Ropes and Gray (2015). *China: China Announces Final Biosimilars Guideline.* https://www.ropesgray.com/news-and-insights/Insights/2015/March/China-Announces-Final-Biosimilars-Guideline.aspx?utm_source=Mondaq&utm_medium=syndication&utm_campaign=View-Original (accessed 20 September 2019).

9 https://www.iqvia.com//media/iqvia/pdfs/nemea/uk/disruption_and_maturity_the_next_phase_of_biologics.pdf (accessed 20 September 2019).

10 https://www.phrma.org/advocacy/research-development/biologics-biosimilars; https://cdn.ymaws.com/www.casss.org/resource/resmgr/cmc_euro_speaker_slides/2018_cmce_JekerleVeronika.pdf (accessed 20 September 2019).

11 https://novel-drugdelivery-systems.pharmaceuticalconferences.com/events-list/biologics-biosimilars (accessed 20 September 2019).

12 https://www.prnewswire.com/news-releases/commercialisation-of-biologics-benchmarking-leading-players-300563293.html (accessed 20 September 2019).

13 https://www.lifescienceleader.com/doc/the-players-in-the-biosimilar-market-archetypes-0001; https://www.biosimilardevelopment.com/doc/biosimilar-pipeline-analysis-market-detailed-industry-report-live-covering-players-biocon-pfizer-0001; https://www.biopharmadive.com/news/7-companies-to-know-in-the-emerging-biosimilars-field/433539; https://www.mckinsey.com/industries/pharmaceuticals-and-medical-products/our-insights/five-things-to-know-about-biosimilars-right-now (accessed 20 September 2019).

14 QI MIDAS Q4 2015, Pharmerging markets only – Pharmerging markets defined as a group of 21 emerging countries adding US$1 billion of cumulative pharmaceutical growth over the next five years.

15 https://www.marketwatch.com/press-release/biologics-market-2019-industry-trends-size-growth-insight-share-emerging-technologies-share-competitive-regional-and-global-industry-forecast-to-2024-2019-05-24 (accessed 20 September 2019).

16 https://mordorintelligence.com:81/industry-reports/biologics-market (accessed 20 September 2019).

17 https://www.businesswire.com/news/home/20180831005463/en/Global-Biologics-Manufacturing-Market-2018-2022-Introduction-New; https://www.iqvia.com/-/media/iqvia/pdfs/nemea/uk/disruption_and_maturity_the_next_phase_of_biologics.pdf (accessed 20 September 2019).

18 https://www.amgenbiotech.com/manufacturing-innovation.html; https://www2.deloitte.com/content/dam/Deloitte/us/Documents/life-sciences-health-care/us-lshc-advanced-biopharmaceutical-manufacturing-white-paper-051515.pdf; https://www.bio.live/visit/news-and-updates/biologics-driving-innovation-drug-packaging-and-devices (accessed 20 September 2019).

19 https://www.iqvia.com/-/media/iqvia/pdfs/nemea/uk/disruption_and_maturity_the_next_phase_of_biologics.pdf; https://www.pharmaceutical-journal.com/news-and-analysis/features/making-drugs-work-better-four-new-drug-deliverynbspmethods/20203530.article?firstPass=false; https://fortune.com/2017/05/03/drug-delivery-breakthroughs (accessed 20 September 2019).

20 QuintilesIMS White paper: cell & gene therapies: innovation to commercialisation – can industry bridge the gap? Published QI 2015; FDA. *Implementation of the Biologics Price Competition and Innovation Act of 2009.* http://www.fda.gov/Drugs/GuidanceComplianceRegulatoryInformation/ucm215089.htm (accessed 20 September 2019).

21 QI MIDAS Q4 2015, Pharmerging markets only – Pharmerging markets defined as a group of 21 emerging countries adding US$1 billion of cumulative pharmaceutical growth over the next five years.

22 FDA (2019). *Considerations in Demonstrating Interchangeability with a Reference Product, Guidance* (May 2019); FDA. *From Our Perspective: Biosimilar Product Labeling.* http://www.fda.gov/Drugs/NewsEvents/ucm493240.htm (accessed 20 September 2019).

23 https://www.mabxience.com/blogs/differences-between-substitution-and-interchangeability; https://www.biopharma-reporter.com/Article/2017/03/21/Biosimilr-switching-interchangeability-and-substitution-the-EU-view; https://safebiologics.org/wp-content/uploads/2016/04/AUS-Onepager-Substitution-FNLe.pdf; https://www.amgenbiosimilars.com/bioengage/prescribing-biosimilars/substitution-and-interchangeability (accessed 20 September 2019).

24 https://www.mckinsey.com/industries/pharmaceuticals-and-medical-products/our-insights/whats-next-for-biosimilars-in-emerging-markets; https://www.beroeinc.com/whitepaper/biosimilars-emerging-markets (accessed 20 September 2019).

25 https://www.thebalance.com/top-biologic-drugs-2663233 (accessed 20 September 2019); QI MIDAS Q4 2015, Pharmerging markets only – Pharmerging markets defined as a group of 21 emerging countries adding US$1 billion of cumulative pharmaceutical growth over the next five years.

3

Status of Biologic Drugs in Modern Therapeutics-Targeted Therapies vs. Small Molecule Drugs

Cody Midlam

Willis Towers Watson, Cincinnati, OH, USA

KEY POINTS

- Biologics are increasingly becoming the focus of pharmaceutical research and development.
- Biologic medicines are able to treat certain diseases more efficiently than SMDs.
- The cost of biologics is resulting in significant and potentially catastrophic strain on payer systems around the globe.
- Orphan drugs will cost more than other drugs and the current development pipeline for these is relatively large.
- Computer modeling techniques will generate an even higher probability of identifying biologic drug targets in the future.

Abbreviations

Abbreviation	Full name
ACE	Angiotensin Converting Enzyme
BLA	Biologic Licensing Application
CA	Carbonic Anhydrase
CBER	Center for Biologics Evaluation and Research
CHO	Chinese Hamster Ovary
COPD	Chronic Obstructive Pulmonary Disease
DBC	Division of Biological Control
DNA	Deoxyribonucleic Acid
EGRF	Epidermal Growth Factor Receptor
ERTs	Enzyme Replacement Therapies
FDA	Food and Drug Administration
GERD	Gastro-oesophageal Reflux Disease
HTA	Health Technology Assessment
ICER	Institute for Clinical and Economic Review
IFN	Interferon
IgG	Immunoglobulin
IL	Interleukin
IP	Intellectual Property
LSD	Lysosomal Storage Disease

Abbreviation	Full name
mAbs	Monoclonal Antibodies
NSCLC	Non-Small Cell Lung Cancer
PD-1	Programmed Death-1
PD-L1	Programmed-Death-Ligand 1
R&D	Research and Development
rDNA	recombinant DNA
RNA	Ribonucleic Acid
TB	Tuberculosis
TNF	Tumor Necrosis Factor
SMDs	Small Molecule Drugs
US	United States

3.1 Biologics in Contemporary Medicine

3.1.1 A History of Drug Development and Commercialization

The use of plants, minerals, and other naturally occurring substances for medicinal purposes dates to ancient times. A trial-and-error paradigm was used to identify which plants or extracts resulted in desired effects, and the end products were non-standardized. In the thirteenth century, drug quality standards were being developed, and in the fourteenth and fifteenth centuries, the first pharmacopeias were being published.[1] Pharmacopoeias list drug effects and directions for their use and laid the foundation for standardized pharmacy practice and drug and drug formulation development. Little changed in the process of drug discovery up until this time as a trial-and-error approach for identifying bioactive compounds with medicinal properties was the norm. This is the foundation for the type of drug discovery known as "irrational design," which resulted in most drug discoveries up until the past century.

At the end of the eighteenth century, digitalis was identified as the active ingredient from the foxglove plant.[2] Scientists were starting to focus on the relationship between the chemical makeup of a plant's active ingredient and the resulting biologic effect in humans. Numerous drugs were developed for treatment during the 1800s including digitalis, quinine, aspirin, and mercury.[3] Modern pharmacology, the study of the properties and reactions of drugs in relation to their therapeutic value, was being developed during this era. The British pharmaceutical company, Beecham, began patenting medicines in 1842 and there were several companies that would go on to manufacture pharmaceuticals in the upcoming century that were founded in the 1800s including Pfizer, Bayer, Solvay S.A., Eli Lilly, Johnson & Johnson, and Hoffmann-La Roche.[4]

As the twentieth century began, the study of pharmacology and the relatively recent technological innovations of the time provided the foundation for the pharmaceutical industry as it exists today. Drug discovery in the 1930s focused on irrational design, screening natural products and finding active ingredients that could be made synthetically. In 1928, *Penicillium* mold was identified as being active against the bacteria staphylococcus. Penicillin was being mass produced and commercially available in the 1940s before the end of World War II.[2] By the end of 1944, Pfizer had perfected a method of production called the deep vat method and was the world's largest penicillin producer. The company was producing enough doses to treat 100 patients each month.[5]

The success of penicillin prompted scientists to research if other soil-dwelling organisms could work better than penicillin for treating infection, or potentially, for treating infections that penicillin could not. Tuberculosis (TB) was then, and is still today, one of the

more prevalent and deadly of bacterial infections. In 1943, a Nobel Prize-winning scientist found a strain of streptomyces that was effective against TB and later dubbed the drug streptomycin. Merck developed streptomycin commercially and by 1949 began administering the drug around the world treating millions of TB sufferers.[5]

Finding, isolating, testing, and marketing a successful treatment for TB began an explosion in antibiotic research by the pharmaceutical industry. Nearly every major manufacturer had an antibiotic research unit by the 1950s with scientists testing a library of soil-based organisms for the next cure.[5] Drugs like vancomycin and erythromycin were developed during this time as scientists worked to find new drugs or ways to tweak the molecular structure of existing drugs. Antibiotic resistance was manifesting itself as a new problem shortly after penicillin was developed, and the need for new antibiotics was high.

While the need for new drugs to treat drug-resistant bacteria remains high today, the interest in this area of research by the pharmaceutical industry is relatively low. By the 1980s, the industry began pulling out of antibiotic R&D; by the 1990s, many of the largest companies were out completely or nearly completely out of this space, and by 2017, 15 of the 18 largest pharmaceutical companies had abandoned the antibiotic market.[5] Pharmaceutical manufacturers would focus their energies from acute treatments to chronic maintenance medications.

3.1.2 The Business Model for Chronic Disease

While drugs to treat chronic disease may be taken for life, antibiotics are commonly taken for only a short course of time until a patient is effectively cured of the infection. This limits how long antibiotics are taken by a patient and subsequently sold by a pharmaceutical company. To further limit the market potential for antibiotics, physicians began hoarding the newest drugs, reserving them only for the very sickest patients with antibiotic-resistant infections. This practice continues today as one component of antibiotic stewardship. The pharmaceutical industry needed more reliable sources of income, and chronic disease was an area of focus and growth.

At the same time as antibiotics were swiftly changing medical practice in the treatment of infection, the Framingham Heart Study was launched in 1948 with the goal of identifying risk factors for cardiovascular disease. It would not be until the early 1960s that reliable data would be reported from this epidemiological study, but drug development that would impact cardiovascular disease was already underway, although initial efforts were not focused there. In the 1950s, Merck was researching carbonic anhydrase (CA) inhibitors for the purpose of reducing blood acidity. The CA inhibitor class of drugs being studied resulted in the side effect of more frequent urination, and soon became known as "diuretics." Eventually, Merck saw potential in this side effect and marketed Diuril® for the treatment of edema in 1958. Soon after, physicians started using it to treat the newly identified risk factor for stroke, now known to be hypertension. Physicians were prescribing Diuril® off-label (meaning it was approved for treating edema but being used for the unapproved indication of hypertension treatment). While it was not the intent of their initial efforts, Merck had just invented the first antihypertensive. Competition soon followed and within a few years more than six thiazide drugs were approved by the United States Food and Drug Administration (FDA).[5]

Antihypertensive drugs had only moderate uptake until the Framingham Heart Study reported potential links to hypertension and stroke and public health epidemiologists were seeing reduced stroke due to antihypertensive drug use in the broader population. There was tremendous market potential, and drug companies were searching for more effective antihypertensives with fewer side effects to increase their competitive advantage. In 1964, the first beta-blocker was identified in Britain and rapidly became one of the best-selling drugs in the world. Beta-blockers would continue to be researched and improved in the upcoming years; however, they still carried concerning side effects such as fatigue and nausea. In the early 1980s, scientists at Squibb started researching pit viper venom as it was known then to lower blood pressure. They found the venom's most active ingredient, teprotide, and identified the enzyme it works on which is angiotensin converting enzyme (ACE). Since teprotide is a peptide and cannot be given orally, the scientists researched the biochemistry and pharmacodynamics and sought to synthesize other compounds that could inhibit ACE while given orally. Little by little they tweaked and tested their new chemical structures based on their insights into how the modified chemicals would affect the enzyme. This process is called "rational design," and in this case it led to the approval of the first commercial ACE-inhibitor, captopril, being approved in the United States in 1981. This drug generated more than one billion dollars in sales in its first full year of being marketed.[5]

Drugs that generate revenue of $1 billion in a given year are commonly referred to as "Blockbuster Drugs." Table 3.1 lists the top 10 all-time best-selling prescription drugs in the United States between 1992 and 2017.[6]

Five of the top 10 drugs are small molecule drugs and the other five are biologics.[6] By 2019, the original patents for all the top 10 small molecule and biologic drugs have expired in the United States and many other countries. Most are available in generic or biosimilar form, respectively, in 2019. All but one of the top 10 drugs can

Table 3.1 Best-selling drugs ranked by sales volume.

	Drug	Lifetime sales (billions)	Clinical uses	Manufacturer	Classification
1	Lipitor® (atorvastatin)	$94.67	Elevated cholesterol	Pfizer	Small molecule
2	Humira® (adalimumab)	$75.72	Inflammatory diseases	AbbVie	Biologic
3	Nexium® (esomeprazole)	$72.45	GERD	AstraZeneca	Small molecule
4	Advair® (fluticasone/salmeterol)	$69.08	COPD	GlaxoSmithKline	Small molecule
5	Enbrel® (etanercept)	$67.78	Inflammatory diseases	Amgen	Biologic
6	Epogen® (epoetin Alfa)	$55.63	Cancer	Amgen	Biologic
7	Remicade® (infliximab)	$54.67	Inflammatory diseases	Johnson & Johnson	Biologic
8	Abilify® (aripiprazole)	$51.34	Mental health	Otsuka/Bristol-Myers Squibb	Small molecule
9	Neulasta® (pegfilgrastim)	$47.40	Cancer	Amgen	Biologic
10	Plavix® (clopidogrel)	$46.48	Cardiovascular	Sanofi/Bristol-Myers Squibb	Small molecule

Source: Adapted from https://www.fiercepharma.com/pharma/
from-old-behemoth-lipitor-to-new-king-humira-u-s-best-selling-drugs-over-25-years.[6]
GERD, gastroesophageal reflux disease; COPD, chronic obstructive pulmonary disease.

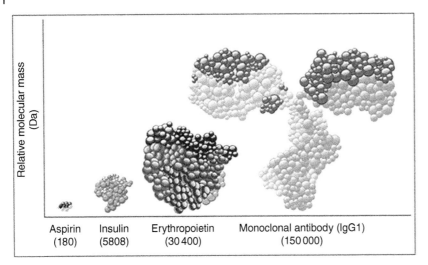

Figure 3.1 Relative molecular mass of small molecule and biologic drugs. *Source:* Mellstedt.[7] (*See insert for color representation of the figure.*)

be self-administered by the patient (meaning they do not require health professionals to administer the drug).

These top drugs have relatively high sales figures in common. There are additional points of similarity, and differences too, that are worth noting. From the context of similarity, all the top 10 drugs except for Neulasta® have indications for chronic disease treatment, meaning they may be taken indefinitely. A key point of differentiation among the top 10 drugs is that half are biologics and half are small molecules. As implied by the naming convention, small molecule drugs have much smaller mass than biologics, which are larger molecular structures as can be seen in Figure 3.1. Biologics are thus generally more complex than traditional, small molecule drugs.

Small molecule drugs held the top positions with blockbuster sales during their patent lifespan, primarily in the late 1990s and through the early 2000s. The biologics, or large molecule drugs, are taking the top sales position in recent years. Additionally, top drugs such as Advair® maintained high sales volume despite losing patents. The common denominator of commercial success among these various products is complexity. It is more challenging to make a generic Advair® inhaler, or infused biologic such as Remicade® than it is to replicate the chemical structure of statin drugs such as Lipitor® or Plavix® for generic production.

3.1.3 Complexity of Biologics

Biological products are typically derived from living cells and include a wide range of product types including vaccines, blood and blood components, gene therapy,

and monoclonal antibodies (mAbs). They can be made of sugars, proteins, nucleic acids, or may be composed of living cells or tissues.[8] Unlike many small molecule drugs that have known chemical structures that can be synthesized and analyzed fully, most biologics are complex and not easily characterized. Most biologics are made using recombinant DNA (rDNA) technology in which living cells are genetically engineered so that they produce the desired proteins.[9]

Biologics are generally more complex to manufacture than small molecule drugs as they are more sensitive to the environment and small variations in production can alter their function. The human body is designed to detect foreign substances, and biopharmaceuticals are no exception. Large biologic products need to be made and administered in such a way that they will not be rejected or neutralized by the body's immune system. The potential for a person's immune system to reject a life-saving drug underscores the importance of safe, reliable, and consistent manufacturing. Some of these challenges are described in Table 3.2, which compares the characteristics of small and large molecule drugs.

The complexity of biologic drugs leads not only to unique challenges with producing novel drugs, but it also makes creating similar versions (biosimilars) more difficult. Unlike small molecules that can typically be well defined and re-created in a laboratory, biologics may have unknown composition, and their synthesis may require thousands of complex steps. In some cases, the exact quantitative components of the finished biologic may be unknown.

Paradoxically, it is the molecular size and relative ambiguity of the large molecule drugs that can make

Table 3.2 Characteristics of small molecule pharmaceuticals vs. biologics.

	Small molecule pharmaceuticals	Biologics
Method of synthesis	Chemical synthesis	Genetically engineered living organisms or cells
Molecular size	Small	Large
Structure	Usually fully known	Complex, frequently partially unknown
Susceptibility to contamination during manufacturing	Low	High
Molecular structure	Relatively simple spatial structures, determined through analytical techniques	Exhibit complex spatial structures, difficult to determine or characterize
Complexity	Relatively pure ingredients	Complex ingredients (impurities, leachables, excipients, by-products)
Sensitivity to physical factors (heat, light)	Low	High
Clinical behavior	Well understood mode(s) of action	Complicated mode(s) of action, not always well understood
Manufacturing process	Straightforward, relatively simple	Highly complex
Species	Interdependent	Specific
Immunogenicity	Nonantigenic (generally)	Antigenic (frequently)

Source: Adapted from https://www.fraserinstitute.org/sites/default/files/biologics-revolution-in-the-production-of-drugs.pdf.[9]

them harder to protect with patents. Their composition is not as easily defined and as such, competitors can develop similar but different drugs. For this reason, both the drug and manufacture/production processes are patented with biologics.[9] In the United States, the FDA requires a biologic licensing application (BLA) and approval for a biologic to be marketed. Supporting the importance for manufacturing patents, the FDA notes the following requirements for a BLA: "Issuance of a biologics license is a determination that the product, the manufacturing process, and the manufacturing facilities meet applicable requirements to ensure the continued safety, purity and potency of the product."[10]

There are additional elements that protect the innovation of innovator firms from generic or biosimilar competitors. In many countries, data exclusivity rights also exist. These data exclusivity rights are granted to innovator firms to ensure that preclinical and clinical trial data, such as safety and efficacy data, are protected from competitors for a defined time period. The data exclusivity period protects clinical trial data for several years after an innovator product reaches the market, during which time competitors cannot use preclinical or clinical trial data in submissions for biosimilar versions. This can prevent a drug from losing patent exclusivity very close to the time it is marketed, but it could also reduce competition.[9] Periods of data exclusivity vary by country as can be seen in Table 3.3.

Table 3.3 Preclinical and clinical trial data exclusivity by nation.

Nation	Small molecule (yr)	Large molecule (yr)
Canada	8	8
European Union	10	10
United States	5	12

Source: Adapted from https://www.fraserinstitute.org/sites/default/files/biologics-revolution-in-the-production-of-drugs.pdf.[9]

3.2 Clinical Characteristics of Biologic Drugs

3.2.1 Biologics vs. Small Molecule Drugs

The unique manufacturing, intellectual property protection, and market potential of biologic drugs has been described previously in this chapter. The remainder of this chapter will review why biologics are gaining a more prominent place in medicine, and what characteristics they possess that small molecule drugs do not or cannot. Small molecule drugs currently make up more than 90% of pharmaceutical therapeutics used for treating disease.[11] They are relatively easy to develop and suit their intended purposes reasonably well; a small molecule can be an effective enzyme inhibitor or receptor ligand, for example. However, there are noteworthy clinical limitations with this therapeutic approach. Typically, for a small molecule to be effective, it needs to be enveloped or be reasonably close to its target protein or transporter; the better the fit, the stronger the affinity and this often relates to how significant the drug's impact is on the human body. Small molecules bind to receptors in similar fashion and when doing so they compete with the natural ligand, which results in agonist, antagonist, or partial agonist/antagonist activity. Most cellular functions in the body, however, are mediated by proteins interacting with other proteins.[11]

A major limitation of traditional small molecule drugs is that they simply fit into a specific destination that results in a biological perturbation, but do not necessarily receive feedback from the body after the drug change has taken place. This can be helpful for inhibiting a bacterial infection or other processes that require one-directional activity, but it can result in unpredictability for more complex biological processes. For example, a small molecule beta-blocker such as propranolol will continue blocking beta-adrenergic receptors regardless of whether a patient is sitting, standing, or exercising. Another issue is that small molecule chemicals are not always site-specific and off-target systems can be affected. Nonselective beta-blockers are contraindicated in some asthmatic patients for this reason; in addition to relaxing smooth muscle in the heart, they can also affect beta receptors in the lung resulting in bronchoconstriction.

Conversely, as biologic drugs can mimic normally existing (endogenous) proteins or protein interactions, they can receive feedback from biological systems in the body while producing their therapeutic activity. Biologic drugs can act more specifically and are often far more potent than small molecules. They more closely resemble naturally occurring proteins within the body and can be more effective in treating diseases. How closely a biologic drug resembles human proteins can have significant impact to its safety and efficacy. Biopharmaceuticals can generally be described by three major areas of their use[2]:

- Prophylactic/ preventive use such as vaccines
- Therapeutic use such as antibodies or enzymes
- Replacement use such as with growth factors or hormones

3.2.2 Vaccines

Vaccines were the first biologic drugs to be developed and regulated by the FDA (Table 3.4). National and global viral pandemics have continued to push research scientists to develop new vaccines throughout the 1900s

Table 3.4 A history of vaccine development and regulation in the United States.

Polio vaccine

In 1938, former US President Franklin D. Roosevelt began a "War on Polio" with the creation of the National Foundation for Infantile Paralysis. In 1954, Jonas Salk's inactivated polio vaccine was tested in 1.8 million children. In 1955, all polio vaccinations were suspended due to ineffective batches of the vaccine circulating in the market and would not resume until all manufacturing facilities were inspected and reviewed to ensure procedures for safety testing were in place.

German measles vaccine

A global epidemic of German measles (rubella) spread to the United States in 1964, infecting about 12.5 million people that year. Rubella is typically a mild virus affecting children and young adults, but it can also pass to an unborn child if pregnant women are infected, resulting in conditions such as mental retardation, blindness, deafness, and heart defects. In 1966, two former Center for Biologics Evaluation and Research (CBER) directors developed the first experimental vaccine, and in 1969, the first vaccines were marketed. By 1988, there were only 225 reported cases of rubella in the United States.

Influenza vaccine

It is estimated that the influenza pandemic of 1918 caused 20 million deaths worldwide. In the 1940s, scientists at the Division of Biological Control – a CBER predecessor – developed the first reliable potency test for flu vaccine so that manufacturers could produce uniform products with desired effectiveness. In 1945, the first flu vaccine was developed, and today CBER works with manufacturers on an annual basis to assist in development and production of annual updated vaccines to keep responding to the ever-evolving influenza virus.

Source: Bren.[12]

3.2.3 Antibodies

Antibodies are immunoglobulins, i.e. proteins with immune functions. Immunoglobulins are categorized into five classes, which identify their structural makeup and the types of immune responses produced: immunoglobulin G (IgG), IgD, IgA, IgM, and IgE. IgG is the most common immunoglobulin in the body, whereas IgE makes up less than 1%.[2] The human body produces antibodies via B cells to attack and destroy antigens. Antigens are foreign substances that the body identifies as invasive and requiring removal. In individuals with autoimmune diseases, such as rheumatoid arthritis or lupus, the body incorrectly identifies itself as being foreign and will attack its own cells. Augmenting this misaligned process can produce clinical benefit and is a target for biological drugs.[2]

There are two predominant ways to classify antibodies based on their source and how they are harvested: polyclonal antibodies and mAbs. As the name suggests, polyclonal antibodies contain many (-poly) antibodies; not only the desired antibody to be used for a clinical effect, but also unwanted antibodies that the immune system may see as foreign. Polyclonal antibodies can be derived from horse blood samples (botulism antitoxin, diphtheria antitoxin, tetanus antitoxin), or from human donors (hepatitis A and B immunoglobulins, or immunoglobulin for measles, rabies, and tetanus, respectively).[2]

In this process, an antigen (foreign matter) is injected in an animal, then the animal develops an immune response mediated by B cells and produces antibodies. The antibodies are harvested and packaged for later clinical use in humans. This type of antibody development has been in existence for decades, but not without its problems. In 1901, a batch of the diphtheria antitoxin became infected with tetanus resulting in the death of 13 children. Subsequently, in 1902, the United States passed The Biologic Control act, which was meant to ensure the safety of biologics and was the predecessor of the FDA's Center for Biologics Evaluation and Research (CBER), which exists today.[12]

mAbs were developed in the 1970s and as the name suggests, would lead to the harvesting of only one (-mono) type of antibody, which leads to more specificity for antigen binding. mAbs are typically produced by injecting a mouse with antigens, resulting in the mouse developing an immune response mediated by B cells. At this point, the B cells are removed and fused with myeloma cells, resulting in what is called a "hybridoma." Myeloma cells are used for their long lifespan and ability to replicate, like cancer cells. This allows mAbs to be produced continuously resulting in an efficient manufacturing machinery. Since mAbs are still produced from animal, the potential for allergic reaction or neutralization by the human body exists.[2]

Advances in science allowed for mAbs to become more humanized over time; first with chimeric antibodies that contained mouse and human proteins (an example is the drug, Reopro®), then with humanized antibodies that further minimized the components made from mice (e.g. Herceptin®), and finally, fully human antibodies (e.g. Humira®, the first fully human antibody approved by the FDA). Antibodies can also be conjugated with other products such as small molecules or radiopharmaceuticals that use the specificity of the antibody-to-antigen to target a site and then release a secondary pharmaceutical agent for therapeutic purposes.[2]

The future of mAbs development may include bispecific antibody development, meaning antibodies that possess two binding specificities. This could be advanta- geous as targeting multiple targets simultaneously could inhibit various receptor-ligand signaling pathways more effectively and limit development of disease-cell resistance. In 30 years since the first therapeutic mAb was approved in 1986, there are more than 294 mAbs being used clinically, with almost 90% of them being humanized mAbs with the remainder being chimeric mAbs.[13]

3.2.4 Enzymes

Enzymes have been a long-standing target for drug development, including small molecules as well as biologics. The biologic enzyme replacement therapies (ERTs) for the treatment of lysosomal storage diseases (LSDs) such as Fabry disease, Gaucher disease, and Pompe disease offer a vantage point into some of the current successes and future potential in this area of biotechnology. There are 50–60 different rare, genetically inherited disorders resulting in deficient lysosomal enzymes. When lysosomal enzymes do not work properly, fats and other enzyme substrates build up throughout the body resulting in widespread cellular complications including death. Enzyme replacement involves the production of enzyme proteins through rDNA technology, then these enzymes are infused into the patient on a recurrent basis as life-long replacement therapy. ERTs have resulted in significant clinical benefit to patients including improved quality of life, walking ability, and respiratory function improvements. However, challenges exist such as neutralizing antibodies and other immune reactions. ERTs are not always able to reach the desired target cells such as those in the central nervous system. Potentially 75% of patients with neurologic dysfunctions may not be adequately treated with ERTs, often due to challenges with drug design and inability to penetrate the blood–brain barrier. Drugs marketed for ERT are some of the most expensive drugs in the world today. Gene therapy is one of many scientific advances being explored to treat these conditions in patients.[14]

3.2.5 Cytokines

Cytokines are proteins that are involved in cell communication and mediating immune system processes. They represent a diverse group of molecules but mainly growth factors and hormones and falling into one of two broad categories: Type I cytokines include interleukins and colony-stimulating factors, and Type II cytokines that are typically interferons. Erythropoietin, thrombopoietin, growth hormone, and prolactin have similar structures and signaling mechanisms as Type I cytokines.[15]

Improperly regulated cytokines can result in a variety of diseases such as autoimmunity and cancer. These characteristics make them attractive targets for therapeutic purposes. For their function, cytokines need to bind to specific receptors. As such, drugs that target the cytokine-mediated immune system can include antibodies that neutralize the cytokine or cytokine receptors, recombinant proteins that are receptor agonists or antagonists, or false receptors that will bind the cytokine itself and neutralize it.[16]

Cytokines are well studied in animal models that are beneficial for developing therapeutic targets; however, there are limitations due to their natural (native) properties. Cytokines have overlapping activity meaning that, when one cytokine is blocked, another may make up for the lost activity. They are also multifunctional in that they affect several processes, potentially in multiple organs in parallel, so affecting or disrupting a specific cytokine may result in unwanted side effects. Inhibiting the natural activity of cytokines can also result in a severely blunted immune system. Some key characteristics of cytokine type of drugs are provided in Table 3.5.

3.2.6 Cytokine-Interferons

Interferon alpha was the first cytokine to be produced using rDNA technology.[18] Interferon is a regulator of growth and differentiation and has clinical efficacy in malignant, viral, immunologic, angiogenic, inflammatory, and fibrotic diseases. Interferon-beta was approved by the FDA in 1993 and is the oldest and most frequently used medication for treating multiple sclerosis to date.[18] Interferon (IFN) works in multiple sclerosis in a variety of ways, decreasing proinflammatory cytokines while also leading to the production of anti-inflammatory cytokines by increasing the activity of suppressor T-cells. The activity of cytokines also makes for a strong candidate in oncology. IFN-alpha was found to have tumor suppressing activity in a rare B-cell neoplasm, which led to expansive research throughout the 1990s on clinical utilization of the molecule for cancer treatment.[19] Intron A® as found in Table 3.5 is approved for five unique cancer types.

3.2.7 Cytokine-Interleukins

Another group of cytokines with established clinical use are interleukins. There have been over three hundred and fifty thousand scientific articles published on interleukin since it was first discovered in 1977. More than sixty cytokines have been designated as interleukins since the initial discovery of monocyte interleukin (IL-1) and lymphocyte interleukin (IL-2). The numbering convention (i.e. IL-___) is based on functional properties and biological structure.[20]

Like other cytokines, the activity of interleukins in the immune system makes them an excellent target to treat immune system-mediated disease such as allergy, asthma, autoimmunity, and chronic infections. For example, describing the interleukin products from Table 3.5, Actemra® binds to IL-6 receptors inhibiting IL-6 mediated signaling. IL-6 is a pro-inflammatory cytokine produced by T-cells, B-cells, lymphocytes, monocytes, and fibroblasts. IL-6 is also produced in synovial cells, which leads to local joint inflammation in rheumatoid arthritis. Cosentyx® is a recombinant human monoclonal antibody that binds to interleukin-17A (IL-17A) cytokine and inhibits its interaction with the IL-17 receptor. Through this interaction, IL-17A inhibits the release of proinflammatory cytokines and chemokines. Dupixent® is a human monoclonal antibody that inhibits IL-4 and IL-13 by binding to a receptor subunit shared by both complexes. Nucala® is an IL-5 antagonist impacting eosinophil activity. Kineret® is an IL-1R antagonist that impacts cartilage degradation and bone resorption, and Zinbryta® binds to IL-2 that is presumed to impact lymphocytes resulting in therapeutic effects in multiple sclerosis.[17]

3.2.8 Tumor Necrosis Factor

Two of the top five best-selling drugs of all time are tumor necrosis factor (TNF) inhibitors, which is an impressive marketing feat despite their relatively specific use for various rare autoimmune diseases.[6] TNF is a major proinflammatory cytokine affecting various aspects of the immune system with a wide range of biologic effects including anti-tumor and antiviral activity. In the body, TNF is a transmembrane protein (tmTNF) that gets cleaved to soluble TNF (solTNF); both forms are biologically active, but with distinct roles that translate to the safety and efficacy profile of TNF inhibitors. For example, inhibiting solTNF results in anti-inflammatory effects, whereas inhibiting tmTNF results in increased susceptibility to infection.[21]

The TNF molecule itself has limited therapeutic activity due to extreme toxicity. The more commonly clinically used agents are the TNF antagonists. Remicade® (infliximab) and Humira® (adalimumab) are both IgG mAbs and work by competitively inhibiting the binding

Table 3.5 Examples of cytokine drugs and diseases treated.

Type of cytokine	Drug name	Manufacturer	Disease(s) treated (per FDA labeling)
Interferon	Actimmune® (interferon gamma-1b)	Horizon Pharma Inc.	Chronic granulomatous disease; Severe, malignant osteopetrosis
	Avonex® (interferon beta-1a)	Biogen Inc.	Multiple sclerosis
	Betaseron® (interferon beta-1b)	Bayer HealthCare	Multiple sclerosis
	Extavia® (interferon beta-1b)	Novartis	Multiple sclerosis
	Intron A® (interferon alfa-2b)	Merck Sharp & Dohme Corp	Hairy cell leukemia; Malignant melanoma; Follicular lymphoma; Condylomata acuminate; AIDS-related Kaposi's sarcoma; Chronic hepatitis C; Chronic hepatitis B
Interleukin	Actemra® (tocilizumab)	Genentech	Rheumatoid arthritis; Giant cell arteritis; Polyarticular juvenile idiopathic arthritis; Systemic juvenile idiopathic arthritis; Cytokine release syndrome
	Cosentyx® (secukinumab)	Novartis	Plaque psoriasis; Psoriatic arthritis; Ankylosing spondylitis
	Dupixent® (dupilumab)	Sanofi-Aventis	Atopic dermatitis; Asthma
	Kineret® (anakinra)	Swedish Orphan Biovitrum AB	Rheumatoid arthritis; Cryopyrin-associated periodic syndromes
	Nucala® (mepolizumab)	GlaxoSmithKline	Severe asthma
	Zinbryta® (daclizumab)	Abbvie	Multiple sclerosis
Tumor necrosis factor	Enbrel® (etanercept)	Amgen	Rheumatoid arthritis; Polyarticular juvenile idiopathic arthritis; Psoriatic arthritis; Ankylosing spondylitis; Plaque psoriasis
	Humira® (adalimumab)	Abbvie	Rheumatoid arthritis; Juvenile idiopathic arthritis; Psoriatic arthritis; Ankylosing spondylitis; Plaque psoriasis; Crohn's disease; Pediatric Crohn's disease; Ulcerative colitis; Hidradenitis suppurativa; Uveitis
	Remicade® (infliximab)	Janssen Biotech	Crohn's disease; Pediatric Crohn's disease; Ulcerative colitis; Pediatric ulcerative colitis; Rheumatoid arthritis; Ankylosing spondylitis; Psoriatic arthritis; Plaque psoriasis
Erythropoietin	Aranesp® (darbepoetin alfa)	Amgen	Anemia due to chronic kidney disease; Anemia due to chemotherapy in patients with cancer
	Procrit® (erythropoietin)	Janssen	Anemia due to chronic kidney disease; Anemia due to zidovudine in patients with HIV-infection; Anemia due to chemotherapy in patients with cancer; Reduction of allogenic red blood cell transfusions in patients undergoing elective, noncardiac surgery; Nonvascular surgery
Colony stimulating growth factor	Neulasta® (pegfilgrastim)	Amgen	Patients with cancer receiving myelosuppressive chemotherapy; Patients with hematopoietic subsyndrome of acute radiation syndrome
Vascular endothelial growth factor	Avastin® (bevacizumab)	Genentech	Metastatic colorectal cancer

Source: Food and Drug Administration (FDA) DailyMed.[17]

of TNF to its receptors. While their mechanism of action is similar, the pharmacokinetic and pharmacodynamic properties of the various drugs in the TNF inhibitor class are different. These unique properties can lead to variations in safety and clinical efficacy for the treatment of autoimmune conditions. It has been noted that up to one-third of RA patients do not respond to anti-TNF therapy due to genetic factors.[21]

3.2.9 Hormones

Biological drugs target a diverse array of hormone-mediated systems. Hormones are intercellular messengers that can be categorized as steroids, polypeptides, or amino acid derivatives. Steroids include estrogens, androgens, and mineral corticoids. Polypeptides include insulin and endorphins, and amino acid derivatives can include epinephrine, norepinephrine, adrenaline, and noradrenaline.[2] The development of a therapeutic insulin has been one of the most important pharmaceutical endeavors to date; a timeline provided in Table 3.6 demonstrates how insulin's discovery and manufacturing have shaped modern medicine and drug development.

One of the most impactful scientific breakthroughs in the development of insulin was the ability to produce human insulin on an industrial scale using rDNA techniques. Prior to this, in the 1970s, diabetic patients were injected with insulins derived from animal

Table 3.6 Insulin: timeline of discovery and manufacture.

1869 – The Islets of Langerhans cells are discovered in the pancreas

1901 – The Islets of Langerhans are identified as cells that produce insulin

1916 – Pancreas extract is found to lower blood sugar when given to diabetic dogs

1921 – Studies show that pancreas removal in dogs results in the symptoms of diabetes; cow pancreas extract demonstrated improvement in the dog's health when given and was termed "insulin"

1922 – A young boy with type 1 diabetes lives for 13 years beyond typical mortality as the first recipient of medical insulin; Eli Lilly becomes the first insulin manufacturer

1936 – The addition of protamine is found to prolong the action of insulin

1950 – Intermediate acting insulin (NPH) is marketed by Novo Nordisk

1955 – Insulin becomes the first protein to be fully genetically sequenced

1963 – Insulin becomes the first human protein to be chemically synthesized

1978 – Genentech makes the first synthetic human insulin via recombinant DNA techniques, which is the first human protein to be manufactured using biotechnology

1982 – Synthetic insulin is renamed as human insulin to better differentiate from animal sources

1996 – Eli Lilly develops the first insulin analogue, which is genetically modified to change pharmacokinetics

Source: Adapted from https://www.diabetes.co.uk/insulin/history-of-insulin.html.[22]

sources. Mass-produced human insulin made from rDNA could be produced uniformly and carried the potential to alleviate shortages inherent in animal-insulin supply. While insulin was the first drug produced using recombinant technology, the capability extends to most of the biological drugs developed to date and continues to progress biotechnological and pharmaceutical sciences today.

3.2.10 Blood Factor Products

Like the scientific developments that improved the production of insulin, blood coagulation factor products for the treatment of hemophilia have also undergone dramatic and important advances. Hemophilia A and B are bleeding disorders caused by dysfunction or deficiencies in coagulation factors VIII and IX that result in impaired blood clot formation and potential hemorrhage. In the 1950s, the only treatment option available was the infusion of human whole blood or plasma. Then in the mid-1960s, scientists discovered how to collect factor products that allowed for collection in a more concentrated form. Donors were now able to donate plasma for production of Factor VIII for replacement. Industries developed around sourcing the material, and patients with hemophilia benefited significantly, being able to treat themselves in their home for the first time.[23]

Unfortunately, the donated pool of plasma used in this process became a host for transmission of blood-borne disease such as hepatitis and HIV. The development of recombinant factor was a way to avoid reliance on human-produced product and to limit viral contamination. In 1984, factor VIII DNA was cloned and in 1992, the first recombinant FVIII product, Recombinate®, was licensed for marketing.[23] Today, there are over a dozen recombinant products developed for hemophilia treatment, with subsequent iterations focusing on using less animal and human-derived elements, limiting the development of inhibiting antibodies and transmission of blood-borne pathogens. Recent drug development efforts have been aimed at extending the half-life of factor products to allow for less frequent dosing, potentially resulting in greater compliance with prophylactic dosing and less reliance on acute bleeding treatment. Human cell lines have been used to produce the latest generation of factor products and genetic therapy is being studied, which could potentially require a single one-time treatment.[23]

3.3 rDNA and Biologic Drug Manufacturing

rDNA is as the name implies, taking DNA from two sources and recombining DNA into one. This allows for modification of DNA and is the basis for much of what is considered genetic engineering. The therapeutic applications have been discussed throughout this chapter; after a protein, messenger system, receptor, or biologic process in the body is understood well, scientists can attempt to augment that system using human-made drug products.

In most cases, to do so specifically, requires the introduction of biologic products, biopharmaceuticals. For example, as history has demonstrated, it is possible to develop such biologic products from inoculating horses with diphtheria bacteria or producing insulin extracts from the pancreas of cows. While scientifically important as well as lifesaving at the time of discovery, these processes are fraught with several limitations, including antibody development following the introduction of agents of animal origin, scalability challenges with mass production from animal stock, lack of uniformity in the end product, and numerous opportunities for potential contamination.

Being able to produce a therapeutic, human protein such as insulin in a controlled environment mitigates many of these challenges. Being able to develop a gene in a laboratory and introduce it into a vector (e.g. bacterium or Chinese hamster ovary [CHO] cells) that will produce millions of copies of a modified, therapeutic protein is a game-changer for drug development and the treatment of human disease.

As genes are the cornerstone of most molecular biology, it is helpful to be able to isolate and amplify specific gene fragments. Through rDNA technology, a gene of interest can be cloned by combining it with another DNA molecule (known as a vector) and inserting this into living cells where replication takes place. To obtain the genes required for this process, DNA needs to be "cut" or "spliced" into smaller fragments. The cutting is mediated by restriction endonucleases, which are enzymes found in bacteria that cut DNA. Through purposeful development, these enzymes are selected for their ability to cut DNA at specific sites. Once DNA fragments are cut as desired, they are then rejoined by ligase, which is an enzyme that catalyzes the joining of two large pieces or molecules.[24]

Consider in this scenario, the specified genes to be cloned produce a useful protein such as insulin. To make many copies of that gene, it needs to be carried in a living cell (a cell that replicates). *E. coli* plasmid vectors and bacteriophage lambda are two of the most commonly used vectors for this purpose. The key difference among the two is that plasmid often lives symbiotically with the host cell and replicates each time the host cell replicates, whereas bacteriophage lambda acts as a virus and kills the host cell, leaving their packaged DNA intact.[25]

Plasmid is a circular, double-stranded DNA molecule. When an important section of DNA (such as the string of DNA that codes for insulin production) is isolated, it can be inserted into a plasmid, which then gets inserted to a bacterial cell such as *E. coli*. As *E. coli* bacteria reproduces, so too does the DNA molecule of interest (the insulin-coding DNA strand in this example). Attaching an antibiotic-resistance gene to the plasmid and exposing the *E. coli* cells to antibiotics ensures that only *E. coli* cells with rDNA inside reproduces. This makes the process more efficient. As the science progressed, it became possible to develop synthetic DNA fragments. These can be useful for making plasmids better suited for cloning or incorporation to study the impact of mutations, for example.[25]

Several more examples beyond insulin can further demonstrate the importance of rDNA technology. The human somatostatin hormone consists of 14 amino acids and inhibits the secretion of somatotropin (growth hormone). It can be used therapeutically to treat acromegaly (excessive somatropin production) and analogs of somatostatin can be used to treat cancer. Somatostatin is produced using plasmid vectors and incubated in *E. coli*. Darbopoetin alfa, a 165-amino acid protein, is a synthetic form of erythropoietin used to increase red blood cell levels. It is produced using rDNA technology, but unlike insulin or somatostatin produced in *E. coli*, darbopoeitin alfa is produced using CHO cells.[24]

CHO cells are used to make nearly 70% of recombinant protein therapeutics today. The first product manufactured using CHO cells was a plasminogen activator called Activase® (r-tPA) in 1987. Rationale for the popularity of CHO cells in biopharmaceutical endeavors include accommodating complex protein folding and post-translational modifications (shaping a protein after it has been made to a desirable structure), adaptive ability to study G-protein coupled receptors in a stable environment, and structural characteristics related to cytoskeletal and microtubule structure, adhesion, and motility. Furthermore, there are logistical benefits to using CHO for protein synthesis; the cells can grow to very high densities in bioreactors, which benefits scalability and CHO cells have been found not to replicate human viruses such as HIV, influenza, polio, herpes, or measles.[26]

3.4 What Does the Future Hold?

3.4.1 Gene Therapy

How genes are identified, isolated, modified, and produced on a large scale using bacteria or CHO cells to generate the desired protein, whether it be, insulin, mAbs, TNF-inhibitors, or others has been described in this chapter so far. Following the production of the therapeutic protein in a manufacturing facility, it is purified through a complex process, and formulated into a form that makes it stable for storage and shipment to the pharmacy, hospital, or clinic where it will be delivered or administered. Finally, the therapeutic protein (biopharmaceutical) is administered by an injection either by a medical professional or by the patient.

The latest breakthroughs in biotechnology research and development (R&D) have been in gene therapies. The initial stages of gene therapy development still require identifying a target gene that is clinically important. There are various ways that gene therapy can be used therapeutically. For example, a patient may be missing a gene that codes for a life-sustaining enzyme, in which case the goal is to introduce that missing gene into the patient, or a patient may have a defective gene that requires modification to restore its normal function. What is unique to gene therapy versus the majority of biological medicines to date is that with gene therapy, the gene is introduced into the patient and the patient makes the protein; the protein is not produced in bacteria or in CHO cells, but rather in the patient.

Like the plasmid vectors described previously for manufacturing biologic drugs like insulin, gene therapies typically require a vector to be successfully administered to a human patient. Viral vectors are the most commonly used vectors in gene therapy representing nearly 70% of clinical trials.[27] Viruses make good vectors due to their natural ability to infect cells. For clinical use, the viruses are modified to avoid causing disease when given to patients. Some viruses, such as retrovirus can incorporate the genetic material into a human cell and chromosome, whereas adenoviruses introduce their DNA into the cell, but the DNA does not get integrated into the chromosome. There exists a broad spectrum of viral vectors for delivering gene therapies and their choice can be influenced by factors such as how much expression is desired and for how long.[27]

3.4.2 Personalized Medicine

Gene therapies are another milestone in the progression from one-size-fits-all medicine to personalized medicine.

Much of this progress has been made in oncology where molecular diagnostics are part of the drug development process and predictive biomarkers are used to guide treatment. By 2018, there were 21 different drugs that had been approved alongside companion diagnostics by the USFDA with testing requirements as part of their labeled approval. Of the drugs with required companion diagnostic tests, nearly half are for treatments of non-small cell lung cancers (NSCLC).[28]

In 2004, epidermal growth factor receptor (EGFR) mutations were identified to have predictive potential. In subsequent years, ALK and ROS mutations would also be identified to further direct cancer treatment based on the types of mutations present in the lung tumors.[29] A decade later, Opdivo® (nivolumab) and Keytruda® (pembrolizumab) were the first programmed death 1 (PD-1) and programed death-ligand 1 (PD-L1) inhibitors approved. PD-1 is a checkpoint protein found on T cells and these drugs cause the patient's immune system to attack cancer. Both these drugs are IgG humanized, mAbs that work by binding to the PD-1 receptor and blocking its interaction with PD-L1 and PD-L2 ligands; blocking PD-1 activity has resulted in decreased tumor growth in clinical trials. There are multiple drugs currently marketed for inhibiting the PD-1 system, and most have companion diagnostics that look for PD-L1 ligand expression to direct use. Additionally, patients with certain cancers are tested for other tumor mutations, which may indicate that a different drug should be tried first before a PD-1 inhibitor. For example, the FDA-labeled indication as a single agent for metastatic NSCLC for Keytruda® includes the following:

> KEYTRUDA®, as a single agent, is indicated for the treatment of patients with metastatic NSCLC whose tumors express PD-L1 (TPS ≥ 1%) as determined by an FDA-approved test, with disease progression on or after platinum-containing chemotherapy. Patients with EGFR or ALK genomic tumor aberrations should have disease progression on FDA-approved therapy for these aberrations prior to receiving KEYTRUDA®.[30]

As this labeling indicates, patients are directed to use an FDA-approved test to determine their PD-L1 expression. If they have EGFR or ALK mutations, they are to try other targeted therapies first (e.g. drugs to treat EGFR or ALK mutations) before using Keytruda®, which targets PD-L1. These types of treatment pathways are becoming synonymous with precision medicine and the impact to the healthcare system is significant. Reimbursement groups welcome the ability to better predict that the medicines being paid for will work, and clinicians and patients welcome the potential for improved health outcomes based on more specific science. Precision medicine, however, will not progress into the future without challenges. The remainder of this chapter will focus on clinical and market-related challenges with biologic drugs in precision medicine.

3.5 Global Biologics Market

According to the USFDA, biologic products are the fastest growing class of therapeutic products in the United States and make up an increasing and substantial share of health care costs.[31] In the past five years, there has been an increase in drug approvals and spending on new drugs globally. The types of drugs being approved are also changing. New drug approvals continue to trend toward biologic, orphan, and oncology products[32]:

- Oncology will make up 30% of new drug approvals.
- Orphan drugs could represent 45% of new drug approvals.

Orphan disease is a term used to describe rare diseases. The definitions vary by country, with some using prevalence rates and others basing the definition on the number of affected individuals. The United States defines a disease as rare when it affects fewer than 200 000 individuals. Europe defines a rare disease as one that affects fewer than 5 individuals per 10 000, whereas Taiwan's definition is fewer than 1 in 10 000. Brazil is in line with the World Health Organization's definition where a disease is considered rare when it affects less than 65 per 100 000 individuals. There are an estimated 5000–8000 rare diseases identified globally, and while individually they are rare, collectively they may impact 6–8% of the population.[33]

As the number of orphan drugs being approved continues to increase, and the use of biomarkers and precision medicine-driven principles continue to gain traction, it is expected that the number of patients treated per new drug will go down, and the price per treatment will go up. Theoretically, a drug that costs $1 million per treatment and treats one person with a rare disease would generate the same return as a drug that costs $1 and treats one million people with a more common disease. There are complex variables associated with either treatment example above, such as costs related to R&D, manufacturing, marketing, regulatory frameworks, and clinical care and monitoring. All these variables impact how much effort and money is spent on R&D by phar-

maceutical companies, which is similar to the issues noted earlier in this chapter with regards to antibiotic R&D. There is an unmet need for treatment of orphan diseases as well as for more effective cancer treatments. Regulatory agencies around the world have taken steps to address development and patient access to rare disease medicines. National plans can encompass funding for orphan disease drug access, research incentives, diagnosis programs, care coordination, and early access programs.[33]

Access to life-saving medications brings immeasurable benefit to patients; however, it comes at a significant cost for payers. Across the world, some of the recently approved biologic drugs, including gene therapies, have placed immense strains on payment systems and put into question their long-term sustainability especially considering the pharmaceutical pipeline that lays ahead. Public and private organizations around the world are working with pharmaceutical companies, insurers, governments, and patient advocacy groups in attempts to better align financial expenditures with clinical benefit. Several organizations perform health technology assessments (HTA) that evaluate new drugs in a systematic fashion to assess clinical safety and efficacy as well as cost-effectiveness. It is expected that independent review of drug pricing by groups like the Institute for Clinical and Economic Review (ICER) may be able to place downward pressure on drug go-to-market prices in the coming years. Outcomes-based contracts that tie reimbursement for a given pharmaceutical to meeting certain clinical metrics are also expected to become more prevalent. It is predicted that by 2022, 30 out of the 50 top drugs will have some form of outcomes-based contracts in place between pharmaceutical manufacturers and payers.[32]

Biosimilars, which are similar versions of innovator biologic drugs, are expected to generate cost-savings compared to the innovator/reference brand that they compete with. The adoption of biosimilars in the United States has been relatively slow compared to other developed markets like the European Union. It is expected that the adoption of biosimilars in Europe will continue to outpace the United States for the upcoming decade; however, major events are on the horizon in the United States including the introduction of the first biosimilars to Humira®, currently the world's top-selling drug.[32] A concerted effort by countries and payers across the globe to educate prescribers and patients on the safety and efficacy of biosimilars will be required to see similar type of savings that small molecule generics generate today.

3.6 Summary

Considering the complexities associated with developing, manufacturing, administering, and paying for biologic drugs, a fundamental question is why the science of drug discovery is moving away from small molecule drugs to biologics? In short, because there is no better alternative that is currently available. In the ongoing development of safer and more efficacious drugs, biologics are the logical next step and this will continue to be the case in the foreseeable future. Most human diseases are caused by signaling imbalances, which involve numerous protein–protein interactions. Biologics afford the capability to alter such complex signaling systems in ways that small molecule drugs, to date, have not been able to. It is estimated that of ~20 000 human proteins, only about 3000 can be regulated by small molecule drugs.[34]

The treatment spectrum of biologic therapeutics is broad. They can limit pro-inflammatory cytokines that erode synovial tissue in rheumatoid arthritis, stimulate red blood cell production in kidney disease, replace the deficient insulin in diabetes, or influence the body to attack cancer using the natural immune system. The umbrella term, biologics, includes the gene therapies used to treat and/or cure diseases that have had little to no treatment options currently. R&D of gene therapies will continue in the upcoming decade and will provide the opportunity for potentially curing certain diseases.

It will be incumbent upon pharmaceutical manufacturers and payers to strike a balance that ensures continued innovation is affordable and accessible on a global scale. Rare diseases that result in dismal health outcomes require therapies that do not exist today and, in many cases, treatment options that exist for diseases today can also be dramatically improved. Precision medicine, including the improved identification of biomarkers and a better understanding of pharmacogenomics, will be one pathway toward finding the nexus of cost-effectiveness and value. While biologics will not cure all diseases or solve all health problems, they will be an integral part of drug therapy in the foreseeable future.

References

1 Sonnedecker, G. (1993). The founding period of the U. S. pharmacopeia: I. European antecedents. *Pharmacy in History* 35 (4): 151–162. www.jstor.org/stable/41112530 (accessed 24 September 2019).

2 Ng, R. (2015). *Drugs: From Discovery to Approval*. Hoboken, NJ: Wiley.

3 Jones, A. (2011). Early drug discovery and the rise of pharmaceutical chemistry. *Drug Testing and Analysis* 3: 337–344. https://doi.org/10.1002/dta.301.

4 Timeline of drug development (2018). Page last edited on 15 June 2018. https://timelines.issarice.com/wiki/Timeline_of_drug_development (accessed 3 August 2019).

5 Kirsch, D. and Ogas, O. (2017). *The Drug Hunters: The Improbable Quest to Discover New Medicines*. New York, NY: Arcade Publishing.

6 Liu, Angus (2018). From old behemoth Lipitor® to new king Humira®: best-selling U.S. drugs over 25 years (14 May) [Online]. FiercePharma. https://www.fiercepharma.com/pharma/from-old-behemoth-lipitor-to-new-king-humira-u-s-best-selling-drugs-over-25-years (accessed 24 September 2019).

7 Mellstedt, H. (2013). Clinical considerations for biosimilar antibodies. *EJC Supplements: EJC: Official Journal of EORTC, European Organization for Research and Treatment of Cancer ... [et al.]* 11 (3): 1–11.

https://doi.org/10.1016/S1359-6349(13)70001-6, https://www.ncbi.nlm.nih.gov/pmc/articles/PMC4048039 (accessed 24 September 2019).

8 The United States Food and Drug Administration. What are "biologics" questions and answers. https://www.fda.gov/about-fda/about-center-biologics-evaluation-and-research-cber/what-are-biologics-questions-and-answers (accessed 24 September 2019).

9 Lybecker, K. (2016). *The Biologics Revolution in the Production of Drugs*. Fraser Institute https://www.fraserinstitute.org/sites/default/files/biologics-revolution-in-the-production-of-drugs.pdf (accessed 24 September 2019).

10 United States Food and Drug Administration (2015). Frequently asked questions about therapeutic biological products (7 July). https://www.fda.gov/drugs/therapeutic-biologics-applications-bla/frequently-asked-questions-about-therapeutic-biological-products (accessed 24 September 2019).

11 Gurevich, E.V. and Gurevich, V.V. (2014). Therapeutic potential of small molecules and engineered proteins. In: *Arrestins – Pharmacology and Therapeutic Potential* (ed. V.V. Gurevich), 1–12. https://doi.org/10.1007/978-3-642-41199-1_1. Heidelberg: Springer-Verlag https://www.ncbi.nlm.nih.gov/pmc/articles/PMC4513659 (accessed 24 September 2019).

12 Bren, L. (2006). The road to the biotech revolution – highlights of 100 years of biologics regulation. FDA Consumer magazine, Centennial Edition (January–February). https://www.fda.gov/files/about%20fda/published/The-Road-to-the-Biotech-Revolution--Highlights-of-100-Years-of-Biologics-Regulation.pdf (accessed 24 September 2019).

13 Chiu, M. and Gilliland, G. (2016). Engineering antibody therapeutics. *Current Opinion in Structural Biology* 38: 163–173. https://www.sciencedirect.com/science/article/pii/S0959440X16300872?via%3Dihub (accessed 24 September 2019).

14 Safary, A., Akbarzadeh Khiavi, M., Mousavi, R. et al. (2018). Enzyme replacement therapies: what is the best option? *BioImpacts: BI* 8 (3): 153–157. https://doi.org/10.15171/bi.2018.17, https://www.ncbi.nlm.nih.gov/pmc/articles/PMC6128977 (accessed 24 September 2019).

15 Ozaki, K. and Leonard, W.J. (2002). Cytokine and cytokine receptor pleiotropy and redundancy. *The Journal of Biological Chemistry* 277: 29355–29358. http://www.jbc.org/content/277/33/29355.full.html#fn-1 (accessed 24 September 2019).

16 Rider, P., Carmi, Y., and Cohen, I. (2016. https://doi.org/10.1155/2016/9259646). Biologics for targeting inflammatory cytokines, clinical uses, and limitations. *International Journal of Cell Biology* 2016: 9259646, 11 pp. https://www.hindawi.com/journals/ijcb/2016/9259646/cta (accessed 24 September 2019).

17 DailyMed [Internet] (2005). Bethesda, MD: U.S. National Library of Medicine. https://dailymed.nlm.nih.gov/dailymed (accessed 18 May 2018). Drug labeling on this Web site is as submitted to the Food and Drug Administration (FDA).

18 Gutterman, J.U. (1994). Cytokine therapeutics: lessons from interferon alpha. *Proceedings of the National Academy of Sciences of the United States of America* 91 (4): 1198–1205. https://doi.org/10.1073/pnas.91.4.1198, https://www.ncbi.nlm.nih.gov/pmc/articles/PMC43124 (accessed 24 September 2019).

19 Saleem, S., Anwar, A., Fayyaz, M. et al. (2019). An overview of therapeutic options in relapsing-remitting multiple sclerosis. *Cureus* 11 (7): e5246. https://doi.org/10.7759/cureus.5246, https://www.cureus.com/articles/21600-an-overview-of-therapeutic-options-in-relapsing-remitting-multiple-sclerosis (accessed 24 September 2019).

20 Akdis, M., Aab, A., Altunbulakli, C. et al. Interleukins (from IL-1 to IL-38), interferons, transforming growth factor β, and TNF-α: receptors, functions, and roles in diseases. *Journal of Allergy and Clinical Immunology* 138 (4): 984–1010. https://www.jacionline.org/article/S0091-6749(16)30715-1/fulltext (accessed 24 September 2019).

21 Lis, K., Kuzawińska, O., and Bałkowiec-Iskra, E. (2014). Tumor necrosis factor inhibitors – state of knowledge. *Archives of Medical Science, AMS* 10 (6): 1175–1185. https://doi.org/10.5114/aoms.2014.47827, https://www.ncbi.nlm.nih.gov/pubmed/25624856 (accessed 24 September 2019).

22 Diabetes.co.uk. History of insulin. Diabetes Digital Media Ltd. https://www.diabetes.co.uk/insulin/history-of-insulin.html (accessed 24 September 2019).

23 Swiech, K., Picanço-Castro, V., and Covas, D.T. (2017). Production of recombinant coagulation factors: are humans the best host cells? *Bioengineered* 8 (5): 462–470. https://doi.org/10.1080/21655979.2017.1279767, https://www.tandfonline.com/doi/full/10.1080/21655979.2017.1279767 (accessed 24 September 2019).

24 Stryjewska, A., Kiepura, K., Librowski, T., and Lochynski, S. (2013). Biotechnology and genetic engineering in the new drug development. Part I. DNA technology and recombinant proteins. *Pharmacological Reports* 65 (5): 1075–1085. http://www.if-pan.krakow.pl/pjp/pdf/2013/5_1075.pdf (accessed 24 September 2019).

25 Lodish, H., Berk, A., Zipursky, S.L. et al. (2000). Section 7.1, DNA cloning with plasmid vectors. In: *Molecular Cell Biology*, 4e. New York: W. H. Freeman https://www.ncbi.nlm.nih.gov/books/NBK21498 (accessed 24 September 2019).

26 Jayapal, K., Wlaschin, K., Hu, W., and Yap, M. *Recombinant Protein Therapeutics from CHO Cells – 20 Years and Counting.* CHO Consortium, SBE Special Section https://pdfs.semanticscholar.org/5f96/12ce9170571f296b75246e80cb671bbb886c.pdf (accessed 24 September 2019).

27 Lundstrom, K. (2018). Viral vectors in gene therapy. *Diseases (Basel, Switzerland)* 6 (2): 42. https://doi.org/10.3390/diseases6020042, https://www.ncbi.nlm.nih.gov/pmc/articles/PMC6023384 (accessed 24 September 2019).

28 Hersom, M. and Jorgensen, J.T. (2018). Companion and complementary diagnostics-focus on PD-L1 expression assays for PD-1/PD-L1 checkpoint inhibitors in non-small cell lung cancer. *Therapeutic Drug Monitoring* 40 (1): 9–16. https://www.ncbi.nlm.nih.gov/pubmed/29084031. (accessed 24 September 2019).

29 Bernicker, E.H., Allen, T.C., and Cagle, P.T. (2019). Update on emerging biomarkers in lung cancer.

Journal of Thoracic Disease 11 (Suppl 1): S81–S88. https://doi.org/10.21037/jtd.2019.01.46, https://www.ncbi.nlm.nih.gov/pmc/articles/PMC6353743/#r36 (accessed 24 September 2019).

30 Keytruda® [package insert] (September 2019). Whitehouse Station, NJ: Merck Sharp & Dohme Corp.. https://www.merck.com/product/usa/pi_circulars/k/keytruda/keytruda_pi.pdf (accessed 24 September 2019).

31 U.S. Food and Drug Administration (2018). Industry information and guidance (last updated 18 July 2018). https://www.fda.gov/drugs/biosimilars/industry-information-and-guidance (accessed 24 September 2019).

32 IQVIA Institute (2019). The global use of medicine in 2019 and outlook to 2023 (29 January). https://www.iqvia.com/insights/the-iqvia-institute/reports/the-global-use-of-medicine-in-2019-and-outlook-to-2023 (accessed 24 September 2019).

33 Dharssi, S., Wong-Rieger, D., Harold, M., and Terry, S. (2017). Review of 11 national policies for rare diseases in the context of key patient needs. *Orphanet Journal of Rare Diseases* 12 (1): 63. https://doi.org/10.1186/s13023-017-0618-0, https://www.ncbi.nlm.nih.gov/pmc/articles/pmid/28359278 (accessed 24 September 2019).

34 Gurevich, E.V. and Gurevich, V.V. (2015). Beyond traditional pharmacology: new tools and approaches. *British Journal of Pharmacology* 172 (13): 3229–3241. https://doi.org/10.1111/bph.13066, https://www.ncbi.nlm.nih.gov/pmc/articles/PMC4500362 (accessed 24 September 2019).

4

Major Classes of Biotherapeutics

Esteban Cruz and Veysel Kayser

Sydney Pharmacy School, Faculty of Medicine and Health, The University of Sydney, Sydney, New South Wales, Australia

---

KEY POINTS

- Current biological therapies include a wide range of classes with different structural features. These include blood-derived products, vaccines, recombinant proteins, engineered-cell therapies, viral therapies, and nucleic acid therapies.
- Monoclonal antibodies have come to dominate the biopharmaceutical market in the last 20 years. The current clinical pipeline suggests that antibody therapeutics will continue to expand their dominance in the near future.
- Novel trends in preclinical development and recent approvals indicate that complex biotherapeutics, such as engineered-cell therapies (e.g. CAR-T cells) and viral therapies (gene therapy and oncolytic viruses), will expand in clinical implementation in the next decade.

Abbreviations

Abbreviation	Full name	Abbreviation	Full name
ADC	Antibody–Drug Conjugate	DTaP	Diphtheria, Tetanus, and Pertussis
ADCC	Antibody Dependent Cellular Cytotoxicity	ECD	Extracellular Domain
ADCP	Antibody Dependent Cellular Phagocytosis	EGFR	Epidermal Growth Factor Receptor
AT	Antithrombin	EMA	European Medicines Agency
BsAb	Bispecific Antibodies	EML	Essential List of Medicines
CBER	Center for Biologics Evaluation and Research	EU	European Union
CD	Cluster of Differentiation	Fab	Fragment Antigen Binding Region
CDC	Complement-Dependent Cytotoxicity	Fc	Fragment Crystallizable Region
CDER	Center for Drug Evaluation and Research	FcRn	Fc Neonatal Receptor
CDR	Complementarity Determining Regions	FDA	United States Food and Drug Administration
CH1	Constant Heavy Chain Domain 1	FFDCA	Federal Food, Drug and Cosmetic Act
CH2	Constant Heavy Chain Domain 2	FIX	Coagulation Factor IX
CH3	Constant Heavy Chain Domain 3	FSH	Follicle-Stimulating Hormone
cHL	Classical Hodgkin Lymphoma	FVIII	Coagulation Factor VIII
CHO	Chinese Hamster Ovary	G-CSF	Granulocyte Colony Stimulating Factor
CKD	Chronic Kidney Disease	GLP-1	Glucagon-Like Peptide 1
CTLA-4	Cytotoxic T-Lymphocyte Associated Protein 4	GLP-2	Glucagon-Like Peptide 2
DNA	Deoxyribonucleic Acid	GM-CSF	Granulocyte and Macrophage Colony Stimulating Factor
DPP-4	Dipeptidyl Peptidase 4		

Biologics, Biosimilars, and Biobetters: An Introduction for Pharmacists, Physicians, and Other Health Practitioners,
First Edition. Edited by Iqbal Ramzan.
© 2021 John Wiley & Sons, Inc. Published 2021 by John Wiley & Sons, Inc.

Abbreviation	Full name
GPIIb/IIIa	Glycoprotein IIb/IIIa
HA	Hemagglutinin
HC	Heavy Chain
HCC	Hepatocellular Carcinoma
hCG	Human Chorionic Gonadotropin
HER-2	Human Epidermal Growth Factor Receptor 2
hGH	Human Growth Hormone
HIV	Human Immunodeficiency Virus
HNSCC	Head and Neck Squamous Cell Cancer
HuCAL	Human Combinatorial Antibody Libraries
ICI	Immune Checkpoint Inhibitors
IFN-β	Interferon Beta
IgE	Immunoglobulin E
IgG	Immunoglobulin G
IL	Interleukin
IL-1RAcP	Human Interleukin-1 Receptor Accessory Protein
kDa	Kilodalton
LAG-3	Lymphocyte Activation Gene 3
LC	Light Chain
LH	Luteinizing Hormone
LMWH	Low Molecular Weight Heparins
M2	Matrix Protein 2
mAb	monoclonal Antibody
MCC	Merkel Cell Carcinoma
MMR	Measles, Mumps and Rubella
mPEG	methoxy Polyethylene Glycol

Abbreviation	Full name
NK	Natural Killer
NSCLC	Non-Small Cell Lung Cancer
PD	Pharmacodynamics
PD-1	Programmed Cell Death Protein 1
PD-L1	Programmed Cell Death Protein 1 Ligand
PEG	Polyethylene Glycol
PHSA	Public Health Service Act
PK	Pharmacokinetics
PMBCL	Primary Mediastinal Large B-cell Lymphoma
RCC	Renal Cell Carcinoma
RNA	Ribonucleic Acid
scFv	single-chain variable Fragments
SHOX	Short Stature Homeobox-Containing Gene
siRNA	small interfering RNA
TIGIT	T-cell Immunoreceptor with Ig and ITIM domains
TIM-3	T-cell Immunoglobulin and Mucin-domain Containing-3
TNF-α	Tumor Necrosis Factor alpha
tPA	tissue Plasminogen Activators
US	United States of America
USD	United State Dollar
VEGF	Vascular Endothelial Growth Factor
VH	Heavy Chain Variable Domain
VISTA	V-domain Ig Suppressor of T-cell Activation
VL	Light Chain Variable Domain
WHO	World Health Organization

4.1 Major Classes of Biotherapeutics

The major classes of biotherapeutics are categorized here according to their pharmacological properties, structural identity, and source of origin, and then further divided into subclasses based on more detailed structural and functional features. Broadly, biological therapies can be divided into recombinant biological macromolecules, plasma-derived therapies, engineered-cell therapies, oncolytic viruses, and vaccines. Table 4.1 provides an outline of a proposed classification system, listing examples of licensed biotherapeutics for each subclass. Due to the complex nature of these therapies, there is naturally an overlap among some classes and subclasses, for instance, plasma-derived therapies contain macromolecules (e.g. proteins), yet the source of these products is fundamentally different to that of recombinant biotherapeutics and are hence assigned to a different class. An alternative example are recombinant vaccines that consist of antigenic proteins produced in recombinant expression systems (e.g. influenza vaccine Flublok Quadrivalent®), where the clinical application of these prophylactic proteins is different in essence to that of most other recombinant proteins.

An excellent online resource on biologics for healthcare professionals in the United States is the Food and Drug Administration (FDA)'s "Purple Book" that lists all licensed biologics in the United States (US) under the Public Health Service Act (PHSA), their date of licensure, and their status as biosimilars or interchangeable biosimilars. The information provided is updated periodically with new approvals and designations. The book issues two separate lists, one containing products licensed by the Center for Drug Evaluation and Research (CDER)[1] and another one with biologics regulated by the Center for Biologics Evaluation and Research

Table 4.1 Major classes of biotherapeutics.

			Examples (INN)	Trade name	Clinical use (broad description)
Biological macromolecules	Recombinant biological macromolecules	Antibody-based therapeutics			
		Full-size IgG	Trastuzumab	Herceptin®	Breast and gastric cancer
			Adalimumab	Humira®	Inflammatory diseases
			Rituximab	Rituxan® / MabThera®	Lymphoma, leukemia, rheumatoid arthritis
		Fc-fusion proteins	Etanercept	Enbrel®	Autoimmune diseases
			Alefacept	Amevive®	Plaque psoriasis
		Antibody–drug conjugates (ADCs)	Ado-trastuzumab emtansine	Kadcyla®	Breast cancer
			Brentuximab vedotin	Adcetris®	Lymphoma
			Polatuzumab vedotin-piiq	Polivy®	Lymphoma
		Bispecific antibodies (BsAbs)	Blinatumomab	Blincyto®	Leukemia (T-cell engager)
			Emicizumab-kxwh	Hemlibra®	Hemophilia A (targets factors IXa and X)
		Immune checkpoint inhibitors (ICIs)	Nivolumab	Opdivo®	Various cancers (e.g. Hodgkin's lymphoma, colorectal, and melanoma)
			Pembrolizumab	Keytruda®	Various cancers (e.g. Hodgkin's lymphoma, Merkel and renal cell carcinomas)
			Ipilimumab	Yervoy®	Melanoma and renal cell carcinoma
	Recombinant proteins	Peptide and protein hormones	Insulin glargine	Lantus®	Diabetes mellitus (type 1 and 2)
			Somatropin (recombinant growth hormone)	Humatrope®	Human growth hormone deficiency in children
		Signaling molecules	Interferon gamma-1b	Actimmune®	Prevent infection in patients with chronic granulomatous disease
			Aldesleukin (human recombinant IL-2)	Proleukin®	Melanoma and renal cell carcinoma
			Pegfilgrastim (human recombinant G-CSF)	Neulasta®	Prevent infections in patients receiving myelosuppressive therapies

(Continued)

Table 4.1 (Continued)

		Examples (INN)	Trade name	Clinical use (broad description)
Plasma-derived Macromolecules	Recombinant clotting factors	Coagulation factor VIII (recombinant)	Jivi®	Hemophilia A (congenital factor VIII deficiency)
		Coagulation factor IX (recombinant)	Rebinyn®	Hemophilia B
	Enzymes	Asparaginase	Elspar®	Acute lymphoblastic leukemia
		Dornase alfa	Pulmozyme®	Cystic fibrosis
	Antisense-oligos	Mipomersen sodium	Kynamro®	Homozygous familial hypercholesterolemia
		Defibrotide	Defitelio®	Hepatic veno-occlusive disease
Oligonucleotides		Eteplirsen	Exondys 51®	Duchenne muscular dystrophy
	siRNA	Patisiran (transthyretin-directed siRNA)	Onpattro®	Polyneuropathy of hereditary transthyretin-mediated amyloidosis
	Aptamers	Pegaptanib (anti-VEGF)	Macugen®	Neovascular macular degeneration
	Clotting factors concentrates	Coagulation factor X	Coagadex®	Hereditary factor X deficiency
		Coagulation factor VIII	Wilate®	Von Willebrand disease and hemophilia A
	Immunoglobulin concentrates	Immune globulin IV, Human-slra	Asceniv®	Primary humoral immunodeficiency for adults and adolescents
		Immune globulin subcutaneous	Cutaquig®	Primary humoral immunodeficiency in adults
	Antitoxins	Crotalidae immune F(ab')2	Anavip®	North American rattlesnake envenomation
		Anthrax immune globulin IV	Anthrasil®	Treatment of inhalational anthrax
Whole cell	Cell therapies	Axicabtagene ciloleucel	Yescarta®	B-cell lymphoma
		Tisagenlecleucel	Kymriah®	B-cell lymphoma and B-cell precursor acute lymphoblasticleukemia
Viral	Oncolytic viruses	Talimogene laherparepvec	Imlygic®	Melanoma
	Viral gene therapy	Voretigene neparvovec	Luxturna®	Retinal dystrophy
Vaccines	Recombinant, inactivated, subunit	Hepatitis B vaccine	Heplisav-B®	Hepatitis B vaccine
		Influenza vaccine, adjuvanted	Fluad®	Influenza vaccine

Complex biologicals

(CBER).[2] The CDER list contains most therapeutic recombinant proteins, including monoclonal antibodies (mAbs), cytokines, growth factors, hormones, enzymes, and fusion proteins. The CBER list, on the other hand, records further biological products such as plasma derivatives, clotting factors, antitoxins, vaccines, probiotics, oncolytic viruses, human cells, and tissue-based products. As the regulatory agencies in Europe (the European Medicines Agency [EMA]) and the US (FDA) often pave the way for subsequent regulatory developments in other regions, these extensive documents are extremely beneficial to gain insight into a broad yet comprehensive overview of the historical implementation of the different classes of biotherapeutics. An important caveat of these lists is that not all biologics are regulated under the PHSA in the United States; insulin, other hormones, some enzymes, and their follow-on biologics are under the Federal Food, Drug and Cosmetic Act (FFDCA). The regulatory status of biologics currently under FFDCA is programmed to change in March 2020 and will be under PHSA regulation.[3] A list of all products scheduled to undergo this transition is available.[4]

Prior to the advent of recombinant DNA technologies, most biologics comprised allergens, vaccines, and plasma-derived products. The first recombinant product to gain regulatory approval was Humulin® (human recombinant insulin) in 1982, in both Europe and the United States. Recombinant therapeutic proteins have since become vastly predominant in development, implementation and market share. Proteins of recombinant origin accounted for more than 70% of FDA (PHSA) biologics approvals (72) in the last five years (102 in total), out of which roughly 65% (i.e. 47) correspond to approvals of novel antibodies or antibody-based therapeutics (i.e. excluding biosimilars). Other major classes with a significant number of first-time approvals in this five-year period include coagulation factors (13), vaccines (9), insulins (9), plasma-derived products (9), and enzymes (7). Signaling molecules excluding insulins (cytokines, hormones, and growth factors) are still widely used in the clinic, yet only two novel FDA approvals (cenegermin and parathyroid hormone) since 2015 reflects a deceleration in clinical development due to market saturation.[1,2] A therapeutic class currently gaining momentum are nucleic acid-based products, which aim to alter gene expression for therapeutic purposes. As such, this class includes viral-gene therapy, antisense oligos, small interfering RNA (siRNA), and engineered-cell therapies. These entities are relatively new, but their enormous therapeutic potential is likely to spur expansive clinical implementation in the future, provided that no significant safety concerns are raised with the newly approved products.

4.2 Antibodies and Antibody-Based Therapeutics

As reflected by recent approvals, total number of available products, and market sales, antibody therapeutics are currently the most important class of biologics.[1,5] Therapeutic mAbs have topped the lists for best-selling pharmaceutical products for several years, featuring 7 out of the 10 best-selling drugs globally in 2018.[6] Even more impressive is that the clinical pipeline of biologics in late stages of clinical development forecasts that mAbs will continue their current market dominance and clinical use in coming years.[7]

Antibody therapeutics experienced remarkable expansion at the turn of the century following the development of antibody humanization techniques. These technologies started in the 1980s and saw fruition in 1998 with the licensure of trastuzumab and palivizumab, the first humanized antibodies to enter the market.[8] Humanization decreased the safety concerns related to the murine origin of the first mAbs and paved the way for the many human and humanized therapeutic mAbs currently available. Prior to 2000, only eight therapeutic (excluding diagnostic mAbs) antibodies had obtained licensure. That number now exceeds 80 products, spanning a variety of clinical applications.[1,9] In light of the relevance of mAbs, the following sections focus primarily on antibody-based therapies.

4.2.1 Monoclonal Antibodies

mAbs have revolutionized the pharmaceutical market in the last two decades.[10] Primary reasons for their clinical success stems from their exquisite molecular specificity toward cognate antigens and design versatility. The advent of high throughput antibody generation techniques (e.g. phage display,[11] yeast display,[12] and human combinatorial antibody libraries[13] [HuCAL®]), alongside a continuous growth in the identification of disease biomarkers, has enabled many clinical applications of therapeutic antibodies since their clinical debut.[14] While oncology and inflammatory indications prevail both currently and in the clinical pipeline, their indications have broadened greatly in recent years. For instance, 2018 saw the approval of antibodies for migraine prevention (erenumab, fremanezumab, and galcanezumab), hypophosphatemia (burosumab), angioedema (lanadelumab), thrombotic thrombocytopenic purpura (caplacizumab), and HIV (ibalizumab).[1,9]

The following section provides a brief description of the structural features of antibodies as an aid to understanding their specificity and versatility.

4.2.2 Structure of Therapeutic mAbs

mAbs are immunoglobulins produced in recombinant expression systems. All currently available mAbs are gamma immunoglobulins (IgG), which consist of approximately 150 kDa heterodimeric proteins containing two heavy (HC) and two light (LC) chains (Figure 4.1)[1,15]. IgGs are further divided into four subclasses – IgG1, IgG2, IgG3, and IgG4, each with different biological properties[16]. Roughly 70% of all therapeutic mAbs are IgG1, with IgG2 and IgG4 representing roughly 16 and 14%, respectively[1,15]. These subtypes differ mostly in the length of the hinge region and the number of disulfide bonds. Antibodies of the IgG3

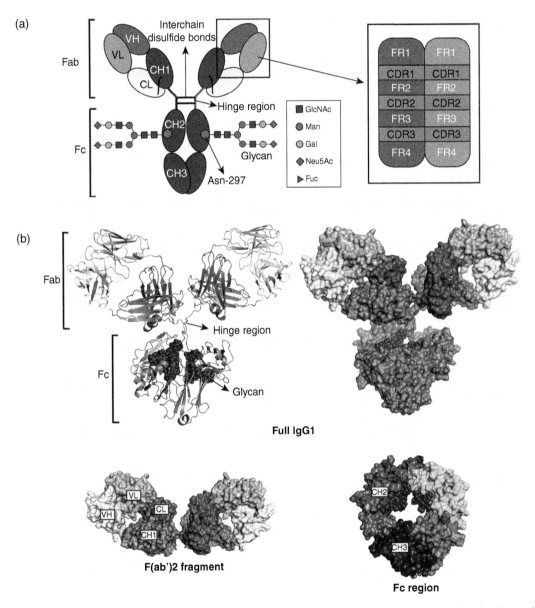

Figure 4.1 Structure of IgG1. (a) Schematic representation of an IgG1 displaying the various domains that make up the heavy chain (VH, CH1, CH2, and CH3) and the light chain (VL and CL). A complex glycan represented as per the standardized Symbol Nomenclature for Glycans (SNFG) is shown bound to the asparagine 297 (Asn-297) residue on the CH2 of the heavy chain. The expanded region provides a closer look to the variable regions of the heavy and light chain, displaying three complementarity-determining regions (CDRs and four framework regions (FR) on each domain. (b) Crystal structure of a murine IgG1 protein (PDB 1IGY) displayed as both cartoon and surface representations. The Fc-glycan is shown in red. The F(ab')2 and Fc region are shown isolated to indicate the position of the domains in the surface representation. Note the orientation of the glycan toward the interior of the protein. (*See insert for color representation of the figure.*)

subclass display high effector function (capacity to elicit an immune response against its antigen); however, their susceptibility to proteolysis has hindered their development[17]. Each antibody chain is segmented into various domains. From N-terminus to C-terminus, heavy chains are made up of one variable domain (VH) and three constant (conserved) domains (CH1, CH2, and CH3). Light chains consist of a variable domain (VL) and a constant domain (CL), with the CL at the C-terminus (Figure 4.1).[18]

Structural units of IgGs can be categorized according to therapeutic function; IgG molecules can be divided into two antigen binding (Fab) regions and one Fc region (Figure 4.1).[19] The two Fab regions in an IgG monomer translates to mAbs being considered bivalent; a single IgG1 can bind to two identical targets expressing the antigen.[20] Each antigen binding region consists of a full LC and the VH and CH1 domains of one of the heavy chains. Light chains are bound to the Fab region of the heavy chain through non-covalent (hydrophobic and electrostatic) interactions and an interchain disulfide bond that is formed proximal to the C-terminus of the light chain and the CH1 domain.[20] Both heavy chains also interact with one another through non-covalent interactions between CH2 and CH3 domains of each respective chain in the Fc region and two disulfide bonds located in the hinge region.[21] The hinge is a highly flexible area of the heavy chain that links the Fc and Fab regions and grants flexibility for the Fab arms and the Fc to move independently.[22] The variable domains of both chains contain seven sections, four framework regions, and three complementarity-determining regions (CDRs) which are hypervariable regions that ultimately determine the specificity of the antibody. The framework regions serve primarily as structural scaffolds for the CDRs[23]. The section of the Fab that binds to the antigen, or the antigen-binding site, is named a paratope. Conversely, the specific region of the antigen that the paratope binds to is called an epitope.[24]

The Fc region plays a key role in the biological function of mAbs, as this region is recognized by Fcγ receptors present in immune effector cells (e.g. natural killer (NK) cells, monocytes, neutrophils, macrophages, and dendritic cells) and can thus trigger antibody-dependent cellular cytotoxicity (ADCC) or antibody-dependent cellular phagocytosis (ADCP) on antigen-coated targets, e.g. tumor cells or pathogens like bacteria or viruses.[25] Alternatively, the Fc region can also bind C1q to initiate the complement cascade and elicit complement-dependent cytotoxicity (CDC).[26] The Fc can further interact in a pH-dependent manner with the neonatal Fc receptor (FcRn) which is responsible for IgG recy-

cling in the lysosomal compartment, where it prevents antibody proteolysis and enables antibody molecules to be recycled back into circulation. Intracellular catabolism is presumed to be the primary elimination route of mAbs, thus FcRn recycling plays a major role in their long elimination half-lives.[27] Prime examples of mAbs with long half-lives are bevacizumab (~20 days),[28] rituximab (~22 days),[29] and trastuzumab (~28 days).[30]

Antibody heavy chains possess an N-glycan on the CH2 domain (thus 2 glycan molecules per monomer) that are important in the biological activity and aggregation propensity of the protein.[31,32] Specifically, these glycans play a key role in the interaction with Fc receptors and complement proteins, hence the particular combinations of glycoforms present on a mAb monomer can determine whether ADCC or CDC is favored in the presence of immune effector cells.[33] Moreover, naturally occurring antibodies can be glycosylated on the Fab region. Roughly 20% of IgGs found in plasma contain an additional N-glycosylation sequon on the Fab region, which can be located either on the HC or the LC. Although not as important as Fc glycosylation, Fab glycosylation can alter binding affinity, pharmacokinetics, and physicochemical stability.[34] Cetuximab (anti-EGFR) is the only therapeutic antibody with Fab glycosylation; however, more mAbs containing this feature may enter the market especially since additional Fab glycosylation has been proposed to increase solubility and reduce aggregation.[35]

4.2.3 Therapeutic Applications of mAbs

The wide-ranging therapeutic applications of mAbs attest to their versatility and unprecedented growth. Historically, mAbs were primarily developed for oncology and inflammatory conditions[1,5]; they were expected to display lower severe side effects compared with conventional small molecule drugs. Most anticancer mAbs target cell membrane proteins overexpressed in tumor cells; the upregulation of these markers is often implicated in tumor development and progression. For instance, the epidermal growth factor receptor (EGFR) (targeted by cetuximab, panitumumab, and necitumumab) and the HER-2 receptor (targeted by trastuzumab and pertuzumab) are both cell surface receptors implicated in signaling pathways that promote exacerbated proliferation in various types of solid tumors.[36] By targeting these receptors, antibodies can inhibit proliferation via blockade of the signaling cascade. Alternatively, mAbs can evoke an immune response against tumor cells overexpressing the antigen by engaging effector cells or the complement system.[37]

Another example of an overexpressed cell-surface molecule targeted by many mAbs is the membrane-embedded CD20 protein (targeted by rituximab, tositumomab, ibritumomab tiuxetan, ofatumumab, obinituzumab, and ocrelizumab). CD20 is a B-lymphocyte antigen expressed in various stages of B-cell development and amplified in certain types of lymphomas and B-cell leukemias.[38] B-cells are involved in multiple pathologies and thus anti-CD20 antibodies can also be used for treatment of autoimmune and inflammatory diseases such as systemic lupus erythematosus, multiple sclerosis, and rheumatoid arthritis.[39,40]

An alternative approach in cancer treatment involves targeting the tumor microenvironment and tumor vasculature, as opposed to the direct depletion of cancer cells. The exquisite specificity of mAbs has been exploited to target soluble proteins that promote angiogenesis; the physiological process by which vascular endothelial cells grow, migrate, and differentiate to form new blood vessels.[41] This process is induced by pro-angiogenic mitogens to provide the vascular growth and vascular remodeling that tumors require to sustain their growth.[41] Two distinct therapeutic mAbs (bevacizumab and rabinizumab) have been developed to target the vascular endothelial growth factor (VEGF), a central regulator of neo-vasculature formation.[1,42] In addition, the antibody ramucirumab binds to the VEGF-2 receptor to disrupt VEGF-mediated signaling.[43] These antiangiogenic agents are not only thought to prevent angiogenesis but also to diminish the interstitial fluid pressure of the tumor, thereby enhancing the capacity of cytotoxic agents to reach the tumor tissue.[44] Immune checkpoint inhibitors (ICIs) are further examples of anticancer mAbs that do not target tumor cells directly (Section 4.3.4).

Antibodies are remarkable tools for disrupting ligand–receptor interactions, either by binding and inactivating soluble ligands, or through interactions with the ligand's receptor that block the ligand-binding site. This signaling-blocking capacity is central to the role of antibody therapeutics in the treatment of inflammatory diseases. Targeting the inflammatory cytokine tumor necrosis factor alpha (TNF-α) and its receptor have been remarkably successful therapeutic strategies, having led to the approval of three distinct full-size antibodies (infliximab, adalimumab, and golimumab), an antibody-fragment (certolizumab pegol), and an Fc-fusion protein (etanercept).[1,9] Remarkably, the anti-TNF-α mAb adalimumab is the best-selling drug in the market as of 2019.[6] Due to the involvement of TNF-α in various inflammatory conditions, these anti-TNF-α therapeutics have been approved for a diverse range of pathologies, including rheumatoid arthritis, Crohn's disease, psoriatic arthritis, ankylosing spondylitis, ulcerative colitis, and plaque psoriasis.[45,46] Other antibodies that target inflammatory cytokine signaling include dupilumab (anti-IL-4α receptor),[47] tocilizumab (anti-IL-6),[48] canakinumab (anti-IL-1β),[48] ixekizumab (anti-IL-17A),[49] and ustekinumab (anti-IL-12 and IL-23).[50] Another example of a mAb used in gastrointestinal inflammatory conditions is vedolizumab, which targets the human α4β7 integrin and provides anti-inflammatory effects in the gut[51].

Therapeutic applications of mAbs have greatly diversified since their introduction. Prior to 2000, most therapeutic mAbs were designed for prophylaxis of transplant rejection, diagnostic purposes, and treatment of cancer and inflammatory diseases.[1,5] Two exceptions were abciximab and pavilizumab; abciximab is an antibody fragment that targets a platelet surface receptor (GPII$_b$/III$_a$) to prevent blood clots, whereas pavilizumab targets a surface epitope of the respiratory syncytial virus to prevent viral infection.[52] Since then, four other antibodies for treatment or prophylaxis of microbial infections have been approved. These are raxibacumab[53] and obiltoxaximab[54] for inhalational anthrax, bezlotuxumab[55] against *C. difficile*, and the recent approval of ibalizumab[56] that targets the human CD4 domain to prevent HIV virus entry into CD4+ cells. Asthma management is another application of mAbs. Omalizumab was the first such antibody to enter the clinic in 2003 (US) and 2005 (EU), for disrupting IgE–IgE receptor signaling in moderate to severe persistent asthma.[57] Three other asthma mAbs (mepolizumab,[58] reslizumab,[59] and benralizumab[60]) have received approval in the last five years, all targeting the cytokine interleukin-5 (IL-5) or its receptor. Other therapeutic indications of mAbs include hypercholesterolemia and dyslipidemia (evolocumab[61] and alirocumab[62]), reversal of anticoagulant effects of dabigatran (idarucizumab[63]), and very recent approvals for X-linked hypophosphatemia (burosumab[64]) and migraine (erenumab[65]).

4.3 Alternative mAb-Based Therapeutics

4.3.1 Fc-Fusion Proteins

Fc-fusion proteins are recombinant chimeric proteins that consist of peptides or proteins genetically fused to an Fc region. The molecular structure of most of these peptide and proteins are considerably smaller than an IgG (Figure 4.2). Most growth factors and cytokines, for

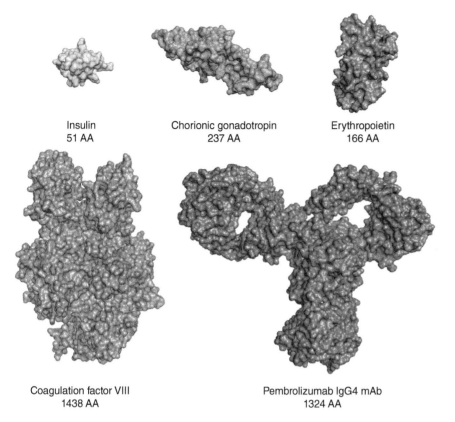

Insulin
51 AA

Chorionic gonadotropin
237 AA

Erythropoietin
166 AA

Coagulation factor VIII
1438 AA

Pembrolizumab IgG4 mAb
1324 AA

Figure 4.2 Structural features of various classes of therapeutic proteins. (*See insert for color representation of the figure.*)

instance, range from 10 to 70 kDa compared with the 150 kDa molecular weight of a full-size antibody. These relatively small sizes lead to increased elimination rates because proteins below ~60 kDa can undergo renal filtration and excretion.[66] While this may be beneficial for certain applications, increased clearance rates diminish efficacy and are not useful for most therapeutic purposes. Fc-fusion proteins consists of an Fc-region fused to the protein of interest to greatly enhance their pharmacokinetic profile via FcRn recycling and the ensuing increase in hydrodynamic size, yielding half-lives of several days.[67] These chimeric proteins generally display a shorter half-life than the parent full-size IgG, but the improvements are significant enough to make them a highly successful therapeutic.[68] A number of Fc-fusion constructs have been designed for research and clinical development with more than 10 reaching the clinic, yet in the last five years, no new product has received licensure.[1, 9]

Fc-fusion proteins can be classified into two classes. The first are immune-adhesins, composed of extracellular domains (ECD) of cell surface receptors fused to an Fc-region. Examples include etanercept (two soluble TNF receptor subunits), rilonacept (human IL-1 receptor [IL-1R] and human IL-1 accessory protein [IL-1RAcP]),

belatacept and abatacept (cytotoxic T-lymphocyte associated protein 4 [CTLA-4]), and aflibercept (vascular endothelial growth factor receptor 1 [VEGFR1] and VEGFR2) all fused to the N-terminus of an Fc-region.[67] These proteins generally work as traps for the receptor's ligand, to prevent the signaling events that the target molecule can evoke. For example, etanercept can sequester TNF to inhibit the inflammatory activity of this cytokine in inflammatory diseases (rheumatoid arthritis, psoriatic arthritis, ankylosing spondylitis, and plaque psoriasis).[69] The second class are constructs comprising other signaling proteins, peptides, or non-natural mimetic peptides bound to an Fc region. Romiplostim[70] (thrombopoietin mimetic peptide construct fused to an Fc region via the C- rather than the N-terminus) and dulaglutide[71] (glucagon-like peptide bound to an Fc region) are both approved agents in this class.

Fc-fusion proteins are structurally complex. A monomer of etanercept, for example, contains six N-linked glycans and multiple intrachain and interchain disulfide bonds and O-linked glycans.[72] Their complexity ensures therapeutic versatility but raises significant manufacturing and regulatory challenges. An excellent reference on the structural features, design considerations, and clinical development of Fc-fusion proteins is available.[67]

4.3.2 Antibody–Drug Conjugates

Antibody–drug conjugates (ADCs) consist of a cytotoxic drug attached to full-size IgGs or mAb derivative through chemical linkers. The ADC concept was conceived several decades ago to selectively deliver potent cytotoxic payloads to a targeted tissue, but it only came to fruition some decades later with the approval of gemtuzumab ozogamicin (Mylotarg®) in 2000.[73] Despite the potential of this concept, ADCs have suffered important setbacks in clinical development; a prime example was the withdrawal of gemtuzumab ozogamicin in 2010 due to its narrow therapeutic index.[74] Mylotarg® was reintroduced into the market in 2017 following a dosing regimen change.[75] ADCs usually carry extremely toxic payloads, many of which would be intolerable if administered alone. Consequently, the chemical linker must have sufficient plasma stability to prevent the drug from being released prematurely and reach untargeted tissues.[14,76] Many initial concerns with ADCs were ascribed to poor linker plasma stability and off-target toxicity of their powerful payloads.[77]

The chemical linker is crucial to determining the site of attachment to the protein and the release mechanism of the drug. Most commonly, the linker binds to the antibody through a chemical reaction with lysine or cysteine residues in a stochastic fashion, i.e. they bind to one or several of these residues on various regions of the mAb.[78] Lysine chemistry, in particular, can lead to considerable heterogeneity as IgG molecules possess many lysines in their primary structure (e.g. trastuzumab contains 88).[79] In contrast, cysteine conjugation requires partial reduction of the interchain disulfide bonds which can lead up to eight localized attachments per monomer, yielding less heterogeneous products.[80] More novel strategies seek to overcome heterogeneity by employing site-specific attachment strategies, such as the insertion of engineered cysteines that can be functionalized without requiring chemical reduction.[81] Other approaches include incorporation of unnatural amino acids in the protein and Fc-glycan attachment.[82,83] Concerning the release mechanism, most early ADC formats were designed with a linker sensitive to either an acidic or reducing environment, both characteristic features of tumor cells, such that the cytotoxic drugs are released upon cellular internalization. Subsequent advancements in linker technology included the development of enzyme-sensitive linkers and non-cleavable linkers.[76] Enzyme-sensitive linkers possess peptide sequences that are cleaved selectively by proteases in the intracellular environment, whereas non-cleavable linkers must undergo protein breakdown by proteolysis inside the cell for the payload to be released.[84]

Further biochemical features of ADCs that have a pivotal impact on the clinical properties are the selection of the molecular target, the antibody format (e.g. IgG subtype, full-size IgG, or antibody fragments), and the cytotoxic payload. Recent advancements in ADC design are further described in Chapter 13 of this book and have also been thoroughly reviewed by others.[14,85] Important advancements in the above-described components of ADCs have led to several recent approvals. Brentuximab vedotin (Adcetris®)[86] and ado-trastuzumab emtansine (Kadcyla®)[87] were approved in 2011 and 2013, respectively, and inotuzumab ozogamicin (Besponsa®, 2017),[88] and polatuzumab vedotin (Polivy®, 2019).[89]

4.3.3 Bispecific Antibodies

Bispecific antibodies (BsAbs) are antibody-based therapeutics that display specificity toward two different targets or epitopes. The history of BsAbs dates to the 1960s with the generation of bispecific antigen binding fragments (Fabs) through reassociation of enzymatic polyclonal F(ab) fragments following disulfide bond reduction.[90] Recombinant technologies have enabled the design of numerous bispecific antibody formats that range from complex IgG molecules with several appended domains to simple single-chain variable fragments (scFv).[91] BsAbs have been mainly developed to treat cancer, but their remarkable versatility translates to treatment of asthma, rheumatoid arthritis, diffuse cutaneous systemic sclerosis, osteoarthritis, neovascular wet age-related macular degeneration, and Sjögren syndrome.[91] The clinical pipeline of BsAbs comprises over 80 distinct formats, with more than 20 commercial technology platforms available.[91,92] However, as it is common with molecules of high complexity, the success of BsAbs in clinical development has been rather elusive and many designs are yet to undergo optimization to realize their full potential. This is reflected in only three bispecifics having received regulatory approval: catumaxomab (Removab®), blinatumomab (Blincyto®), and emicizumab (Hemlibra®), with catumaxomab being voluntarily withdrawn from the market for commercial reasons.[1,9] Only one bispecific (faricimab) is in Phase III trials as of 2019.[91]

4.3.4 Immune Checkpoint Inhibitors

ICIs are a relatively new class of biologics and discussed in detail in Chapters 5 and 13. Their market history started in 2011 with the approval of ipilimumab,[93] a full-size IgG1 antibody that targets the CTLA-4 receptor. Structurally, current ICIs are conventional full-size antibodies, yet their mechanism of action sets them in a distinct therapeutic class. There are currently seven ICI therapeutics approved, four of them are IgG1 and three of them IgG4 immunoglobulins.[1,9] As their name suggests, these proteins act by inhibiting suppressive signals responsible for preventing an immune response against tumor cells, thus "releasing the brakes" of the immune system to trigger a potent anti-cancer response.[14,94] Two inhibitory molecular pathways have been targeted in the clinic; CTLA-4 and the adaptive regulatory interaction between the programmed cell death protein 1 (PD-1) and its ligand PD-L1. Currently, ipilimumab is the only approved ICI that targets CTLA-4. All other available drugs target PD-1 (pembrolizumab, nivolumab, and cemiplimab) or PD-L1 (durvalumab, avelumab, and atezolizumab). The mechanistic description of the events that lead to an enhanced antitumor immune response are highly complex and are yet to be fully elucidated. Extensive reviews on the topic are available.[95] ICIs have already had tremendous success in the clinic. Nivolumab (Opdivo®) and pembrolizumab (Keytruda®) were among the top 10 best-selling drugs in 2018 by total sales, despite only entering the market in 2014.[6]

A crucial advantage of immune checkpoint inhibition, compared with other antibody-based applications in oncology, is that ICI can potentially be employed in a wide variety of tumors and hematological malignancies.[96] This is exemplified by pembrolizumab, which is currently indicated in melanoma, non-small cell lung cancer (NSCLC), head and neck squamous cell cancer (HNSCC), classical Hodgkin lymphoma (cHL), primary mediastinal large B-cell lymphoma (PMBCL), urothelial carcinoma, microsatellite instability-high cancer, gastric cancer, cervical cancer, hepatocellular carcinoma (HCC), Merkel cell carcinoma (MCC), and renal cell carcinoma (RCC).[97] In addition, ICIs have shown long-term antitumor effects in patients that respond to this therapy.[98] Conversely, low objective response rates and emergence of acquired resistance are still important clinical limitations.[99] However, better understanding of the immune system in tumor control and identification of relevant biomarkers that predict efficacy are likely to overcome such drawbacks. Several biologics targeting alternative pathways for immune checkpoint inhibition (e.g. LAG-3, TIM-3, TIGIT, and VISTA) are undergoing clinical development.[100]

4.4 Therapeutic Signaling Molecules

Signaling molecules are defined here as those that can regulate cellular processes through an interaction between the signaling molecule and a cell surface receptor. These cell surface receptors can communicate with intracellular effector molecules, in most cases kinases, wherein ligand-mediated activation of the receptor triggers a downstream signaling cascade that alters the function of intracellular proteins and gene expression. As such, we include hormones, growth factors, cytokines, and their corresponding analogs in the category of therapeutic signaling molecules.

In general, signaling molecules possess relatively simpler structures compared with other biotherapeutics (Figure 4.2). Many endogenous hormones and growth factors consist of short peptides, often containing two subunits covalently bound through disulfide bonds.[101–103] The smaller size of signaling molecules can be beneficial for their mode of action as coordinators of intricate homeostatic processes, yet they also have disadvantages as biotherapeutics given their increased clearance rates, as previously discussed for Fc-fusion proteins. This shortcoming has been the focus of many protein engineering strategies as part of novel therapeutic signaling molecule developments.[67] Another technique to increase plasma stability is protein PEGylation; the chemical attachment of polyethylene glycol (PEG) polymers on residue side chains confers higher hydrophilicity to proteins and increases their hydrodynamic size to prevent glomerular filtration. PEG attachment can also prevent enzymatic digestion of the protein.[104]

Different classes of biotherapeutics display significant differences in structural features with important implications in biological activity and pharmacokinetic profiles. Most evident is the prolonged biological half-life of therapeutic proteins with molecular weights above the glomerular cutoff filtration barrier (~60 kDa) and/or with an Fc-region that enables FcRn-mediated recycling. Therapeutic proteins such as Factor VIII and mAbs are considerably larger than most other therapeutics proteins, e.g. hormones, cytokines, growth factors, and enzymes. Human insulin (PDB 3I40), human chorionic gonadotropin (PDB 1HCN), human erythropoietin (PDB 1BUY), human coagulation factor VIII (PDB 3CDZ), and pembrolizumab (PDB 5DK3) are displayed as surface representations in PyMol.

4.4.1 Peptide and Protein Hormones

4.4.1.1 Insulins and Glucagon-like Peptide Analogs

Hormones are regulatory molecules produced and secreted by glands that travel via circulation and reach distant tissues to regulate physiological functions. Hormones were among the first biologic medicines to be used in the clinic, best exemplified by the history of therapeutic insulin.[105]

Structurally, human insulin is a 5.8 kDa protein comprising two peptide chains (A-chain and B-chain) bound together by two interchain disulfide bonds.[106] Important segments of insulin are highly conserved among species, hence porcine and bovine insulin were used in therapy before the introduction of recombinant human insulin.[105] The first recombinant insulin analog (insulin lispro 1996) conferred a faster onset of action for rapid postprandial adjustments. This was achieved by reversing proline-28 and lysine-29 present on the C-terminus of the B-chain to allow faster dissolution of the insulin hexamers into dimers and monomers that are absorbed more rapidly after subcutaneous injection.[107] A similar rationale was employed for the development of rapid-acting insulins, insulin aspart (proline B28 to aspartic acid), and insulin glulisine (arginine B3 to lysine, and lysine B29 to glutamic acid).[108] Long-acting insulins followed similar principles but guided toward favoring aggregated states as a way to provide a depot, where the aggregate slowly releases monomer as the equilibrium shifts from the absorption of the monomeric form.[109] The primary structure of the long-acting insulin glargine was modified to be made soluble at lower pH (4) but insoluble at physiological pH (7.4), where it forms micro-precipitates in the subcutaneous tissue causing slow release of the monomer. Insulin detemir was endowed with a long-acting action through a different design, wherein lysine B29 possesses myristic acid (14-carbon fatty acid) that promotes binding to albumin.[109] The ultra-long acting insulin degludec was similarly modified to have a hexadecanedioic acid on lysine B29. The added functional group leads to the formation of large multi-hexamers that serve as a depot for prolonged release. Conversely, the modified release profile of insulin NPH is achieved by complexing insulin with protamine.[110]

Insulin has also been formulated as a combination of insulins with different pharmacokinetic profiles and in conjunction with glucagon-like peptide 1 (GLP-1) analogs. The biphasic Novomix®, approved in 2000 (EU) and 2001 (USA, Novolog® Mix), contains 30% soluble insulin aspart mixed with 70% protamine-crystalized insulin aspart. This product is also available as a 50% mixture. Ryzodeg® 70/30 combines insulin degludec (70%) with insulin aspart (30%).[111] Novo Nordisk and Sanofi-Aventis have both recently developed combination products containing an insulin and a GLP-1 analog. Xultophy® (Novo Nordisk)[112] combines insulin degludec with liraglutide (GLP-1 analog), and Suliqua® (Sanofi-Aventis)[113] combines insulin glargine with lixisenatide (GLP-1 analog).

Glucagon-like peptides (GLP-1 and GLP-2) have seen an important increase in clinical market approval and clinical use in the last decade.[9] Recombinant glucagon was first approved in 1998 for treatment of hypoglycemia. Endogenous glucagon is mainly secreted by the alpha-cells of the islets of Langerhans and can induce gluconeogenesis and glycogenolysis to regulate low glucose levels. Conversely, GLP-1 can promote insulin secretion in a glucose-dependent manner and this function can thus be used to reduce blood glucose concentrations in type 2 diabetes.[114] Endogenous GLP-1, however, suffers from high clearance rates due to its intrinsic susceptibility to enzymatic degradation, particularly by dipeptidyl peptidase 4 (DPP-4).[115] Incretin-mimetic GLP-1 receptor agonists (GLP-1 analogs) were then developed to provide more suitable pharmacokinetic profiles by using analogs more resistant to DPP-4 degradation and several have been approved.[9, 115] Exenatide, the first GLP-1 analog in the clinic (2005), is a synthetic version of Exendin-4, a peptide derived from the venom of *Heloderma suspectum* that has sufficient homology to human GLP-1 (53%) and is resistant to DPP-4 degradation.[116] The next recombinant GLP-1 analog to enter the clinic was liraglutide, approved in 2009 (EU) and 2010 (US), consisting of a fatty acid-derivatized human GLP-1 peptide analog.[117] Other structural modifications implemented are derivatization with a C18 fatty acid (semaglutide),[118] fc-fusion (dulaglutide),[119] and fusion to albumin (albiglutide).[120] The GLP-2 analog teduglutide has also been approved for short bowel syndrome.[121]

4.4.1.2 Growth Hormone, Gonadotropins, and Other Hormones

4.4.1.2.1 Human Growth Hormone
Human growth hormone (hGH) (somatotropin) is a 22 kDa mitogenic protein (191 amino acids) predominantly produced and secreted by somatotropic cells in the hypothalamus anterior pituitary gland.[122] Somatotropin plays a key role in human development, stimulating protein synthesis and cell proliferation and regeneration to promote muscle, cartilage, and bone growth.[123] Somatotropin has been used since the 1950s for growth hormone deficiencies, originally of bovine origin or purified from cadaveric human pituitary glands.[124] The first recombinant therapeutic hGH (Somatrem®, 1985) was a modified version of the human somatotropin with an extra methionine on the N-terminal produced in *E. coli*.[125] Eli Lilly obtained approval of an unmodified somatotropin (Humatrope®) two years later (1987). Humatrope® is currently indicated for pediatric patients with growth hormone deficiency, Turner Syndrome,

Idiopathic Short Stature, short stature homeobox-containing gene (SHOX) deficiency, and children classified as being born small for their gestational age and in adults with growth hormone deficiency that meet certain clinical criteria.[126] Other hGH products include the biosimilar Omnitrope®.[9] Novel hGH in clinical development include somapacitan,[127] a long-acting fatty-acid derivatized analog designed for increased albumin binding and TransCon® hGH,[128] an hGH prodrug bound to an methoxy polyethylene glycol (mPEG) carrier through an autohydrolyzed linker. A PEGylated recombinant hGH receptor antagonist (pegvisomant) has also been developed for treatment of acromegaly.[129]

4.4.1.2.2 Gonadotropins

Gonadotropins are hormones produced and secreted by gonadotrope cells of the anterior pituitary that stimulate the physiological function of the gonads, used therapeutically in assisted reproductive therapies.[130] Therapeutic gonadotropins include follicle-stimulating hormone (FSH), human chorionic gonadotropin (hCG), and luteinizing hormone (LH).[130] Structurally, gonadotropins are heavily glycosylated heterodimeric proteins possessing a conserved alfa-chain comprised of 93 amino acids and a unique beta structure for each variant. The beta subunits of these hormones are made up of 111 (FSH), 145 (hCG), and 121 (LH) amino acids.[131]

The first therapeutic FSH and LH were mostly extracted and purified from menopausal urine.[132] hCG on the other hand is mainly produced in early stages of pregnancy, thus the first therapeutic hCGs were obtained from the urine of pregnant women. Recombinant FSH (follitropin alfa, Merck Serono) was first approved in 1995 (EU) and 1997 (US), produced in CHO cells.[133] hCG (Ovitrelle® [EU] or Ovidrel® [US], Merck Serono) and LH (lutropin alfa, EMD Serono) were both approved in 2000 in the EU, and in 2001 and 2004 in the United States, respectively. Other isoforms of FSH have since entered the clinic, including follitropin beta (Puregon®, Follistim®, Fertavid®), follitropin delta (Rekovelle®), and Pergoveris®, a product that combines recombinant FSH and recombinant LH. Corifollitropin alfa[134] (Elonva®), approved in 2010 (EU), is a fusion protein where the C-terminal peptide of the beta subunit of the hCG was fused to the beta chain of the FSH, granting the molecule enhanced biological half-life. All recombinant gonadotropins are produced in CHO cells with the exception of Rekovelle® (follitropin delta), produced in the human PER.C6 cell line obtained from human embryonic retinal cells.[9]

4.4.2 Hematopoietins and Related Molecules

4.4.2.1 Colony-Stimulating Factors

Recombinant granulocyte colony-stimulating factor (G-CSF) and granulocyte macrophage colony-stimulating factor (GM-CSF) are used to promote clonal expansion and differentiation of granulocytes and monocytes in patients suffering from neutropenia due to myeloid leukemia, congenital neutropenia, and administration of myelosuppressive or myeloablative therapy.[1] Endogenous G-CSF and GM-CSF are predominantly secreted by immunocompetent cells and cells present in the hematopoietic microenvironment upon stimulation by an inflammatory response to a pathogen.

Human G-CSG is secreted as a 174 amino acid 19.6 kDa glycoprotein with an O-glycosylation site and two disulfide bonds.[103] In contrast to the endogenous human protein, recombinant G-CSF lacks O-linked carbohydrates due to the expression system (*E. coli*). Endogenous GM-CSF is a glycosylated protein made up of 127 amino acid residues, with varying molecular weight (14–35 kDa) that depends on the degree of glycosylation and the glycoforms present. Remarkably, G-CSF and GM-CSF share a conserved structural architecture with other growth factors used therapeutically despite displaying low sequence similarity, which consists in a four-alfa helix bundle motif. Other therapeutic signaling molecules possessing this structural feature include hGH, erythropoietin, interleukin-2 (IL-2), and interferon beta (IFN-β).[103]

The first recombinant G-CSF (filgrastim) and GM-CSF (sargramostim) were both approved in 1991, for management of chemotherapy-induced neutropenia and autologous bone marrow transplantation, respectively. Another version of G-CSF produced in CHO cells (lenograstim) was approved in the EU in 1993.[135] Filgrastim was later engineered to possess a polyethylene glycol group on the N-terminal amine through a site-specific reductive alkylation reaction. The PEGylated version of filgrastim (pegfilgrastim) was approved in 2002 (US and EU), and it extended biological half-life of unmodified filgrastim from 3.5–4 h to approximately 42 h, enabling once-per-cycle administration.[136] Lipegfilgrastim, another long-acting CSF, was later developed by Teva Pharma and approved in 2013 (EU).[137] Several biosimilars of filgrastim and pegfilgrastim have also been approved in the EU, with the first one (filgrastim ratiopharm) obtaining approval in 2008.[9] Importantly, Zarxio® (filgrastim-sndz) was the very first biosimilar to be granted marketing authorization by the FDA in 2015.[138]

4.4.2.2 Erythropoietin

Erythropoietin is a growth factor with an extensive and successful clinical track record. Recombinant erythropoietin first entered the market in 1989, with the FDA approval of Epogen® (Amgen) for use in anemia.[139] Compared with insulin and other peptide hormones, human erythropoietin is significantly more complex in structural features, comprising 165 amino acid residues, 3 N-glycosylation sites, 1 O-glycosylation site, and 2 disulfide bonds (~30–34 kDa).[140] Erythropoietin production in the kidney and in the liver is dramatically increased as a response to cellular hypoxia, being subsequently secreted to inhibit apoptosis of erythroid precursor cells and promote erythropoiesis. Functionally, erythropoietin can also be classified as a growth factor, as it can induce cellular proliferation and differentiation.[141] Epogen® is currently indicated for treatment of anemia caused by chronic kidney disease (CKD), zidovudine treatment, and myelosuppressive chemotherapy, and for reduction of the need to provide allogeneic red-blood cell transfusions in patients undergoing nonvascular, noncardiac surgery that are at risk of perioperative blood loss.[139]

Since the approval of Epogen®, other recombinant variants have been marketed, including isoforms epoetin beta, epoetin delta, epoetin zeta, epoetin theta, and darbepoetin alfa and methoxy polyethylene glycol epoetin beta. Epoetin isoforms bear the same primary structure, yet they vary in their glycosylation profile, which grants them marked differences in stability, solubility, and pharmacokinetic profile.[142] Darbepoetin alfa was engineered to possess two extra N-glycosylation sites to extend serum half-life.[143] Mircera® (methoxy polyethylene glycol epoetin beta) is a PEGylated version of epoetin alfa that alters receptor binding kinetics and confers the protein a profoundly longer half-life.[144]

4.5 Blood-Related Products

Some of the first biologics were plasma extracts or plasma-purified proteins; examples are human albumin and human immunoglobulin, used clinically since the 1940s.[2] One of the first biological products used in blood clotting disorders was Hemophil M® (human antihemophilic factor), a plasma-purified lyophilized concentrate of coagulation factor VIII (FVIII) obtained through affinity chromatography using a murine monoclonal antibody, approved by FDA in 1966 for hemophilia A.[145] Factor VIII would later become part of the WHO Model List of Essential Medicines (EML).[146] Recombinant biopharmaceuticals have revolutionized the treatment of blood clotting disorders and have become essential in the treatment of hemophilia, venous thrombosis, angioedema, myocardial infarction, and von Willebrand disease.[147] These biotherapies are classified into three main categories based on their pharmacology: clotting or coagulation factors, anticoagulants, and thrombolytics. While the use of serum-derived protein concentrates for blood disorders is still prevalent, clinical development is currently dominated by recombinant products.

4.5.1 Recombinant Coagulation Factors

Recombinant coagulation factors replace deficient clotting factors in clotting disorders. The first recombinant clotting factor was a human factor VIII glycoprotein (Recombinate®) produced in CHO cells.[148] Human factor VIII is a complex heterodimeric globular protein made up of a heterogeneous heavy chain (90–200 kDa) and a light chain (80 kDa). The protein possesses various domains arranged as A1-A2-B-A3-C1-C2 from N-terminal to C-terminal.[149] Thrombin cleavage then converts FVIII into its activated form FVIIIa.[150] FVIII has been extremely successful as a replacement therapy in hemophilia A.[151]

A number of recombinant versions of factor VIII are available and the protein remains the most important coagulation factor in the clinic. Recently we have seen the approval of several alternative versions of FVIII. Refacto® (moroctog alfa), approved in 1999 (EU) and 2000 (US), is a B-domain deleted recombinant version of factor VIII.[152] The B-domain is heavily glycosylated, containing 19 potential glycosylation sites; consequently, the expression of B-domain deleted or truncated versions represents an advantage from a manufacturing perspective.[153] Turoctocog alfa (NovoEight®) is a similar FVIII analog with a truncated B-domain.[154] Other B-domain deleted versions include susoctocog alfa (Obizur®),[155] simoctocog alfa (Nuwiq®),[156] and lonoctocog alfa (Afstyla®).[157] Susoctocog alfa is a porcine VFIII indicated for patients with acquired hemophilia A as a result of the generation of human anti-FVIII antibodies.[158] Efmoroctocog alfa is an Fc-fusion protein containing a B-domain deleted FVIII fused to an Fc region with improved pharmacokinetic profile.[159] Adynovate®, a PEGylated full-length FVIII, is another product with extended biological half-life.[160] Furthermore, a recombinant von Willebrand factor – Vonvendi® – received regulatory approval in 2015 (US) for treatment of von Willebrand disease.[161]

Additional recombinant coagulation factors used therapeutically include factor IX, factor VIIa, thrombin (factor IIa), factor Xa, and a subunit of factor XIII.[162] Recombinant factor IX (FIX) is used in the treatment of hemophilia B, FIX deficiency, and has also been engineered for improved therapeutic properties. Specifically, the biological half-life of therapeutic FIX has been improved via PEGylation (Rebinyn® – nonacog beta pegol),[163] Fc-fusion (Alprolix® – eftrenonacog alfa),[164] and albumin-fusion (Idelvion® – albutrepenonacog alfa).[165] As discussed in the section on bispecifics, the anti-FIXa/anti-FX bispecific antibody emicizumab (Hemlibra®) entered the market in 2017 for prevention or reduction of the frequency of bleeding in patients with hemophilia A.[166]

4.5.2 Anticoagulants and Thrombolytic Agents

Anticoagulants are drugs that can disrupt molecular events in the coagulation cascade following platelet aggregation. This is in contrast to antiplatelet and thrombolytic therapies, which prevent thromboembolic events by acting at different stages of the formation of the blood clot. Anticoagulant drugs include a wide range of drugs of synthetic and biological origin, including coumarins (e.g. warfarin), direct oral coagulants (e.g. dabigatran, rivaroxaban), heparin and related products (e.g. heparin, lower molecular weight heparins), synthetic inhibitors of factor Xa (e.g. fondaparinux), antithrombin drugs (ATryn®), plasma kallikrein inhibitors (e.g. ecallantide), and complement component 1 (C1) esterase inhibitors (e.g. conestat alfa).[167] This section covers agents of recombinant origin.

The first heparin biologic was marketed in the US (1939) and was derived from beef lung tissue, requiring a complex manufacturing process.[168] Porcine intestinal mucosa later became and remains the most prevalent source of heparin.[169] Mucosal heparins comprise a complex and heterogeneous mixture of glycosaminoglycans, ranging from 5 to 25 kDa in molecular weight. Heparins exert their anticoagulant effect via interaction with antithrombin III (AT), which causes a conformational change that markedly increases the capacity of AT to bind and inactivate thrombin, factor Xa, and other coagulation factors. The size of the oligosaccharide affects the capacity of heparin to inhibit thrombin, where inactivation requires a heparin length of at least 18 saccharides. In contrast, a particular heparin pentasaccharide is sufficient to inhibit FXa.[170] Based on this size-dependency

of the pharmacodynamics of heparin, low-molecular weight heparins (LMWH) were later developed (1970s and 1980s) as alternatives to inhibit preferentially FXa and eliciting more predictable anticoagulant effects and pharmacokinetics, conferring less risk of bleeding.[171] Currently available LMWH include enoxaparin, parnarin, tinzaparin, nadroparin, and dalteparin. LMWH are prepared by depolymerization of a heparin precursor using various methods, among which deaminative cleavage with an organic nitrite or with nitrous acid are the most common.[172] At least 60% of the polysaccharide content in LMWH must be below 8 kDa.[173] LMWH have replaced heparins in many therapeutic applications due to improved pharmacokinetic profiles, reduced hemorrhagic effect, and patient convenience. ATryn®, a recombinant antithrombin protein, is a unique example of an anticoagulant biologic purified from milk of transgenic goats.[174] The hirudins, lepirudin, and desirudin were also approved recombinant products but they have been withdrawn from the market.[9]

Thrombolytics are a therapeutic class dominated by recombinant tissue plasminogen activators (tPA) used to break down blood clots; indications include acute ischemic stroke, acute myocardial infarction, and pulmonary embolism.[175] Human tPA is a ~70 kDa serine protease responsible for converting plasminogen into its activated form, the fibrinolytic protease plasmin.[176] Recombinant tPA (alteplase) was first (1980) cloned and expressed at Genentech, and would later become one of the first (1987) biotherapeutic produced in a mammalian expression system, second only to the monoclonal antibody muromomab CD3 (1986).[177] Following the clinical success of alteplase, analogs reteplase[178] (approved in 1996 in US and EU) and tenecteplase[179] (approved in 2000 in US and 2001 in EU) were subsequently developed. Reteplase is an *E. coli* expressed analog that lacks the three N-terminal domains of tPA, granting the protein more reversible binding to fibrin and a longer half-life.[180] Tenecteplase is a variant of tPA bearing several mutations that enhance resistance to plasminogen activator inhibitor-1 and fibrin specificity.

4.6 Biosimilars

The first antibodies that entered the market (many of which became blockbuster drugs) have recently come off patent or will do so over the next few years.[1] This has prompted tremendous commercial interest in the industry to produce "highly similar" versions of these blockbuster drugs once the reference product exclusivity expires. These products are designated biosimilars, or follow-on biologics. Alternatively, modifications may be performed to existing biologics to enhance their clinical performance, yielding so-called biobetters (see definitions in Chapter 1). A more in-depth discussion of the regulatory framework and approval pathways for biosimilars (and biobetters) can be found in Chapters 2, 7, and 10.

References

1 FDA, Center for Drug Evaluation and Research. (2019). List of licensed biological products with (1) reference product exclusivity and (2) biosimilarity or interchangeability evaluations to date. https://www.fda.gov/media/89589/download (accessed 19 November 2019).

2 FDA, Center for Biologic Evaluation and Research. (2019). List of licensed biological products with (1) reference product exclusivity and (2) biosimilarity or interchangeability evaluations to date. https://www.fda.gov/media/89426/download (accessed 14 August 2019).

3 George, K. and Woollett, G. (2019). Insulins as drugs or biologics in the USA: what difference does it make and why does it matter? *BioDrugs* 33 (5): 447–451.

4 FDA. (2019). Preliminary list of approved NDAs for biological products that will be deemed to be BLAs on March 23, 2020. https://www.fda.gov/media/119229/download (updated 31 August 2019; accessed 23 August 2019).

5 Elgundi, Z., Reslan, M., Cruz, E. et al. (2017). The state-of-play and future of antibody therapeutics. *Advanced Drug Ddelivery Reviews* 122: 2–19.

6 Urquhart, L. (2019). Top drugs and companies by sales in 2018. *Nature Reviews Drug Discovery* https://doi.org/10.1038/d41573-019-00049-0.

7 Kaplon, H. and Reichert, J.M. (2019). Antibodies to watch in 2019. *mAbs* 11 (2): 219–238.

8 Slamon, D.J., Leyland-Jones, B., Shak, S. et al. (2001). Use of chemotherapy plus a monoclonal antibody against HER2 for metastatic breast cancer that overexpresses HER2. *New England Journal of Medicine* 344 (11): 783–792.

9 Walsh, G. (2018). Biopharmaceutical benchmarks 2018. *Nature Biotechnology* 36 (12): 1136–1145.

10 Rajewsky, K. (2019). The advent and rise of monoclonal antibodies. *Nature* 575 (7781): 47–49.

11 McCafferty, J., Griffiths, A.D., Winter, G., and Chiswell, D.J. (1990). Phage antibodies: filamentous phage displaying antibody variable domains. *Nature* 348 (6301): 552–554.

12 Boder, E.T. and Wittrup, K.D. (1997). Yeast surface display for screening combinatorial polypeptide libraries. *Nature Biotechnology* 15 (6): 553–557.

13 Knappik, A., Ge, L., Honegger, A. et al. (2000). Fully synthetic human combinatorial antibody libraries (HuCAL) based on modular consensus frameworks and CDRs randomized with trinucleotides. *Journal of Molecular Biology* 296 (1): 57–86.

14 Cruz, E. and Kayser, V. (2019). Monoclonal antibody therapy of solid tumors: clinical limitations and novel strategies to enhance treatment efficacy. *Biologics* 13: 33–51.

15 Reichert, J.M.; The Antibody Society (2019). Antibody therapeutics approved or in regulatory review in the EU or US. https://www.antibodysociety.org/resources/approved-antibodies (accessed 23 October 2019).

16 Salfeld, J.G. (2007). Isotype selection in antibody engineering. *Nature Biotechnology* 25 (12): 1369–1372.

17 Saito, S., Namisaki, H., Hiraishi, K. et al. (2019). A stable engineered human IgG3 antibody with decreased aggregation during antibody expression and low pH stress. *Protein Science* 28 (5): 900–909.

18 Harris, L.J., Larson, S.B., Hasel, K.W. et al. (1992). The three-dimensional structure of an intact monoclonal antibody for canine lymphoma. *Nature* 360 (6402): 369–372.

19 Janeway, C.A., Travers, P., Walport, M. et al. (2001). The structure of a typical antibody molecule. In: *Immunobiology: The Immune System in Health and Disease*, 5e. New York: Garland Science.

20 Stanfield, R.L. and Wilson, I.A. (2009). Antibody molecular structure. In: *Therapeutic Monoclonal Antibodies: From Bench to Clinic* (ed. Z. An), 51–66. Wiley.

21 Rispens, T., Davies, A.M., Ooijevaar-de Heer, P. et al. (2014). Dynamics of inter-heavy chain interactions in human immunoglobulin G (IgG) subclasses studied by kinetic Fab arm exchange. *The Journal of Biological Chemistry* 289 (9): 6098–6109.

22 Harris, L.J., Skaletsky, E., and McPherson, A. (1998). Crystallographic structure of an intact IgG1 monoclonal antibody. *Journal of Molecular Biology* 275 (5): 861–872.

23 Foote, J. and Winter, G. (1992). Antibody framework residues affecting the conformation of the hypervariable loops. *Journal of Molecular Biology* 224 (2): 487–499.

24 Van Oss, C.J. (1995). Hydrophobic, hydrophilic and other interactions in epitope-paratope binding. *Molecular Immunology* 32 (3): 199–211.

25 Jefferis, R., Lund, J., and Pound, J.D. (1998). IgG-Fc-mediated effector functions: molecular definition of interaction sites for effector ligands and the role of glycosylation. *Immunological Reviews* 163 (1): 59–76.

26 Leatherbarrow, R.J., Rademacher, T.W., Dwek, R.A. et al. (1985). Effector functions of a monoclonal aglycosylated mouse IgG2a: binding and activation of complement component C1 and interaction with human monocyte Fc receptor. *Molecular Immunology* 22 (4): 407–415.

27 Roopenian, D.C. and Akilesh, S. (2007). FcRn: the neonatal Fc receptor comes of age. *Nature Reviews Immunology* 7 (9): 715–725.

28 Han, K., Peyret, T., Marchand, M. et al. (2016). Population pharmacokinetics of bevacizumab in cancer patients with external validation. *Cancer Chemother Pharmacol* 78 (2): 341–351.

29 Li, J., Levi, M., Charoin, J.-E. et al. (2007). Rituximab exhibits a long half-life based on a population pharmacokinetic analysis in non-Hodgkin's lymphoma (NHL) patients. *Blood* 110 (11): 2371–2371.

30 Bruno, R., Washington, C.B., Lu, J.F. et al. (2005). Population pharmacokinetics of trastuzumab in patients with HER2+ metastatic breast cancer. *Cancer Chemother Pharmacol* 56 (4): 361–369.

31 Kayser, V., Chennamsetty, N., Voynov, V. et al. (2011). Glycosylation influences on the aggregation propensity of therapeutic monoclonal antibodies. *Biotechnology Journal* 6 (1): 38–44.

32 Zheng, K., Bantog, C., and Bayer, R. (2011). The impact of glycosylation on monoclonal antibody conformation and stability. *MAbs* 3 (6): 568–576.

33 Jefferis, R. (2009). Recombinant antibody therapeutics: the impact of glycosylation on mechanisms of action. *Trends in Pharmacological Sciences* 30 (7): 356–362.

34 van de Bovenkamp, F.S., Hafkenscheid, L., Rispens, T., and Rombouts, Y. (2016). The emerging importance of IgG Fab glycosylation in immunity. *The Journal of Immunology* 196 (4): 1435.

35 Reslan, M., Sifniotis, V., Cruz, E. et al. (2020). Enhancing the stability of adalimumab by engineering additional glycosylation motifs. *International Journal of Biological Macromolecules* 158: 189–196. https://doi.org/10.1016/j.ijbiomac.2020.04.147.

36 Sigismund, S., Avanzato, D., and Lanzetti, L. (2018). Emerging functions of the EGFR in cancer. *Molecular Oncology* 12 (1): 3–20.

37 Cragg, M.S. and Glennie, M.J. (2004). Antibody specificity controls in vivo effector mechanisms of anti-CD20 reagents. *Blood* 103 (7): 2738–2743.

38 Cragg, M.S., Walshe, C.A., Ivanov, A.O., and Glennie, M.J. (2005). The biology of CD20 and its potential as a target for mAb therapy. *Current Directions in Autoimmunity* 8: 140–174.

39 Hauser, S.L., Bar-Or, A., Comi, G. et al. (2016). Ocrelizumab versus interferon beta-1a in relapsing multiple sclerosis. *New England Journal of Medicine* 376 (3): 221–234.

40 Eisenberg, R. and Albert, D. (2006). B-cell targeted therapies in rheumatoid arthritis and systemic lupus erythematosus. *Nature Clinical Practice Rheumatology* 2 (1): 20–27.

41 Kerbel, R.S. (2008). Tumor angiogenesis. *New England Journal of Medicine* 358 (19): 2039–2049.

42 Hicklin, D.J. and Ellis, L.M. (2005). Role of the vascular endothelial growth factor pathway in tumor growth and angiogenesis. *Journal of Clinical Oncology* 23 (5): 1011–1027.

43 Fuchs, C.S., Tomasek, J., Yong, C.J. et al. (2014). Ramucirumab monotherapy for previously treated advanced gastric or gastro-oesophageal junction adenocarcinoma (REGARD): an international, randomised, multicentre, placebo-controlled, phase 3 trial. *The Lancet* 383 (9911): 31–39.

44 Jain, R.K., Duda, D.G., Clark, J.W., and Loeffler, J.S. (2006). Lessons from phase III clinical trials on anti-VEGF therapy for cancer. *Nature Clinical Practice Oncology* 3 (1): 24–40.

45 van Dullemen, H.M., van Deventer, S.J.H., Hommes, D.W. et al. (1995). Treatment of Crohn's disease with anti-tumor necrosis factor chimeric monoclonal antibody (cA2). *Gastroenterology* 109 (1): 129–135.

46 Chaparro, M., Burgueno, P., Iglesias, E. et al. (2012). Infliximab salvage therapy after failure of ciclosporin in corticosteroid-refractory ulcerative colitis: a multicentre study. *Alimentary Pharmacology & Therapeutics* 35 (2): 275–283.

47 Wenzel, S., Ford, L., Pearlman, D. et al. (2013). Dupilumab in persistent asthma with elevated eosinophil levels. *New England Journal of Medicine* 368 (26): 2455–2466.

48 De Benedetti, F., Brunner, H.I., Ruperto, N. et al. (2012). Randomized trial of tocilizumab in systemic juvenile idiopathic arthritis. *New England Journal of Medicine* 367 (25): 2385–2395.

49 Leonardi, C., Matheson, R., Zachariae, C. et al. (2012). Anti–interleukin-17 monoclonal antibody ixekizumab in chronic plaque psoriasis. *New England Journal of Medicine* 366 (13): 1190–1199.

50 Sandborn, W.J., Gasink, C., Gao, L.-L. et al. (2012). Ustekinumab induction and maintenance therapy in refractory Crohn's disease. *New England Journal of Medicine* 367 (16): 1519–1528.

51 Feagan, B.G., Rutgeerts, P., Sands, B.E. et al. (2013). Vedolizumab as induction and maintenance therapy for ulcerative colitis. *New England Journal of Medicine* 369 (8): 699–710.

52 Feltes, T.F., Cabalka, A.K., Meissner, H.C. et al. (2003). Palivizumab prophylaxis reduces hospitalization due to respiratory syncytial virus in young children with hemodynamically significant congenital heart disease. *The Journal of Pediatrics* 143 (4): 532–540.

53 Migone, T.-S., Subramanian, G.M., Zhong, J. et al. (2009). Raxibacumab for the treatment of inhalational anthrax. *New England Journal of Medicine* 361 (2): 135–144.

54 Greig, S.L. (2016). Obiltoxaximab: first global approval. *Drugs* 76 (7): 823–830.

55 Wilcox, M.H., Gerding, D.N., Poxton, I.R. et al. (2017). Bezlotoxumab for prevention of recurrent *Clostridium difficile* infection. *New England Journal of Medicine* 376 (4): 305–317.

56 Emu, B., Fessel, J., Schrader, S. et al. (2018). Phase 3 study of ibalizumab for multidrug-resistant HIV-1. *New England Journal of Medicine* 379 (7): 645–654.

57 Strunk, R.C. and Bloomberg, G.R. (2006). Omalizumab for asthma. *New England Journal of Medicine* 354 (25): 2689–2695.

58 Haldar, P., Brightling, C.E., Hargadon, B. et al. (2009). Mepolizumab and exacerbations of refractory eosinophilic asthma. *New England Journal of Medicine* 360 (10): 973–984.

59 Castro, M., Zangrilli, J., Wechsler, M.E. et al. (2015). Reslizumab for inadequately controlled asthma with elevated blood eosinophil counts: results from two multicentre, parallel, double-blind, randomised, placebo-controlled, phase 3 trials. *The Lancet Respiratory Medicine* 3 (5): 355–366.

60 Nair, P., Wenzel, S., Rabe, K.F. et al. (2017). Oral glucocorticoid–sparing effect of benralizumab in severe asthma. *New England Journal of Medicine* 376 (25): 2448–2458.

61 Sabatine, M.S., Giugliano, R.P., Keech, A.C. et al. (2017). Evolocumab and clinical outcomes in patients with cardiovascular disease. *New England Journal of Medicine* 376 (18): 1713–1722.

62 Robinson, J.G., Farnier, M., Krempf, M. et al. (2015). Efficacy and safety of alirocumab in reducing lipids and cardiovascular events. *New England Journal of Medicine* 372 (16): 1489–1499.

63 Pollack, C.V., Reilly, P.A., Eikelboom, J. et al. (2015). Idarucizumab for dabigatran reversal. *New England Journal of Medicine* 373 (6): 511–520.

64 Carpenter, T.O., Whyte, M.P., Imel, E.A. et al. (2018). Burosumab therapy in children with X-linked hypophosphatemia. *New England Journal of Medicine* 378 (21): 1987–1998.

65 Dodick, D.W., Ashina, M., Brandes, J.L. et al. (2018). ARISE: a phase 3 randomized trial of erenumab for episodic migraine. *Cephalalgia* 38 (6): 1026–1037.

66 Maack, T. (2011). Renal handling of proteins and polypeptides. In: *Comprehensive Physiology* (ed. R. Terjung), 2039–2082. Wiley for American Physiological Society.

67 Ultee, M.E. (2014). Therapeutic Fc-fusion proteins: Edited by Steven M Chamow, Thomas Ryll, Henry B Lowman and Deborah Farson. *mAbs* 6 (4): 810–811.

68 Liu, L. (2018). Pharmacokinetics of monoclonal antibodies and Fc-fusion proteins. *Protein & Cell* 9 (1): 15–32.

69 FDA. (2019). Enbrel (Etanercept) label. https://www.accessdata.fda.gov/drugsatfda_docs/label/2012/103795s5503lbl.pdf (accessed 23 September 2019).

70 Kuter, D.J., Rummel, M., Boccia, R. et al. (2010). Romiplostim or standard of care in patients with immune thrombocytopenia. *New England Journal of Medicine* 363 (20): 1889–1899.

71 Dungan, K.M., Povedano, S.T., Forst, T. et al. (2014). Once-weekly dulaglutide versus once-daily liraglutide in metformin-treated patients with type 2 diabetes (AWARD-6): a randomised, open-label, phase 3, non-inferiority trial. *The Lancet* 384 (9951): 1349–1357.

72 Houel, S., Hilliard, M., Yu, Y.Q. et al. (2014). N- and O-glycosylation analysis of etanercept using liquid chromatography and quadrupole time-of-flight mass spectrometry equipped with electron-transfer dissociation functionality. *Analytical Chemistry* 86 (1): 576–584.

73 Bross, P.F., Beitz, J., Chen, G. et al. (2001). Approval summary: gemtuzumab ozogamicin in relapsed acute myeloid leukemia. *Clinical Cancer Research: An Official Journal of the American Association for Cancer Research* 7 (6): 1490–1496.

74 Hughes, B. (2010). Antibody–drug conjugates for cancer: poised to deliver? *Nature Reviews Drug Discovery* 9 (9): 665–667.

75 Jen, E.Y., Ko, C.W., Lee, J.E. et al. (2018). FDA approval: gemtuzumab ozogamicin for the treatment of adults with newly diagnosed CD33-positive acute myeloid leukemia. *Clinical Cancer Research: An*

Official Journal of the American Association for Cancer Research 24 (14): 3242–3246.

76 Lu, J., Jiang, F., Lu, A., and Zhang, G. (2016). Linkers having a crucial role in antibody-drug conjugates. *International Journal of Molecular Sciences* 17 (4): 561.

77 Tolcher, A.W. (2016). Antibody drug conjugates: lessons from 20 years of clinical experience. *Annals of Oncology* 27 (12): 2168–2172.

78 Jain, N., Smith, S.W., Ghone, S., and Tomczuk, B. (2015). Current ADC linker chemistry. *Pharmaceutical Research* 32 (11): 3526–3540.

79 Chen, L., Wang, L., Shion, H. et al. (2016). In-depth structural characterization of Kadcyla® (ado-trastuzumab emtansine) and its biosimilar candidate. *mAbs* 8 (7): 1210–1223.

80 Behrens, C.R., Ha, E.H., Chinn, L.L. et al. (2015). Antibody–drug conjugates (ADCs) derived from Interchain cysteine cross-linking demonstrate improved homogeneity and other pharmacological properties over conventional heterogeneous ADCs. *Molecular Pharmaceutics* 12 (11): 3986–3998.

81 Tumey, L.N., Li, F., Rago, B. et al. (2017). Site selection: a case study in the identification of optimal cysteine engineered antibody drug conjugates. *The AAPS Journal* 19 (4): 1123–1135.

82 Axup, J.Y., Bajjuri, K.M., Ritland, M. et al. (2012). Synthesis of site-specific antibody-drug conjugates using unnatural amino acids. *Proceedings of the National Academy of Sciences of the United States of America* 109 (40): 16101–16106.

83 van Geel, R., Wijdeven, M.A., Heesbeen, R. et al. (2015). Chemoenzymatic conjugation of toxic payloads to the globally conserved N-glycan of native mAbs provides homogeneous and highly efficacious antibody–drug conjugates. *Bioconjugate Chemistry* 26 (11): 2233–2242.

84 Rock, B.M., Tometsko, M.E., Patel, S.K. et al. (2015). Intracellular catabolism of an antibody drug conjugate with a noncleavable linker. *Drug Metabolism and Disposition* 43 (9): 1341–1344.

85 Tsuchikama, K. and An, Z. (2018). Antibody-drug conjugates: recent advances in conjugation and linker chemistries. *Protein & Cell* 9 (1): 33–46.

86 Younes, A., Bartlett, N.L., Leonard, J.P. et al. (2010). Brentuximab vedotin (SGN-35) for relapsed CD30-positive lymphomas. *New England Journal of Medicine* 363 (19): 1812–1821.

87 Verma, S., Miles, D., Gianni, L. et al. (2012). Trastuzumab emtansine for HER2-positive advanced breast cancer. *New England Journal of Medicine* 367 (19): 1783–1791.

88 Kantarjian, H.M., DeAngelo, D.J., Stelljes, M. et al. (2016). Inotuzumab ozogamicin versus standard therapy for acute lymphoblastic leukemia. *New England Journal of Medicine* 375 (8): 740–753.

89 Deeks, E.D. (2019). Polatuzumab vedotin: first global approval. *Drugs* 79 (13): 1467–1475.

90 Nisonoff, A. and Rivers, M.M. (1961). Recombination of a mixture of univalent antibody fragments of different specificity. *Archives of Biochemistry and Biophysics* 93: 460–462.

91 Labrijn, A.F., Janmaat, M.L., Reichert, J.M., and Parren, P.W.H.I. (2019). Bispecific antibodies: a mechanistic review of the pipeline. *Nature Reviews Drug Discovery* 18 (8): 585–608.

92 Kontermann, R.E. and Brinkmann, U. (2015). Bispecific antibodies. *Drug Discovery Today* 20 (7): 838–847.

93 Hodi, F.S., O'Day, S.J., McDermott, D.F. et al. (2010). Improved survival with ipilimumab in patients with metastatic melanoma. *New England Journal of Medicine* 363 (8): 711–723.

94 Wei, S.C., Duffy, C.R., and Allison, J.P. (2018). Fundamental mechanisms of immune checkpoint blockade therapy. *Cancer Discovery* 8 (9): 1069–1086.

95 Nguyen, L.T. and Ohashi, P.S. (2015). Clinical blockade of PD1 and LAG3 – potential mechanisms of action. *Nature Reviews Immunology* 15 (1): 45–56.

96 Fan, L., Li, Y., Chen, J.-Y. et al. (2019). Immune checkpoint modulators in cancer immunotherapy: recent advances and combination rationales. *Cancer Letters* 456: 23–28.

97 FDA. (2019). Keytruda (pembrolizumab) label. https://www.accessdata.fda.gov/drugsatfda_docs/label/2018/125514s034lbl.pdf (accessed 23 August 2019).

98 Pons-Tostivint, E., Latouche, A., Vaflard, P. et al. (2019). Comparative analysis of durable responses on immune checkpoint inhibitors versus other systemic therapies: a pooled analysis of phase III trials. *JCO Precision Oncology* 3: 1–10.

99 Haslam, A. and Prasad, V. (2019). Estimation of the percentage of US patients with cancer who are eligible for and respond to checkpoint inhibitor immunotherapy drugs. *JAMA Network Open* 2 (5): e192535–e192535.

100 Marin-Acevedo, J.A., Dholaria, B., Soyano, A.E. et al. (2018). Next generation of immune checkpoint therapy in cancer: new developments and challenges. *Journal of Hematology & Oncology* 11 (1): 39.

101 Timofeev, V.I., Chuprov-Netochin, R.N., Samigina, V.R. et al. (2010). X-ray investigation of gene-engineered human insulin crystallized from a solution containing polysialic acid. *Acta Crystallographica Section F, Structural Biology and Crystallization Communications* 66 (Pt 3): 259–263.

102 Ultsch, M.H., Somers, W., Kossiakoff, A.A., and de Vos, A.M. (1994). The crystal structure of affinity-matured human growth hormone at 2 A resolution. *Journal of Molecular Biology* 236 (1): 286–299.

103 Hill, C.P., Osslund, T.D., and Eisenberg, D. (1993). The structure of granulocyte-colony-stimulating factor and its relationship to other growth factors. *Proceedings of the National Academy of Sciences of the United States of America* 90 (11): 5167–5171.

104 Veronese, F.M. and Mero, A. (2008). The impact of PEGylation on biological therapies. *BioDrugs* 22 (5): 315–329.

105 Lapolla, A. and Dalfra, M.G. (2020). Hundred years of insulin therapy: purified early insulins. *American Journal of Therapeutics* 27 (1): e24–e29. https://doi.org/10.1097/MJT.0000000000001081.

106 Scapin, G., Dandey, V.P., Zhang, Z. et al. (2018). Structure of the insulin receptor–insulin complex by single-particle cryo-EM analysis. *Nature* 556 (7699): 122–125.

107 Chance, R.E., Glazer, N.B., and Wishner, K.L. (1999). Insulin Lispro (Humalog®). In: *Biopharmaceuticals, and Industrial Perspective* (eds. G. Walsh and B. Murphy). Dordrecht: Springer.

108 Cahn, A., Miccoli, R., Dardano, A., and Del Prato, S. (2015). New forms of insulin and insulin therapies for the treatment of type 2 diabetes. *The Lancet Diabetes & Endocrinology* 3 (8): 638–652.

109 Peterson, G.E. (2006). Intermediate and long-acting insulins: a review of NPH insulin, insulin glargine and insulin detemir. *Current Medical Research and Opinion* 22 (12): 2613–2619.

110 Hermansen, K., Madsbad, S., Perrild, H. et al. (2001). Comparison of the soluble basal insulin analog insulin detemir with NPH insulin: a randomized open crossover trial in type 1 diabetic subjects on basal-bolus therapy. *Diabetes Care* 24 (2): 296–301.

111 Fulcher, G.R., Christiansen, J.S., Bantwal, G. et al. (2014). Comparison of insulin degludec/insulin aspart and biphasic insulin aspart 30 in uncontrolled, insulin-treated type 2 diabetes: a phase 3a, randomized, treat-to-target trial. *Diabetes Care* 37 (8): 2084–2090.

112 Chaplin, S. and Patel, V. (2015). Xultophy®: combination therapy for the treatment of type 2 diabetes. *Prescriber* 26 (17): 32–34.

113 Deeks, E. (2019). Insulin glargine/lixisenatide in type 2 diabetes: a profile of its use. *Drugs & Therapy Perspectives* 35 (10): 470–480.

114 De Leon, D.D., Crutchlow, M.F., Ham, J.Y., and Stoffers, D.A. (2006). Role of glucagon-like peptide-1 in the pathogenesis and treatment of diabetes mellitus. *The International Journal of Biochemistry & Cell Biology* 38 (5–6): 845–859.

115 Tudurí, E., López, M., Diéguez, C. et al. (2016). Glucagon-like peptide 1 analogs and their effects on pancreatic islets. *Trends in Endocrinology & Metabolism* 27 (5): 304–318.

116 Davidson, M.B., Bate, G., and Kirkpatrick, P. (2005). Exenatide. *Nature Reviews Drug Discovery* 4 (9): 713–714.

117 Knudsen, L.B. and Lau, J. (2019). The discovery and development of liraglutide and semaglutide. *Frontiers in Endocrinology* 10: 155.

118 Lau, J., Bloch, P., Schäffer, L. et al. (2015). Discovery of the once-weekly glucagon-like peptide-1 (GLP-1) analogue semaglutide. *Journal of Medicinal Chemistry* 58 (18): 7370–7380.

119 Umpierrez, G., Tofe Povedano, S., Perez Manghi, F. et al. (2014). Efficacy and safety of dulaglutide monotherapy versus metformin in type 2 diabetes in a randomized controlled trial (AWARD-3). *Diabetes Care* 37 (8): 2168–2176.

120 Pratley, R.E., Nauck, M.A., Barnett, A.H. et al. (2014). Once-weekly albiglutide versus once-daily liraglutide in patients with type 2 diabetes inadequately controlled on oral drugs (HARMONY 7): a randomised, open-label, multicentre, non-inferiority phase 3 study. *The Lancet Diabetes & Endocrinology* 2 (4): 289–297.

121 Vipperla, K. and O'Keefe, S.J. (2014). Targeted therapy of short-bowel syndrome with teduglutide: the new kid on the block. *Clinical and Experimental Gastroenterology* 7: 489–495.

122 Backeljauw, P. and Hwa, V. (2016). Growth hormone physiology. In: *Growth Hormone Deficiency: Physiology and Clinical Management* (ed. L.E. Cohen), 7–20. Cham: Springer International Publishing.

123 Brinkmann, J.E. and Sharma, S. (2019). Physiology, growth hormone. In: *StatPearls [Internet]*. Treasure Island, FL: StatPearls Publishing.

124 Blizzard, R.M. (2012). History of growth hormone therapy. *Indian Journal of Pediatrics* 79 (1): 87–91.

125 Vicens-Calvet, E., Potau, N., Carracosa, A. et al. (1986). Clinical experience with somatrem in growth hormone deficiency. *Acta Paediatrica* 75 (s325): 33–40.

126 FDA (2019). Humatrope (somatropin rDNA origin) label. https://www.accessdata.fda.gov/drugsatfda_docs/label/2014/019640s092lbl.pdf (accessed 12 September 2019).

127 Juul, R.V., Rasmussen, M.H., Agerso, H., and Overgaard, R.V. (2019). Pharmacokinetics and pharmacodynamics of once-weekly somapacitan in children and adults: supporting dosing rationales with a model-based analysis of three phase I trials. *Clinical Pharmacokinetics* 58 (1): 63–75.

128 Sprogøe, K., Mortensen, E., Karpf, D.B., and Leff, J.A. (2017). The rationale and design of TransCon Growth Hormone for the treatment of growth hormone deficiency. *Endocrine Connections* 6 (8): R171–R181.

129 Freda, P., Gordon, M., Kelepouris, N. et al. (2015). Long-term treatment with pegvisomant as monotherapy in patients with acromegaly: experience from acrostudy. *Endocrine Practice* 21 (3): 264–274.

130 Leão, R.B.F. and Esteves, S.C. (2014). Gonadotropin therapy in assisted reproduction: an evolutionary perspective from biologics to biotech. *Clinics (Sao Paulo)* 69 (4): 279–293.

131 Lustbader, J.W., Pollak, S., Lobel, L. et al. (1996). Three-dimensional structures of gonadotropins. *Molecular and Cellular Endocrinology* 125 (1): 21–31.

132 Daya, S. and Gunby, J. (1999). Recombinant versus urinary follicle stimulating hormone for ovarian stimulation in assisted reproduction. *Human Reproduction* 14 (9): 2207–2215.

133 Bassett, R.M. and Driebergen, R. (2005). Continued improvements in the quality and consistency of follitropin alfa, recombinant human FSH. *Reproductive BioMedicine Online.* 10 (2): 169–177.

134 The Corifollitropin Alfa Ensure Study Group (2010). Corifollitropin alfa for ovarian stimulation in IVF: a randomized trial in lower-body-weight women. *Reproductive BioMedicine Online* 21 (1): 66–76.

135 Gisselbrecht, C., Prentice, H.G., Bacigalupo, A. et al. (1994). Placebo-controlled phase III trial of lenograstim in bone-marrow transplantation. *The Lancet* 343 (8899): 696–700.

136 Molineux, G. (2004). The design and development of pegfilgrastim (PEG-rmetHuG-CSF, Neulasta). *Current Pharmaceutical Design* 10 (11): 1235–1244.

137 Guariglia, R., Martorelli, M.C., Lerose, R. et al. (2016). Lipegfilgrastim in the management of chemotherapy-induced neutropenia of cancer patients. *Biologics* 10: 1–8.

138 Raedler, L.A. (2016). Zarxio® (filgrastim-sndz): first biosimilar approved in the United States. *American Health & Drug Benefits* 9 (Spec Feature): 150–154.

139 FDA (2019). Epogen® (epoetin alfa) label. https://www.accessdata.fda.gov/drugsatfda_docs/label/2017/103234s5363s5366lbl.pdf. (accessed 15 August 2019).

140 Syed, R.S., Reid, S.W., Li, C. et al. (1998). Efficiency of signalling through cytokine receptors depends critically on receptor orientation. *Nature* 395 (6701): 511–516.

141 Jelkmann, W. (2013). Physiology and pharmacology of erythropoietin. *Transfusion Medicine and Hemotherapy* 40 (5): 302–309.

142 Deicher, R. and Hörl, W.H. (2004). Differentiating factors between erythropoiesis-stimulating agents. *Drugs* 64 (5): 499–509.

143 Egrie, J.C., Dwyer, E., Browne, J.K. et al. (2003). Darbepoetin alfa has a longer circulating half-life and greater in vivo potency than recombinant human erythropoietin. *Experimental Hematology* 31 (4): 290–299.

144 Berns, J.S., Wong, T.C., and Dawson, S. (2019). 9 – Anemia in chronic kidney disease. In: *Chronic Kidney Disease, Dialysis, and Transplantation*, 4e (eds. J. Himmelfarb and T.A. Ikizler), 136–144.e136. Philadelphia: Elsevier.

145 Addiego, J.E. Jr., Gomperts, E., Liu, S.L. et al. (1992). Treatment of hemophilia A with a highly purified factor VIII concentrate prepared by anti-FVIIIc immunoaffinity chromatography. *Thrombosis and Haemostasis* 67 (1): 19–27.

146 WHO (2019). Wordl Health Organization model list of essential medicines, 21st list 2019. https://apps.who.int/iris/bitstream/handle/10665/325771/WHO-MVP-EMP-IAU-2019.06-eng.pdf?ua=1 (accessed 8 October 2019).

147 Adivitiya, K.Y.P. (2017). The evolution of recombinant thrombolytics: current status and future directions. *Bioengineered* 8 (4): 331–358.

148 Bray, G., Gomperts, E., Courter, S. et al. (1994). A multicenter study of recombinant factor VIII (recombinate): safety, efficacy, and inhibitor risk in previously untreated patients with hemophilia A. The Recombinate Study Group. *Blood* 83 (9): 2428–2435.

149 Ngo, J.C.K., Huang, M., Roth, D.A. et al. (2008). Crystal structure of human factor VIII: implications for the formation of the factor IXa-factor VIIIa complex. *Structure (London, England: 1993)* 16 (4): 597–606.

150 Fay, P.J. (2004). Activation of factor VIII and mechanisms of cofactor action. *Blood Reviews* 18 (1): 1–15.

151 Aledort, L., Ljung, R., Mann, K., and Pipe, S. (2014). Factor VIII therapy for hemophilia A: current and future issues. *Expert Review of Hematology* 7 (3): 373–385.

152 Pittman, D.D., Alderman, E.M., Tomkinson, K.N. et al. (1993). Biochemical, immunological, and in vivo functional characterization of B-domain-deleted factor VIII. *Blood* 81 (11): 2925–2935.

153 Gruppo, R.A., Brown, D., Wilkes, M.M., and Navickis, R.J. (2003). Comparative effectiveness of full-length and B-domain deleted factor VIII for prophylaxis – a meta-analysis. *Haemophilia: The Official Journal of the World Federation of Hemophilia* 9 (3): 251–260.

154 Ahmadian, H., Hansen, E.B., Faber, J.H. et al. (2016). Molecular design and downstream processing of turoctocog alfa (NovoEight®), a B-domain truncated factor VIII molecule. *Blood Coagul Fibrinolysis* 27 (5): 568–575.

155 Lillicrap, D., Schiviz, A., Apostol, C. et al. (2016). Porcine recombinant factor VIII (Obizur®; OBI-1; BAX801): product characteristics and preclinical profile. *Haemophilia: The Official Journal of the World Federation of Hemophilia* 22 (2): 308–317.

156 Klukowska, A., Szczepański, T., Vdovin, V. et al. (2016). Novel, human cell line-derived recombinant factor VIII (human-cl rhFVIII, Nuwiq®) in children with severe haemophilia A: efficacy, safety and pharmacokinetics. *Haemophilia: The Official Journal of the World Federation of Hemophilia* 22 (2): 232–239.

157 Raso, S. and Hermans, C. (2018). Lonoctocog alfa (rVIII-SingleChain) for the treatment of haemophilia A. *Expert Opinion on Biological Therapy* 18 (1): 87–94.

158 Burness, C.B. and Scott, L.J. (2016). Susoctocog alfa: a review in acquired haemophilia A. *Drugs* 76 (7): 815–821.

159 Mahlangu, J., Powell, J.S., Ragni, M.V. et al. (2014). Phase 3 study of recombinant factor VIII Fc fusion protein in severe hemophilia A. *Blood* 123 (3): 317–325.

160 Konkle, B.A., Stasyshyn, O., Chowdary, P. et al. (2015). Pegylated, full-length, recombinant factor VIII for prophylactic and on-demand treatment of severe hemophilia A. *Blood* 126 (9): 1078–1085.

161 Franchini, M. and Mannucci, P.M. (2016). Von Willebrand factor (Vonvendi®): the first recombinant product licensed for the treatment of von Willebrand disease. *Expert Review of Hematology* 9 (9): 825–830.

162 Arruda, V.R., Doshi, B.S., and Samelson-Jones, B.J. (2017). Novel approaches to hemophilia therapy: successes and challenges. *Blood* 130 (21): 2251–2256.

163 Young, G., Collins, P.W., Colberg, T. et al. (2016). Nonacog beta pegol (N9-GP) in haemophilia B: a

multinational phase III safety and efficacy extension trial (paradigm™4). *Thrombosis Research* 141: 69–76.

164 Hoy, S.M. (2017). Eftrenonacog alfa: a review in haemophilia B. *Drugs* 77 (11): 1235–1246.

165 Lyseng-Williamson, K.A. (2017). Coagulation factor IX (recombinant), albumin fusion protein (albutrepenonacog alfa; Idelvion®): a review of its use in haemophilia B. *Drugs* 77 (1): 97–106.

166 Oldenburg, J., Mahlangu, J.N., Kim, B. et al. (2017). Emicizumab prophylaxis in hemophilia A with inhibitors. *New England Journal of Medicine* 377 (9): 809–818.

167 Harter, K., Levine, M., and Henderson, S.O. (2015). Anticoagulation drug therapy: a review. *Western Journal of Emergency Medicine* 16 (1): 11–17.

168 Barrowcliffe, T.W. (2012). History of heparin. In: *Heparin – A Century of Progress* (eds. R. Lever, B. Mulloy and C.P. Page), 3–22. Berlin, Heidelberg: Springer.

169 van der Meer, J.-Y., Kellenbach, E., and van den Bos, L.J. (2017). From farm to pharma: an overview of industrial heparin manufacturing methods. *Molecules* 22 (6): 1025.

170 Thunberg, L., Backstrom, G., and Lindahl, U. (1982). Further characterization of the antithrombin-binding sequence in heparin. *Carbohydrate Research* 100: 393–410.

171 Hirsh, J. (1991). Rationale for development of low-molecular-weight heparins and their clinical potential in the prevention of postoperative venous thrombosis. *American Journal of Surgery* 161 (4): 512–518.

172 Sadowski, R., Gadzała-Kopciuch, R., and Buszewski, B. (2019). Recent developments in the separation of low molecular weight heparin anticoagulants. *Current Medicinal Chemistry* 26 (1): 166–176.

173 Gray, E., Mulloy, B., and Barrowcliffe, T.W. (2008). Heparin and low-molecular-weight heparin. *Thrombosis and Haemostasis* 99 (5): 807–818.

174 Gavin, W. (2014). ATryn®: 1st GE (genetically engineered) animal success story for production of a human recombinant pharmaceutical. *BMC Proceedings* 8 (4): O4.

175 Gurman, P., Miranda, O.R., Nathan, A. et al. (2015). Recombinant tissue plasminogen activators (rtPA): a review. *Clinical Pharmacology and Therapeutics* 97 (3): 274–285.

176 Bachmann, F. and Kruithof, I.E. (1984). Tissue plasminogen activator: chemical and physiological aspects. *Seminars in Thrombosis and Hemostasis* 10 (1): 6–17.

177 Collen, D. and Lijnen, H.R. (2004). Tissue-type plasminogen activator: a historical perspective and personal account. *Journal of Thrombosis and Haemostasis* 2 (4): 541–546.

178 Noble, S. and McTavish, D. (1996). Reteplase. *Drugs* 52 (4): 589–605.

179 Davydov, L. and Cheng, J.W.M. (2001). Tenecteplase: a review. *Clinical Therapeutics* 23 (7): 982–997.

180 Smalling, R.W. (1997). Pharmacological and clinical impact of the unique molecular structure of a new plasminogen activator. *European Heart Journal* 18 Suppl F: F11–F16.

5

Drug Targets for Biologics

Felcia Lai, Bryson A. Hawkins, Esteban Cruz, Jane R. Hanrahan,
Paul W. Groundwater, and David E. Hibbs

Sydney Pharmacy School, Faculty of Medicine and Health, The University of Sydney, Sydney, New South Wales, Australia

---

KEY POINTS

- Biologic medicines are designed rationally based on an understanding of the complex biochemical and physiologic pathways in a disease state.
- Understanding the structure and function of key protein molecules in these pathways (drug targets) has transformed the discovery and development paradigm of biologic medicines.
- The discovery of biologic drugs is expanding exponentially as new potential endogenous targets/biochemical pathways are identified and exploited therapeutically.

Abbreviations

Abbreviation	Full name
ADAM-17; TACE	TNF-α Converting Enzyme
ADCC	Antibody-Dependent Cell-mediated Cytotoxicity
Akt	Protein Kinase B
AMD	Age-related Macular Degeneration
APCs	Antigen-Presenting Cells
B7-DC; CD273; PD-L2	Programmed Cell Death Ligand 2
B7-H1; CD274; PD-L1	Programmed Cell Death Ligand 1
BMSCs	Bone Marrow Stromal Cells
CD	Cluster of Differentiation
CD120a; p55/60; TNF-R1	TNF-α Receptor Type 1
CD120b; p75/80; TNF-R2	TNF-α Receptor Type 2
CD152; CTLA-4	Cytotoxic T-Lymphocyte-Associated Antigen 4
CD279; PD-1	Programmed Cell Death Protein 1
CHD	Coronary Heart Disease

Abbreviation	Full name
Da	Dalton
DNA	Deoxyribonucleic Acid
ECD	Extracellular Domain
EGF	Epidermal Growth Factor
EGF-A	Epidermal Growth Factor Repeat A
EGFR; ErbB-1; HER1	Epidermal Growth Factor Receptor
EMA	European Medicines Agency
epigen	Epithelial Mitogen
ErbB; HER	Epidermal Growth Factor Family
ErbB2; HER2; HER2/Neu	Human Epidermal Growth Factor Receptor 2
ERK	Extracellular Signal-Regulated Kinases
ESCC	Esophageal Squamous Cell Carcinoma
Fab	Fragment Antigen Binding Region
Fc	Constant Binding Fragment
FDA	United States Food and Drug Administration
Flt	FMS-like Tyrosine Kinase

Abbreviation	Full name		Abbreviation	Full name
HB-EGF	Heparin-Binding EGF-Like Growth Factor		PDB	Protein Data Bank
HIF-1	Hypoxia-Inducible Factor-1		PEG	Polyethylene Glycol
HMG-CoA	3-Hydroxy-3-Methylglutaryl Coenzyme A		PI3K	Phosphoinositide 3-Kinase
HNSCC	Head and Neck Squamous Cell Carcinoma		PIGF	Placental Growth Factor
IacP	Interleukin-1 accessory Protein		PKC	Protein Kinase C
IFN	Interferon		PLC	Phospholipase C
IgE	Immunoglobulin E		RANK; TNFRSF11	Receptor Activator of Nuclear Factor-κB
IGF-IR	Insulin-like Growth Factor-1 Receptor		RANKL; TNFSF11	Receptor Activator of Nuclear Factor-κB Ligand
IgV	Immunoglobulin Variable		RIPK-1	Receptor Interacting Serine/Threonine Protein Kinase-1
IHC	Immunohistochemical			
IL	Interleukin		RTKs	Tyrosine Kinase Receptors
IL-12Rβ1	IL-12 Receptor-β1		SHP-2	Src Homology Region 2 Domain-Containing Phosphatase-2
IL-12Rβ2	IL-12 Receptor-β2			
IL-17RA	IL-17 Receptor A		SREBP-2	Sterol-Responsive Element-Binding Protein 2
IL-6R	IL-6 Receptor			
IL-6st	IL-6 Signal Transducer		STAT3	Signal Transducer and Activator of Transcription 3
Jak1	Janus kinase 1			
Jak2	Janus kinase 2		STATs	Signal Transducer and Activator of Transcription Proteins
KDR	Kinase Insert Domain Receptor			
LDL	Low-Density Lipoprotein		sTNF-α	soluble TNF-α
LDL-C	Low-Density Lipoprotein Cholesterol		TCR	T Cell Receptor
LDL-R	Low-Density Lipoprotein Receptor		TGF-α	Transforming Growth Factor-α
MAPK	Mitogen-Activated Protein Kinase		TGF-β	Transforming Growth Factor-β
MEK	Mitogen-Activated Protein Kinase Kinase		Th	Helper T-cell
MHC	Major Histocompatibility Complex		TILs	Tumor-Infiltrating Lymphocytes
mTOR	mammalian Target of Rapamycin		tmTNF-α	transmembrane TNF-α
NFAFc1	Nuclear Factor of Activated t-cell Cytoplasmic 1		TNF	Tumor Necrosis Factor
			TNF-α	Tumor Necrosis Factor-α
NF-κB	Nuclear Factor-κB		TRADD	TNFR type 1-Associated-Death-Domain
NIK	NF-κB-Inducing Kinase		TRAF6	TNF Receptor Associated Factor-6
NP-1	Neuropilin-1		TRAFs	TNF Receptor Associated Factors
NRG	Neuregulins		Tregs	Regulatory T cells
NSCLC	Non-Small-Cell Lung Carcinoma		Tyk2	Tyrosine Kinases 2
OPG	Osteoprotegerin		VEGF	Vascular Endothelial Growth Factor
PCSK9	Proprotein Convertase Subtilisin/Kexin type 9		VEGFR	VEGF Receptor

5.1 Introduction

Since 1982, when the first biologic drug, a recombinant human insulin protein, was given approval,[1] closely followed in the same year by the approval of the first monoclonal antibody, Muromonab CD3,[2] the development of biologic drugs has continued to grow at a rapid rate. Biologic drugs have been a game changer for conditions such as arthritis, psoriasis, inflammatory bowel diseases, and various forms of cancer. As in the conventional small molecule drug design protocol, the identification and validation of the physiological target are two of the key stages in the process. In this chapter, the biochemical, physiologic, and pharmacological basis of biologic medicines will be discussed, based on the most common major targets of biologics as adopted during the drug discovery process.

5.2 Immune Checkpoints

The discovery of immune checkpoints has significantly changed the landscape of cancer immunotherapy in the last decade, improving survival rates of advanced and metastatic tumors significantly. Immune checkpoints are inhibitory receptors that regulate the immune system and prevent autoimmunity. They cause T-cell dysfunction and exhaustion by transducing (extracellular to intracellular transmission) signals that inhibit T-cell proliferation, cytokine production, and cytolytic function.[3]

The two immune checkpoints that have been extensively studied so far are programmed cell death protein 1 (PD-1; also known as CD279) and cytotoxic T-lymphocyte-associated antigen 4 (CTLA-4; also known as CD152). Ipilimumab, a CTLA-4 inhibitor, is the first United States (US) Food and Drug Administration (FDA) approved immune checkpoint inhibitor (2011),[4] followed by nivolumab which inhibits PD-1 in 2014.[5] There are currently seven FDA-approved (five European Medicines Agency [EMA] approved) immune checkpoints inhibitors, including six inhibitors of PD-1 or its ligand, programmed cell death ligand 1 (PD-L1; also known as CD274 and B7-H1) as well as one CTLA-4 inhibitor, for treating a wide range of tumors.

PD-1 is a transmembrane protein that contains an extracellular N-terminal Immunoglobulin Variable set-like (IgV-like) domain, a transmembrane domain, and a C-terminal cytoplasmic domain.[6] PD-1 is expressed on a broad range of immune cells, including activated T-cells, B-cells, macrophages, and tumor-infiltrating lymphocytes (TILs).[7] The interaction between PD-1 and PD-L1 plays a crucial role in maintaining peripheral immune cell tolerance. Murine (mouse) studies demonstrated that PD-1-deficient mice developed lupus-like glomerulonephritis and arthritis, revealing the important role of the PD-1/PD-L1 axis in immune homeostasis.[8] PD-1 has two ligands, PD-L1 and programmed cell death ligand 2 (PD-L2; also known as CD273 and B7-DC). PD-L1 is responsible for most of the cellular responses that we know and is more clinically relevant in cancer immunotherapy. Its expression is induced by proinflammatory cytokines such as interferon-γ (IFN-γ) and interleukin (IL)-4[9] and it is expressed on hematopoietic cells such as T-cells, B-cells, dendritic cells, macrophages as well as non-hematopoietic cells like pancreatic islet cells and vascular endothelial cells.[10] PD-1 also binds to PD-L2 to send immune suppressive signals but PD-L2 has a much more restricted expression and is far less prevalent compared with PD-L1.[11]

To activate T cells, T-cell receptor (TCR) identifies foreign antigen that are bound to major histocompatibility complex (MHC) on the surface of antigen-presenting cells (APCs) together with co-stimulating signals from CD86/CD28 interaction (see paragraph below). The interaction mediates IFN-γ release and prompts the upregulation of PD-L1 in order to balance immune response and limit tissue damage caused by excessive immune cell activity. Upon PD-L1/PD-1 binding, Src homology region 2 domain-containing phosphatase-2 (SHP-2) is recruited and binds to the cytoplasmic C-terminal of PD-1, which inhibits TCR signaling cascades (e.g. PI3K/Akt/mTOR pathway (an intracellular signaling pathway important in regulating the cell cycle) and the Ras-Raf-MEK/ERK pathway, which is a series of proteins that communicate a signal from a receptor on the surface of the cell to DNA in the nucleus), ultimately the signals for T cells activation is suppressed.[6] However, this mechanism is often exploited in cancer. Regulation of PD-L1 expression is complex and happens at different stages, especially in tumor cells. Signaling pathways, transcriptional factors like hypoxia-inducible factor α (HIF-1), signal transducer and activator of transcription 3 (STAT3), and nuclear factor-κB (NF-κB) as well as other post-translational factors all play a part in regulating the expression of PD-L1 in cancer cells. Cancer cells overexpress PD-L1 to protect the tumor by evading immunologic surveillance, preventing the proper generation of an immune response to the tumor.[12,13] Anti-PD-1/PD-L1 monoclonal antibodies (mAbs) induce anti-tumor immunity by inhibiting the binding between PD-L1 and PD-1. They shift the immune system balance toward T-cell activation, which works against cancer cells to destroy the tumor.[14]

CTLA-4, like PD-1, also sends suppressive signals and inhibits T-cell proliferation and expansion. However, CTLA-4 attenuates T cell activation at different stages of immune responses, mainly blocking T-cell activation at the earlier stage in lymphoid tissues. In conventional T-cells, CTLA-4 is rapidly expressed on cell surface when induced after TCR signaling. However, it is expressed constitutively (i.e. producing a biological response in the absence of a bound ligand) in regulatory T-cells (Tregs) at high levels.[15] CTLA-4 and its homologue CD28 share a pair of ligands, B7-1 (CD80) and B7-2 (CD86), expressed on APCs and compete to bind them. The binding of B7 to CD28 induces positive activating signal and upregulates the production of cytokines such as interleukin-2 (IL-2), which increases T cell proliferation and survival. In contrast, the binding of CTLA-4 to B7 results in the inhibition of IL-2 and downregulation of peripheral T-cell response. Though

CD28 has a much lower affinity for both B7-1 and B7-2 compared with CTLA-4, CD28 expression is more abundant and less restricted.[16] Mice lacking CTLA-4 develop lymphoproliferative disease and die within three to four weeks of birth.[17] There are several models hypothesizing how CTLA-4 inhibits T-cells but the exact downstream signaling mechanism of CTLA-4 remains controversial and unclear.[18] Anti-CTLA-4 blockade with ipilimumab enhances tumor immunity. When used in conjunction with PD-1/PD-L1 inhibitors, it provides synergistic effects and strengthens tumor specific immunity. This combination therapy has shown improved efficacy compared with either monotherapy and resulted in significantly longer progression-free survival[19,20] (Figure 5.1).

Biomarkers are specific chemicals that correlate to disease prognosis or treatment response. High PD-L1 expression are observed in various solid tumors, including breast cancer, gastric cancer, hepatocellular carcinoma, non-small cell lung carcinoma (NSCLC), and renal cell carcinoma, and has been shown to be closely related to poor prognosis.[22–24] Therefore, assessment of PD-L1 expression offers great potential in guiding patient selection and predicting treatment response to PD-1/PD-L1 inhibitors. Unfortunately, the association between PD-L1 expression and treatment response is not straightforward. Furthermore, patients with tumors of low or negative PD-L1 expression may also benefit from PD-1/PD-L1 therapy.[25] Currently, companion PD-L1 testing is required by the US FDA for the use of pembrolizumab in certain tumor types including

NSCLC, gastric cancer, cervical cancer, head and neck squamous cell carcinoma (HNSCC), and urothelial carcinoma and esophageal squamous cell carcinoma (ESCC). The four approved immunohistochemical (IHC) assays to measure PD-L1 expressions utilize four different PD-L1 antibodies (22C3, 28-2, SP263, and SP142), each have their own scoring systems and the sample types used for each assay are also different.[26] The lack of standardization between these IHC assays make comparison and interpretation of the results and their clinical utility problematic.[27] To maximize treatment response and avoid unnecessary toxicities in immune checkpoints blockade therapies, harmonization of PD-L1 IHC testing and validation of other predictive biomarkers are required.

5.3 Human Epidermal Growth Factor Family

The human epidermal growth factor family (ErbB; HER) is a group of receptor tyrosine kinases (RTKs) that comprise four members: HER1 (EGFR; ErbB1; HER1), HER2 (HER2/neu; ErbB2), HER3 (ErbB3), and HER4 (ErbB4). EGFR is constitutively expressed in a wide range of epithelial tissues and is involved in a wide range of cellular activities, particularly cell proliferation, survival, and differentiation. Mice without functional EGFR develop abnormalities in skin, kidney, brain, liver, and gastrointestinal tract.[28] All members of the family share a similar structure which consists of

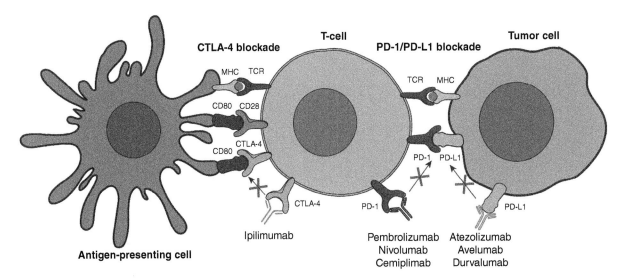

Figure 5.1 CTLA-4 and PD-1/PD-L1 blockade using immune checkpoint inhibitors. Tumor antigens bind to MHC and interact with TCR on T cells. CTLA-4 binds to CD80 to suppress T cell activation. Ipilimumab prevents CTLA-4 from binding to CD80 and activates immune response to tumor cells. Similarly, the immune-suppressing signal sent by PD-1/PD-L1 binding is blocked by anti-PD-1 and anti-PD-L1 antibodies. *Source:* Adapted from Cruz and Kayser[21]. (*See insert for color representation of the figure.*)

three domains: (1) N-terminal extracellular domain (ECD), (2) transmembrane domain, and (3) C-terminal intracellular protein kinase domain. The ECD is further subdivided into four subdomains, based on the binding of different signaling molecules and receptor dimerization.[29]

To activate these receptors and transduce signals, the process begins with a ligand binding to the ECD, which will trigger a conformational change in the ECD from its "tethered" conformation (where a loop needed for dimerization is hidden), following which the dimerization domain will be unmasked, allowing the receptor to dimerize and form either a homodimer or heterodimer with a different member from the ErbB family. The dimerization activates the receptor and leads to phosphorylation of specific tyrosine residues in the cytoplasmic tail which creates binding site for downstream signaling proteins, resulting in activation of signaling cascades such as Ras-Raf-MEK/ERK and PI3K/Akt/mTor intracellular pathways (see Section 5.2), leading to cell migration, proliferation, adhesion, and cell survival.[30]

The type of signaling pathway that is activated is determined by the ligand-type bound to ECD and the dimerization partner. More than 10 ligands have been identified so far, allowing the EGF receptors to transduce a wide range of signals. They are divided into three groups based on their receptor specificity. First group binds specifically to EGFR and comprises epidermal growth factor (EGF), transforming growth factor-α (TGF-α), epigen (epithelial mitogen), and amphiregulin (an autocrine growth factor). The second group binds to both EGFR and HER4, including heparin-binding EGF-like growth factor (HB-EGF), epiregulin, and betacellulin (proteins that play roles in cell differentiation). The last group are neuregulins (NRG), NRG 1-4, and they bind to HER4. NRG-1 and NRG-2 also bind to HER3.[31,32]

Two members of the ErbB family, EGFR and HER4, are fully functional RTKs, capable of signaling as homodimers and heterodimers following the binding of signaling molecules. Although HER2 is an orphan receptor and has no identifiable ligand, it is constitutively present at cell surface in active conformation, allowing HER2 homodimer to induce ligand-independent signals. HER2 can also dimerize with another member of the EGF family and is considered the preferred heterodimerization partner (see Figure 5.2). HER3 can accept a wide range of ligands but lacks intrinsic kinase activity; it can be activated and mediates signaling through heterodimerization.[35] Homodimers tend to transduce weaker signals, whereas HER2-containing heterodimers, especially HER2-HER3 dimer, produce potent signaling leading to cell proliferation and cell survival.[33,36]

5.3.1 EGFR

Overexpression of EGFR and/or its ligands in cancer cells can promote tumor growth by promoting cancer cell proliferation, metastases, angiogenesis, and inhibition of

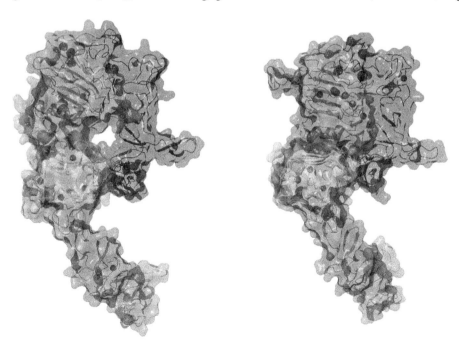

Figure 5.2 Extracellular domain structures of EGFR bound to EGF (PDB ID: 3NJP) (left) and HER2 (PDB ID: 6J71). Red – domain I; blue – domain II; green – domain III; purple – domain IV; yellow – EGF. *Source:* Adapted from Lu et al.[33] and Wang et al.[34] (*See insert for color representation of the figure.*)

apoptosis. EGFR is a strong prognostic indicator in head and neck, ovarian, cervical, bladder, and esophageal cancers.[37] There are currently three anti-EGFR mAbs, cetuximab, necitumumab, and panitumumab, as well as a range of small molecule tyrosine kinase inhibitors.

Although all three of the mAbs bind to domain III of the ECD, their mechanisms are different. Cetuximab inhibits both ligand-dependent and ligand-independent actions of EGFR by partially blocking the ligand binding region on domain III, as well as sterically preventing the receptor to adopt the active conformation needed for dimerization.[38] A clinical study revealed that a patient who was under cetuximab treatment acquired a point mutation (S492R) in the ECD and developed cetuximab resistance while still responding to panitumumab, suggesting that the panitumumab epitope might overlap with cetuximab or is in close proximity but it is not identical.[39,40] Nimotuzumab inhibits ligand binding but still allows conformational change of the receptor.[41] Nevertheless, all of them ultimately inhibit cancer cell proliferation, angiogenic growth factor production, tumor-induced angiogenesis, and cancer cell invasion by inhibiting EGFR activation. Cetuximab and nimotuzumab also mediate antibody-dependent cellular cytotoxicity (ADCC) *in vitro*, which can enhance anti-tumor effects.[42,43]

5.3.2 HER2

Overexpression of HER2 has been observed in certain cancers including ovarian,[44,45] lung adenocarcinoma,[46] uterine cancer,[47] gastric cancer,[48] endometrial cancer,[49] and salivary duct carcinomas.[50] It is however best known for its strong association in the development and progression of breast cancer[45]; the overexpression of HER2 occurs in approximately 25–30% of breast cancers and is associated with poor prognosis and higher rates of recurrence, rendering HER2 an important molecular target for HER2-positive breast cancer therapy.[45]

Trastuzumab was first approved by the FDA in 1998.[51] It binds to domain IV of HER2's ECD, preventing the activation of its intracellular tyrosine kinase thus blocking downstream signaling. Trastuzumab also induces immune response *via* ADCC and inhibits angiogenesis.[52–54] Unfortunately, many patients do not respond to therapy or develop refractory disease within one year of trastuzumab treatment.[55] Molecular mechanisms to trastuzumab resistance that were hypothesized include (1) altered HER2 molecular structure, (2) HER2 shedding, (3) activation of other HER receptors to compensate for the loss of HER2 signaling through increased ligand production, (4) compensatory signaling through

overexpression of other protein such as insulin-like growth factor-1 receptor (IGF-1R), and (5) constitutive activation of downstream signaling cascade through different pathways.[56]

Novel agents or combination strategies were explored to overcome trastuzumab resistance. Pertuzumab is the second anti-HER2 antibody approved by the FDA (2012) for the treatment of HER2-positive metastatic breast cancer.[57] It prevents the formation of HER2 dimers (both homodimers and heterodimers) by binding to domain II of the ECD, which is essential for dimerization. Like trastuzumab, it diminishes ligand-dependent HER2 signaling and induces ADCC, which inhibits the growth of breast cancer.[58]

5.4 Vascular Endothelial Growth Factor

Vascular endothelial growth factor (VEGF), also known as vascular permeability factor, is a mediator of the process of vasculogenesis, defined as the differentiation of precursor cells (angioblasts) into endothelial cells and the development of a vascular network, and angiogenesis, which is the growth of new capillaries from pre-existing blood vessels, as well as an inducer of vessel permeability.[59] The human VEGF family comprises of five members, VEGF-A (commonly known as VEGF), VEGF-B, VEGF-C, VEGF-D, and placental growth factor (PlGF); VEGF is the most studied and clinically relevant. VEGF-A regulates the proliferation, migration, and survival of endothelial cells in embryonic development, wound healing, and female reproductive cycle in adults.[60] Several VEGF-A isoforms exist due to alternative splicing (where a single gene codes for multiple proteins). VEGF-A$_{165}$ has intermediate properties among all VEGF-A isoforms and is believed to be the most important isoform for VEGF-A cellular response.[61]

Cellular hypoxia is one of the major triggers of VEGF-A expression. The lack of oxygen increases the production of hypoxia-inducible factor-1 (HIF-1) in the cell, and consequently upregulates the production of VEGF-A.[62] Inflammatory cytokines (e.g. IL-1α and IL-6) and growth factors (e.g. EGF, TGF-α) are also known to upregulate VEGF-A expression.[63]

Dimerized VEGF-A binds to tyrosine kinase receptors found on endothelial cell surface; the receptor dimerizes, activates through phosphorylation of the tyrosine kinase domain, and triggers downstream signaling cascade. Members of the VEGF family bind to three RTKs, fms-like tyrosine kinase-1 (Flt-1; also known as

VEGF receptor, VEGFR-1),[64] kinase insert domain receptor (KDR; VEGFR-2),[65] and Flt-4 (VEGFR-3).[66]

VEGF-A binds and activates two of the three receptors, VEGFR-1 and VEGFR-2; its signaling through these two receptors are also regulated by co-receptors neuropilin-1 (NP-1).[67] Even though VEGFR-1 has a higher affinity for VEGF-A, VEGFR-1 possesses weak kinase activity. Therefore, VEGFR-2 is responsible for most of the cellular responses through several signaling cascades, such as PI3K/Akt, the Src family, and PLCγ-PKC-MAPK pathway,[68–70] promoting proliferation, migration, and survival of endothelial cells as well as angiogenesis. The biological function of VEGFR-1 is not fully understood. Soluble form of VEGFR-1 has been shown to play a negative role in angiogenesis by acting as a decoy receptor to VEGF-A and regulates its signaling in vascular endothelial cells.[71,72] High expression levels of VEGFR-1 can be detected in a variety of cancer cells and contributes to tumor angiogenesis and growth.[73]

Strategies have been developed to interfere with VEGF-A/VEGFR interaction and hence reduce excessive angiogenesis, a common characteristic observed in numerous pathophysiological conditions, such as rheumatoid arthritis (RA), psoriasis, proliferative retinopathies, and age-related macular degeneration (AMD).[74] Currently, FDA-approved VEGF-A/VEGFR antibodies and VEGF-A-trap (aflibercept) are indicated for two major diseases, cancer and eye disorders.[75–80]

In some types of tumors, an increased amount of pro-angiogenic factors are secreted, including VEGF-A, to rapidly develop a vascular network to support its high proliferative rate.[81] The result is the formation of a structurally and functionally abnormal vascular network, leading to poor tumor perfusion and increased hypoxia, and consequently disrupting the diffusion and effects of cytotoxic drugs and radiotherapy, as well as immune cell infiltration. VEGF-A inhibitor, in combination with cytotoxic agents, could provide synergistic effects by restoring normal tumor vascular network and increasing tumor perfusion, allowing the diffusion of these chemotherapeutic agents and radiation treatment to reach the target site and exert their anti-cancer effects. When administered as monotherapy, VEGF-A antibody reduces cell proliferation, resulting in decreased tumor blood supply and thus inhibiting tumor growth.[82] Angiogenesis is also associated with ocular diseases such as AMD and proliferative retinopathies. Abnormal blood vessel growth, choroidal neovascularization, in the retina and choroid lead to fluid and blood leakage below the macular, resulting in distortion of normal retinal structure, scarring, and ultimately vision loss.[83] High VEGF-A expression can be found in the vitreous of AMD patients and is involved in subretinal angiogenesis.[84,85]

5.5 Interleukins

Interleukins consist of more than 60 potent pro-inflammatory pleiotropic (having multiple effects) cytokines.[86] Thirty-eight of these endocrine-mediated secretory proteins are well known to play a direct or communicative role in the pathogenesis of chronic inflammatory and proliferative conditions and diseases.[87] As with other cytokines, interleukins cannot usually be detected in healthy individuals; they are defense mediators that rapidly appear after biological stress. In certain autoimmune conditions such as Crohn's disease and RA, interleukins' levels are elevated, giving rise to their pathogenicity.[88] Their pathogenetic action is either direct or *via* forwarding cytokinetic signaling to alternate cytokines, for example, tumor necrosis factor-α (TNF-α).[89,90] Interleukins are produced by various cells, most commonly activated macrophages and T-lymphocytes.[86] Interleukins can be categorized into subfamilies based on shared sequence homology; this often determines their quaternary structure. As already mentioned, there are a large number of interleukins; however, only five have licensed biologic therapies targeting them: IL-1, IL-6, IL-12, IL-17, and IL-23.

5.5.1 IL-1

The IL-1 family represents a group of pleiotropic cytokines consisting of IL-1α, IL-1β, and IL-1Ra (antagonist), which are located on chromosome 2q13 and are heterodimers with a mean molecular weight of 18kDa. The IL-1 family interacts with two receptor subgroups (IL-1-type I and IL-1-type II).[86,91] Excluding IL-1Ra, the IL-1 family of cytokines share activated macrophages as a common source for the precursor molecule pro-IL-1. Pro-IL-1 is an inactive form of IL-1α and IL-1β.[91] Pro-IL-1 secretion is triggered by pro-inflammatory events, although interestingly IL-1 can self-induce pro-IL-1 by autocrine induction (where a cell secretes a hormone or chemical messenger that binds to autocrine receptors on the same cell, producing changes in the cell itself).[92] Once released, pro-IL-1 undergoes activation *via* intracellular caspase-1 enzyme or extracellular neutrophilic proteases to form active IL-1α or IL-1β that promote the phenotypic helper T-cell 17 (T$_\text{h}$17) when preforming active cytokine functions.[93,94] IL-1-type I is an active cell line located on the cell membrane, IL-1-type I has an accessory co-receptor (a receptor binds to a signaling molecule in addition to the primary receptor), IL-1-type *IacP* is expressed in brain cells which is shown to downregulate IL-1 cytokine forward signaling; IL-1-type II does not have a transmembrane domain and therefore lacks the ability to forward signal cellular communication and acts as a decoy receptor site.[95,96] Although not well understood, the signal induction of IL-1 results in the release of nuclear factor kappa-light-chain, enhancing activated B cells (NF-κB) and mitogen-activated protein kinase (MAPK), which contribute to pro-inflammatory response.[97]

5.5.2 IL-6

IL-6, previously known as interferon *β*-2, is a single glycoprotein cytokine with a mean molecular weight of 23kDa. The primary sources of IL-6 are T and B lymphocytes and to a lesser extent, monocytes, fibroblasts, and activated macrophages.[87] IL-6 has a number of pro-inflammatory functions; most important is activation of T and B lymphocytes in the adaptive immune system, which trigger the production of many acute-phase proteins.[87] Similar to IL-1, IL-6 is required for the development of T helper cells and tumor necrosis factor (TNF, see Section 5.6); it is also required for Th-17 helper T cell creation, which is a vital component of interleukin production required for neutrophil mobilization.[98] IL-6 belongs to a large group of cytokines, which express their cellular outcome *via* interactions with the IL-6 receptor (IL-6R) and/or IL-6R coupled with the glycoprotein130 subunit (IL-6 signal transducer (IL-6st)). IL-6 binds to the functional IL-6 receptor which is composed of two chains and two of above-mentioned IL-6st, creating an IL-6/IL-6st complex. The IL-6/IL-6st complex is soluble and is also located on the cell membrane to allow transmembrane signaling.[99] Binding of IL-6 to the complex causes homodimerization of IL-6st that activates janus kinase 1 (Jak1), janus kinase 2 (Jak2), and to a lesser extent tyrosine kinases 2 (Tyk2).[99] As a result of kinase stimulation, there is enhanced signal transduction and activation of transcription proteins (STATs), specifically transcription-3 (STAT3).[99] STAT3 upregulation increases chemokine and other cytokine release, which lead to the overall pro-inflammatory response.[99] Interestingly, IL-6ST is a shared transduction chain for other cytokines; this may explain its multimodal effects and number of indications for IL-6 inhibitors like tocilizumab.

5.5.3 IL-17

IL-17 is a group of seven cytokines (IL-17A to F) of which two, IL-17A/F, have a well-documented cellular function. They are predominately secreted by activated T$_h$17. IL-17A/F are pro-inflammatory like other interleukins causing neutrophilia and activation of other cellular targets.[100] There are several IL-17 receptors to mirror the number of cytokines of the family; these receptors are cysteine knots (a complex structural loop involving three disulphide bridges), homodimeric/heterodimeric in nature.[101] Due to the similarity of IL-17A and F, both bind to IL-17 receptor A (IL-17RA), although IL-17A has a higher binding affinity. IL-17A/F bind to a coupled IL-17RA and IL-17 receptor C (IL-17RC) complexed with an intracellular Actinin alpha 1 moiety (Act1) (IL-17RA/C-Act1) to allow further downstream signaling.[101] The activation of the IL-17RA/C-Act1 protein complex is not fully understood; what is known is that the receptors are in various tissues and upon binding IL-17A/F, a cellular signal causes downstream MAPK and NF-κB regulation *via* a ubiquitination/de-ubiquitination equilibrium of various TRAFs (TNF receptor associated factors), which are a family of proteins involved in regulation of inflammation, apoptosis, and immune responses. TRAFs have both up- and downstream regulatory roles in IL-17 mediated pro-inflammatory responses; this results in a differential response to the initiating stimuli.[102] The signaling pathway for pro-inflammatory responses is *via* NF-κB and MAPK activation, which is highly dependent on TRAF6 levels.[102]

5.5.4 IL-12 and IL-23

IL-12 and IL-23 play important pro-inflammatory and immunoregulatory effects *via* stimulation of the adaptive immune system.[103,104] IL-12 and IL-23A are predominantly induced by exogenous agents like bacteria or parasite causing activation of peripheral blood mononuclear cells; however, IL-12 is released prior to IL-23 and therefore is needed for IL-23 release.[105,106] Both these cytokines bind to a receptor that consists of two chains; one shared chain is IL-12 receptor-*β*1 (IL-12R*β*1) and the second is a complement chain, IL-12 receptor-*β*2 (IL-12R*β*2) or IL-23 receptor (IL-23R), respectively.[107] Binding results in cytokine-induced auto-phosphorylation and trans-phosphorylation of both Jak1 and Tyk2 kinases, which in turn activate STATs, which further induce pro-inflammatory and immunoregulatory regulators such as IFN-*γ*. The adaptive immune system is activated by IL-12 or IL-23 stimulating IFN-*γ* production, which leads to a cascaded formation of activated helper T cells 1 (T$_h$1) and inhibition helper T cell 2 (T$_h$2). This dampens the hormonal immune system resulting in lower immunoglobulin E (IgE) levels and thus amplifying T$_h$1 cellular pro-inflammatory response to stimulus.[107,108]

5.5.5 Interleukin-Directed Therapies

As outlined above, IL-1, IL-6, IL-17, IL-12, and IL-23 are potent pro-inflammatory pleiotropic cytokines, and as such strategies have been developed to interfere with the interleukin pathways. The reduction of these cellular interactions would be favorable for numerous pro-inflammatory pathophysiological diseases. Currently, FDA-approved interleukin antibodies are indicated for an array of conditions. The first interleukin biologic approved by the FDA in 2001 was anakinra, an *E. coli* sourced recombinant non-glycosylated human IL-1 antagonist (17 kDa) and was found to be used in the treatment of RA.[109] Subsequently, anakinra gained FDA approval for other autoimmune conditions. In 2009, a new interleukin biologic was given FDA approval, ustekinumab, a human IgG1κ monoclonal antibody (149 kDa) targeting the IL-12/IL-23 receptor cascade was licensed for moderate to severe plaque psoriasis in 2009.[110] Since 2009, many further indications have been added, most recently being ulcerative colitis in 2019.[110] Tocilizumab was the third FDA-approved interleukin biologic. Tocilizumab is a recombinant humanized IL-6 antibody (145 kDa); it targets both membrane-bound and soluble IL-6R and IL-6R/IL-6st, and it was first licensed in 2010 for RA.[111] Tocilizumab has been the most intensely studied resulting in the most FDA-approved indications.[111] Sarilumab is another biologic that targets membrane-bound and soluble IL-6R and IL-6R/IL-6st; it is a human IL-6 antibody (150 kDa). It is the most recent of the interleukin biologics to gain approval, for the treatment of RA in 2017.[112] Sarilumab may be used as monotherapy or in combination with methotrexate, the recommended dose is 200 mg once every two weeks; however, dosage modifications are required in specific situations: neutropenia, thrombocytopenia, or elevated liver function.[112]

The remaining FDA-approved interleukin directed therapy is secukinumab, a human IgG1κ monoclonal antibody (151 kDa), which inhibits the IL-17A cellular signaling by selectively binding to the cytokine, thus inhibiting IL-17A binding ability at the receptor.[113] Secukinumab first gained FDA approval for plaque psoriasis in 2015, in the year following, it gained two further indications of use.[113] Secukinumab has condition-specific initiation and maintenance doses, which are administered *via* subcutaneous methods using either formulated vials or pre-filled syringe devices.[113] It is important to note that there are also numerous downstream kinase inhibitors approved; however, as outlined above, cytokine cellular pathways are complex, multifactorial in regulation, and further studies are required for quantification of the key cellular targets for the management of interleukin mediated pro-inflammatory conditions.

5.6 Tumor Necrosis Factor

Inflammation is characterized by vasodilation and recruitment of immune cells to a site of cellular injury; one of the central players in this process is the TNF protein. TNF (or TNF-α) is a potent pro-inflammatory pleiotropic cytokine that plays a role in the pathogenesis of many chronic inflammatory, proliferative, or infectious diseases.[114] TNF-α cannot usually be detected in healthy individuals, hence TNF-α is regarded as a defense molecule, being the first cytokine to rapidly appear in the bloodstream after injury or stress, with other pro-inflammatory mediators (IL-1 and IL-6, see Section 5.5) dependent on its prior release.[89,115] TNF-α is produced mainly by activated macrophages and T lymphocytes and exists as two variants, the precursory cell-bound transmembrane TNF-α (tmTNF-α) and its unbound soluble form (sTNF-α).[90,116,117] TNF-α has two receptors, Type 1 and 2, TNF-R1 (or CD120a, p55/60) and TNF-R2 (or CD120b, p75/80) and these are present on almost all mononucleated cells, albeit differentially regulated in different cell types and in normal versus disease tissues.[118]

TmTNF-α is expressed as a type II cell surface homotrimer peptide with subsequent proteolytic cleavage by TNF-α converting enzyme (TACE or ADAM-17) to release the soluble, smaller homotrimer form.[119]

TACE also processes several other membrane proteins, including the TNF receptors, which it may also liberate into their soluble forms, to neutralize circulating TNF-α, thus acting in both a pro- and anti-inflammatory fashion.

TmTNF-α acts bimodally, as a molecule that transmits signals both as a ligand and as a receptor in a cell-to-cell contact fashion.[120] TmTNF-α and sTNF-α bind to both TNF-R1 and -R2; however, the biological action of tmTNF-α is primarily *via* interaction with TNF-R2.[121] After binding, TNF-α induces a pro-inflammatory signaling cascade that is highly complex and not fully understood although there is activation of the MAPKs and NF-κB pathways, which induce upregulation of several pro-inflammatory cytokines, including TNF-α (Figure 5.3).

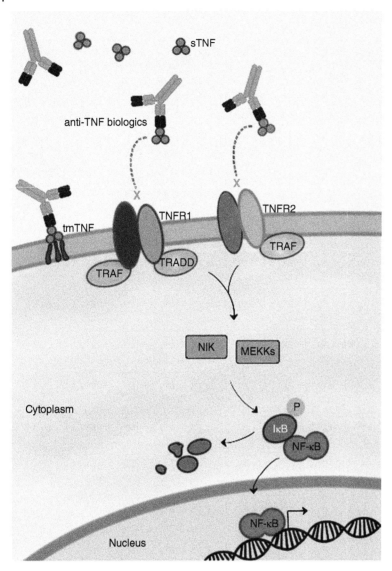

Figure 5.3 Mechanism of action of TNF-α biologics; infliximab is shown binding to tmTNF-α and sTNF-α. *Source:* Adapted from Pedersen et al.[122] (*See insert for color representation of the figure.*)

The signaling mechanism of TNF-R1 is initiated by the recruitment of a protein, TNFR type 1-associated-death-domain (TRADD). TRADD recruits two additional proteins, importantly, TNF receptor associated factor (TRAF) and receptor interacting serine/threonine protein kinase-1 (RIPK-1); this TRADD-TRAF-RIPK-1 trimer is released from the TNF-R1 complex.[114] RIPK-1 recruits mitogen activated kinase kinase kinases (MEKKs, MAP3Ks, Raf), leading to phosphorylation of the inhibitor of NF-κB kinase (IαB) and ubiquitination and subsequent degradation. This process allows NF-κB to enter the nucleus and induce expression of pro-inflammatory genes encoding for a number of chemokines and cytokines.[123] TNF-R2 does not interact with TRADD, due to the lack of a death domain, but can interact with TRAF, leading to the recruitment of NF-κB-inducing kinase (NIK) and inhibition of nuclear factor kappa (IKK) to phosphorylate IαB, degradation, and subsequent pro-inflammatory gene encoding.[114]

A number of synthetic anti-TNF-α biologics target TNF-α, for chronic inflammatory conditions. It was thought that all TNF-α blockers share common mechanisms of action, acting as decoys, binding to either/or tmTNF-α or sTNF-α, or preventing binding to TNF-R1/2 (Figure 5.3); however, the mechanisms of action are far more complex.[124]

Originally, approved by the FDA in 1998, infliximab, a 25% murine (variable region), 75% human (constant region), chimeric IgG1 κ dimeric antibody was initially useful in Crohn's disease[125] and subsequently for other autoimmune disorders. Antichimeric immune responses to infliximab (and the need for use in combination with methotrexate) prompted the removal of the murine sequence and the subsequent fusion of two human TNF-R2 to the IgG1 antibody to produce the dimeric fusion protein etanercept, approved in 1998. Adalimumab, approved in 2002, was designed so that there were no nonhuman components; it is indistinguishable from natural human IgG1, blocks both TNF-R1 and -2 receptors, has low immunogenicity,[126] and is on the WHO List of Essential Medicines. Certolizumab pegol was next to be approved in 2008 for Crohn's disease that did not respond to first-line therapies. Certolizumab pegol is composed of the antibody binding fragment (Fab) of the monoclonal antibody for TNF-α, conjugated with a 40 kDa polyethylene glycol (PEG); however, it lacks the constant binding fragment (Fc) that is present in infliximab and adalimumab. Certolizumab pegol has a distinct mechanism of action when compared with other TNF-α blockers,[127] has low immunogenicity, with the PEGylation giving it a half-life comparable to other blockers. Golimumab received approval in 2009 as the first once-monthly treatment of RA, psoriatic arthritis, and ankylosing spondylitis.

5.7 Receptor Activator of Nuclear Factor-KB Ligand (RANKL)

Bone remodeling occurs constantly in the human body, repairing damage and renewing skeletal integrity and strength. Bone homeostasis is maintained by osteocytes that regulate the balance between bone-building osteoblasts and bone-resorbing osteoclasts. The imbalance between bone formation and bone resorption may be influenced by several factors, including inflammatory cytokines, growth factors, and hormones, leading to bone disorders such as osteoporosis and osteopetrosis.[128]

Osteoclasts play an important role in the development of osteoporosis, a skeletal disorder characterized by low bone density and compromised bone strength that leads to frequent bone fractures.[128] Osteoclast activity is stimulated by the presence of receptor activator of nuclear factor-κB ligand (RANKL; also known as TNFSF11). RANKL is a type II transmembrane protein expressed by several cell types such as osteoblasts, lymphocytes, bone marrow stromal cells (BMSCs), and predominantly by osteocytes in the bone.[129] It is part of the TNF superfamily and also exists in a biologically active soluble form.[130]

For bone resorption, RANKL binds to receptor activator of nuclear factor-κB (RANK; also known as TNFRSF11A) present on the surface of osteoclasts and osteoclast precursors. The complex then binds to TNF receptor associated factor-6 (TRAF6), which triggers downstream signaling cascade, activating MAPK, NF-κB, and nuclear factor of activated T-cell cytoplasmic 1 (NFAFc1), and eventually to differentiation and activation of osteoclasts. *RANK* and *RANKL* knockout mice have defects in osteoclast function, resulting in bone resorption and osteopetrosis.[131,132] RANKL also binds to osteoprotegerin (OPG), a decoy receptor of RANKL that hinders RANKL/RANK binding. OPG knockout mice exhibited osteoporosis due to increased osteoclastogenesis,[133] indicating the importance of OPG in balancing RANK/RANKL activity.

Expression of RANKL is affected by numerous factors and one of these is hormones. Women approaching menopause often experience bone loss and fragility due to reduced levels of oestrogen, which causes an increased expression of RANKL and reduced expression of OPG, thus increasing bone resorption.[134,135] Patients receiving hormone ablation therapy for breast cancer (e.g. oestrogen suppression therapy) or prostate cancer (testosterone suppression through chemical or surgical castration) experience bone loss through the same mechanism.[136–138]

Apart from osteoporosis, RANKL is also involved in bone metastases, which are often seen in cancer, especially in solid tumors like breast and prostate cancer. Tumor cells spread and metastasize to the bone, causing either osteolytic or osteoblastic bone metastases, or a mix of the two. Fractures, bone pain, and spinal cord compression are all signs of bone metastases.[139] In certain cancers, tumor cells can either produce RANKL directly or secrete cytokines and growth factors, including interleukin (IL)-1, IL-6, IL-8, IL-11, TNF-α, and parathyroid hormone-related peptide, causing the secretion of RANKL by stroma cells and osteoblasts and/or reduction of OPG expression, resulting in an increase of osteoclast activity and bone resorption. Consequently, bone destruction causes cytokines such as transforming growth factor beta (TGF-β) and insulin-like growth factor release from the bone matrix leading to tumor growth and tumor survival. The cycle repeats and forms a vicious cycle, causing osteolytic destruction of the bone matrix and the development of bone metastasis.[140]

The involvement of RANKL in different systems means that the RANK/RANKL/OPG axis is an attractive target for the prevention of bone loss by inhibiting the development and activity of osteoclasts.

Denosumab, a fully human mAb against RANKL, is used for the treatment of different bone disorders.[141]

5.8 Proprotein Convertase Subtilisin/Kexin Type 9 (PCSK9)

Low-density lipoprotein (LDL) cholesterol is a modifiable risk factor for atherosclerosis and coronary heart disease (CHD). In humans, cholesterol is carried around in combination with different lipoproteins. When a cell requires additional cholesterol, the cholesterol in the bloodstream binds to LDL, the LDL-C ligand then binds to LDL receptors (LDL-R) that are present on the surface of hepatocytes. The LDL-C/LDL-R complex is internalized to the endosomes and dissociates due to the acidic endosome environment. LDL-R recycles to the cell surface by recycling vesicles and endosomes containing LDL-C fuse with lysosome, leading to the degradation of LDL-C to release cholesterol. The process keeps repeating and results in lower circulating LDL-C levels.[142]

Cholesterol homeostasis is regulated by proprotein convertase subtilisin/kexin type 9 (PCSK9), which degrades LDL-R to control the levels of LDL-R present on cell surface *via* two major mechanisms. (1) Extracellular PCSK9 binds to LDL-R, the complex is internalized in the hepatocyte which subsequently degrades LDL-R, affecting the recycling and availability of LDL-R; (2) intracellular PCSK9 binds to LDL-R and directs lysosomal delivery for degradation before LDL-R is transported to the cell surface.[143]

PCSK9 mutation was first discovered in 2003 in patients with autosomal-dominant familial hypercholesterolemia. Increase in PCSK9 function leads to lower LDL-R and higher levels of LDL-C[144] and subsequent loss of PSCK9 function and reduced LDL-C levels and conferred protection against CHD.[145] PCSK9 is expressed abundantly in the liver and is predominantly regulated by transcription factor sterol-responsive element-binding protein 2 (SREBP-2),[146] which also upregulates transcription of other genes involved in cholesterol metabolism including 3-hydroxy-3-methylglutaryl coenzyme A (HMG-CoA) reductase and LDL-R.[147] Decreased intracellular cholesterol activates the expression of SREBP-2, which increases the expression of LDL-R to encourage the uptake of LDL-C from the circulation. SREBP-2 also promotes the expression of PCSK9, which degrades LDL-R to balance hepatic uptake of cholesterol and maintain lipid homeostasis.[146]

PCSK9 is synthesized as a zymogen (an enzyme requiring activation by another enzyme) consisting of (1) signal peptide, (2) prosegment, (3) catalytic domain, and (4) C-terminal domain. Inactive PCSK9 first loses the signal peptide, then it undergoes co-translational autocatalytic cleavage in the endoplasmic reticulum to cleave the prosegment and the mature PCSK9 protein, which contains the catalytic domain and the C-terminal domain. However, the prosegment remains closely associated with the mature PCSK9 after cleavage and facilitates the translocation of the protein from the endoplasmic reticulum to the Golgi apparatus as well as blocking access of potential substrates to the catalytic domain. Mature PCSK9 is then released into the circulation by hepatocytes and binds its catalytic domain to the epidermal growth factor repeat A (EGF-A) region of LDL-R, the complex is internalized and enters the endosomes. Instead of dissociating like the LDL-C/LDL-R complex, the acidic environment inside the endosomes strengthens the association between PCSK9 and LDL-R which prevents the dissociation of the complex which is then routed to the lysosome where PCSK9 and LDL-R are both degraded.[148]

PSCK9 inhibitors offer a novel approach to LDL-C reduction; by binding to PCSK9, the inhibitors prevent the binding of PCSK9 to LDL-R, which increases the number of LDL-R that are available to remove LDL circulating in the blood, thereby lowering LDL-C levels and reducing the risk of CHD. Approved PCSK9 inhibitors, alirocumab and evolucumab, are both fully humanized mAbs used for homozygous familial hypercholesterolemia and primary hypercholesterolemia, especially in patients who are statin intolerant or when other treatments have not provided sufficient clinical benefits.[149] PCSK9 is also closely related to inflammation, endothelial apoptosis, blood pressure regulation, glucose metabolism, and adipogenesis; this suggests PCSK9 inhibitors may have a role in other metabolic disorders.[150,151]

5.9 Concluding Remarks

As outlined in this chapter, biologic drugs have assumed huge prominence in modern therapeutics and healthcare. Biologics are clearly very effective and offer treatment options for many patients with debilitating chronic diseases or life-threatening illnesses like cancer for which there have not been any other viable treatment options. The choice of a suitable endogenous macromolecular target is key to this success of biologic medicines. The development of biologic drugs relies upon the deep understanding of the complex biochemical signaling pathways that are involved in many disease pathologies. In addition, increasingly, genetic differences have been identified in individuals in processes that are responsible for either upregulation or downregulation of particular proteins involved in signaling pathways which may lead to various diseases. Not always, but often, many of the fine details of the role (and interconnectivity) of the various pathways of the chosen target in biological systems remain unclear, leading to treatment failure and/or clinically significant side effects. With the rapidly expanding understanding of basic biochemistry and endogenous signaling pathways, and the consequent identification of new pathways, and hence new potential drug targets, the number, efficacy, and safety profiles of new biologics are likely to be the future of modern drug design and therapeutics.

References

1 Chance, R.E. and Frank, B.H. (1993). Research, development, production, and safety of biosynthetic human insulin. *Diabetes Care* 16 (Suppl. 3): 133.

2 Todd, P.A. and Brogden, R.N. (1989). Muromonab CD3. *Drugs* 37 (6): 871–899.

3 Pardoll, D.M. (2012). The blockade of immune checkpoints in cancer immunotherapy. *Nature Reviews Cancer* 12 (4): 252–264.

4 FDA (2011). Yervoy® (ipilimumab) (updated September 2019). www.fda.gov (accessed 16 June 2020).

5 FDA (2014). Opdivo® (nivolumab) (updated September 2019). www.fda.gov (accessed 16 June 2020).

6 Okazaki, T., Maeda, A., Nishimura, H. et al. (2001). PD-1 immunoreceptor inhibits B cell receptor-mediated signaling by recruiting src homology 2-domain-containing tyrosine phosphatase 2 to phosphotyrosine. *Proceedings of the National Academy of Sciences* 98 (24): 13866–13871.

7 Agata, Y., Kawasaki, A., Nishimura, H. et al. (1996). Expression of the PD-1 antigen on the surface of stimulated mouse T and B lymphocytes. *International Immunology* 8 (5): 765–772.

8 Nishimura, H., Nose, M., Hiai, H. et al. (1999). Development of lupus-like autoimmune diseases by disruption of the PD-1 gene encoding an ITIM motif-carrying immunoreceptor. *Immunity* 11 (2): 141–151.

9 Eppihimer, M.J., Gunn, J., Freeman, G.J. et al. (2002). Expression and regulation of the PD-L1 immunoinhibitory molecule on microvascular endothelial cells. *Microcirculation: The Official Journal of the Microcirculatory Society, Inc.* 9 (2): 133–145.

10 Sharpe, A.H., Wherry, E.J., Ahmed, R., and Freeman, G.J. (2007). The function of programmed cell death 1 and its ligands in regulating autoimmunity and infection. *Nature Immunology* 8 (3): 239.

11 Latchman, Y., Wood, C.R., Chernova, T. et al. (2001). PD-L2 is a second ligand for PD-1 and inhibits T cell activation. *Nature Immunology* 2 (3): 261.

12 Blank, C., Gajewski, T.F., and Mackensen, A. (2005). Interaction of PD-L1 on tumor cells with PD-1 on tumor-specific T cells as a mechanism of immune evasion: implications for tumor immunotherapy. *Cancer Immunology, Immunotherapy: CII* 54 (4): 307–314.

13 Chen, J., Jiang, C., Jin, L., and Zhang, X. (2015). Regulation of PD-L1: a novel role of pro-survival signalling in cancer. *Annals of Oncology* 27 (3): 409–416.

14 Robainas, M., Otano, R., Bueno, S., and Ait-Oudhia, S. (2017). Understanding the role of PD-L1/PD1 pathway blockade and autophagy in cancer therapy. *OncoTargets and Therapy* 10: 1803.

15 Jain, N., Nguyen, H., Chambers, C., and Kang, J. (2010). Dual function of CTLA-4 in regulatory T cells and conventional T cells to prevent multiorgan autoimmunity. *Proceedings of the National Academy of Sciences of the United States of America* 107 (4): 1524–1528.

16 Wolchok, J.D. and Saenger, Y. (2008). The mechanism of anti-CTLA-4 activity and the negative regulation of T-cell activation. *The Oncologist* 13 (Supplement 4): 2–9.

17 Waterhouse, P., Penninger, J.M., Timms, E. et al. (1995). Lymphoproliferative disorders with early lethality in mice deficient in Ctla-4. *Science (New York, NY)* 270 (5238): 985–988.

18 Walker, L.S. and Sansom, D.M. (2015). Confusing signals: recent progress in CTLA-4 biology. *Trends in Immunology* 36 (2): 63–70.

19 Larkin, J., Chiarion-Sileni, V., Gonzalez, R. et al. (2015). Combined nivolumab and ipilimumab or monotherapy in untreated melanoma. *The New England Journal of Medicine* 373 (1): 23–34.

20 Wolchok, J.D., Kluger, H., Callahan, M.K. et al. (2013). Nivolumab plus ipilimumab in advanced melanoma. *The New England Journal of Medicine* 369 (2): 122–133.

21 Cruz, E. and Kayser, V. (2019). Monoclonal antibody therapy of solid tumors: clinical limitations and novel strategies to enhance treatment efficacy. *Biologics: Targets & Therapy* 13: 33.

22 Muenst, S., Schaerli, A., Gao, F. et al. (2014). Expression of programmed death ligand 1 (PD-L1) is associated with poor prognosis in human breast cancer. *Breast Cancer Research and Treatment* 146 (1): 15–24.

23 Mu, C.-Y., Huang, J.-A., Chen, Y. et al. (2011). High expression of PD-L1 in lung cancer may contribute to poor prognosis and tumor cells immune escape through suppressing tumor infiltrating dendritic cells maturation. *Medical Oncology* 28 (3): 682–688.

24 Wang, A., Wang, H., Liu, Y. et al. (2015). The prognostic value of PD-L1 expression for non-small cell lung cancer patients: a meta-analysis. *European Journal of Surgical Oncology: the Journal of the European Society of Surgical Oncology and the British Association of Surgical Oncology* 41 (4): 450–456.

25 Taube, J.M., Klein, A., Brahmer, J.R. et al. (2014). Association of PD-1, PD-1 ligands, and other features of the tumor immune microenvironment with response to anti–PD-1 therapy. *Clinical Cancer Research: An Official Journal of the American Association for Cancer Research* 20 (19): 5064–5074.

26 FDA (2019). List of cleared or approved companion diagnostic devices (in vitro and imaging tools). https://www.fda.gov/medical-devices/vitro-diagnostics/list-cleared-or-approved-companion-diagnostic-devices-vitro-and-imaging-tools (accessed 16 June 2020).

27 Udall, M., Rizzo, M., Kenny, J. et al. (2018). PD-L1 diagnostic tests: a systematic literature review of scoring algorithms and test-validation metrics. *Diagnostic Pathology* 13 (1): 12.

28 Threadgill, D.W., Dlugosz, A.A., Hansen, L.A. et al. (1995). Targeted disruption of mouse EGF receptor: effect of genetic background on mutant phenotype. *Science* 269 (5221): 230–234.

29 Yarden, Y. and Sliwkowski, M.X. (2001). Untangling the ErbB signalling network. *Nature Reviews. Molecular Cell Biology* 2 (2): 127.

30 Cho, H.-S., Mason, K., Ramyar, K.X. et al. (2003). Structure of the extracellular region of HER2 alone and in complex with the Herceptin Fab. *Nature* 421 (6924): 756.

31 Singh, B., Carpenter, G., and Coffey, R.J. (2016). EGF receptor ligands: recent advances. *F1000Research* 5: F1000. Faculty Rev-2270.

32 Hynes, N.E. and MacDonald, G. (2009). ErbB receptors and signaling pathways in cancer. *Current Opinion in Cell Biology* 21 (2): 177–184.

33 Lu, C., Mi, L.-Z., Grey, M.J. et al. (2010). Structural evidence for loose linkage between ligand binding and kinase activation in the epidermal growth factor receptor. *Molecular and Cellular Biology* 30 (22): 5432–5443.

34 Wang, Z., Cheng, L., Guo, G. et al. (2019). Structural insight into a matured humanized monoclonal antibody HuA21 against HER2-overexpressing cancer cells. *Acta Crystallographica Section D: Structural Biology* 75 (6): 554–563.

35 Riese, D.J. and Stern, D.F. (1998). Specificity within the EGF family/ErbB receptor family signaling network. *BioEssays* 20 (1): 41–48.

36 Mitri, Z., Constantine, T., and O'Regan, R. (2012). The HER2 receptor in breast cancer: pathophysiology, clinical use, and new advances in therapy. *Chemotherapy Research and Practice* 2012: 743193. https://doi.org/10.1155/2012/743193.

37 Nicholson, R., Gee, J., and Harper, M. (2001). EGFR and Cancer Prognosis. *European Journal of Cancer* 37: 9–15.

38 Li, S., Schmitz, K.R., Jeffrey, P.D. et al. (2005). Structural basis for inhibition of the epidermal growth factor receptor by cetuximab. *Cancer Cell* 7 (4): 301–311.

39 Montagut, C., Dalmases, A., Bellosillo, B. et al. (2012). Identification of a mutation in the extracellular domain of the epidermal growth factor receptor conferring cetuximab resistance in colorectal cancer. *Nature Medicine* 18 (2): 221.

40 Voigt, M., Braig, F., Göthel, M. et al. (2012). Functional dissection of the epidermal growth factor receptor epitopes targeted by panitumumab and cetuximab. *Neoplasia* 14 (11): 1023–1031.

41 Talavera, A., Friemann, R., Gómez-Puerta, S. et al. (2009). Nimotuzumab, an antitumor antibody that targets the epidermal growth factor receptor, blocks ligand binding while permitting the active receptor conformation. *Cancer Research* 69 (14): 5851–5859.

42 Kurai, J., Chikumi, H., Hashimoto, K. et al. (2007). Antibody-dependent cellular cytotoxicity mediated by cetuximab against lung cancer cell lines. *Clinical Cancer Research* 13 (5): 1552–1561.

43 Mazorra, Z., Lavastida, A., Concha-Benavente, F. et al. (2017). Nimotuzumab induces NK cell activation, cytotoxicity, dendritic cell maturation and expansion of EGFR-specific T cells in head and neck cancer patients. *Frontiers in Pharmacology* 8: 382.

44 Berchuck, A., Kamel, A., Whitaker, R. et al. (1990). Overexpression of HER-2/neu is associated with poor survival in advanced epithelial ovarian cancer. *Cancer Research* 50 (13): 4087–4091.

45 Slamon, D.J., Godolphin, W., Jones, L.A. et al. (1989). Studies of the HER-2/neu proto-oncogene in human breast and ovarian cancer. *Science* 244 (4905): 707–712.

46 Shigematsu, H., Takahashi, T., Nomura, M. et al. (2005). Somatic mutations of the HER2 kinase domain in lung adenocarcinomas. *Cancer Research* 65 (5): 1642–1646.

47 Slomovitz, B.M., Broaddus, R.R., Burke, T.W. et al. (2004). Her-2/neu overexpression and amplification in uterine papillary serous carcinoma. *Journal of Clinical Oncology* 22 (15): 3126–3132.

48 Gravalos, C. and Jimeno, A. (2008). HER2 in gastric cancer: a new prognostic factor and a novel therapeutic target. *Annals of Oncology* 19 (9): 1523–1529.

49 Rolitsky, C.D., Theil, K.S., McGaughy, V.R. et al. (1999). HER-2/neu amplification and overexpression in endometrial carcinoma. *International journal of Gynecological Pathology: Official Journal of the International Society of Gynecological Pathologists* 18 (2): 138–143.

50 Press, M.F., Pike, M.C., Hung, G. et al. (1994). Amplification and overexpression of HER-2/neu in carcinomas of the salivary gland: correlation with poor prognosis. *Cancer Research* 54 (21): 5675–5682.

51 FDA (1998). Herceptin® (trastuzumab). (updated November 2018). www.fda.gov (accessed 16 June 2020).

52 Hudis, C.A. (2007). Trastuzumab – mechanism of action and use in clinical practice. *The New England Journal of Medicine* 357 (1): 39–51.

53 Clynes, R.A., Towers, T.L., Presta, L.G., and Ravetch, J.V. (2000). Inhibitory Fc receptors modulate *in vivo* cytoxicity against tumor targets. *Nature Medicine* 6 (4): 443.

54 Kumar, R. and Yarmand-Bagheri, R. (2001). The role of HER2 in angiogenesis. *Seminars in Oncology* 28: 27–32.

55 Gajria, D. and Chandarlapaty, S. (2011). HER2-amplified breast cancer: mechanisms of trastuzumab resistance and novel targeted therapies. *Expert Review of Anticancer Therapy* 11 (2): 263–275.

56 Nahta, R. and Esteva, F.J. (2006). HER2 therapy: molecular mechanisms of trastuzumab resistance. *Breast Cancer Research: BCR* 8 (6): 215.

57 FDA (2012). Perjeta® (pertuzumab). (updated December 2018) www.fda.gov (accessed 16 June 2020).

58 Metzger-Filho, O., Winer, E.P., and Krop, I. (2013). Pertuzumab: optimizing HER2 blockade. *Clinical Cancer Research: An Official Journal of the American Association for Cancer Research* 19 (20): 5552–5556.

59 Dvorak, H.F., Brown, L.F., Detmar, M., and Dvorak, A.M. (1995). Vascular permeability factor/vascular endothelial growth factor, microvascular hyperpermeability, and angiogenesis. *The American Journal of Pathology* 146 (5): 1029–1039.

60 Ferrara, N., Gerber, H.P., and LeCouter, J. (2003). The biology of VEGF and its receptors. *Nature Medicine* 9 (6): 669–676.

61 Maes, C., Carmeliet, P., Moermans, K. et al. (2002). Impaired angiogenesis and endochondral bone formation in mice lacking the vascular endothelial growth factor isoforms VEGF164 and VEGF188. *Mechanisms of Development* 111 (1-2): 61–73.

62 Büchler, P., Reber, H.A., Büchler, M. et al. (2003). Hypoxia-inducible factor 1 regulates vascular endothelial growth factor expression in human pancreatic cancer. *Pancreas* 26 (1): 56–64.

63 Neufeld, G., Cohen, T., Gengrinovitch, S., and Poltorak, Z. (1999). Vascular endothelial growth factor (VEGF) and its receptors. *The FASEB Journal* 13 (1): 9–22.

64 Shibuya, M., Yamaguchi, S., Yamane, A. et al. (1990). Nucleotide sequence and expression of a novel human receptor-type tyrosine kinase gene (flt) closely related to the fms family. *Oncogene* 5 (4): 519–524.

65 Terman, B.I., Dougher-Vermazen, M., Carrion, M.E. et al. (1992). Identification of the KDR tyrosine kinase as a receptor for vascular endothelial cell growth factor. *Biochemical and Biophysical Research Communications* 187 (3): 1579–1586.

66 Karkkainen, M.J., Makinen, T., and Alitalo, K. (2002). Lymphatic endothelium: a new frontier of metastasis research. *Nature Cell Biology* 4 (1): E2–E5.

67 Herzog, B., Pellet-Many, C., Britton, G. et al. (2011). VEGF binding to NRP1 is essential for VEGF stimulation of endothelial cell migration, complex formation between NRP1 and VEGFR2, and signaling via FAK Tyr407 phosphorylation. *Molecular Biology of the Cell* 22 (15): 2766–2776.

68 Takahashi, T., Ueno, H., and Shibuya, M. (1999). VEGF activates protein kinase C-dependent, but Ras-independent Raf-MEK-MAP kinase pathway for DNA synthesis in primary endothelial cells. *Oncogene* 18 (13): 2221–2230.

69 Guo, D., Jia, Q., Song, H.Y. et al. (1995). Vascular endothelial cell growth factor promotes tyrosine phosphorylation of mediators of signal transduction that contain SH2 domains. Association with endothelial cell proliferation. *The Journal of Biological Chemistry* 270 (12): 6729–6733.

70 Eliceiri, B.P., Paul, R., Schwartzberg, P.L. et al. (1999). Selective requirement for Src kinases during VEGF-induced angiogenesis and vascular permeability. *Molecular Cell* 4 (6): 915–924.

71 Kendall, R.L. and Thomas, K.A. (1993). Inhibition of vascular endothelial cell growth factor activity by an endogenously encoded soluble receptor. *Proceedings of the National Academy of Sciences* 90 (22): 10705–10709.

72 Hiratsuka, S., Minowa, O., Kuno, J. et al. (1998). Flt-1 lacking the tyrosine kinase domain is sufficient for normal development and angiogenesis in mice. *Proceedings of the National Academy of Sciences of the United States of America* 95 (16): 9349–9354.

73 Wu, Y., Zhong, Z., Huber, J. et al. (2006). Anti-vascular endothelial growth factor receptor-1 antagonist antibody as a therapeutic agent for cancer. *Clinical Cancer Research* 12 (21): 6573–6584.

74 Carvalho, J.F., Blank, M., and Shoenfeld, Y. (2007). Vascular endothelial growth factor (VEGF) in autoimmune diseases. *Journal of Clinical Immunology* 27 (3): 246–256.

75 FDA (2004). Avastin® (bevacizumab) (updated June 2018). www.fda.gov (accessed 16 June 2020).

76 FDA (2014). Cyramza® (ramucirumab) (updated May 2019). www.fda.gov (accessed 16 June 2020).

77 FDA (2006). Lucentis® (ranibizumab) (updated March 2018). www.fda.gov (accessed 16 June 2020).

78 FDA (2011). Eylea® (aflibercept) (updated August 2019). www.fda.gov (accessed 16 June 2020).

79 FDA (2012). Zaltrap® (ziv-aflibercept). (updated June 2016). www.fda.gov (accessed 16 June 2020).

80 FDA (2019). Beovu® (brolucizumab-dbll) (updated October 2019). www.fda.gov (accessed 16 June 2020).

81 Reinmuth, N., Parikh, A.A., Ahmad, S.A. et al. (2003). Biology of angiogenesis in tumors of the gastrointestinal tract. *Microscopy Research and Technique* 60 (2): 199–207.

82 Viallard, C. and Larrivee, B. (2017). Tumor angiogenesis and vascular normalization: alternative therapeutic targets. *Angiogenesis* 20 (4): 409–426.

83 Barakat, M.R. and Kaiser, P.K. (2009). VEGF inhibitors for the treatment of neovascular age-related macular degeneration. *Expert Opinion on Investigational Drugs* 18 (5): 637–646.

84 Kliffen, M., Sharma, H.S., Mooy, C.M. et al. (1997). Increased expression of angiogenic growth factors in age-related maculopathy. *The British Journal of Ophthalmology* 81 (2): 154–162.

85 Wells, J.A., Murthy, R., Chibber, R. et al. (1996). Levels of vascular endothelial growth factor are elevated in the vitreous of patients with subretinal neovascularisation. *The British Journal of Ophthalmology* 80 (4): 363–366.

86 Dembic, Z. (2015). *The Cytokines of the Immune System: The Role of Cytokines in Disease Related to Immune Response*. Burlington: Elsevier Science.

87 Akdis, M., Aab, A., Altunbulakli, C. et al. (2016). Interleukins (from IL-1 to IL-38), interferons, transforming growth factor β, and TNF-α: receptors, functions, and roles in diseases. *Journal of Allergy and Clinical Immunology* 138 (4): 984–1010.

88 Coppin, H., Roth, M.-P., and Liblau, R.S. (2003). Cytokine and cytokine receptor genes in the susceptibility and resistance to organ-specific autoimmune diseases. *Cytokines and Chemokines in Autoimmune Disease*: 33–65.

89 Fong, Y., Tracey, K.J., Moldawer, L.L. et al. (1989). Antibodies to cachectin/tumor necrosis factor reduce interleukin 1 beta and interleukin 6 appearance during lethal bacteremia. *The Journal of Experimental Medicine* 170 (5): 1627–1633.

90 Pennica, D., Nedwin, G.E., Hayflick, J.S. et al. (1984). Human tumour necrosis factor: precursor structure, expression and homology to lymphotoxin. *Nature* 312 (5996): 724–729.

91 Dinarello, C.A. (1997). Interleukin-1. *Cytokine & Growth Factor Reviews* 8 (4): 253–265.

92 Dinarello, C.A., Ikejima, T., Warner, S. et al. (1987). Interleukin 1 induces interleukin 1. I. Induction of circulating interleukin 1 in rabbits in vivo and in human mononuclear cells in vitro. *The Journal of Immunology* 139 (6): 1902–1910.

93 Netea, M.G., Nold-Petry, C.A., Nold, M.F. et al. (2009). Differential requirement for the activation of the inflammasome for processing and release of IL-1β in monocytes and macrophages. *Blood* 113 (10): 2324–2335.

94 Andrei, C., Margiocco, P., Poggi, A. et al. (2004). Phospholipases C and A2 control lysosome-mediated IL-1β secretion: implications for inflammatory processes. *Proceedings of the National Academy of Sciences* 101 (26): 9745–9750.

95 Smith, D.E., Lipsky, B.P., Russell, C. et al. (2009). A central nervous system-restricted isoform of the interleukin-1 receptor accessory protein modulates neuronal responses to interleukin-1. *Immunity* 30 (6): 817–831.

96 Colotta, F., Re, F., Muzio, M. et al. (1993). Interleukin-1 type II receptor: a decoy target for IL-1 that is regulated by IL-4. *Science* 261 (5120): 472–475.

97 Weber, A., Wasiliew, P., and Kracht, M. (2010). Interleukin-1 (IL-1) pathway. *Science Signaling* 3 (105): cm1.

98 Scheller, J., Ohnesorge, N., and Rose-John, S. (2006). Interleukin-6 trans-signalling in chronic inflammation and cancer. *Scandinavian Journal of Immunology* 63 (5): 321–329.

99 O'Reilly, S., Ciechomska, M., Cant, R., and van Laar, J.M. (2014). Interleukin-6 (IL-6) trans signaling drives a STAT3-dependent pathway that leads to hyperactive transforming growth factor-β (TGF-β) signaling promoting SMAD3 activation and fibrosis via gremlin protein. *Journal of Biological Chemistry* 289 (14): 9952–9960.

100 Gu, C., Wu, L., and Li, X. (2013). IL-17 family: cytokines, receptors and signaling. *Cytokine* 64 (2): 477–485.

101 Ely, L.K., Fischer, S., and Garcia, K.C. (2009). Structural basis of receptor sharing by interleukin 17 cytokines. *Nature Immunology* 10 (12): 1245–1251.

102 Kao, C.-Y., Kim, C., Huang, F., and Wu, R. (2008). Requirements for two proximal NF-κB binding sites and IκB-ζ in IL-17A-induced human β-defensin 2 expression by conducting airway epithelium. *Journal of Biological Chemistry* 283 (22): 15309–15318.

103 Wu, C.-Y., Gadina, M., Wang, K. et al. (2000). Cytokine regulation of IL-12 receptor β2 expression: differential effects on human T and NK cells. *European Journal of Immunology* 30 (5): 1364–1374.

104 Trinchieri, G. (2003). Interleukin-12 and the regulation of innate resistance and adaptive immunity. *Nature Reviews Immunology* 3 (2): 133–146.

105 Trinchieri, G. (1998). Interleukin-12: a cytokine at the interface of inflammation and immunity. In: *Advances in Immunology*, vol. 70 (ed. F.J. Dixon), 83–243. Academic Press.

106 Gerosa, F., Baldani-Guerra, B., Lyakh, L.A. et al. (2008). Differential regulation of interleukin 12 and interleukin 23 production in human dendritic cells. *The Journal of Experimental Medicine* 205 (6): 1447–1461.

107 Watford, W.T., Hissong, B.D., Bream, J.H. et al. (2004). Signaling by IL-12 and IL-23 and the immunoregulatory roles of STAT4. *Immunological Reviews* 202 (1): 139–156.

108 Mullen, A.C., High, F.A., Hutchins, A.S. et al. (2001). Role of T-bet in commitment of TH1 cells before IL-12-dependent selection. *Science* 292 (5523): 1907.

109 Kineret F. physician packet insert: Kineret (anakinra) prescribing information (2001). Rockville, MD: US Food and Drug Administration.

110 FDA (2019). Stelara (ustekinumab). Janssen-Ortho Inc., Toronto, Ontario. www.accessdata.fda.gov (accessed 16 June 2020).

111 FDA (2019). Actemra. Genentech. www.accessdata.fda.gov (accessed 16 June 2020).

112 FDA (2018). KEVZARA® (sarilumab). Sanofi Biotechnology ©2018 Regeneron Pharmaceuticals, Inc.www.accessdata.fda.gov (accessed 16 June 2020).

113 FDA (2018). COSENTYX® (secukinumab). www.accessdata.fda.gov. Novartis Pharmaceuticals Corporation.

114 Bradley, J.R. (2008). TNF-mediated inflammatory disease. *The Journal of Pathology* 214 (2): 149–160.

115 Monaco, C., Nanchahal, J., Taylor, P., and Feldmann, M. (2015). Anti-TNF therapy: past, present and future. *International Immunology* 27 (1): 55–62.

116 Kriegler, M., Perez, C., DeFay, K. et al. (1988). A novel form of TNF/cachectin is a cell surface cytotoxic transmembrane protein: ramifications for the complex physiology of TNF. *Cell* 53 (1): 45–53.

117 Luettig, B., Decker, T., and Lohmann-Matthes, M.L. (1989). Evidence for the existence of two forms of membrane tumor necrosis factor: an integral protein and a molecule attached to its receptor. *The Journal of Immunology.* 143 (12): 4034.

118 Al-Lamki, R.S., Wang, J., Skepper, J.N. et al. (2001). Expression of tumor necrosis factor receptors in normal kidney and rejecting renal transplants. *Laboratory Investigation* 81 (11): 1503–1515.

119 Tang, P., Hung, M.-C., and Klostergaard, J. (1996). Human pro-tumor necrosis factor is a homotrimer. *Biochemistry* 35 (25): 8216–8225.

120 Horiuchi, T., Mitoma, H., Harashima, S.-I. et al. (2010). Transmembrane TNF-alpha: structure, function and interaction with anti-TNF agents. *Rheumatology (Oxford, England)* 49 (7): 1215–1228.

121 Grell, M., Douni, E., Wajant, H. et al. (1995). The transmembrane form of tumor necrosis factor is the prime activating ligand of the 80 kDa tumor necrosis factor receptor. *Cell* 83 (5): 793–802.

122 Pedersen, J., Coskun, M., Soendergaard, C. et al. (2014). Inflammatory pathways of importance for management of inflammatory bowel disease. *World Journal of Gastroenterology* 20 (1): 64–77.

123 Liu, T., Zhang, L., Joo, D., and Sun, S.-C. (2017). NF-κB signaling in inflammation. *Signal Transduction and Targeted Therapy* 2: 17023.

124 Nielsen, O.H. and Ainsworth, M.A. (2013). Tumor necrosis factor inhibitors for inflammatory bowel disease. *New England Journal of Medicine* 369 (8): 754–762.

125 Present, D.H., Rutgeerts, P., Targan, S. et al. (1999). Infliximab for the treatment of fistulas in patients with Crohn's disease. *New England Journal of Medicine* 340 (18): 1398–1405.

126 Rau, R. (2002). Adalimumab (a fully human anti-tumour necrosis factor α monoclonal antibody) in the treatment of active rheumatoid arthritis: the initial results of five trials. *Annals of the Rheumatic Diseases* 61 (suppl 2): ii70.

127 Pasut, G. (2014). Pegylation of biological molecules and potential benefits: pharmacological properties of certolizumab pegol. *BioDrugs* 28 (1): 15–23.

128 Raisz, L.G. (2005). Pathogenesis of osteoporosis: concepts, conflicts, and prospects. *Journal of Clinical Investigation* 115 (12): 3318–3325.

129 Nakashima, T., Hayashi, M., Fukunaga, T. et al. (2011). Evidence for osteocyte regulation of bone homeostasis through RANKL expression. *Nature Medicine* 17 (10): 1231.

130 Boyle, W.J., Simonet, W.S., and Lacey, D.L. (2003). Osteoclast differentiation and activation. *Nature* 423 (6937): 337.

131 Li, J., Sarosi, I., Yan, X.-Q. et al. (2000). RANK is the intrinsic hematopoietic cell surface receptor that controls osteoclastogenesis and regulation of bone mass and calcium metabolism. *Proceedings of the National Academy of Sciences of the United States of America* 97 (4): 1566–1571.

132 Kong, Y.Y., Yoshida, H., Sarosi, I. et al. (1999). OPGL is a key regulator of osteoclastogenesis, lymphocyte development and lymph-node organogenesis. *Nature* 397 (6717): 315–323.

133 Mizuno, A., Amizuka, N., Irie, K. et al. (1998). Severe osteoporosis in mice lacking osteoclastogenesis inhibitory factor/osteoprotegerin. *Biochemical and Biophysical Research Communications* 247 (3): 610–615.

134 Hofbauer, L.C., Khosla, S., Dunstan, C.R. et al. (1999). Estrogen stimulates gene expression and protein production of osteoprotegerin in human osteoblastic cells. *Endocrinology* 140 (9): 4367–4370.

135 Streicher, C., Heyny, A., Andrukhova, O. et al. (2017). Estrogen regulates bone turnover by targeting RANKL expression in bone lining cells. *Scientific Reports* 7 (1): 6460.

136 Cheung, A.M., Tile, L., Cardew, S. et al. (2012). Bone density and structure in healthy postmenopausal women treated with exemestane for the primary prevention of breast cancer: a nested substudy of the MAP.3 randomised controlled trial. *The Lancet Oncology* 13 (3): 275–284.

137 Aapro, M.S. (2004). Long-term implications of bone loss in breast cancer. *Breast (Edinburgh, Scotland)* 13 (Suppl 1): S29–S37.

138 Diamond, T.H., Higano, C.S., Smith, M.R. et al. (2004). Osteoporosis in men with prostate carcinoma receiving androgen-deprivation therapy: recommendations for diagnosis and therapies. *Cancer* 100 (5): 892–899.

139 Mundy, G.R. (2002). Metastasis to bone: causes, consequences and therapeutic opportunities. *Nature Reviews. Cancer* 2 (8): 584.

140 Kozlow, W. and Guise, T.A. (2005). Breast cancer metastasis to bone: mechanisms of osteolysis and implications for therapy. *Journal of Mammary Gland Biology and Neoplasia* 10 (2): 169–180.

141 FDA (2010). Prolia® (denosumab) (updated July 2019). www.fda.gov

142 Leren, T.P. (2014). Sorting an LDL receptor with bound PCSK9 to intracellular degradation. *Atherosclerosis* 237 (1): 76–81.

143 Horton, J.D., Cohen, J.C., and Hobbs, H.H. (2007). Molecular biology of PCSK9: its role in LDL metabolism. *Trends in Biochemical Sciences* 32 (2): 71–77.

144 Abifadel, M., Varret, M., Rabès, J.-P. et al. (2003). Mutations in PCSK9 cause autosomal dominant hypercholesterolemia. *Nature Genetics* 34 (2): 154.

145 Cohen, J.C., Boerwinkle, E., Mosley, T.H. Jr., and Hobbs, H.H. (2006). Sequence variations in PCSK9, low LDL, and protection against coronary heart disease. *New England Journal of Medicine* 354 (12): 1264–1272.

146 Jeong, H.J., Lee, H.S., Kim, K.S. et al. (2008). Sterol-dependent regulation of proprotein convertase subtilisin/kexin type 9 expression by sterol-regulatory element binding protein-2. *Journal of Lipid Research* 49 (2): 399–409.

147 Horton, J.D., Goldstein, J.L., and Brown, M.S. (2002). SREBPs: activators of the complete program of cholesterol and fatty acid synthesis in the liver. *The Journal of Clinical Investigation* 109 (9): 1125–1131.

148 Bergeron, N., Phan, B.A.P., Ding, Y. et al. (2015). Proprotein convertase subtilisin/kexin type 9 inhibition: a new therapeutic mechanism for reducing cardiovascular disease risk. *Circulation* 132 (17): 1648–1666.

149 FDA (2015). Repatha® (evolocumab) (updated Feburary 2019). www.fda.gov

150 Urban, D., Pöss, J., Böhm, M., and Laufs, U. (2013). Targeting the proprotein convertase subtilisin/kexin type 9 for the treatment of dyslipidemia and atherosclerosis. *Journal of the American College of Cardiology* 62 (16): 1401–1408.

151 Momtazi-Borojeni, A.A., Sabouri-Rad, S., Gotto, A.M. et al. (2019). PCSK9 and inflammation: a review of experimental and clinical evidence. *European Heart Journal Cardiovascular Pharmacotherapy* 5 (4): 237–245.

6

Pivotal Biology, Chemistry, Biochemistry, and Biophysical Concepts of Biologics and Biosimilars

Veysel Kayser and Mouhamad Reslan

Sydney Pharmacy School, Faculty of Medicine and Health, The University of Sydney, Sydney, New South Wales, Australia

---

KEY POINTS

- Biologics constitute a wide range of products including peptides, hormones, antibodies and other recombinant proteins, vaccines, cell, and gene therapies.
- Due to their complex nature and complex formulation, characterization of their structural features and formulation attributes is of utmost importance. This is only possible by employing a variety of different biophysical/analytical tools including chromatography, spectroscopy, separation, calorimetry, and microscopy methods.
- Monoclonal antibody (mAb) products have become top-selling medicines in the last couple of decades. As a result of their remarkable versatility, new modalities of mAb therapies such as antibody–drug conjugates (ADCs) are gaining more prominence.
- Successful production of biologics is highly dependent on their post-translational modifications (PTMs), structural and formulation stabilities, and degradation especially by protein aggregation under long-term storage conditions.

Abbreviations

Abbreviation	Full name
3D	Three-Dimensional
ADC	Antibody–Drug Conjugate
ADCC	Antibody-Dependent Cellular Cytotoxicity
ADCP	Antibody-Dependent Cellular Phagocytosis
AFM	Atomic Force Microscope
ANS	8-Anilinonaphthalene-1-Sulfonic Acid
AUC	Analytical Ultracentrifugation
CD	Circular Dichroism
CE-SDS	Capillary Electrophoresis-Sodium Dodecyl Sulfate
CH1	Constant Heavy Chain Domain 1
CH2	Constant Heavy Chain Domain 2
CH3	Constant Heavy Chain Domain 3
CHO	Chinese Hamster Ovary (cells)
ΔCp	Heat Capacity Change
DIM	Dynamic Imaging Microscopy

Abbreviation	Full name
DLS	Dynamic Light Scattering
DNA	Deoxyribonucleic Acid
DSC	Differential Scanning Calorimetry
E. Coli	*Escherichia Coli*
EMA	European Medicines Agency
Fab	Fragment Antigen Binding Region
Fc	Fragment Crystallizable Region
FcγRIIIA	Human Leukocyte Receptor IIIA
FDA	United States Food and Drug Administration
FTIR	Fourier-Transform Infrared Spectroscopy
HA or H	Hemagglutinin
HEK	Human Embryonic Kidney Cells
HPLC	High Performance Liquid Chromatography
IEX	Ion-Exchange Chromatography
IgG	Immunoglobulin G
IgG1	Immunoglobulin G Subclass 1

Biologics, Biosimilars, and Biobetters: An Introduction for Pharmacists, Physicians, and Other Health Practitioners,
First Edition. Edited by Iqbal Ramzan.
© 2021 John Wiley & Sons, Inc. Published 2021 by John Wiley & Sons, Inc.

Abbreviation	Full name
IgG4	Immunoglobulin G Subclass 4
IM	Intramuscular
IP	Intellectual Property
IR	Infrared
ITC	Isothermal Titration Calorimetry
IV	Intravenous
kDa	kiloDalton
LAIV	Live Attenuated Influenza Vaccine
LC-MS	Liquid Chromatography–Mass Spectrometry
LMWH	Low Molecular Weight Heparin
LO	Light Obscuration
mAb	monoclonal Antibody
MDCK	Madin-Darby Canine Kidney Cells
MS	Mass Spectrometry
NA or N	Neuraminidase
NMR	Nuclear Magnetic Resonance
PEG	Polyethylene Glycol
PFP	Pre-Filled Pen
PFS	Pre-Filled Syringe

Abbreviation	Full name
PTM	Post-Translational Modification
RNA	Ribonucleic Acid
RP	Reverse Phase
SAP	Spatial-Aggregation-Propensity
SC	Subcutaneous
SDS-PAGE	Sodium Dodecyl Sulfate-Polyacrylamide Gel Electrophoresis
SEC	Size-Exclusion Chromatography
STED	Stimulated Emission Depletion
SV	Sedimentation Velocity
TEM	Transmission Electron Microscope
TFF	Tangential Flow Filtration
TIRF	Total Internal Reflection Fluorescence
Tm	Melting Temperature
US	United States of America
USD	United States Dollar
USP	The United States Pharmacopeia
UV-Vis	Ultraviolet-Visible Spectrophotometry
VEGF	Vascular Endothelial Growth Factor
WHO	World Health Organization

6.1 Definitions: Biologics vs. Small Molecule Drugs

Biologics or biopharmaceuticals[1] can be broadly described as medicinal products manufactured or derived from a variety of living organisms to prevent, treat, or diagnose diseases and their preparations being made up of a type of protein, mixture of proteins, or other biological substances such as DNA, RNA, viruses, and their components. These products include antibodies, vaccines, therapeutic serum, toxins, antitoxins, blood components or derivatives, allergenic products, and polypeptides (excluding chemically synthesized entities). Cell- and gene-therapy may also be referred to as "biological"; however, they are beyond the scope of this chapter, due to their unique applications and are discussed in Chapter 13. Biologics can vary in size and complexity; short peptides may be as small as a few kDa, while complex proteins can reach sizes greater than

100 kDa, a thousand-fold larger than common small molecule drugs.[1,2] For example, monoclonal antibodies (mAbs) are ~150 kDa, and are composed of more than 1000 amino acids, while ibuprofen is only 206 Da, approximately the same size of one tryptophan residue (Figure 6.1). Due to the complexity of biologics, chemical synthesis is not a feasible manufacturing option, as it is for small molecule drugs. Instead, biologics are either produced in bioreactors of living cells (human or animal) or microorganisms such as bacteria or yeast and purified for use as medicinal products or harvested from blood serum/plasma. This major difference between small molecule drugs and biologics introduces several unique challenges[3,4] for the production, characterization, and cost of biopharmaceuticals and their use as medicinal products (Table 6.1).

Biologics offer several advantages compared to small molecule drugs. Firstly, biologics have similar or shorter development cycles[6] – they can be developed with one or more disease target(s) or perhaps no specific disease target, e.g. many types of solid cancers overexpress vascular endothelial growth factor (VEGF) and a drug that targets VEGF can be used for the treatment of a variety of cancers. Biologics can be developed and refined with maximal specificity and binding affinity to a target using recombinant technologies such as phage display

---

1 Different terms are used in the literature interchangeably: biologics, biotherapeutics, biopharmaceutics, biopharmaceuticals, biologicals, biologic drugs, biomedicines, or biological medicines. Although there might be minor differences in their meanings, different jurisdictions and countries have adopted one term over another.

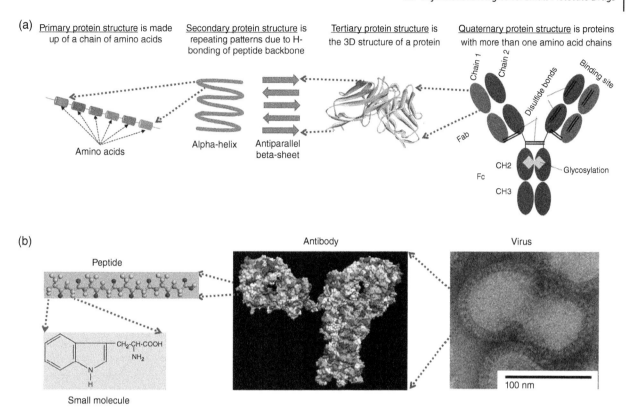

Figure 6.1 (a) Primary, secondary, tertiary, and quaternary protein structures. The quaternary structure depicts different domains of an IgG antibody and shows some of the disulfide bonds and its glycosylation site. It has two light chains (Chain 1) and two heavy chains (Chain 2). (b) Size comparison of small molecule, peptide, mAb, and influenza virus. (*See insert for color representation of the figure.*)

Table 6.1 Comparison of common characteristics of biologics and small molecule drugs.

Characteristics	Biologics	Small molecule drugs
Molecular weight	2000 to more than 150 000 Da	<500 Da
Physical/chemical/biological properties	Complex	Less complex
Production method	Using living organisms/cells	Chemical synthesis
Critical process steps in manufacture	Many	Significantly less
Characterization	Difficult, often not well characterized	Relatively simple, well characterized
Physical/chemical stability	Poor – requires stabilizing excipients and low temperature storage	Relatively stable at room temperature
Administration	Limited – usually parenteral	Flexible – usually oral
Half-life	Days to weeks – dosing daily to monthly	Relatively short – dosing every few hours is common
Specificity/affinity for target	Very high	Relatively low – off-target effects more common
Safety profile	Less common side effects, but can be immunogenic	More common side effects, but less likely to be immunogenic
Cost	High production and treatment costs	Relatively lower production and treatment costs
Heterogeneity	High	Low or none

Source: Adapted from Baldo.[5]

and protein engineering, as opposed to the more limited manipulation of small molecules as potential agonists or antagonists as targeted therapies. For example, several blood disorders caused by protein deficiencies are easily treated by using recombinant technology to produce replacement proteins. The rapid, exponential growth of biologics in the last 50 years is a testament to their revolutionary role in disease prevention, treatment, and diagnosis. Therapeutic antibodies and antibody-based products are particularly successful, dominating the top 10 list of best-selling medicines worldwide throughout the last decade.[7] This trend is likely to continue in the future as there are thousands of antibody products in clinical trials and many more in different stages of research and development of pharmaceutical industry pipelines.

In recent years, "biosimilars" have emerged as well. A biosimilar is a follow-on biologic product produced by a substantially similar process, is highly comparable to the originator product, and used for same indications. The most critical aspect of a biosimilar product is its similarity to the reference product – with no clinically meaningful differences in terms of purity, safety, and potency (safety and effectiveness). This involves extensive characterization and demonstration of similarity of its structure, function, chemical identity, purity, and bioactivity compared to the reference product. If there are minor differences between the biosimilar and originator product in clinically inactive components, they might be acceptable. Currently, there are 23 biosimilar products approved by the US Food and Drug Administration (FDA) and more than 60 by the European Medicines Agency (EMA) as of October 2019[8]; however, less than 10 of them are in the US market due mainly to various patent protections that may cause intellectual property (IP) disputes between manufacturers or due to settlements between originator and biosimilar product manufacturers. For example, AbbVie (the manufacturer of the originator adalimumab) made deals with every biosimilar manufacturer of adalimumab, delaying all biosimilars of this mAb to be marketed in the United States until 2023.[9] This is different to pay-for-delay deals between originator and generic manufacturers that we have seen some years ago, in particular for small molecule drugs; however, such deals still cause delays for biosimilar products to be marketed, even after they have received FDA approvals. These deals are due to various patent litigation issues.[10] For example, AbbVie has numerous add-on patents for adalimumab and each and every one of the biosimilar manufacturers has signed licensure deals with AbbVie,

which in turn prevents their own marketing soon after approvals. Nevertheless, the number of biosimilars will continue to increase in the near future as many biosimilar products are in clinical trials and patent protections of some top-selling innovator biologics are ending soon, enabling the development of their biosimilars. It is highly likely that these biosimilar products will continue to be "skinny labeled" – smaller indication lists, because of exclusivities and IPs that originator manufacturers have secured previously for their reference product.

Another term, "biobetter," is often used for a follow-on biologic that implies some improvement over an existing biologic. Biobetters, sometimes referred as "next-generation biologics," usually have the same target of action as an existing biologic, but display improved attributes in manufacturing, disposition, efficacy, and safety. Although it has no official description yet, products that display better shelf-life due to improved formulation characteristics (e.g. preparations of subcutaneous (SC) formulations instead of intravenous formulations), longer serum half-lives compared to existing biologics, or novel design formats such as antibody–drug conjugates (ADCs), antibody–nanoparticle complexes, hyperglycosylated mAbs, or bispecifics can all be categorized as biobetters. More information on some of the biobetters such as ADCs and bispecifics and how to achieve such enhanced properties via, for example, hyperglycosylation are summarized in Sections 13.2–13.4.

Another important aspect of biologics and one of the critical differences between biologics and small molecules is their differentiations in clinical trials and pathways from Phase I to approval. Traditionally, safety of a product, along with other important parameters such as dose variations, is first determined in the Phase I trial with a small set of healthy volunteers, and efficacy and side effects are determined in Phase II and III. According to regulatory agency guidelines, biosimilars still have to go through clinical trials but their clinical tests may be significantly shortened, i.e. abbreviated new drug applications. Biosimilars of low molecular weight heparins (LMWHs) do not even need to be tested clinically as both the FDA and EMA indicate that physicochemical characterization would be sufficient for their approval if these tests show high similarity for the biosimilar and the originator product in addition to their similar production methods.[11] Interestingly, some new products such as immune checkpoint inhibitors (see Sections 4.3.4 and 13.6 for more information) are tested directly in patients rather than healthy volunteers in Phase I, due to their invasive effects, especially for various late-stage oncology patients. In other words, the efficacy of a biologic is tested along its safety aspect in Phase I and hence approval of the product may be obtained from the regulatory agency even after extended Phase I or Phase II trials. These new developments suggest that the approval pathway, aims of clinical tests in each phase, or trial designs for some biologics might be varied in the future and reliance of a biologic drug's approval on obtaining a positive Phase III outcome may not be necessary, as an encouraging Phase I/II outcome might be sufficient. Some vaccine candidates such as Zika, Ebola or COVID-19 vaccines could also benefit from such recent regulatory developments and decisions.

There are also fundamental differences between biologics and small molecules in terms of product interchangeability, switching, and substitution, which are discussed separately in other chapters of this book.

Another difference between biologics and small molecules is the nomenclature used, which can be quite complex and unique for some biologics, in particular mAbs (Table 6.2): after the initial prefix, which carries no special meaning, a two-letter syllable shows the indication that the product will be used for (i.e. if the mAb

Table 6.2 Nomenclature/naming of therapeutic monoclonal antibodies.

Syllables	What it means	Example
-mab	Monoclonal antibody	Cetuximab (Ce-tu-xi-mab): tumor-directed chimeric monoclonal antibody
-u-mab	Fully human monoclonal antibody	Adalimumab (ada-lim-u-mab): immune system-directed fully human monoclonal antibody
-zu-mab	Humanized monoclonal antibody	Trastuzumab (tras-tu-zu-mab): tumor-directed humanized monoclonal antibody
-xi-mab	Chimeric monoclonal antibody	Infliximab (inf-li-xi-mab): immune system-directed chimeric monoclonal antibody
-o-mab	Mouse monoclonal antibody	Racotumomab (Raco-tum-o-mab): tumor-directed mouse monoclonal antibody
-tu(m)-xx-mab	Tumor-directed monoclonal antibody	Alemtuzumab (alem-tu-zu-mab): tumor-directed humanized monoclonal antibody
-ci-xx-mab	Circulatory-directed monoclonal antibody	Idarucizumab (idaru-ci-zu-mab): circulatory-directed humanized antibody

Source: Adapted from Refs. 12–14.

is for a tumor in general (e.g. *-tu(m)-*), what type of tumor (e.g. *-co(l)-* for colon); *-ci(r)-* if it is for circulatory). The subsequent syllable is related to the source of the mAb (i.e. *-u-* for human, *-zu-* for humanized, or *-xi-* for chimeric). The last three syllables for mAbs are always *mab*. This is completely different to the nomenclature used for small molecule medicines.

Influenza viruses also have a designated naming convention since 1979.[15] Influenza viruses are identified based on the type of their surface glycoproteins, hemagglutinins (HA or H), and neuraminidases (NA or N). There are four different types of influenza, A, B, C, and D. Influenza A, B, and C cause respiratory illness in humans but only type A and B cause major issues while C only induces mild reactions. Type D does not cause any illness in humans but mainly causes illness in cattle. Therefore, current influenza vaccines are prepared against only type A and B. Influenza A is further divided into subtypes based on its HA and NA content; there are 18 different HA subtypes and 11 different NA subtypes.[15] Naming of influenza viruses involves defining which subtype HA and NA are present on the virus. For example, if subtype HA 1 and subtype NA 1 are present, the virus is called H1N1. Many different subtypes of influenza A can exist when different subtypes of HA and NA are mixed. Luckily however, not all subtypes infect humans and mainly only H1N1 and H3N2 are observed to infect humans, although there are exceptions such as H5N1 and H9N5 have also been detected to cause illnesses in some cases. It is also worth noting that to further complicate things, there are more than one subtype of a similar strain that can be observed, e.g. more than one type of H1N1 exist and the pandemic H1N1 that started in Mexico was a different strain to the H1N1 that was circulating around the globe previously. Consequently, one can estimate that there are hundreds, or even thousands, of different influenza A subtypes existing at any given time and birds tend to be the main host for almost all different kinds of influenza viruses. Each year, WHO determines which three or four influenza virus strains should be included in the seasonal influenza vaccines based on the previous year's circulation.[16]

In contrast, influenza B is not as diverse as influenza A and hence it is divided into different lineages and strains such as influenza B/Texas or influenza B/Brisbane.

Naming convention for influenza viruses comprises of antigenic type (i.e. influenza A or B), if it is not human origin, then host of origin, geographical location where the virus was isolated, strain number, year it was isolated; if it is an influenza A, then type of HA and NA in parenthesis; all these information is put together but separated by a forward slash such as: A/California/7/2009 (H1N1) or B/Brisbane/60/2008.

6.2 Biochemical and Biophysical Properties

6.2.1 Protein Structure

Proteins must adopt a 3D shape in order to exert their functions. Although there are chaperones to help some proteins to fold into a correct form, many proteins fold into their unique shape on their own. Currently, it is impossible to predict a protein's 3D shape from its amino acid sequence. This amino acid sequence (primary protein structure) determines the shape and function of the protein (Figure 6.1).

Many proteins have some repeating patterns such as alpha-helices or beta-sheets that are held by an array of hydrogen bonds, and these patterns provide basic structural features of a protein (secondary protein structure) (Figure 6.1). Secondary structures are held together by additional amino acid sequences forming a unique 3D shape from a single amino acid chain (tertiary protein structure). If more than one amino acid chain is part of a protein, it gives rise to a quaternary protein structure (Figure 6.1).

An IgG antibody is represented as an example to demonstrate primary, secondary, tertiary, and quaternary protein structure in Figure 6.1. IgGs are multidomain glycoproteins having two light and two heavy chains. Disulfide bonds connect light and heavy chains, and different domains of each chain together. The CH2 domain of the heavy chain is typically glycosylated. Each protein has a different glycosylation pattern which along with other post-translational modifications (PTMs), introduces heterogeneity in protein formulations.

6.2.2 Post-Translational Modifications

Heterogeneity in biologic products can arise due to PTMs during production. The types of PTMs are dependent on the host system expressing the protein (i.e. bacterial, plant, fungal, and mammalian) and the growth conditions of the cells.[3,17] In fact, even if the same cells are used to express a protein, variability within the batch and batch-to-batch may still occur, although variability is minimized by controlling the nutrients fed to the cells and the cell growth conditions such as temperature and CO_2 levels. PTMs of relevance to therapeutic proteins include the addition of an N- or

O-linked glycan (sugar group – type can vary), carboxylation, hydroxylation, proteolysis, amidation, sulfation, and disulfide linking.[18,19] Glycosylation is of major importance to therapeutic proteins as the presence and type of glycan structure can have significant effects on the biochemical and physical properties of the protein, including its stability and function.[20,21] In any given batch of a therapeutic protein product, several different glycosylation patterns may exist, with a high proportion of one or two types of glycan structures leading to heterogeneity within the same batch.[19,20,22]

6.2.2.1 Glycosylation

Glycosylation is one of the most common and important PTMs of biopharmaceuticals due its significant impact on the stability, immunogenicity, and function of proteins such as therapeutic antibodies.[4,21,23] Glycosylation involves the enzymatic addition of a carbohydrate group or glycan to a specific residue within a unique sequence. For example, N-linked glycosylation occurs when a glycan attaches to the nitrogen atom of an asparagine (Asn) residue. The Asn residue must be positioned in a particular sequence such as Asn-X-Ser or Asn-X-Thr to be recognized as an attachment site (X is any amino acid residue except proline).[18] Additionally, the Asn residue must be conformationally accessible (i.e. solvent-exposed) for enzymatic addition of the glycan. O-linked glycosylation, which is less common in therapeutic antibodies, occurs when a glycan attaches to the oxygen atom of serine or threonine.[22] Glycosylation is a common source of product heterogeneity due to the many types of glycan groups and combinations of sugars that may be attached to each protein. For example, therapeutic IgG1 antibodies, which are commonly produced via mammalian host systems, contain a glycosylation site in the heavy chain (Asn 297), which may be attached to different combinations of N-acetylglucosamine, N-acetylneuraminic (sialic acid), mannose, fucose, or galactose groups (Figure 6.2). The resulting glycosylation patterns affect the conformational stability, half-life, and biological activity of the antibody and result in product heterogeneity. For instance, if the attached glycan lacks a fucose core, antibody affinity to the FcγRIIIA – a protein receptor expressed on the surface of immune cells such as natural killer cells and macrophages – is improved, resulting in higher antibody-dependent cell-mediated cytotoxicity (ADCC).[4] Antibody such as obinituzumab is an example of biopharmaceutical engineered to have defucosylated glycans (Figure 6.2) to enhance ADCC activity.[4] Additionally, sialic acid content of glycans has been shown to affect the rate of clearance of proteins after

Figure 6.2 Types of N-glycans observed in therapeutic proteins. All N-glycans are composed of the core structure highlighted in red. (*See insert for color representation of the figure.*)

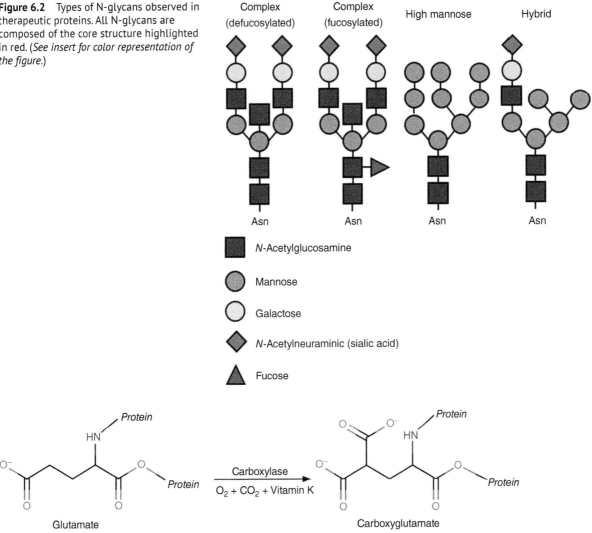

Complex (defucosylated)　Complex (fucosylated)　High mannose　Hybrid

Asn　Asn　Asn　Asn

■ *N*-Acetylglucosamine

● Mannose

○ Galactose

◆ *N*-Acetylneuraminic (sialic acid)

▲ Fucose

Glutamate

Carboxylase
$O_2 + CO_2 + Vitamin K$

Carboxyglutamate

Figure 6.3 Post-translational γ-carboxylation of glutamate residues in proteins. (*See insert for color representation of the figure.*)

injection, with less sialic acid resulting in faster clearance or shorter half-life.[20] Glycosylation of antibodies increases the conformational stability of the hinge region, an interdomain sequence of amino acids that connects the two heavy chains of some antibodies and protects it against proteolysis. Aglycosylated antibodies (i.e. no glycan) have been shown to have very poor conformational and colloidal stability leading to aggregation.[21] Colloidal stability refers to the resistance of protein molecules in solution to self-association, aggregation, and/or precipitation.

6.2.2.2　γ-Carboxylation and β-Hydroxylation

γ-Carboxylation refers to the conversion of glutamate residues to carboxyglutamate via specific carboxylase enzymes (Figure 6.3). β-hydroxylation refers to the conversion of aspartate residues to hydroxyaspartate via

specific hydroxylase enzymes. Carboxylation and hydroxylation are important for the functioning of several proteins, such as blood coagulation factors VII, IX, and X (including their recombinant counterparts, e.g. NovoSeven®). Carboxyglutamate and hydroxyaspartate residues act as calcium-binding sites, which are essential for protein activation. Binding of calcium ions induces conformational changes, which are required for the correct folding of the protein domain responsible for its functional activity.[18]

6.2.2.3　Amidation

Amidation refers to the modification of a peptide's C-terminal carboxyl group with an amide functional group. It is more common in bioactive peptides and contributes to increased stability against enzymatic proteolytic degradation, which leads to increased half-life, and

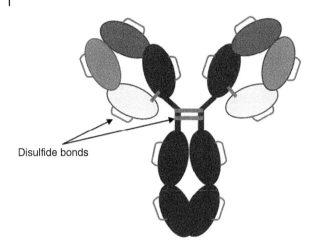

Figure 6.4 An IgG1 antibody with 4 interchain (green) and 12 intra-chain disulfide bonds (red). (*See insert for color representation of the figure.*)

enhanced activity, i.e. improved binding affinity to target G-protein coupled receptors.[18] This is an enzyme-catalyzed process and is particularly important in many hormones and growth factors.[24]

6.2.2.4 Sulfation

Sulfation involves the covalent addition of a sulfo group to solvent-exposed tyrosine residues in a protein by enzymes called tyrosylprotein sulfotransferases. Sulfation is not a common PTM in marketed biopharmaceuticals; however, it can play a role in increasing protein–protein interactions and consequently increasing the activity of some proteins. For example, ALPROLIX®, a recombinant Factor VIII fused to a crystallizable antibody fragment (Fc), has six sulfated tyrosines, which are essential to its activity.[25] Different expression systems affect the degree of PTMs such as sulfation; for example, expression in human embryonic kidney (HEK) cells has been shown to result in higher levels of sulfation and carboxylation than expression in Chinese hamster ovary (CHO) cells.[25]

6.2.2.5 Disulfide Linking

A disulfide linkage is a covalent bond between two thiol groups in two cysteine (Cys) residues. Disulfide linking is critical for the stabilization of the tertiary and quaternary structure of multi-domain proteins such as antibodies (Figure 6.4).[18] In a typical IgG1 or IgG4 monoclonal antibody, there are 4 interchain disulfide bonds in the hinge region and 12 intra-chain disulfide bonds between the domains (Figure 6.4). Because of the impact of disulfide bonds on the structural stability of proteins, disulfide-related modifications such as disulfide bond cleavage or scrambling must be monitored closely and avoided.

Cleavage of disulfide bonds can occur as a result of the presence of radicals, metal ions, or redox systems (e.g. iron, ascorbate and oxygen) in the cell culture media.[26] Incomplete processing of disulfide bonds can also occur in the endoplasmic reticulum of eukaryote[2] host systems resulting in "open" disulfide bonds.[26] Some prokaryote organisms such as *E. Coli* may not facilitate correct disulfide bond formation, especially if high levels of protein expression are required. Disulfide bond scrambling is another interesting phenomenon whereby interchain bonds in some antibodies become rearranged leading to antibody heterogeneity.[26]

6.3 Chemical and Physical Stability

Biologics are sensitive products that may undergo physical or chemical degradation to light, agitation, temperature, acidity/alkalinity, oxygen levels, presence of divalent metal ions, and various other stressors that are encountered during manufacturing, transport, and storage.[3,4,27,28] Degradation of a biological product is of significant concern as it can result in loss of potency and efficacy and increased likelihood of severe side effects, including immunogenic reactions. Furthermore, degradation of biologics contributes to product heterogeneity. Some of the most common types of degradation are summarized below:

Physical

1) Aggregation (clumping together)
2) Unfolding/denaturation

Chemical

1) Fragmentation
2) Oxidation
3) Deamidation/isomerization
4) Pyroglutamate formation

6.3.1 Aggregation

One of the primary pathways of protein degradation is protein aggregation, i.e. the clumping or association of protein molecules or monomers together to form larger structures or aggregates. Aggregation leads to loss of potency and is associated with increased immunogenicity.[29–32] Additionally, growth of aggregates to a certain size

---

2 Eukaryotes are organisms with cells containing a membrane-bound nucleus. Prokaryotes are single cell organisms with no membrane-bound organelles.

Figure 6.5 Native and non-native protein aggregation. (*See insert for color representation of the figure.*)

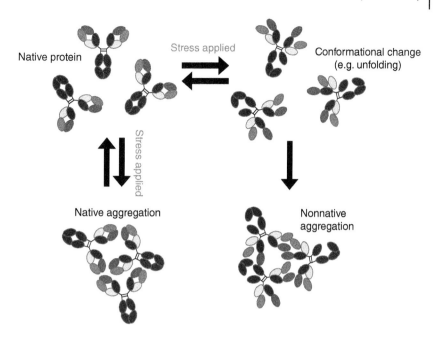

can lead to the precipitation of the protein and the appearance of visible particulates, which are not safe for injection.[29] Protein aggregation is driven by: (i) irreversible covalent interactions between protein monomers such as the formation of inter-disulfide bonds and (ii) non-covalent interactions such as electrostatic or hydrophobic interactions, which are occasionally reversible. Aggregation may be preceded by chemical or structural changes, such as unfolding or oxidation. This type of aggregation is termed non-native state aggregation[33] (Figure 6.5).

Protein aggregation is one of the most challenging degradation pathways to control. There are many stages in the manufacturing process that can induce protein aggregation including (i) production of the protein in the host organism, (ii) exposure to high salt concentrations during purification steps, (iii) exposure to low pH during viral inactivation, (iv) agitation during mixing of excipients, (v) shear stress and adsorption to membrane during tangential flow filtration[3] to increase protein concentration above desired target, (vi) filtration of the product to remove particulates, (vii) freeze-thaw steps between manufacturing processes, and (viii) transport and storage of the final product.[4,30–32,34]

Aggregate content must typically be controlled to less than 5% by the end of the product shelf-life period to be approved by regulatory authorities. There are two main approaches to control protein aggregation: (i) stabilizing excipients[35] such as salts, amino acids, sugars, (polyols) and surfactants; (ii) protein engineering, often guided by predictive computational tools such as molecular dynamics simulations or studies of the mechanism of protein aggregation.[36–39] These computational tools predict protein aggregation regions either based on amino acid sequence or spatial information from crystal structure. The latter methods generally provide better information since an aggregation-prone motif could be formed by amino acids that are far apart; however, placed closely to each other in the protein's 3D structure. Numerous tools are available including SAP and AGGRESCAN 3D.[38–41]

In this context, many viral vaccines require preservation of full virus structure in order to induce immunization. The rabies vaccine is a good example of this,[42] but many other viral vaccines are also formulated with full virus that is inactivated or attenuated, e.g. Polio vaccine. If structural integrity is not retained due to various formulation conditions, or due to external stress such as elevated heat or shear stress during manufacturing, some of these viruses are destroyed or "split" and hence the vaccine may lose its ability to produce an immunize response. If the virus is split, enhanced protein aggregation is observed because membrane proteins as well as other viral proteins that have large hydrophobic surfaces are now water-exposed and become aggregation-prone. Such protein aggregation is usually overlooked

---

3 Tangential flow filtration (TFF) or Cross-flow filtration is a filtration process whereby the solvent and any small molecules in a solution, such as salts, are filtered through small pores in the membrane, while molecules larger than the membrane pores such as proteins are retained and pass along the membrane surface. The retained molecules are returned to the starting container and can thus become more concentrated in the solution. TFF may also be used to exchange the solvent or solutes in solution.

in vaccines if aggregation is not extensive; however, an extensive formation of large aggregates at submicron levels may cause product quality and safety issues as was the case with influenza vaccines from bioCSL and Novartis in 2010 and 2012, respectively.[43,44]

6.3.2 Unfolding/Denaturation

Denaturation or unfolding is a significant structural or conformational change in a protein that is beyond "normal" or expected protein fluctuations.[45,46] Unfolding may be reversible or irreversible and is often induced by exposure to excessive heat or acidic/alkaline conditions. Some processing steps such as filtration and shear effects induced by pumps and other equipment may also lead to protein denaturation. Protein structural changes such as unfolding can trigger or accelerate the aggregation process by exposing "sticky" hydrophobic regions typically buried in the native protein structure.[4,46] Aggregation caused by unfolding is usually irreversible and is more likely to induce immunogenicity or other hypersensitivity reactions as the protein has undergone significant structural change and is more likely to be recognized by the body as foreign.[4] The melting temperature(s) of the protein or its separate domains is often used to understand the thermal or conformational stability of the protein to ensure unfolding does not occur during manufacturing processes such as freeze- or spray-drying, and to ensure it is protected from unfolding at storage and transport temperatures.

6.3.3 Fragmentation

Fragmentation is the cleavage of proteins into smaller units or peptides and most commonly affects multimeric (two or more subunits) and multidomain proteins such as therapeutic antibodies.[5,47] Enzymatic fragmentation is facilitated by the action of proteases, which may remain in the purified protein harvest after production.[48] Nonenzymatic fragmentation occurs via hydrolysis of peptide bonds between amino acids. For example, fragmentation of therapeutic antibodies may occur in the hinge region and between domains (e.g. CH2-CH3).[4,47] Peptide bonds between aspartic acid (Asp) and certain amino acids such as proline (Pro), glycine (Gly), tryptophan (Trp), or serine (Ser) are more susceptible to hydrolysis.[47] Fragmentation of biopharmaceuticals can result in loss of potency due to loss of structure and binding affinity to the target. Fragment content must typically be controlled to less than 5% by the end of the product shelf-life period to be approved by regulatory authorities.

6.3.4 Oxidation

Oxidation is caused by exposure of amino acids to oxygen, which may be present in the form of reactive oxygen species or free radicals. Although all amino acids may undergo oxidation, Cys, His, Met, Tyr, Phe, or Trp residues are most commonly affected, resulting in increased protein polarity, conformational changes, aggregation, or loss of potency if residues in binding regions are affected.[4,49] Reduction and oxidation of Cys residues in lyophilized insulin has been shown to promote covalent aggregation during storage.[50] Oxidation is more likely to occur in solvent-exposed regions and is dependent on solution conditions such as pH and buffer type. Exposure to hypoxic conditions (e.g. during cell culture) can promote oxidation through the formation of reactive oxygen species.[4] Exposure to metals such as Fe or Cu as impurities or during downstream processes can also catalyze oxidation of proteins.[49] For instance, Cu-based metal affinity chromatography has been shown to promote oxidation of lactate dehydrogenase.[51] Long-term storage of proteins and their exposure to visible and ultraviolet light also promotes oxidation. Oxidation of biopharmaceuticals is controlled by adding antioxidants such as methionine and ascorbic acid, or nitrogen filling to remove oxygen from product containers.

6.3.5 Deamidation

Deamidation occurs when the amide side chains of Asn or Gln transform into carboxylic acid groups resulting in their conversion to Asp and Glu, respectively. Deamidation is more common for Asn residues and is initiated by nucleophilic attack of the nitrogen of the neighboring amino acid on the C-terminal end.[4,52] This results in loss of ammonia and formation of a succinimide intermediate, which then undergoes hydrolysis to form Asp or its isomer, IsoAsp. The succinimide intermediate can also undergo racemization before hydrolysis, forming D-Asp or D-IsoAsp.[53] Deamidation increases the net negative charge of the protein resulting in charge heterogeneity. Deamidation of residues in specific regions such as binding regions, may also affect the potency and stability of the protein.[54]

6.3.6 Pyroglutamate Formation

Glu or Gln can undergo cyclization at the N-terminal end of a polypeptide or protein to form pyroglutamate. Transformation of a Gln residue results in a loss of an amine group making the protein more acidic or negatively charged. However, conversion of a Glu to a PyroGlu has no effect on the net charge.[54,55]

Pyroglutamate formation is common in therapeutic antibodies and can be catalyzed *in vivo*; however, it has not been shown to impact the potency or stability of any therapeutic product.[56] Nevertheless, each protein is unique and the impact of pyroglutamate formation should be investigated for each product.

6.4 Formulation Considerations and Devices

The presence of proteases in most human body cavities and tracts severely limits the flexibility of administration of biologic therapies. Nearly all biotherapies are currently administered parenterally to bypass or avoid protease digestion, harsh conditions in the stomach, and first-pass metabolism in the liver. Parenteral administration also "bypasses" other biologic drug delivery obstacles such as poor absorption of large proteins from the stomach or intestine. Traditionally, the most common parenteral route was intravenous (IV); however, due to issues with patient acceptance and subsequent compliance, most biopharmaceutical manufacturers are shifting toward small-volume subcutaneous (SC) injections, which can be self-administered. Developing a SC formulation is more challenging for several reasons[57]:

1) There is lower bioavailability via the SC route (compared to IV).
2) SC formulations are limited to small volumes such as 0.5–2 mL.
3) Due to (1) and (2), significantly higher concentrations of protein are required to administer the same dose as an IV infusion.
4) Protein aggregation is accelerated with increasing protein concentration.
5) SC syringes/devices may not support viscous solutions greater than 20–50 centipoise.
6) At concentrations typically required for SC administration (e.g. >100 mg/mL), solution viscosity is increased almost exponentially with increasing protein concentrations.[57]

The two most common types of devices for SC administration of biologics are prefilled syringes (PFSs) and prefilled pens (PFPs) or auto-injectors.[4] PFSs are the most basic "device" for SC delivery of biologics; they are composed of a glass barrel, which contains the drug product, a needle with a shield, finger flanges to support injection,

a stopper, and a plunger rod, which is manually pushed to inject the product.[58] PFPs or auto-injectors often have a PFS contained within the device but are activated automatically through a single click or press of a button. They also have other features including audible clicks when the product is injected or the injection is complete, hidden needles, and viewing windows to observe injection progress.[58,59] More sophisticated PFP devices have also been developed with advanced electronic features including on-screen or voice instructions, skin sensors and an emergency stop mechanism, injection logs, and a medicine information chip.[59] PFPs or auto-injectors are preferred by most patients as they make the injection process easier, especially for patients who may have poor manual dexterity due to arthritis or other medical complications. They may also reduce injection-site pain as there is a reduced chance for incorrect injection technique (e.g. injecting too quickly), increasing patient confidence and adherence.[59]

Noninvasive administration routes are most desirable as they alleviate issues such as injection-site injuries and inflammation and phobias to needles.[60] Pulmonary delivery is the most promising noninvasive administration route for biopharmaceuticals and will likely become more commonplace in the future, as it allows targeted treatment of lung diseases[61]; it can also deliver proteins systemically,[60] bypassing some of the challenges with oral delivery including the harsh pH of the stomach and hepatic first-pass metabolism. The lungs also offer a large surface area and thin alveolar epithelium for rapid drug absorption.[60] A limited number of biopharmaceuticals have been developed as nebulized solutions or inhalers (Pulmozyme® and Affreza®). Affreza® is one of the first marketed inhalable protein powders that delivers insulin through the lungs and into the systemic circulation for patients with type I and II diabetes. Overcoming absorption issues with larger proteins such as mAbs and other barriers such as macrophage clearance and protease activity is currently a major hurdle for systemic delivery of biologics through the lungs.[60,62]

6.5 Analytical Methods/Tools

Chemical, biochemical, biophysical, PTMs, and other properties of biologics need to be characterized robustly in order to ensure their structural and formulation stabilities correspond to the specified guidelines so they are efficacious and safe for human usage.

Each biologic needs to be characterized starting from conformational properties (i.e. primary, secondary, tertiary, and quaternary structure) of proteins including

---

4 Auto-injectors differ slightly from some types of PFPs – some PFPs require manual needle insertion while auto-injectors do not. However, the terms are often used interchangeably.

Table 6.3 Summary of commonly used biophysical methods used in biologics research and primary information obtained from each method (in *italics*).

Optical	Separation	Spectroscopic	Calorimetric
• TEM • Optical • AFM • LO/DIM • *Primary information: size, shape, protein-protein interactions, particulates*	• AUC • HPLC • Electrophoresis • *Primary information: molecular weight, number of species, concentration*	• UV-Vis • Fluorescence • Light scattering • MS • NMR • CD and FTIR • *Primary information: molecular interactions and structure*	• DSC • ITC • *Primary information: melting temperature, molecular interaction, and structure*

correct amino acid sequence, any unfolding intermediates (i.e. tertiary and quaternary structure), to type of aggregates, small-size aggregates including dimers or trimers to large aggregates and even submicron range particulate formation, to the formulation stability and accelerated studies, and finally to the efficacy of the drug product.

Many methods are employed for such characterization at different stages of development, formulation, and in the final drug product. These methods are summarized in Table 6.3 and are categorized as spectroscopic (UV-Vis, fluorescence, LC-MS, DLS, etc.), separation (HPLC, centrifugation, SDS-PAGE, etc.), optical/imaging (TEM, light obscuration, etc.), and calorimetric (DSC, etc.). Some of these methods are used on a routine basis and can be considered as the industry workhorse (e.g. HPLC) while others would be used rarely due to the difficulties involved in accessing them as well as the exorbitant purchase price, required technical expertise, and low-throughput (e.g. TEM).

Some commonly used methods are summarized in the following sections and listed in Table 6.3.

6.5.1 Spectroscopic Methods

There is a diverse range of spectroscopic methods and generally they are preferred methods because of their sensitivity, ease of use, and high-throughput potential. Crucial information such as amino acid sequence, glycosylation profile, tertiary and higher order structural information, aggregate formation, etc., can all be tracked using spectroscopic methods.

6.5.1.1 UV-Vis

UV-Vis spectroscopy (UV-Vis) may be used to measure the total protein concentration of soluble proteins. This is achieved by measuring the absorbance of aromatic amino acids present in proteins (tryptophan, tyrosine, and phenylalanine) between 250 and 300 nm. Absorption spectra of these amino acids somewhat overlap but Trp absorbance is much higher than for other amino acids and thus generally only Trp absorbance is used in experimental studies. Trp absorbance is more prevalent around 280 nm and majority of protein concentration calculations are carried out via protein absorption at this wavelength.

6.5.1.2 Fluorescence Spectroscopy

Intrinsic protein fluorescence occurs due to aromatic residues (tryptophan, tyrosine, and phenylalanine); however, Trp has the highest quantum yield ($\phi = 0.14$) and therefore generally only Trp fluorescence is observed in proteins. Trp fluorescence is very sensitive to its environment and its emission spectrum depends on whether Trp is located in a hydrophilic (i.e. aqueous) or a hydrophobic environment. Due to this sensitivity of its vicinity, Trp fluorescence is often used to detect tertiary/quaternary conformational changes of a protein. For example, if a protein has only one Trp residue that is located inside the protein in a hydrophobic environment and restricted in motion, its fluorescence would be well below 350 nm, which is the emission maximum of free Trp in an aqueous environment. Upon unfolding, if this Trp residue is completely solvent-exposed and can move around freely, then it is expected to display an emission maximum around 350 nm in the absence of any quenching moiety. Consequently, protein tertiary/quaternary structure can be probed by observing the change in the emission maximum of this Trp residue. Nevertheless, most proteins, especially therapeutic protein products, have more than one protein and due to fluorescence being additive, protein fluorescence is somewhat more complicated than this particular example. For instance, mAbs may have more than 20 Trp residues that are located all around the protein, some located on the surface of the protein while others are buried or partially buried in the hydrophobic pockets inside the protein. Furthermore, energy transfer between Trp–Trp residues and Trp fluorescence quenching by adjacent amino acid residues can further complicate the observed protein fluorescence.[63,64] Nevertheless, any change in the Trp emission generally is related to structural stability of the protein or sometimes in rare cases to protein–protein interactions. Therefore, one can track conformational stability of a therapeutic protein using Trp fluorescence.

Another fluorescence technique that is often employed in biologics is external dye-binding method where a hydrophobic dye is used to detect the presence of protein unfolding species or more commonly to detect protein aggregation. Most commonly used dyes include ANS, Thioflavin T, Congo red, Nile red, and Bodipy. A common feature of these dyes is that they are not highly soluble in water because they are hydrophobic in nature and they do not fluoresce well in aqueous environment, sometimes due to intramolecular charge transfer. Protein aggregates tend to possess hydrophobic environments, which attract such hydrophobic molecules and thus dyes intercalate within aggregates. For example, thioflavin T binds to any protein aggregates nonspecifically but has a high affinity for beta-sheet aggregates and is often used in protein aggregation studies. Upon binding, fluorescence intensities of these dyes increase and very often their fluorescence blue-shifts. Such a change is used to probe aggregate formation as well as determining protein unfolding, depending on experimental conditions.[65,66]

6.5.1.3 Mass Spectrometry

Mass spectrometry (MS) is a technique used to detect and measure changes in molecules based on their mass-to-charge ratios and is widely used in proteomics. This includes studying protein structure and function, identification of a protein's amino acid sequence, and detecting PTMs and chemical modifications to specific amino acid residues that may affect the stability of the biologic. MS is often preceded by various chromatographic techniques, most commonly liquid chromatography (LC-MS) to separate molecules of interest (e.g. degradation products). MS analysis typically requires digestion of the protein into smaller units (peptides) to determine the protein sequence. This allows the identification of chemical modifications such as oxidation or deamidation of specific amino acids. Other similar techniques identify and characterize PTMs such as the glycosyla-

tion profile of a protein. LC-MS or LC-MS/MS is the most accurate approach for the detection of minor changes in protein primary sequence; however, MS analysis is time-consuming.

6.5.1.4 CD and FTIR

Circular dichroism (CD) and Fourier-transform infrared spectroscopy (FTIR) are spectroscopic techniques that characterize the secondary structure of a protein and other biologics. CD uses a circularly polarized light to determine the absorbance differences between left- and right-circular light, and in combination with Beer's law, molar ellipticity (θ) is determined and plotted against wavelength (λ). Commonly observed different secondary structures of proteins, i.e. α-helices, β-sheets, or random coils, have characteristic θ vs. λ traces, and hence, highly valuable information about a protein's secondary structure may be obtained via CD.[67]

Similarly, structural analysis may be performed via the FTIR method, which also provides information about secondary structures of proteins by measuring certain vibrational (absorption) bands of the peptide backbone. The most prominent infrared (IR) bands are amide I bands that are located at 1700–1600 cm^{-1}; however, there are eight other IR bands that are related to protein secondary structure, with varying sensitivities.[68]

6.5.2 Microscopy Methods

Many microscopy methods exist with varying resolution for the study of biologics. Traditionally, optical microscopy has inadequate resolution to study biologics (resolution is related to diffraction limit, which can be estimated as $\sim \lambda/2$). However, in the last several decades, this has changed with the advancement of several novel methods such as the stimulated emission depletion (STED) microscopy, which detects particles beyond the diffraction limit. Super-resolution microscopy also exists to study protein–protein interactions and other related phenomena like total-internal reflection fluorescence (TIRF), confocal, spinning disc, etc.

Other than optical microscopy methods, electron microscopy is frequently employed to study biologics; transmission electron microscopy (TEM) is the most common. Atomic force microscopy (AFM) is another method that is sometimes used to study biologics. Although microscopy methods provide crucial structural, size, and shape information, unfortunately the majority of microscopy methods are slow, cannot be used in a high-throughput manner, require technical expertise, and are expensive. There are many other types of microscopies, of most relevance to this chapter is TEM.

6.5.2.1 TEM

TEM provides information on the size and shape of particles over a wide range of sizes from small aggregates (a few nanometers) to large aggregates or viruses/bacteria (submicron/micron). Biological samples are usually negatively stained with uranyl acetate or other commonly used staining techniques for better visualization. TEM provides invaluable information about the product; however, the equipment is quite expensive, imaging with TEM requires technical expertise, and TEM is quite slow as imaging cannot be performed in a high-throughput manner. Thus, it is not widely used. Products with large structures or aggregates such as vaccines benefit from TEM imaging as no other method can provide such detailed structural information. An example of a TEM image with an influenza vaccine is shown in Figure 6.6.

6.5.2.2 Light Obscuration and Dynamic Imaging Microscopy

Subvisible particles (such as large aggregates) above 1 μm in diameter are analyzed using light obscuration (LO) particle counters (USP method) or flow imaging particle analysis systems. It is crucial to monitor subvisible particles in a biologic sample as they can act as precursors to visible particles, which are highly dangerous if injected into a patient. When a protein sample containing subvisible particles is passed through a laser light source and detector, the light blocked or obscured by the particles reduces the light intensity that is detected, which is then processed to determine the size of the particle (LO). In contrast, dynamic imaging microscopy (DIM) is a particle imaging analysis method that utilizes digital imaging to measure and count sub-

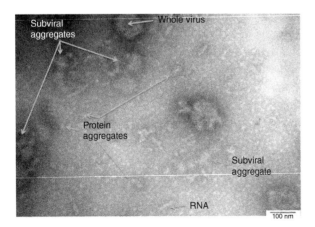

Figure 6.6 Different size particles present in an influenza vaccine sample as visualized by a negatively stained TEM image. (*See insert for color representation of the figure.*)

visible particles in a liquid sample as they pass through the flow cell. DIM visualizes particles in a protein sample, which can assist in identifying their type and source (i.e. proteinaceous or extraneous).

6.5.3 Separation Methods

Molecules may be separated according to their size, charge, hydrophobicity, and other characteristics. These properties are specific to different types of biologics and sometimes undesired biological moieties can be detected, quantified, or purified using one or more of these separation methods. For example, size-exclusion chromatography (SEC) can identify and quantify aggregates, monomers, and fragments in a therapeutic monoclonal antibody formulation; this method is the industry workhorse. Similarly, a vaccine formulation may be characterized using SEC but because large particles cannot traverse the chromatographic column, other types of separation methods such as analytical ultracentrifugation (AUC) or sucrose gradient methods are used instead. Such methods use centrifugal force rather than a column to separate different entities, and hence, a large range of particle sizes can be detected. While some chromatographic methods are used in a high-throughput manner and automated (e.g. SEC), other separation methods such as AUC are extremely laborious and slow and require specific expertise and equipment. Nevertheless, if each component can be separated and quantified, it can assist with elucidating degradation pathways such as protein aggregation and product characterization; separation methods therefore will always play a major role in the development and characterization of biologics.

6.5.3.1 Chromatography

Chromatographic techniques such as SEC, ion-exchange (IEX), and reverse-phase (RP) chromatography are used widely to characterize biologics and their chemical and physical stability over time. SEC is the gold-standard for separating and quantifying monomers and high molecular weight species such as protein aggregates. IEX is used to separate and quantify charge variants of a protein, termed acidic and basic variants, which often include oxidized and deamidated species. RP is used to determine the degree of protein oxidation, by separating protein variants based on their hydrophobicity (protein may be digested to aid identification of a region of interest). IEX and RP are both often used to monitor chemical degradation as they are higher throughput methods compared to MS techniques. However, they are not sensitive enough to determine the exact nature or quantity of charged or oxidized variants which require MS analysis.

6.5.3.2 AUC

AUC may be used to separate and quantify different types of aggregates and monomers as well as for structure and macroscopic characterization of various products including vaccines. With the AUC analysis software, one can analyze sedimentation velocity (SV). AUC accurately characterizes different types of aggregates at high resolution and SV data provide information on the strength of discrete and overall macromolecular interactions.

6.5.3.3 Electrophoresis Methods (SDS-PAGE, CE-SDS)

Electrophoresis methods such as sodium dodecyl sulfate-polyacrylamide gel electrophoresis (SDS-PAGE) and capillary electrophoresis-sodium dodecyl sulfate (CE-SDS) utilize an electric current to separate proteins in the non-native state (denatured), and in either reduced (intact) or non-reduced form (split into smaller chains/fragments) based on their molecular weight. These techniques allow identification of the purity of the protein sample and quantify contaminants such as host cell proteins and degraded protein fragments, which may no longer be functional.

6.5.4 Calorimetric Methods

Calorimetry is based on measuring the small amount of heat exchange in the system that occurs upon protein unfolding or other molecular interactions. Subsequently, one can calculate important thermodynamic parameters such as enthalpy, entropy, and heat capacity change (ΔCp) of the system, which are all related to physical and chemical properties of biologics. For example, temperature-induced protein unfolding depends on the experimental conditions, such as protein concentration, buffer component, solution pH, or additives. At low protein concentrations, e.g. 1 mg/mL, if the temperature is increased slowly over time, proteins unfold at a given temperature (temperature where 50% of protein unfolds is called melting temperature, Tm). Using a differential scanning calorimeter (DSC), one can measure this Tm value, calculate ΔCp and other thermodynamic parameters, and obtain an understanding of protein's structural stability, i.e. the higher the Tm, the more stable is the protein. If the protein is engineered to accommodate mutations including glycosylation, conjugated with other moieties like PEG (PEGylation), or simply formulated differently with additional excipients, the Tm parameter would give information on the conformational stability of the protein. Similarly, isothermal titration calorimetry (ITC) is informative when studying enthalpy changes, particle–protein interactions, binding affinities, and stoichiometry of a biologic.

6.5.5 Accelerated Studies

Accelerated studies are generally performed at elevated temperatures to determine both structural stability as well as aggregation propensity of products. The outcomes from such studies are also extrapolated to predict a product's behavior under long-term storage conditions. In addition, rapid formulation screenings may also be performed via accelerated studies. For example, the effect of an additive on protein aggregation are tested in the presence or absence of such an additive by incubating the protein at higher temperatures (<Tm) for various time periods to induce aggregation. The nature of the aggregates formed at high protein concentrations can also be investigated and then monomer content is measured using size exclusion-high performance liquid chromatography (SEC). Light scattering, external dye-binding, and perhaps TEM may also be used to determine aggregate formation patterns. Similarly, a biologic product's susceptibility to fragmentation, oxidation, and other types of degradation may also be determined rapidly in different formulations.

Subsequently, using Arrhenius equation/plot, the results are extrapolated to lower temperatures and in turn, the long-term behavior of the product is estimated. If the protein or vaccine degradation does not exhibit Arrhenius behavior, then a non-Arrhenius model is employed.[28]

6.6 Influenza Vaccines

Immunization has been very successful, particularly in reducing deaths in children from infectious diseases. Among infectious diseases, influenza has a special place because of its extensive viral drift and shift, and thus requiring annual vaccination.[69] Nevertheless, morbidity and mortality from seasonal influenza are still extremely high; between 290 000 and 600 000 deaths and millions of hospitalizations worldwide annually.[69-71] Influenza causes major problems even in developed countries; e.g. more than 35 000 deaths in the United States each year.[72] Its effects in developing countries, however, is more detrimental; for children <5 years, 99% of deaths are due to influenza-related infections.[73] High risk groups are young children, elderly, and the pregnant but all age groups are at risk as exemplified during the A(H1N1pdm09) pandemic (the swine flu) in 2009.

The majority of the influenza vaccines are egg-based (produced in embryonated chicken eggs) and inactivated, although cell-based and attenuated influenza vaccines also exist. Major egg-based vaccines include AGRIFLU®, FLUAD®, FLUVIRAL®, FLUZONE®, INFLUVAC®, VAXIGRIP®, FLULAVAL TETRA®, VAXIGRIP®, AFLURIA®, and FLUAD®. The cell-based influenza vaccine FLUCELVAX QUADRIVALENT® is produced in Madin-Darby canine kidney (MDCK) cells, and FLUBLOK® QUADRIVALENT and FLUBLOK® HIGH DOSE are manufactured using insect cell-baculovirus production technology. Lastly, FLUMIST® QUADRIVALENT is a live attenuated influenza vaccine (LAIV) that is delivered via a nasal spray. Some of the abovementioned vaccines (FLUZONE® and VAXIGRIP®) are manufactured by the same manufacturer but are only available in some countries.

To address current and potential concerns associated with vaccine development and manufacturing, innovative methods are needed to characterize not only the final vaccine product, but also intermediate key processing steps,[74,78] from inoculation of the eggs/cells until the fill-finish step.[79,80] More critically, to develop vaccines rationally rather than the current empirical trial-and-error approach, the following is required: (i) what are the critical parameters in vaccine production (V. Kayser (in preparation). Enhancing the stability of influenza vaccine formulations with DoE),[81] (ii) how can these be controlled during key manufacturing steps, and (iii) how they synergistically affect vaccine formulation stability and the final product?

Other unanswered questions in influenza vaccine development include:

- What parameters directly affect virus yield (e.g. virus strain, type of glycosylation, growth factors, nutrients, immunosuppressants, temperature, and inoculate concentration)?
- How do titer and surfactant concentrations affect formulation stability?
- How do other processing steps (e.g. number of filtration steps) influence formulation stability?

Answers to these questions should provide a mechanistic understanding (and control) of virus yield and formulation stability. This is specifically important in preparing for the next influenza pandemic.

References

1 Lagassé, H.A.D., Alexaki, A., Simhadri, V.L. et al. (2017). Recent advances in (therapeutic protein) drug development. *F1000Research* 6: 113.

2 Morrow, T. and Felcone, L.H. (2004). Defining the difference: what makes biologics unique. *Biotechnology Healthcare* 1 (4): 24–29.

3 Sifniotis, V., Cruz, E., Eroglu, B., and Kayser, V. (2019). Current advancements in addressing key challenges of therapeutic antibody design, manufacture, and formulation. *Antibodies* 8 (2): 36.

4 Elgundi, Z., Reslan, M., Cruz, E. et al. (2017). The state-of-play and future of antibody therapeutics. *Advanced Drug Delivery Reviews* 122: 2–19.

5 Baldo, B.A. (2016). Approved biologics used for therapy and their adverse effects. In: *Safety of Biologics Therapy: Monoclonal Antibodies, Cytokines, Fusion Proteins, Hormones, Enzymes, Coagulation Proteins, Vaccines, Botulinum Toxins* (ed. B.A. Baldo), 1–27. Cham: Springer International Publishing.

6 Beall, R.F., Hwang, T.J., and Kesselheim, A.S. (2019). Pre-market development times for biologic versus small-molecule drugs. *Nature Biotechnology* 37 (7): 708–711.

7 Lindsley, C.W. (2019). Predictions and statistics for the best-selling drugs globally and in the United States in 2018 and a look forward to 2024 projections. *ACS Chemical Neuroscience* 10 (3): 1115.

8 FDA (2019). FDA-approved biosimilar products (updated 5 September 2019). https://www.fda.gov/drugs/biosimilars/biosimilar-product-information (accessed 16 June 2020).

9 GaBi Online (2018). AbbVie makes more deals delaying adalimumab biosimilars in the US. http://www.gabionline.net/Pharma-News/AbbVie-makes-more-deals-delaying-adalimumab-biosimilars-in-the-US (accessed 16 June 2020).

10 GaBi Online (2019). Boehringer Ingelheim finally signs licensing deal for Humira® biosimilar. http://www.gabionline.net/layout/set/print/Pharma-News/Boehringer-Ingelheim-finally-signs-licensing-deal-for-Humira-biosimilar (accessed 16 June 2020).

11 Lee, S., Raw, A., Yu, L. et al. (2013). Scientific considerations in the review and approval of generic enoxaparin in the United States. *Nature Biotechnology* 31: 220.

12 WHO (2019). General policies for monoclonal antibodies. https://www.who.int/medicines/services/inn/generalpoliciesmonoclonalantibodiesjan10.pdf (accessed 16 June 2020).

13 WHO (2019). The use of stems in the selection of International Nonproprietary Names (INN) for pharmaceutical substances. https://www.who.int/medicines/services/inn/StemBook2009.pdf (accessed 16 June 2020).

14 Grothey, A. (2019). Nomenclature of monoclonal antibodies (-mab). https://www.grepmed.com/images/4599 (accessed 16 June 2020).

15 CDC (2019). Types of influenza viruses https://www.cdc.gov/flu/about/viruses/types.htm (accessed 16 June 2020).

16 WHO (2019). WHO recommendations on the composition of influenza virus vaccines https://www.who.int/influenza/vaccines/virus/recommendations/en (accessed 16 June 2020).

17 Ghaderi, D., Zhang, M., Hurtado-Ziola, N., and Varki, A. (2012). Production platforms for biotherapeutic glycoproteins. Occurrence, impact, and challenges of non-human sialylation. *Biotechnology and Genetic Engineering Reviews* 28: 147–175.

18 Walsh, G. (2010). Post-translational modifications of protein biopharmaceuticals. *Drug Discovery Today* 15 (17): 773–780.

19 Walsh, G. and Jefferis, R. (2006). Post-translational modifications in the context of therapeutic proteins. *Nature Biotechnology* 24 (10): 1241–1252.

20 Higel, F., Seidl, A., Sörgel, F., and Friess, W. (2016). N-glycosylation heterogeneity and the influence on structure, function and pharmacokinetics of monoclonal antibodies and Fc fusion proteins. *European Journal of Pharmaceutics and Biopharmaceutics* 100: 94–100.

21 Kayser, V., Chennamsetty, N., Voynov, V. et al. (2011). Glycosylation influences on the aggregation propensity of therapeutic monoclonal antibodies. *Biotechnology Journal* 6 (1): 38–44.

22 Zheng, K., Yarmarkovich, M., Bantog, C. et al. (2014). Influence of glycosylation pattern on the molecular properties of monoclonal antibodies. *MAbs* 6 (3): 649–658.

23 Cymer, F., Beck, H., Rohde, A., and Reusch, D. (2018). Therapeutic monoclonal antibody N-glycosylation – structure, function and therapeutic potential. *Biologicals* 52: 1–11.

24 Bradbury, A.F. and Smyth, D.G. (1991). Peptide amidation. *Trends in Biochemical Sciences* 16: 112–115.

25 Lalonde, M.-E. and Durocher, Y. (2017). Therapeutic glycoprotein production in mammalian cells. *Journal of Biotechnology* 251: 128–140.

26 Moritz, B. and Stracke, J.O. (2017). Assessment of disulfide and hinge modifications in monoclonal antibodies. *Electrophoresis* 38 (6): 769–785.

27 Manning, M.C., Chou, D.K., Murphy, B.M. et al. (2010). Stability of protein pharmaceuticals: an update. *Pharmaceutical Research* 27 (4): 544–575.

28 Kayser, V., Chennamsetty, N., Voynov, V. et al. (2011). Evaluation of a non-arrhenius model for therapeutic monoclonal antibody aggregation. *Journal of Pharmaceutical Sciences* 100 (7): 2526–2542.

29 Wang, W., Singh, S.K., Li, N. et al. (2012). Immunogenicity of protein aggregates – concerns and realities. *International Journal of Pharmaceutics* 431 (1): 1–11.

30 Moussa, E.M., Panchal, J.P., Moorthy, B.S. et al. (2016). Immunogenicity of therapeutic protein aggregates. *Journal of Pharmaceutical Sciences* 105 (2): 417–430.

31 Uchino, T., Miyazaki, Y., Yamazaki, T., and Kagawa, Y. (2017). Immunogenicity of protein aggregates of a monoclonal antibody generated by forced shaking stress with siliconized and nonsiliconized syringes in BALB/c mice. *The Journal of Pharmacy and Pharmacology* 69 (10): 1341–1351.

32 Vazquez-Rey, M. and Lang, D.A. (2011). Aggregates in monoclonal antibody manufacturing processes. *Biotechnology and Bioengineering* 108 (7): 1494–1508.

33 Roberts, C.J. (2006). *Nonnative Protein Aggregation. Misbehaving Proteins: Protein (Mis)Folding, Aggregation, and Stability*, 17–46. New York, NY: Springer New York.

34 Cromwell, M.E.M., Hilario, E., and Jacobson, F. (2006). Protein aggregation and bioprocessing. *The AAPS Journal* 8 (3): E572–E579.

35 Reslan, M. and Kayser, V. (2018). Ionic liquids as biocompatible stabilizers of proteins. *Biophysical Reviews* 10 (3): 781–793.

36 Roberts, C.J. (2007). Non-native protein aggregation kinetics. *Biotechnology and Bioengineering* 98 (5): 927–938.

37 Voynov, V., Chennamsetty, N., Kayser, V. et al. (2009). Predictive tools for stabilization of therapeutic proteins. *MAbs* 1 (6): 580–582.

38 Kuyucak, S. and Kayser, V. (2017). Biobetters from an integrated computational/experimental approach. *Computational and Structural Biotechnology Journal* 15: 138–145.

39 Agrawal, N.J., Kumar, S., Wang, X. et al. (2011). Aggregation in protein-based biotherapeutics: computational studies and tools to identify aggregation-prone regions. *Journal of Pharmaceutical Sciences* 100 (12): 5081–5095.

40 Chennamsetty, N., Voynov, V., Kayser, V. et al. (2010). Prediction of aggregation prone regions of therapeutic proteins. *The Journal of Physical Chemistry B* 114 (19): 6614–6624.

41 Zambrano, R., Jamroz, M., Szczasiuk, A. et al. (2015). AGGRESCAN3D (A3D): server for prediction of aggregation properties of protein structures. *Nucleic Acids Research* 43 (W1): W306–W313.

42 Kayser, V., Françon, A., Pinton, H. et al. (2017). Rational design of rabies vaccine formulation for enhanced stability. *Turkish Journal of Medical Sciences.* 47: 987–995.

43 swissinfo.ch (2012). Immunisation trouble: Novartis flu vaccine deliveries stopped (28 October). https://www.swissinfo.ch/eng/immunisation-trouble_novartis-flu-vaccine-deliveries-stopped/33805730 (accessed 16 June 2020).

44 FDA (2014). BioCSL Fluvax – investigations into the cause of fevers in children under 5 years. https://www.tga.gov.au/media-release/biocsl-fluvax-investigations-cause-fevers-children-under-5-years (accessed 16 June 2020).

45 Souza, V.P., Ikegami, C.M., Arantes, G.M., and Marana, S.R. (2016). Protein thermal denaturation is modulated by central residues in the protein structure network. *The FEBS Journal* 283 (6): 1124–1138.

46 Day, R., Bennion, B.J., Ham, S., and Daggett, V. (2002). Increasing temperature accelerates protein unfolding without changing the pathway of unfolding. *Journal of Molecular Biology* 322 (1): 189–203.

47 Vlasak, J. and Ionescu, R. (2011). Fragmentation of monoclonal antibodies. *MAbs* 3 (3): 253–263.

48 Gao, S.X., Zhang, Y., Stansberry-Perkins, K. et al. (2011). Fragmentation of a highly purified monoclonal antibody attributed to residual CHO cell protease activity. *Biotechnology and Bioengineering* 108 (4): 977–982.

49 Torosantucci, R., Schöneich, C., and Jiskoot, W. (2014). Oxidation of therapeutic proteins and peptides: structural and biological consequences. *Pharmaceutical Research* 31 (3): 541–553.

50 Costantino, H.R., Langer, R., and Klibanov, A.M. (1994). Moisture-induced aggregation of lyophilized insulin. *Pharmaceutical Research* 11 (1): 21–29.

51 Krishnamurthy, R., Madurawe, R.D., Bush, K.D., and Lumpkin, J.A. (1995). Conditions promoting metal-catalyzed oxidations during immobilized Cu-iminodiacetic acid metal affinity chromatography. *Biotechnology Progress* 11 (6): 643–650.

52 Zhang, K., Zhu, X., and Lu, Y. (2018). 6 – The proteome of cataract markers: focus on crystallins. In: *Advances in Clinical Chemistry*, vol. 86 (ed. G.S. Makowski), 179–210. Elsevier.

53 Shire, S.J. (2015). 3 – Stability of monoclonal antibodies (mAbs). In: *Monoclonal Antibodies* (ed. S.J. Shire), 45–92. Woodhead Publishing.

54 Leblanc, Y., Ramon, C., Bihoreau, N., and Chevreux, G. (2017). Charge variants characterization of a monoclonal antibody by ion exchange chromatography coupled on-line to native mass spectrometry: case study after a long-term storage at +5°C. *Journal of Chromatography B* 1048: 130–139.

55 Liu, Y.D., Goetze, A.M., Bass, R.B., and Flynn, G.C. (2011). N-terminal glutamate to pyroglutamate conversion in vivo for human IgG2 antibodies. *Journal of Biological Chemistry* 286 (13): 11211–11217.

56 Beyer, B., Schuster, M., Jungbauer, A., and Lingg, N. (2018). Microheterogeneity of recombinant antibodies: analytics and functional impact. *Biotechnology Journal* 13 (1): 1700476.

57 Shire, S.J., Shahrokh, Z., and Liu, J. (2004). Challenges in the development of high protein concentration formulations. *Journal of Pharmaceutical Sciences* 93 (6): 1390–1402.

58 Shire, S.J. (2015). 8 – Development of delivery device technology to deal with the challenges of highly viscous mAb formulations at high concentration. In: *Monoclonal Antibodies* (ed. S.J. Shire), 153–162. Woodhead Publishing.

59 van den Bemt, B.J.F., Gettings, L., Domanska, B. et al. (2019). A portfolio of biologic self-injection devices in rheumatology: how patient involvement in device design can improve treatment experience. *Drug Delivery* 26 (1): 384–392.

60 Uchenna Agu, R., Ikechukwu Ugwoke, M., Armand, M. et al. (2001). The lung as a route for systemic delivery of therapeutic proteins and peptides. *Respiratory Research* 2 (4): 198.

61 Bodier-Montagutelli, E., Mayor, A., Vecellio, L. et al. (2018). Designing inhaled protein therapeutics for topical lung delivery: what are the next steps? *Expert Opinion on Drug Delivery* 15 (8): 729–736.

62 Lip Kwok, P.C. and Chan, H.-K. (2011). 2 – Pulmonary delivery of peptides and proteins. In: *Peptide and Protein Delivery* (ed. C. Van Der Walle), 23–46. Boston: Academic Press.

63 Kayser, V., Chennamsetty, N., Voynov, V. et al. (2011). Tryptophan-tryptophan energy transfer and classification of tryptophan residues in proteins using a therapeutic monoclonal antibody as a model. *Journal of Fluorescence* 21 (1): 275–288.

64 Kayser, V., Turton, D.A., Aggeli, A. et al. (2004). Energy migration in novel pH-triggered self-assembled β-sheet ribbons. *Journal of the American Chemical Society* 126 (1): 336–343.

65 Kayser, V., Chennamsetty, N., Voynov, V. et al. (2012). A screening tool for therapeutic monoclonal antibodies: identifying the most stable protein and its best formulation based on thioflavin T binding. *Biotechnology Journal* 7 (1): 127–132.

66 Kayser, V., Chennamsetty, N., Voynov, V. et al. (2011). Conformational stability and aggregation of therapeutic monoclonal antibodies studied with ANS and thioflavin T binding. *MAbs* 3 (4): 408–411.

67 Nakanishi, K., Berova, N., and Woody, R. (eds.) (1994). *Circular Dichroism: Principles and Applications.* New York, NY: VCH.

68 Yang, H., Yang, S., Kong, J. et al. (2015). Obtaining information about protein secondary structures in aqueous solution using Fourier transform IR spectroscopy. *Nature Protocols* 10 (3): 382–396.

69 World Health Organization (2018). Influenza (seasonal). http://www.who.int/mediacentre/factsheets/fs211/en (accessed 16 June 2020).

70 Soema, P.C., van Riet, E., Kersten, G., and Amorij, J.-P. (2015). Development of cross-protective influenza A vaccines based on cellular responses. *Frontiers in Immunology* 6: 1–9.

71 WHO (2019). Influenza: burden of disease https://www.who.int/influenza/surveillance_monitoring/bod/en (accessed 16 June 2020).

72 Newall, A.T. and Scuffham, P.A. (2008). Influenza-related disease: the cost to the Australian healthcare system. *Vaccine* 26 (52): 6818–6823.

73 Nair, H., Brooks, W.A., Katz, M. et al. (2011). Global burden of respiratory infections due to seasonal influenza in young children: a systematic review and meta-analysis. *The Lancet* 378 (9807): 1917–1930.

74 Lee, K.K.H., Sahin, Y.Z., Neeleman, R. et al. (2016). Quantitative determination of the surfactant-induced split ratio of influenza virus by fluorescence spectroscopy. *Human Vaccines & Immunotherapeutics* 12 (7): 1757–1765.

75 Sahin, Z., Akkoc, S., Neeleman, R. et al. (2017). Nile red fluorescence spectrum decomposition enables rapid screening of large protein aggregates in complex biopharmaceutical formulations like influenza vaccines. *Vaccine* 35 (23): 3026–3032.

76 Sahin, Z., Neeleman, R., Haines, J., and Kayser, V. (2019). Preparation-free method can enable rapid surfactant screening during industrial processing of influenza vaccines. *Vaccine* 37 (8): 1073–1079.

77 Tay, T., Agius, C., Hamilton, R. et al. (2015). Investigation into alternative testing methodologies for characterization of influenza virus vaccine. *Human Vaccines & Immunotherapeutics* 11 (7): 1673–1684.

78 Clénet, D. (2018). Accurate prediction of vaccine stability under real storage conditions and during temperature excursions. *European Journal of Pharmaceutics and Biopharmaceutics* 125: 76–84.

79 Cruz, E., Cain, J., Crossett, B., and Kayser, V. (2018). Site-specific glycosylation profile of influenza A (H1N1) hemagglutinin through tandem mass spectrometry. *Human Vaccines & Immunotherapeutics* 14 (3): 508–517.

80 She, Y.-M., Li, X., and Cyr, T.D. (2019). Remarkable structural diversity of N-glycan sulfation on influenza vaccines. *Analytical Chemistry* 91 (8): 5083–5090.

81 Ahl, P.L., Mensch, C., Hu, B. et al. (2016). Accelerating vaccine formulation development using design of experiment stability studies. *Journal of Pharmaceutical Sciences* 105 (10): 3046–3056.

7

Biosimilarity and Interchangeability of Biologic Drugs-General Principles, Biophysical Tests, and Clinical Requirements to Demonstrate Biosimilarity

Hans C. Ebbers

Scientific Affairs Biosimilars, Biogen International GmbH, Baar, Switzerland

---

KEY POINTS

- Biosimilars are approved based on a robust biosimilarity exercise, involving comprehensive assessments of quality attributes, functional testing, and comparative clinical studies.
- Different regulatory jurisdictions/regions have taken different approaches toward switching and the concept of "interchangeability."
- Accumulated evidence on switching biosimilars does not suggest negative effects with respect to immunogenicity, pharmacokinetics, or effectiveness.
- Nocebo effects have been observed in patients switched to biosimilars.

Abbreviations

Abbreviations	Full name
ADCC	Antibody-Dependent Cellular Cytotoxicity
ADCP	Antibody-Dependent Cellular Phagocytosis
BPCI	Biologics Price Competition and Innovation
CQA	Critical Quality Attribute
ELISA	Enzyme Linked Immunosorbent Assay
EMA	European Medicines Agency
EU	European Union
FDA	Food and Drug Administration
IBD	Inflammatory Bowel Disease
PBAC	Pharmaceutical Benefits Advisory Committee
PD	Pharmacodynamic

Abbreviations	Full name
PK	Pharmacokinetic
QTPP	Quality Target Product Profile
RA	Rheumatoid Arthritis
RBP	Reference Biotherapeutic Product
RCT	Randomized Controlled Trial
SBPs	Similar Biotherapeutic Products
SE-HPLC	Size Exclusion High Performance Liquid Chromatography
TGA	Therapeutic Goods Administration
TNF	Tumor Necrosis Factor
WHO	World Health Organization

7.1 Introduction

7.1.1 The Biosimilar Development Paradigm

When biosimilars are approved as such by a competent regulatory authority, these products have been demonstrated to be "similar in terms of quality, safety and efficacy to an already licensed biotherapeutic product (the reference product, or innovator) for which substantial evidence exists of safety and efficacy."[1]

As we have seen in the previous chapters, biosimilars are approved based on a robust biosimilarity exercise. During this biosimilarity exercise, a manufacturer demonstrates that there are no clinically relevant differences

---

Biologics, Biosimilars, and Biobetters: An Introduction for Pharmacists, Physicians, and Other Health Practitioners,
First Edition. Edited by Iqbal Ramzan.
© 2021 John Wiley & Sons, Inc. Published 2021 by John Wiley & Sons, Inc.

between the biosimilar candidate and its reference product. The theoretical framework for approving biosimilars is based on the comparability exercise that is performed when a manufacturer changes its manufacturing process, which has been applied for innovator products for many years.[2] During a comparability exercise, several batches of a biological product produced before and after the change are compared to ensure that the manufacturing change did not have a clinically relevant effect on the product quality. The level of data that needs to be presented to regulatory bodies in order to establish comparability between the pre- and post-change material depends on the nature of the change. Characterization of a biotechnological/biological product by appropriate techniques includes the determination of physicochemical properties, biological activity, immunochemical properties (if any), purity, impurities, contaminants, and quantity.[3] Additional nonclinical and (in rare cases) clinical data may be required, depending on the nature of the change and the level of comparability pre- and post-change. This regulatory approach has been successfully applied to innovator biologics for many years before the introduction of the first biosimilars in the European Union (EU) around 2005.

The basic approach and paradigm of establishing biosimilarity has been pioneered by the EU where legislation allowing for the registration of biosimilars was introduced in 2004, followed by the release of several guidance documents by the European Medicines Agency (EMA) in 2005. The United States introduced legislation to allow for the approval of biosimilars in 2009, which was also followed by the release of several guidance documents.

The pathway to establish biosimilarity became the global standard following the adoption of the WHO Guidelines on evaluation of similar biotherapeutic products (SBPs) in 2009, which set out the scientific principles, including the stepwise approach, which should be applied for the demonstration of similarity between an SBP and the reference biotherapeutic product (RBP):

> High similarity at the quality level is regarded as a prerequisite for enabling the use of a tailored nonclinical and clinical programme for licensure. The goal of the clinical comparability exercise is to confirm the similarity established at previous stages of development and to demonstrate that there are no clinically meaningful differences between the SBP and the RBP – and not to re-establish safety and efficacy, as this has been done already for the RBP. The decision on licensure of the SBP should be based upon evaluation of the totality of evidence from quality, nonclinical and clinical parameters. It should be noted that clinical studies cannot be used to resolve substantial differences in physicochemical characteristics and biological activity between the RBP and the SBP. If substantial differences in quality attributes are present, a stand-alone licensing approach may be considered.[1]

The basic approach toward demonstrating biosimilarity is very similar across the globe and all major regulatory regions require a stepwise approach, starting with an extensive quality comparison and followed by confirmatory nonclinical and clinical studies (Figure 7.1).

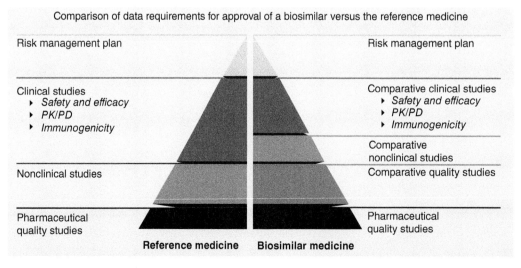

Figure 7.1 Development paradigm of biosimilars versus innovators. *Source:* Adapted from https://ec.europa.eu/docsroom/documents/22924/attachments/1/translations/en/renditions/native.[4] (*See insert for color representation of the figure.*)

7.1.2 Establishing Analytical and Functional Similarity

The foundation to establishing biosimilarity are the comparative quality studies that are aimed to demonstrate similarity between the reference product and the proposed biosimilar. It is expected that the manufacturer of a biosimilar candidate uses state-of-the-art analytical and other methods to demonstrate similarity regarding several aspects of the product:

- Primary structure
- Higher order structures
- Post-translational modifications
- Impurities
- Biological activity (potency)

Biosimilars are expected to have the same primary structure, i.e. an identical amino acid sequence as their reference products. The only exception being differences in C-terminal lysine variants, which may be acceptable, if duly justified that these do not affect biological activity.[5] C-terminal lysine variants are clipped rapidly *in vivo* and have been shown to have no clinical consequence.[6,7] For other attributes of the product, it is not possible to demonstrate identicality. Since we are dealing with proteins that are produced in living systems, there will always be a degree of variability (microheterogeneity) of quality attributes that could potentially affect the clinical performance of a product.[8] Microheterogeneity for both the reference product and the biosimilar means that we are dealing with complex mixtures that always contain many closely related molecules. This is the result of different sources of variability. First, the inherent variability that is associated with expression system. Proteins are expressed in living cells and are modified by multiple enzymatic processes that affect post-translational modifications such as glycosylation, deamidation, oxidation, and methylation.[9] Well-known examples are monoclonal antibody glycosylation patterns that show considerable variability between different cell lines, and even between different clones of a given cell line.[10] Post-translational modifications are also very susceptible to growth conditions, including the use of nutrients, pH, and oxygen. Therefore, the manufacturing process itself can have a profound impact on the end product and therefore needs to be tightly controlled. Even if a manufacturer would use the same cell line as the manufacturer of the reference product, the manufacturing process is proprietary information and thus cannot be copied and every manufacturer has its own manufacturing process, using different reagents and methods to control the quality. In addition to the previously mentioned microheterogeneity, there is also a degree of variability in assays that are used to determine certain quality attributes, particularly in the biological assays that are used to determine the biological activity of biological products. Due to the various sources of variability, we are dealing with relatively wide quality ranges, as compared to synthetic small molecule drugs.

In order to establish biosimilarity, a manufacturer needs to investigate many different aspects of its product to form a complete and holistic picture of its product at several different levels of analysis. Before developing a biosimilar, a manufacturer must know as much as possible about the reference product and therefore must begin by thoroughly characterizing many batches of the reference product for a multitude of quality attributes to establish the range of variation for each of these attributes. Based on these analyses, the product's quality target product profile (QTPP) is determined.[9,11] Figure 7.2 shows a hypothetical example of what such a QTPP might look like, for selected quality attributes.

A biosimilar is then developed that matches the QTPP as closely as possible. What follows is a comprehensive, head-to-head comparability exercise (i.e. performing direct comparisons between the biosimilar candidate and its reference product) using state-of-the-art methods to evaluate the proposed product and the reference product. The FDA describes this in its (draft) guidance[13]:

> "A meaningful comparative analytical assessment depends on, among other things, the capabilities of available state-of-the-art analytical assays to assess, for example, the molecular weight of the protein, complexity of the protein (higher order structure and posttranslational modifications), degree of heterogeneity, functional properties, impurity profiles, and degradation profiles denoting stability. Physicochemical and functional characterization studies should be sufficient to establish relevant quality attributes, including those that define a product's identity, quantity, safety, purity, and potency. The product-related impurities and product-related substances should be identified, characterized as appropriate, quantified, and compared using multiple lots of the proposed biosimilar product and multiple lots of the reference product, to the extent feasible and relevant, as part of an assessment of the potential impact on the safety, purity, and potency of the product."

Manufacturers need to perform side-by-side comparisons using an appropriate number of lots and methods of appropriate (analytical) sensitivity and specificity to

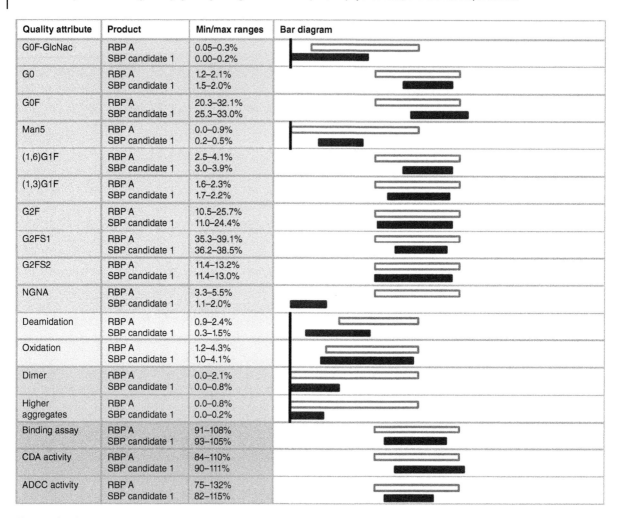

Quality attribute	Product	Min/max ranges	Bar diagram
G0F-GlcNac	RBP A	0.05–0.3%	
	SBP candidate 1	0.00–0.2%	
G0	RBP A	1.2–2.1%	
	SBP candidate 1	1.5–2.0%	
G0F	RBP A	20.3–32.1%	
	SBP candidate 1	25.3–33.0%	
Man5	RBP A	0.0–0.9%	
	SBP candidate 1	0.2–0.5%	
(1,6)G1F	RBP A	2.5–4.1%	
	SBP candidate 1	3.0–3.9%	
(1,3)G1F	RBP A	1.6–2.3%	
	SBP candidate 1	1.7–2.2%	
G2F	RBP A	10.5–25.7%	
	SBP candidate 1	11.0–24.4%	
G2FS1	RBP A	35.3–39.1%	
	SBP candidate 1	36.2–38.5%	
G2FS2	RBP A	11.4–13.2%	
	SBP candidate 1	11.4–13.0%	
NGNA	RBP A	3.3–5.5%	
	SBP candidate 1	1.1–2.0%	
Deamidation	RBP A	0.9–2.4%	
	SBP candidate 1	0.3–1.5%	
Oxidation	RBP A	1.2–4.3%	
	SBP candidate 1	1.0–4.1%	
Dimer	RBP A	0.0–2.1%	
	SBP candidate 1	0.0–0.8%	
Higher aggregates	RBP A	0.0–0.8%	
	SBP candidate 1	0.0–0.2%	
Binding assay	RBP A	91–108%	
	SBP candidate 1	93–105%	
CDA activity	RBP A	84–110%	
	SBP candidate 1	90–111%	
ADCC activity	RBP A	75–132%	
	SBP candidate 1	82–115%	

Figure 7.2 Quality attribute ranges of a biosimilar candidate and its reference product. A, targeting a cell membrane bound target and where Fc functionality is an important part of the clinical mode of action. The lengths of the bars show the relative boundaries of the quality attribute ranges. Attributes are shown as a black line on the left; this black line represents the point of origin (i.e. 0%). *Source:* From Schiestl et al.[12] (*See insert for color representation of the figure.*)

provide meaningful information as to whether the proposed product and the reference product are highly similar, meaning that the methods used should be able to identify very small differences between products and should be able to distinguish between different analytes or features of interest.[13] The result of this head-to-head comparability exercise should be a detailed overview of physicochemical attributes, such as glycan structures, disulfide bond analysis, higher order structure analysis, aggregate levels that should be highly similar, as is detailed in the hypothetical example from Figure 7.2. Not all differences between a biosimilar and its reference are relevant for the benefit–risk profile of a given product, so manufacturers should focus on those quality attributes that matter, so-called "critical quality attributes," which have been defined by the International Conference for Harmonization as: *"a physical, chemical, biological, or microbiological property or characteristic*

that should be within an appropriate limit, range, or distribution to ensure the desired product quality. CQAs are generally associated with the drug substance, excipients, intermediates (in-process materials), and drug product."[14] So, it is very important for a biosimilar developer to understand the product very well to justify any differences that may be identified between a biosimilar candidate and its reference product. For example, certain glycan structures may be present in different quantities than a reference product, but this may not be of any clinical relevance. Other quality attributes may differ from the reference product if it can be justified they are of no clinical consequence, or may be considered desirable, for example, lower levels of impurities such as aggregates, that may be beneficial, as a higher content of aggregates has been linked to immunogenicity.[15]

An example of which quality attributes are compared for monoclonal antibodies and their rationale is provided

Table 7.1 Representative example of quality attributes for a monoclonal antibody.

Quality attributes	Criticality	Impact on biological activity or clinical function	Analytical methods
Primary structures			
Amino acid sequence	+++	Might influence potency	Peptide mapping with LC/MS or LC/MS/MS
N-terminal pyroglutamate	++	Might influence *in vivo* pharmacokinetics	Edman degradation
C-terminal lysine removal	+	Might not influence *in vitro* biological activity	
Disulfide bonds	+++	Might influence potency	
Higher-order structures	++	Might influence the receptor or antigen binding	CD, fluorescence, DSC FT-IR
		Might influence potency	
Charged variants			
Deamidated form	+	Might influence *in vitro* biological activity	LC/MS, IEX, CE, IEF, RP-HPLC
Oxidized form	++	Might lead to highly immunogenic aggregates	HPAEC-PAD
Sialylated form	+++	Might influence *in vivo* clearance	LC/DMB labeling
Mass variants			
Aggregates	+++	Might induce immunogenicity	MS, SEC with UV or MALLS, HIC, CGE, AUC, SDS-PAGE
Truncated form	+	Might maintain some biological activity	
Monomer	+	Might maintain some biological activity	
Pegylation	+++	Might influence *in vivo* clearance	
Oligosaccharides			
Fucose or galactose	+++	Might influence Fc effector functions	HPAEC-PAD
Non-human glycans	+++	Might induce immunogenicity	LC or CE with labeling (e.g. 2AA or 2AB)
High-Man glycans	++	Might induce immunogenicity	LC/MS
		Might influence *in vivo* clearance	LC/MS/MS
		Might influence ADCC	
Non-glycosylated form	++	Might influence Fc effector functions	
Biological activities			
ADCC	+++	Might influence mode of action	Cell-based assay, ELISA, SPR
ADCP	++	Might influence mode of action	
CDC	+++	Might influence mode of action	
Apoptosis	++	Might influence mode of action	
Antigen/antibody binding	+++	Might influence mode of action	
FcγR binding	++	Might influence ADCC	
C1q binding	++	Might influence CDC	
FcRn binding	++	Might influence *in vivo* clearance	
Content	+++	Might influence *in vivo* pharmacokinetics	UV spectrometry

Source: From Kwon et al.[5]
AA, anthranilic acid; AUC, analytical ultracentrifugation; CD, circular dichroism; CE, capillary electrophoresis; CGE, capillary gel electrophoresis; DMB, 1,2-dia,ino-4,5-methylenedioxy-benzene; DSC, differential scanning calorimetry; ELISA, enzyme-linked immunosorbent assay; FT-IR, Fourier transform infrared spectroscopy; HIC, hydrophobic interaction chromatography; HPAEC-PAD, high-performance anion-exchange chromatography with pulsed amperometric detection; IEF, isoelectric focusing; IEX, ion-exchange chromatography; LC, liquid chromatography; MALLS, multi-angle laser light scattering; MS, mass spectrometry; RP-HPLC, reversed-phase high-performance liquid chromatography; SEC, size-exclusion chromatography; SPR, surface plasmon resonance; 2AB, 2-aminobenzamide.

in Table 7.1. Developers of biosimilars are requested to use different complementary ("orthogonal") analytical methods based on different physicochemical principles to investigate a given quality attribute of the product, for example, determining aggregation levels using size exclusion high performance chromatography (SE-HPLC) and sedimentation velocity analytical ultracentrifugation, or analyzing charge variants using both affinity chromatographic methods and electrophoresis.[13,16,17] Furthermore, complementary analytical techniques such as peptide mapping (i.e. digesting proteins using enzymes such as trypsin that cleave proteins at very specific enzyme- and protein-dependent sites generating parent protein-dependent peptide profiles "fingerprinting"), coupled to separation techniques (like liquid chromatography) and mass spectrometry are powerful and very sensitive methods to identify potential differences between two products. A detailed discussion on the methods applied to characterize pharmaceutical proteins is beyond the scope of this chapter, but several excellent reviews and books are available on this topic and the reader is referred to these resources.[18–20]

Functional testing is important to establish similarity in biological activity. This usually involves receptor binding studies (e.g. using surface plasmon resonance or ELISA-based methods), or cell-based assays (e.g. TNF neutralization assays or antibody-dependent cellular cytotoxicity [ADCC] assays).[18] Such functional assays may also be useful to address any differences observed with other methods to justify that these are of no clinical consequence. A good example of the latter is the infliximab biosimilar, CT-P13, which showed slightly lower levels of so-called a-fucosylated glycans than the reference product. Antibodies that contain glycans that lack fucose have a higher affinity for the FcγRIIIa (CD 16) receptor, which is expressed on macrophages, natural killer (NK) cells, and monocytes, and is an important determinant of ADCC, which in turn is believed to play an important role in the efficacy of infliximab in its gastro-enterologic indications, but not for its rheumatologic and dermatologic indications.[21] Indeed, it was shown that CT-P13 demonstrated lower ADCC activity than the reference product in a very sensitive ADCC assay utilizing only purified NK cells. However, these assays were very sensitive and not very representative for actual physiological conditions. The biosimilar company that had filed the marketing auhtorisation application was able to show that under more physiological conditions, i.e. using peripheral blood mononuclear cells, rather than purified NK cells, or by adding low levels of serum to the assays, these differences were no longer apparent.[22] These additional data convinced the EMA that the observed differences were not clinically meaningful, which was later confirmed in clinical studies.[23] The above provides a good example of how a product is analyzed at several levels and that a conclusion is based on the "totality of evidence" that is provided by a biosimilar developer. Analytical and functional methods are used in combination ensuring the robustness of the comparative quality assessments.

Because of the complexity of biological products and the different sources of variability that are inherent to their production and the relatively wide ranges observed for their quality attributes, a conclusion on biosimilarity needs to be based on a multitude of methods. This is the reason why we are talking about biosimilars and not "biogenerics." However, it is often misinterpreted that biosimilars are "merely" similar and not identical, suggesting that lower quality standards apply to biosimilars, or that there is still considerable question whether these products are of a "sufficient likeliness."[24] The comparability exercise aims to demonstrate that the two products do not show any relevant differences in terms of structure and function. From the above, it should be clear that the quality standards that apply to biosimilars are as rigorous as for their reference products and based on a very thorough similarity assessment.

7.1.3 Demonstration of Clinical Similarity

The quality and functional assessment form the foundation for establishing that two products are similar to an extent that any differences are of no clinical consequence. The clinical part of the biosimilarity exercise intends to confirm that what was seen in the analytical test program in the laboratory, indeed does not translate into clinically meaningful differences when the products are administered to healthy volunteers and/or patients.

Following a successful demonstration of similarity in terms of physicochemical and functional characteristics, a demonstration of similarity in terms of pharmacokinetics (i.e. disposition or concentration–time profiles in blood/serum) and clinical efficacy and safety is required. It should be noted that analytical studies are much more sensitive in identifying differences between two products, and demonstrating similarity in clinical studies may not justify substantial differences in quality attributes.[25] A good example is the EMA refusal of Solumarv®, a biosimilar insulin candidate, where the clinical studies might have been supportive of biosimilarity, but shortcoming in the analytical comparability testing methods, as well as other concerns on the use of material other than what was planned to be used commercially, led to the decision that the product was not approvable.[26] Unlike clinical studies for new innovator biologic molecules, the aim of clinical studies performed with biosimilars is to address slight differences that may have been identified in the quality comparison and to confirm comparable clinical performance of the biosimilar and the reference product.[25] The objective is not to re-establish a benefit–risk profile, which has already been demonstrated for the reference product, but to confirm similarity in a sensitive clinical model.

Demonstrating similarity in terms of pharmacokinetics is of key importance for a successful biosimilarity exercise and a key determinant for a successful demonstration of biosimilarity.[27] A single-dose crossover study (i.e. each subject/patient receives both the reference and biosimilar product in a randomized sequence) is preferred by regulators. However, this is not always feasible due to the long half-life of many biologicals, particularly monoclonal antibodies, and the fact that antidrug–antibody formation may occur that could affect the product's serum concentrations. Therefore, randomized single dose, parallel group studies (i.e. one group of subjects/patients receives the reference product and the other group receives the biosimilar) are carried out for these products. Often, these studies are performed in healthy volunteers as they are considered to be more sensitive than patients, as they are usually not receiving concomitant medications that may affect the half-life of the products under investigation. PK studies, however, are not always feasible in healthy volunteers, for ethical reasons. For example, PK studies for rituximab biosimilars were performed in patients.

A final step is to confirm clinical similarity in a sensitive clinical model. In principle, biosimilars may be approved on the basis of PK/PD studies (i.e. involving a measure of response to a drug, PD, in addition to serum concentration measurements) only, but this may not be appropriate for most products. For molecules for which there are well-established pharmacodynamic markers and no known concerns of immunogenicity, this may be acceptable. Examples are insulin or filgrastim, which are relatively simple molecules and PD responses to both drugs can measured by clamping studies (in case of insulins), absolute neutrophil counts (in case of filgrastim) in either patients or volunteers.[27] However, for more complicated products that lack a validated PD marker to assess the efficacy of the drug and immunogenicity is a concern, e.g. TNF inhibitors, similar efficacy of the biosimilar will usually have to be demonstrated in adequately powered, randomized and controlled clinical trial(s).[1] Such trials should in principle have an "equivalence design," i.e. capable of showing that the biosimilar is neither worse or better than its reference product in

terms of a predefined (efficacy) endpoint, by a prespecified margin in a head-to-head comparative study. This margin should be justified on the basis of clinical relevance and statistical grounds and represent the largest difference in efficacy that would not be considered clinically relevant.[1,28] These trials should be adequately powered (the number of subjects/patients required is adequate and statistically justifiable) and usually enroll several hundreds of patients. Also, pre-licensing safety data should be obtained in a sufficient number of patients, with a sufficient length of exposure to characterize the safety profile of the biosimilar. It should be noted that most studies are only powered to be able to draw conclusions on the primary efficacy endpoint, but they provide a general reassurance that the safety profile is in line with the known safety profile of the reference product. Further monitoring in the post-marketing phase is usually required and in the EU, as for any newly approved drug, biosimilars receive a black triangle (▼) to indicate that they should be closely monitored for adverse events for five years following approval. Furthermore, clinical studies should always collect immunogenicity data. All therapeutic proteins have the potential to elicit immune responses.[29] Immunogenicity may be influenced by many factors including the primary sequence of the product, process-related impurities, excipients, stability, route of administration, dosing, and patient- or disease-related factors.[1,29] Therefore, the clinical trials should also collect sufficient immunogenicity data to rule out clinically relevant differences between a biosimilar and its reference product.

As part of the package to demonstrate similarity, a manufacturer may provide additional testing using *in vitro* models to support similarity in indications that are not studied in the comparative clinical trials, to support extrapolation of indications not studied as part of the clinical development program. [22]

7.2 Interchangeability

The main promise of biosimilars is to reduce drug expenditure without compromising the quality of care that patients are receiving. This may lead to better access to treatments and potentially to better treatment outcomes due to timely initiation of therapy and reduced need to discontinue treatment because of cost constraints. The concept of establishing biosimilarity has received widespread acceptance across the globe and treating patients naive to a given biologic product with its biosimilar is no longer subject of discussion. However, the discussion persists whether patients can

> **Scheme 7.1 Terminology Relating to Interchangeability**
>
> Based on the EMA definition, interchangeability refers to the property of a medicine to be exchanged with another one, which is expected to have the same clinical effects. This may mean replacing a reference product with a biosimilar or vice versa or replacing one biosimilar with another. Such replacing may be done by
>
> - Switching, if the prescriber decides to exchange one medicine for another one with the same therapeutic intent.
> - Automatic substitution, which is the practice of dispensing one medicine instead of another equivalent and interchangeable one at the pharmacy level, without previously consulting the prescriber.[32]

be safely switched from a reference biologic product to a biosimilar, or between two biosimilars.[30]

To realize the full promise of cost saving without compromising quality of care, there is clearly a need to switch patients treated with a reference product to a biosimilar. Indeed, there are ample examples of switching between reference biologic and biosimilar products for nonmedical reasons including selection in hospital formularies based on hospital purchase policies or national tenders, or at the request of patients.[31] Nevertheless, the question whether biosimilars can be safely and effectively switched has been highly contested and it remains a topic of fierce debate when discussing biosimilars. Also, different regulatory regions have taken different approaches toward switching and the concept of "interchangeability" (Scheme 7.1).

7.2.1 Global Approaches to Interchangeability

7.2.1.1 The European Approach

The EU, with the longest history of approving biosimilars, has restrained from providing any guidance on interchangeability.[4] In the European model, the EMA evaluates a medicinal product, following a positive opinion of the EMA's scientific committee, the Committee for Medicinal Products for Human Use (CHMP), the European Commission grants a marketing authorization, which means a product may be marketed across the EU. However, the individual member states decide on reimbursement of medicines, which means that the decision on interchangeable use and substitution of the biosimilars is taken at the national level. This

has sometimes been misinterpreted to mean that the EMA does not consider biosimilars to be interchangeable. However, individual members of the EMA's Biosimilars Working Party published an article in 2017 with the very clear conclusion:

> "On the basis of current knowledge, it is unlikely and very difficult to substantiate that two products, comparable on a population level, would have different safety or efficacy in individual patients upon a switch. Our conclusion is that biosimilars licensed in the EU are interchangeable."[33]

Unlike the US situation discussed below, in the EU "interchangeability" is a scientific term, which has no legal meaning. In European guidance, for most products there has never been a requirement to investigate the act of switching in pre-authorization clinical studies, the exception being epoetin, where a maintenance study in patients on stable treatment with the reference product is recommended. For some of the early biosimilars (e.g. filgrastim), switching is not likely to happen often, as patients are treated for short periods of time, so the question of interchangeability has never been a topic of major discussion.

When the first biosimilars were introduced onto the EU market, most countries were cautious about advising on switching to biosimilars. Also many medical societies or physician speciality groups were cautious and generally advocated that switching should not take place unless approved by the prescriber.[34] Over the past 13 years the experience with and overall acceptance of biosimilars has increased substantially in the EU and the general consensus at this moment across the EU is that switching should always involve the prescriber. In most European countries, automatic substitution (i.e. without prescriber consultation or consent) at the level of the pharmacist is not permitted.[35]

7.2.1.2 The United States Approach

Of all the regulatory regions, only the United States has a legal statute defining interchangeable products.

The United States Congress passed the Biologics Price Competition and Innovation (BPCI) Act of 2009 as part of the Patient Protection and Affordable Care Act.[36] The BPCI Act amended the Public Health and Service Act, thereby establishing a regulatory approval pathway for biosimilar products in the United States.[37] Unlike the European situation, interchangeability is a legal designation. This legal designation is of importance, because this means products can be substituted for the reference product by a pharmacist without the intervention of the prescribing healthcare practitioner, if so permitted by state laws. Interchangeability is defined in the BPCI Act to mean that the "interchangeable biological product is biosimilar to the reference product," and it can be "expected to produce the same clinical result as the reference product in any given patient." In addition, "for a product that is administered more than once to an individual, the risk in terms of safety or diminished efficacy of alternating or switching between use of the [interchangeable] product and its reference product is not greater than the risk of using the reference product without such alternation or switch."[36,37]

So, unlike the European situation, the United States has a legal definition of what it means to be an interchangeable product, which clearly mandates a *clinical* proof of interchangeability. The FDA was charged with implementing a regulatory framework around these statutory provisions (BPCI Act). It took several years before the final guidance was released describing the FDA's recommendations for obtaining the legal status of interchangeability.[38]

One of the key questions was how to design a clinical study that would meet FDA's requirements. The FDA recommends a dedicated switch study *"with a lead-in period of treatment with the reference product, followed by a randomized two-arm period – with one arm incorporating switching between the proposed interchangeable product and the reference product (switching arm) and the other remaining as a non-switching arm receiving only the reference product (non-switching arm) – may be appropriate when designing a switching study."* The primary endpoint of such a study should assess the impact of switching or alternating between the use of the proposed interchangeable product and the reference product on clinical PK and PD (if available).[38] To date, no product has sought for an "interchangeable designation" and thus no interchangeable products have been approved. Also, several questions are still open, for example, how to deal with multiple interchangeable products; would it suffice to demonstrate interchangeability vs the reference product only?

7.2.1.3 Approaches in Other Regions

The Australian TGA has adopted the European framework and guidelines for the approval of biosimilars. Interchangeability and substitutability is determined by the Australian Pharmaceutical Benefits Advisory Committee (PBAC), which has taken a case by case approach to determine whether a biosimilar is substitutable with its reference product (i.e. "a" flagged) based on three criteria[39]:

- The Therapeutic Goods Administration (TGA) has determined that the product is a biosimilar of the reference medicine as evidenced by ARTG registration documentation.
- Availability of supportive data relating to the effects of switching between the reference product and the biosimilar product/s.
- Practical considerations relating to substitution by the pharmacist at the point of dispensing. This includes strength of formulation, number of units per pack, and maximum quantities between the brands, which may make substitution at the pharmacy level difficult from a practical perspective.

Currently, at the time of writing this chapter, several biosimilars are considered interchangeable and substitutable at the level of the pharmacy without the requirement of prescriber involvement, including adalimumab, infliximab, and etanercept. In the Australian situation, even if a medicine is substitutable, the doctor can tick the "brand substitution not permitted" box when writing a prescription. If this box is ticked, by law the pharmacist cannot dispense a brand other than that prescribed.[40]

7.2.2 Experience from Switching Biosimilars

Since the authorization of the first biosimilar somatropin in the EU in 2006, considerable evidence has been gathered for over 50 approved biosimilar products in the EU as of April 2019 and over 20 in the United States (Figure 7.3).[41]

7.2.2.1 Data from Registration Trials

For somatropin biosimilars, the registration studies enrolled either volunteers or patients naive to somatropin treatment, which led to questions from prescribers whether these products can be safely switched.[42,43] The early epoetin studies did investigate clinical efficacy and safety in patients on stable treatment with the reference product, thus providing additional information on the effects of switching to a biosimilar.

Also, for regulatory registration studies for monoclonal antibodies, patients with prior use of the reference product were excluded. The main reason for this is that one of the key outstanding questions to be addressed in clinical studies is whether a biosimilar has the same immunogenicity profile as its reference product, which has led to the recommendation to enroll patients naive to the reference product.[28] At the time the monoclonal antibody

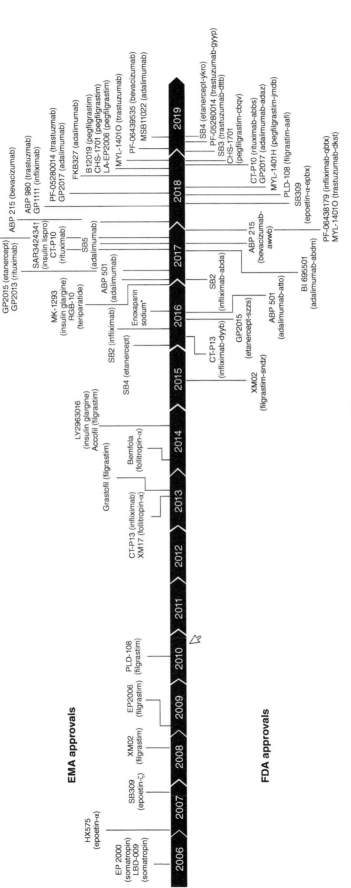

Figure 7.3 Time line of EMA and FDA biosimilar approvals. Source: From Ebbers and Schellekens.[30]

guidance was released by the EMA, it did include a paragraph that switching data may be a relevant part of the post-approval lifecycle management of a biosimilar:

> "Depending on the handling of biosimilars and reference medicinal products in clinical practice at national level, "switching" and "interchanging" of medicines that contain a given monoclonal antibody might occur. Thus, applicants are recommended to follow further development in the field and consider these aspects as part of the risk management plan."[28]

The discussions on switching patients intensified when biosimilars of TNF inhibitors came to the market, that were indicated for several chronic immune mediated inflammatory diseases, including rheumatoid arthritis (RA), inflammatory bowel disease (IBD), and psoriasis.

There was no requirement to investigate switching during the double-blind period and different study designs were employed. The first biosimilars, CT-P13 (infliximab) and SB4 (etanercept), did not investigate switching during the double-blind phase of the clinical efficacy studies, but switching was investigated in open-label extensions where a subset of patients who received the reference product during the double-blind period of the study transitioned to receive the biosimilar for a period up to one year after transitioning.[44–46] The products that followed SB5 and ABP 501 (both adalimumabs) included a double-blind switching arm in their pivotal studies. Some more recently approved products also adopted an "alternating design," probably with the aim of meeting US regulatory requirements with regard to establishing interchangeability.[47,48]

None of the double-blind studies published to date have reported a detrimental effect of switching to a biosimilar. Also alternating between biosimilars and their reference products did not lead to reports of negative effects on efficacy, safety, or of discontinuation rates as compared to continuous treatment with either the biosimilar or reference product (Figure 7.4).

Optimal switching study design

1. Randomized design with appropriate control arms
2. At least 1-way switch from originator to biosimilar
3. Assessment of immunogenicity
4. Sufficient washout between treatment (multiple switching)
5. Sufficient power to assess efficacy and safety (equivalence)
6. Sufficient follow-up period

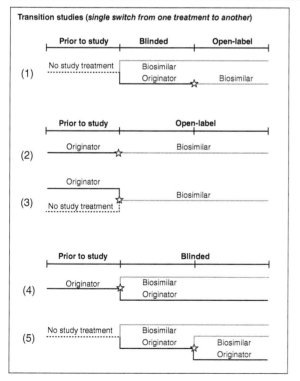

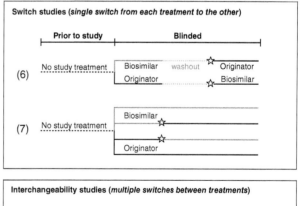

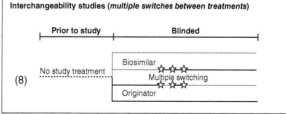

Figure 7.4 Elements of the optimal switching study and study designs employed by switching studies of reference biologics and biosimilars. *Source:* From Moots et al.[49]

7.2.2.2 Real-World Evidence

As discussed earlier, biosimilars are approved based on a tailored regulatory package that is deeply rooted in demonstrating analytical and functional similarity in the laboratory and a reduced clinical package to demonstrate similar PK and address any residual uncertainty in comparative clinical studies. These studies usually involve a limited number of patients, usually followed up for one to two years. After the introduction of the first biosimilars, several learned physician societies were not convinced that switching to biosimilars could be recommended because of questions around potential increases in immunogenicity following a switch, long-term safety, lack of switching data in actual patients, and concerns about extrapolation of indications.[34,50] This may have been partly due to an unfamiliarity with the biosimilarity concept. Also, the fact that only naive patients were enrolled in most randomized clinical trials (RCTs) for some frequently used products, particularly TNF inhibitors, raised questions on whether data from registration studies could be extrapolated to patients on stable reference biologic treatment. This has led to calls for "real-world data" for the use of biosimilars in the real patient populations to provide additional data and reassurance that biosimilars can be safely and effectively used in clinical practice. In some cases, such studies have been mandated by regulatory authorities, but also biosimilar manufacturers have conducted and reported on additional studies to provide additional reassurances for prescribers that biosimilars can be safely and effectively switched. Furthermore, real-world data may provide additional evidence to guide decision-making on some outstanding questions such as effectiveness in indications not studied during the clinical development program (indication extrapolation), the effects of switching from a reference product to a biosimilar, but also between biosimilars, or long-term effects of treatment with biosimilars. A substantial body of evidence has been collected on the performance of biosimilars after marketing authorization, particularly in Europe and several comprehensive reviews on the accumulated data have been published in recent years.[31,51,52] The conclusions based on the total body of evidence are clear. Based on the currently available substantial body of evidence, there is no signal that switching to biosimilars is associated with a loss of effectiveness and no evidence has been found that switching leads to increased immunogenicity. However, it must be noted that most, if not all, studies were not designed around safety endpoints and most of them were uncontrolled open-label studies.

A second finding was that several studies reported higher discontinuation rates in patients receiving biosimilars, something that was not seen in blinded studies.[53,54] It was noted in some studies that differences in terms of discontinuation rates as compared to patients being treated with the reference product were mostly explained by subjective adverse events, meaning adverse events that were experienced by patients, but could not be verified by the treating physician. Most data are available for TNF inhibitors and several studies found that these observed differences in discontinuation rates may be explained by so-called nocebo effects (Scheme 7.2).[55,56]

Scheme 7.2 Nocebo Effects

The term "nocebo" derives from the Latin verb nocere ("I shall harm") and was first introduced in the 1960s to designate noxious effects produced by a placebo during clinical trials.[57] The term nocebo effect was later more generally applied to describe "[negative effects] of a pharmacological or non-pharmacological medical treatment that is induced by patients' expectations, and that is unrelated to the physiological action of the treatment."[58] It is the negative equivalent of the better known "placebo effect" when positive expectations have been observed in many clinical situations, such as pain medication and epilepsy and switching to generic drugs. Nocebo effects have been described as "Nonspecific adverse effects are generally non-serious symptoms that are idiosyncratic, not clearly attributable to the pharmacological action of the drug involved, and not dose dependent" and have been observed for a variety of diseases.[57]

Closely related to nocebo effects are so-called incorrect attribution effects, where patients attribute certain events to their medication that may not necessarily be caused by this medication. The introduction of biosimilars has brought renewed attention to this phenomenon as they may be perceived as "cheap copies." Nocebo effects can have several negative consequences including lowering patients' quality of life and negatively affecting treatment adherence.[59]

Nocebo phenomena are mainly a result of physician–patient communication and inappropriate patient expectations, and communication strategies such as "positive framing" have been shown to reduce nocebo effects and increased acceptance of patients to switch to a biosimilar.[60,61] Increased understanding at the level of prescribers of the biosimilarity concept may foster confidence that may be transferred to patients, which may reduce nocebo effects.[62]

Data from the Danish DANBIO registry found no difference in terms of effectiveness following a nationwide switch of patients suffering from several inflammatory rheumatic diseases from reference etanercept to the biosimilar SB4. However, a higher rate of discontinuation was observed as compared to an historic cohort. Some patients were switched back to the reference product due to a lack of effectiveness. However, looking in more detail at these patients, at the time of switching back, objective measures of disease activity such as the number of swollen joints and the inflammatory blood marker C-reactive protein were unchanged, but a patient-reported measure of disease activity, the "patient global score," was considerably worse. Again, this suggests that nocebo effects appear to play a major role in patients being switched to a biosimilar. So, although current evidence does not indicate that switching to biosimilars negatively affects effectiveness and safety, nocebo effects are common and may lead to increased discontinuation of biosimilars. Healthcare providers have an important role in mitigating these effects by facilitating acceptance of biosimilars and minimizing/limiting the risks of nocebo effects.

7.2.2.3 Biosimilar to Biosimilar Switching

While there is now considerable evidence that switching to biosimilars does not affect effectiveness, most of these studies concern switches from a reference product to a biosimilar. The question has been raised whether the same can be said for switching between two biosimilars.[63] Concerns again are centered around the potential of immunogenicity, following a switch between two biosimilars. Limited data are currently available for biosimilar to biosimilar switching, at the time of writing, only on switching between different brands of infliximab, namely from CT-P13 to SB2. The limited studies performed so far do not report any concerns in terms of safety, immunogenicity, or reduced effectiveness following a switch between biosimilars in patients with IBD, rheumatic diseases, or psoriasis, but more data are awaited.[64–67]

7.3 Conclusions

The general approach toward demonstrating biosimilarity has become a well-established regulatory pathway across the globe. A framework pioneered in the EU has been in place for 13 years and is expanding, adapting and evolving continuously, and has become increasingly harmonized globally following the adoption of the WHO guidelines on evaluating biosimilars. Biosimilars are approved on a robust data package that ensures that there are no clinically relevant differences between a biosimilar and its reference product. However, different regulatory approaches have been taken across the globe toward interchangeability. The body of evidence gathered today is reassuring that no untoward effects occur when switching from a reference product to a biosimilar. Establishing interchangeability is first and foremost a demonstration of biosimilarity, while in the United States, additional clinical requirements may be required to fulfill the legal requirements with respect to interchangeability. Currently available data from clinical trials and real-world evidence support that switching to a biosimilar is safe and effective, but nocebo effects are valid and should be considered. While there is limited data available on switching between biosimilars or switching back to a reference product, there is no reason to assume that this will have clinical consequences.

References

1 World Health Organisation (2009). *Guidelines on Evaluation of Similar Biotherapeutic Products (SBPs), Annex 2.* Technical Report Series No. 977.

2 Vezer, B., Buzas, Z., Sebeszta, M., and Zrubka, Z. (2016). Authorized manufacturing changes for therapeutic monoclonal antibodies (mAbs) in European Public Assessment Report (EPAR) documents. *Current Medical Research and Opinion* 32 (5): 829–834.

3 International Conference on Harmonisation (2004). *Comparability of Biotechnological/Biological Products Subject to Changes in Their Manufacturing Process Q5E.* https://www.ich.org/fileadmin/Public_Web_Site/ ICH_Products/Guidelines/Quality/Q5E/Step4/Q5E_ Guideline.pdf (accessed 30 April 2019).

4 European Medicines Agency, European Commission. *Biosimilars in the EU. Information Guide for Healthcare Professionals.* https://ec.europa.eu/docsroom/ documents/22924/attachments/1/translations/en/ renditions/native (accessed 13 May 2019).

5 Kwon, O., Joung, J., Park, Y. et al. (2017). Considerations of critical quality attributes in the analytical comparability assessment of biosimilar products. *Biologicals: Journal of the International Association of Biological Standardization* 48: 101–108.

6 van den Bremer, E.T., Beurskens, F.J., Voorhorst, M. et al. (2015). Human IgG is produced in a pro-form that requires clipping of C-terminal lysines for maximal complement activation. *MAbs* 7 (4): 672–680.

7 Jung, S.K., Lee, K.H., Jeon, J.W. et al. (2014). Physicochemical characterization of Remsima. *MAbs* 6 (5): 1163–1177.

8 Pivot, X., Pegram, M., Cortes, J. et al. (2019). Three-year follow-up from a phase 3 study of SB3 (a trastuzumab biosimilar) versus reference trastuzumab in the neoadjuvant setting for human epidermal growth factor receptor 2-positive breast cancer. *European Journal of Cancer (Oxford, England: 1990)* 120: 1–9.

9 Vulto, A.G. and Jaquez, O.A. (2017). The process defines the product: what really matters in biosimilar design and production? *Rheumatology (Oxford)* 56 (suppl. 4): iv14–iv29.

10 van Berkel, P.H., Gerritsen, J., Perdok, G. et al. (2009). N-linked glycosylation is an important parameter for optimal selection of cell lines producing biopharmaceutical human IgG. *Biotechnology Progress* 25 (1): 244–251.

11 Bui, L.A., Hurst, S., Finch, G.L. et al. (2015). Key considerations in the preclinical development of biosimilars. *Drug Discovery Today* 20 (suppl. 1): 3–15.

12 Schiestl, M., Li, J., Abas, A. et al. (2014). The role of the quality assessment in the determination of overall biosimilarity: a simulated case study exercise. *Biologicals: Journal of the International Association of Biological Standardization* 42 (2): 128–132.

13 Food and Drug Administration (2019). *Development of Therapeutic Protein Biosimilars: Comparative Analytical Assessment and Other Quality-Related Considerations Guidance for Industry (draft)*. https://www.fda.gov/media/125484/download (accessed 17 July 2019).

14 International Conference on Harmonisation (2009). *Pharmaceutical Development Q8(R2)*. https://database.ich.org/sites/default/files/Q8_R2_Guideline.pdf (accessed 6 November 2019).

15 Ratanji, K.D., Derrick, J.P., Dearman, R.J., and Kimber, I. (2014). Immunogenicity of therapeutic proteins: influence of aggregation. *Journal of Immunotoxicology* 11 (2): 99–109.

16 European Medicines Agency (2014). *Guideline on Similar Biological Medicinal Products Containing Biotechnology-Derived Proteins as Active Substance: Quality Issues (Revision 1)*. EMA/CHMP/BWP/247713/2012. https://www.ema.europa.eu/en/documents/scientific-guideline/guideline-similar-biological-medicinal-products-containing-biotechnology-derived-proteins-active_en-0.pdf (accessed 16 August 2019).

17 Carpenter, J.F., Randolph, T.W., Jiskoot, W. et al. (2010). Potential inaccurate quantitation and sizing of protein aggregates by size exclusion chromatography: essential need to use orthogonal methods to assure the quality of therapeutic protein products. *Journal of Pharmaceutical Sciences* 99 (5): 2200–2208.

18 Koulov, A. (2019). Protein stability and characterization. In: *Pharmaceutical Biotechnology: Fundamentals and Applications* (eds. D.J.A. Crommelin, R.D. Sindelar and B. Meibohm), 33–56. Cham: Springer International Publishing.

19 Jiskoot, W. and Crommelin, D.J.A. (2019). Biophysical and biochemical characteristics of therapeutic proteins. In: *Pharmaceutical Biotechnology: Fundamentals and Applications* (eds. D.J.A. Crommelin, R.D. Sindelar and B. Meibohm), 19–32. Cham: Springer International Publishing.

20 Fekete, S., Guillarme, D., Sandra, P., and Sandra, K. (2016). Chromatographic, electrophoretic, and mass spectrometric methods for the analytical characterization of protein biopharmaceuticals. *Analytical Chemistry* 88 (1): 480–507.

21 Food and Drug Administration (2016). FDA briefing document: arthritis advisory committee meeting (9 February). https://www.fda.gov/media/95987/download (accessed 6 November 2019).

22 Weise, M., Kurki, P., Wolff-Holz, E. et al. (2014). Biosimilars: the science of extrapolation. *Blood* 124 (22): 3191–3196.

23 Ye, B.D., Pesegova, M., Alexeeva, O. et al. (2019). Efficacy and safety of biosimilar CT-P13 compared with originator infliximab in patients with active Crohn's disease: an international, randomised, double-blind, phase 3 non-inferiority study. *Lancet* 393 (10182): 1699–1707.

24 Webster, C.J., Wong, A.C., and Woollett, G.R. (2019). An efficient development paradigm for biosimilars. *BioDrugs* 33 (6): 603–611.

25 European Medicines Agency (2014). *Guideline on Similar Biological Medicinal Products*. CHMP/437/04 Rev 1. http://www.ema.europa.eu/docs/en_GB/document_library/Scientific_guideline/2014/10/WC500176768.pdf (accessed 17 June 2020).

26 European Medicines Agency (2015). *Assessment Report Solumarv*. EMA/596513/2015. https://www.ema.europa.eu/en/documents/assessment-report/solumarv-epar-public-assessment-report_en.pdf (accessed 15 November 2019).

27 Wolff-Holz, E., Tiitso, K., Vleminckx, C., and Weise, M. (2019). Evolution of the EU biosimilar framework: past and future. *BioDrugs* 33 (6): 621–634. https://doi.org/10.1007/s40259-019-00377-y.

28 European Medicines Agency. *Guideline on Similar Biological Medicinal Products Containing Monoclonal Antibodies – Non-clinical and Clinical Issues*. EMA/CHMP/BMWP/403543/2010. https://www.ema.europa.eu/en/documents/scientific-guideline/guideline-similar-biological-medicinal-products-containing-monoclonal-antibodies-non-clinical_en.pdf (accessed 2 September 2019).

29 Schellekens, H. (2002). Bioequivalence and the immunogenicity of biopharmaceuticals. *Nature Reviews. Drug Discovery* 1 (6): 457–462.

30 Ebbers, H.C. and Schellekens, H. (2019). Are we ready to close the discussion on the interchangeability of biosimilars? *Drug Discovery Today* 24 (10): 1963–1967. https://doi.org/10.1016/j.drudis.2019.06.016.

31 Cohen, H.P., Blauvelt, A., Rifkin, R.M. et al. (2018). Switching reference medicines to biosimilars: a systematic literature review of clinical outcomes. *Drugs* 78 (4): 463–478.

32 Trifiro, G., Marciano, I., and Ingrasciotta, Y. (2018). Interchangeability of biosimilar and biological reference product: updated regulatory positions and pre- and post-marketing evidence. *Expert Opinion on Biological Therapy* 18 (3): 309–315.

33 Kurki, P., van Aerts, L., Wolff-Holz, E. et al. (2017). Interchangeability of biosimilars: a European perspective. *BioDrugs: Clinical Immunotherapeutics, Biopharmaceuticals and Gene Therapy* 31 (2): 83–91.

34 Danese, S. and Gomollon, F. (2013). ECCO position statement: the use of biosimilar medicines in the treatment of inflammatory bowel disease (IBD). *Journal of Crohn's & Colitis* 7 (7): 586–589.

35 Medicines for Europe. Positioning statements on physician-led switching for biosimilar medicines (updated April 2018). https://www.medicinesforeurope.com/wp-content/uploads/2017/03/M-Biosimilars-Overview-of-positions-on-physician-led-switching.pdf (accessed 17 July 2020).

36 US Congress. Biologics Price Competition and Innovation Act of 2009. Public Law 111–148.

37 Christl, L.A., Woodcock, J., and Kozlowski, S. (2017). Biosimilars: the US regulatory framework. *Annual Review of Medicine* 68: 243–254.

38 Food and Drug Administration (2019). *Considerations in Demonstrating Interchangeability with a Reference Product Guidance for Industry*. https://www.fda.gov/media/124907/download (accessed 17 June 2020); https://www.fda.gov/downloads/Drugs/GuidanceComplianceRegulatoryInformation/Guidances/UCM537135.pdf (accessed 17 June 2020).

39 Australian Government (2018). Pharmaceutical benefits scheme. Public Summary Document – March 2018 PBAC Meeting. https://www.pbs.gov.au/industry/listing/elements/pbac-meetings/psd/2018-03/files/biosimilar-medicines-considering-brand-equivalence-substitution-psd-march-2018.docx (accessed 6 October 2019).

40 Australian Government, Department of Health (2019). *Biosimilar Medicines: The Basics for Health Care Professionals*. https://www1.health.gov.au/internet/main/publishing.nsf/content/biosimilar-awareness-initiative/$File/Biosimilar-medicines-the-basics-for-healthcare-professionals-Brochure.pdf (accessed 6 October 2019).

41 European Medicines Agency. European Public Assessment reports. https://www.ema.europa.eu/en/medicines (accessed 17 June 2020).

42 Declerck, P.J., Darendeliler, F., Goth, M. et al. (2010). Biosimilars: controversies as illustrated by rhGH. *Current Medical Research and Opinion* 26 (5): 1219–1229.

43 Ranke, M.B. (2008). New preparations comprising recombinant human growth hormone: deliberations on the issue of biosimilars. *Hormone Research* 69 (1): 22–28.

44 Emery, P., Vencovsky, J., Sylwestrzak, A. et al. (2017). Long-term efficacy and safety in patients with rheumatoid arthritis continuing on SB4 or switching from reference etanercept to SB4 [published online ahead of print, 9 August 2017]. *Annals of the Rheumatic Diseases* 76 (12): 1986–1991. https://doi.org/10.1136/annrheumdis-2017-211591.

45 Park, W., Yoo, D.H., Miranda, P. et al. (2017). Efficacy and safety of switching from reference infliximab to CT-P13 compared with maintenance of CT-P13 in ankylosing spondylitis: 102-week data from the PLANETAS extension study. *Annals of the Rheumatic Diseases* 76 (2): 346–354.

46 Yoo, D.H., Prodanovic, N., Jaworski, J. et al. (2017). Efficacy and safety of CT-P13 (biosimilar infliximab) in patients with rheumatoid arthritis: comparison between switching from reference infliximab to CT-P13 and continuing CT-P13 in the PLANETRA extension study. *Annals of the Rheumatic Diseases* 76 (2): 355–363.

47 Griffiths, C.E.M., Thaci, D., Gerdes, S. et al. (2017). The EGALITY study: a confirmatory, randomized, double-blind study comparing the efficacy, safety and immunogenicity of GP2015, a proposed etanercept biosimilar, vs. the originator product in patients with

moderate-to-severe chronic plaque-type psoriasis. *The British Journal of Dermatology* 176 (4): 928–938.

48 Blauvelt, A., Lacour, J.P., Fowler, J.F. Jr. et al. (2018). Phase III randomized study of the proposed adalimumab biosimilar GP2017 in psoriasis: impact of multiple switches. *The British Journal of Dermatology* 179 (3): 623–631.

49 Moots, R., Azevedo, V., Coindreau, J.L. et al. (2017). Switching between reference biologics and biosimilars for the treatment of rheumatology, gastroenterology, and dermatology inflammatory conditions: considerations for the clinician. *Current Rheumatology Reports* 19 (6): 37.

50 O'Callaghan, J., Barry, S.P., Bermingham, M. et al. (2019). Regulation of biosimilar medicines and current perspectives on interchangeability and policy. *European Journal of Clinical Pharmacology* 75 (1): 1–11.

51 McKinnon, R.A., Cook, M., Liauw, W. et al. (2018). Biosimilarity and interchangeability: principles and evidence: a systematic review. *BioDrugs: Clinical Immunotherapeutics, Biopharmaceuticals and Gene Therapy* 32 (1): 27–52.

52 Inotai, A., Prins, C.P.J., Csanadi, M. et al. (2017). Is there a reason for concern or is it just hype? – a systematic literature review of the clinical consequences of switching from originator biologics to biosimilars. *Expert Opinion on Biological Therapy* 17 (8): 915–926.

53 Odinet, J.S., Day, C.E., Cruz, J.L., and Heindel, G.A. (2018). The biosimilar nocebo effect? A systematic review of double-blinded versus open-label studies. *Journal of Managed Care & Specialty Pharmacy* 24 (10): 952–959.

54 Numan, S. and Faccin, F. (2018). Non-medical switching from originator tumor necrosis factor inhibitors to their biosimilars: systematic review of randomized controlled trials and real-world studies. *Advances in Therapy* 35 (9): 1295–1332.

55 Tweehuysen, L., Huiskes, V.J., van den Bemt, B.J. et al. (2018). Open-label non-mandatory transitioning from originator etanercept to biosimilar SB4: 6-month results from a controlled cohort study. *Arthritis & Rheumatology (Hoboken, NJ)* 70: 1408–1418.

56 Tweehuysen, L., van den Bemt, B.J.F., van Ingen, I.L. et al. (2018). Subjective complaints as the main reason for biosimilar discontinuation after open-label transition from reference infliximab to biosimilar infliximab. *Arthritis & Rheumatology (Hoboken, NJ)* 70: 60–68.

57 Planes, S., Villier, C., and Mallaret, M. (2016). The nocebo effect of drugs. *Pharmacology Research & Perspectives* 4 (2): e00208.

58 Pouillon, L., Danese, S., Hart, A. et al. (2019). Consensus report: clinical recommendations for the prevention and management of the nocebo effect in biosimilar-treated IBD patients. *Alimentary Pharmacology & Therapeutics* 49 (9): 1181–1187.

59 Kristensen, L.E., Alten, R., Puig, L. et al. (2018). Non-pharmacological effects in switching medication: the nocebo effect in switching from originator to biosimilar agent. *BioDrugs* 32 (5): 397–404.

60 Barnes, K., Faasse, K., Geers, A.L. et al. (2019). Can positive framing reduce nocebo side effects? Current evidence and recommendation for future research. *Frontiers in Pharmacology* 10: 167.

61 Gasteiger, C., Jones, A.S.K., Kleinstäuber, M. et al. (2019). The effects of message framing on patients' perceptions and willingness to change to a biosimilar in a hypothetical drug switch. *Arthritis Care & Research* https://doi.org/10.1002/acr.24012.

62 Rezk, M.F. and Pieper, B. (2017). Treatment outcomes with biosimilars: be aware of the nocebo effect. *Rheumatology and Therapy* 4 (2): 209–218.

63 Danese, S. and Peyrin-Biroulet, L. (2017). IBD: to switch or not to switch: that is the biosimilar question. *Nature Reviews. Gastroenterology & Hepatology* 14 (9): 508–509.

64 Harris, C., Harris, R., Young, D. et al. (2019). IBD biosimilar to biosimilar infliximab switching study: preliminary results. Abstract P0419. *UEG Journal* 7 (suppl. 8): 391.

65 Lauret, A., Moltó, A., Abitbol, V. et al. (2019). OP0227 effects of successive switches to different biosimilars infliximab on immunogenicity in chronic inflammatory diseases in daily clinical practice. *Annals of the Rheumatic Diseases* 78 (suppl. 2): 190–191.

66 Bouhnik, Y., Fautrel, B., Desjeux, G. et al. (2019). PERFUSE: a French prospective/retrospective non-interventional cohort study of infliximab-naive and transitioned patients receiving infliximab biosimilar SB2; 1st interim analysis. *UEG Journal* 7 (suppl. 8): 626.

67 Gisondi, P., Virga, C., and Girolomoni, G. (eds.) (2019). Cross-switch from CT-P13 to sb2 infliximab biosimilars in patients with chronic plaque psoriasis. *Journal of the European Academy of Dermatology and Venereology* 33: 29–29.

8

Pharmacokinetics of Biologics

Andrew J. McLachlan and Jeffry Adiwidjaja

Sydney Pharmacy School, Faculty of Medicine and Health, The University of Sydney, Sydney, New South Wales, Australia

---

KEY POINTS

- The pharmacokinetic properties of biologic medicines, including monoclonal antibodies, are significantly different from drugs that are small chemical entities.
- Molecular size and charge, differences in tertiary structure, glycosylation, and folding all influence their pharmacokinetic behavior.
- Formulation and route of administration have a major impact on the disposition and immunogenicity of therapeutic proteins.
- Many biologics display nonlinear pharmacokinetic behavior depending on the nature of the molecule, dose, and the expression of their molecular target.
- Factors that contribute to variability in monoclonal antibody pharmacokinetic behavior are different to those that affect small molecules.
- Anti-drug antibodies may significantly increase the clearance and reduce the efficacy of biological medicines.
- Mechanistic pharmacokinetic models and therapeutic drug monitoring are new tools to guide optimal dosing of monoclonal antibodies.

Abbreviations

Abbreviation	Full name
ADAs	Anti-Drug Antibodies
CL	Clearance
DDI	Drug–Drug Interaction
FcRn	Neonatal Fc Receptor
IgG	Immunoglobulin G
kDa	Kilodaltons

Abbreviation	Full name
mAb	monoclonal Antibody
PBPK	Physiologically Based Pharmacokinetics
PK/PD	Pharmacokinetics/ Pharmacodynamics
RES	Reticuloendothelial System
TMDD	Target-Mediated Drug Disposition

Biologics, Biosimilars, and Biobetters: An Introduction for Pharmacists, Physicians, and Other Health Practitioners,
First Edition. Edited by Iqbal Ramzan.
© 2021 John Wiley & Sons, Inc. Published 2021 by John Wiley & Sons, Inc.

8.1 Introduction

Biological medicines (or biologics) contain active substances from biological sources, such as living cells or organisms. Biologics comprise proteins, sugar moieties and/or nucleic acids, enzymes, or may be living cells (tissues). This broad class of drugs includes vaccines, blood components, gene therapy drugs, somatic cells, tissue-engineered medicines, and protein therapeutics. The latter subclass includes a broad range of macromolecules (polypeptides or proteins) that range from peptide molecules, human insulin, growth hormones, erythropoietin, enzymes, or large and complex molecules such as monoclonal antibodies (mAbs; 150 000 Da) which are typically the product of biotechnology processes (such as recombinant DNA techniques and manufactured in bacterial or mammalian cell culture).[1, 2] Recombinant human insulin was the first approved biotechnologically derived drug product (in 1982) which was followed by many protein therapeutics during the past four decades.[3] As biotechnological techniques and cell-based production methods have become available, the ability to produce human versions of therapeutic proteins has significantly expanded. Whereas once such proteins for therapeutic use would have been isolated from animal (e.g. bovine or porcine insulin) or human tissues (e.g. human growth hormone isolated from cadaveric pituitary glands) or plasma (e.g. clotting factors), proteins can now be manufactured to the highest standards of purity and safety (e.g. reduced risk for Creutzfeldt–Jakob disease [CJD] from cadaveric growth hormone or immunogenicity from animal extracted insulins).

Of the biologic medicines, it is the monoclonal antibodies which have revolutionized healthcare and the medicines industry with blockbusters that treat some of the most significant chronic (inflammatory bowel disease, rheumatoid arthritis, psoriasis) and life-threatening diseases (such as cancers and infection).[4] At present, there are approximately 150 monoclonal antibodies and peptide therapeutics approved for marketing by the United States Food and Drug Administration (FDA)[1,5,6] and European Medicines Agency (EMA)[7] across a diverse range of therapeutic areas with over 600 biologics, mostly monoclonal antibodies, in various clinical stages of development.[5] Among the top 10 selling drugs in 2018, 7 were monoclonal antibodies, most of which have been approved for treatment of cancers (pembrolizumab, trastuzumab, bevacizumab, nivolumab, and rituximab).

Adalimumab (Humira®), an anti-tumor necrosis factor (TNF) drug used for various types of arthritis has topped the global sales of drugs in several previous years and likely to do so in the coming years until its patent expiry and the introduction of the biosimilar products in the United States in 2023.[8]

Monoclonal antibodies are therapeutic proteins that target specific epitopes to achieve their therapeutic effects. These proteins can be partly (-xumab) or fully (-umab) humanized monoclonal antibodies or generated from a variety of animal cell systems such as murine cells (-omab) or be chimeric (-ximab). The antibody therapeutics available in the market today belong to the immunoglobulin G (IgG) class either as naked monoclonal antibodies (antibodies that work by themselves) or antibody derivatives, e.g. Fc-fusion proteins (Fc domain of IgG is genetically linked to a peptide or protein of interest), antibody–drug conjugates (a potent cytotoxic small molecule drug is conjugated to an antibody), and bispecific antibodies (an antibody engineered to bind to two different antigens or two different epitopes on the same antigen). It is noteworthy that Fc-fusion proteins are often not categorized as antibodies due to a lack of fragment that binds to antigens (Fab) in their structures. Some examples of different types of monoclonal antibodies (and their nomenclature) are presented in Table 8.1 (see also, Chapters 4 and 13).

Advances in biotechnology that have allowed reliable and efficient/enhanced production of human proteins as therapeutic agents have facilitated a greater understanding of the clinical pharmacology of therapeutic proteins, especially with respect to the relationship between the administered dose and the concentrations achieved in the body (pharmacokinetics) and the factors that influence pharmacokinetic behavior. Clinical pharmacological principles, including pharmacokinetics (relationship between dose, concentration, and time) and pharmacodynamics (relationship between dose/concentration and response), are the foundation of therapeutics. These fundamental concepts have an essential place in drug development, medicines regulation/approval, and clinical therapeutics to guide decision-making around dosing of medicines to patients. By their nature, these therapeutic proteins are complex, especially more complex than medicines that are synthetic or naturally occurring small molecules and their pharmacokinetics are influenced by a range of molecule-specific, disease-related, and physiological factors. The manner in which these medicines are produced, formulated, and administered can all have a significant impact on their

Table 8.1 Monoclonal antibody nomenclature, humanized component, and examples.

Nomenclature	Human component (%)	Generic suffix	Example (type)	Indication
Murine	0	-omab	Ibritumomab tiuxetan ZEVALIN® (Murine IgG1)	Non-Hodgkin's lymphoma
Chimeric	65	-ximab	Infliximab REMICADE® (Chimeric mouse/human IgG1)	Rheumatoid arthritis, Crohn's disease
Humanized	>90	-zumab	Trastuzumab HERCEPTIN® (CDR-grafted mouse/human IgG1)	Breast cancer
Fully human	100	-umab	Adalimumab HUMIRA® (Human IgG1)	Rheumatoid arthritis

IgG, immunoglobulin G.

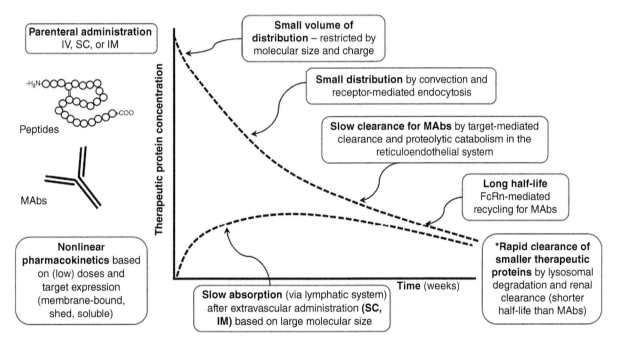

Figure 8.1 Graph summarizing the pharmacokinetic features of therapeutic proteins after parenteral administration.

pharmacokinetic and pharmacodynamic behavior in the body[4] (Figure 8.1). Indeed, many of the monoclonal antibodies in drug development failed to progress in clinical development and regulatory approval due to issues related to their pharmacokinetics.[5]

There is no better example of this pharmacokinetic and pharmacodynamic complexity (and subtlety) than insulin, one of the first biologics used in medicine.[9] The onset and duration of action of insulins vary significantly depending on subtle changes in structure and formulation from insulin lispro (onset, 15 minutes,

duration three to four hours) to insulin glargine (onset one to two hours, duration 24 hours).[9]

8.1.1 Focus of This Chapter

This chapter describes the general pharmacokinetics of biologics with a strong focus on monoclonal antibodies, as the most common and complex biological medicines. The aim is to provide an understanding of the pharmacokinetics processes that impact on the disposition of therapeutic proteins, such as monoclonal antibodies, and the unique factors that lead to pharmacokinetic variability. This will be discussed in the context of such information being used by health professionals and applied to drug development, regulatory decision-making, and optimizing patient care through dose individualization. This chapter also compares (and contrasts) the pharmacokinetic properties of biologicals (with a focus on monoclonal antibodies) and small molecule medicines as well as between monoclonal antibodies and other therapeutic protein drugs. The role of mechanistic pharmacokinetic modeling and therapeutic drug monitoring (TDM) to guide biologic dosing is also explored.

Pharmacokinetic and pharmacodynamic considerations play an important part of consideration of biosimilarity and interchangeability of biologic drugs. These issues are also briefly discussed here and more widely explored in Chapter 7.

8.2 Pharmacokinetics and Pharmacodynamics of Biological Medicines

Biological medicines, including monoclonal antibodies, have unique pharmacokinetics and pharmacodynamics properties (Figure 8.1), especially when compared to small molecule medicines. The "pharmacological" strengths and limitations of small molecules when compared to biologics are comprehensively summarized in Table 8.2.

The ability of monoclonal antibodies to specifically bind and target specific antigen ligands with high affinity gives these therapeutic agents clear advantage over small molecules, which lack this same specificity for their therapeutic target in the body. Yet, these biologics have some limitations when it comes to other pharmacological or pharmacokinetic properties. Biological medicines are absorbed, distributed, and eliminated via very different mechanisms when compared to medicines that are smaller synthetic or naturally occurring molecules, which in itself provides substantial barriers to the manner in which these medicines are administered and can access their pharmacological target. The complex interplay of factors that influence the pharmacokinetics of therapeutic proteins[2,4,5] are presented in Figures 8.1 and 8.2.

There are few notable differences between pharmacokinetic characteristics of mAbs and other (non-antibody) protein drugs, e.g. differences in the major pathway of absorption based on molecular sizes (i.e. via lymphatic or blood circulation), lack of FcRn-mediated recycling in non-antibody therapeutic proteins that protects them from lysosomal degradation (catabolism) and hence a relatively shorter half-life compared to that of mAbs, and different contribution of renal clearance depending on molecular size and charge.

8.2.1 Absorption

Oral administration is not possible due to the large molecular size and the degradation of proteins in the gastrointestinal tract, so most biologics are administered via the parenteral route typically intravenous (IV), subcutaneous (SC), or intramuscular (IM) administration.[2,4] Bioavailability (fraction of administered dose that reaches systemic circulation) following SC and IM administration may be erratic ranging from 20 to 95%. The absorption after these routes of administration can also be very slow (with peak plasma concentrations observed over one to eight days post-dose) and typically occurs via the lymphatic system. This slow systematic absorption after SC administration leads to absorption rate-limited elimination from the body. This can lead to low flat concentration–time profiles of monoclonal antibodies after SC administration. This is particularly the case for large therapeutic proteins (molecular weight larger than 16 kDa).[3] Conversely, those therapeutic proteins with molecular weights under this cutoff are predominantly well absorbed into the systemic circulation, e.g. recombinant human insulin and insulin aspart with time to reach peak plasma concentrations (t_{max}) in the order of hours.[10] Insulin glargine is an analog of human insulin that exploits the absorption rate-limited elimination by having a different solubility to native insulin at biological pH.[9] When administered SC, the glargine forms microprecipitates in the SC tissue slowly releasing the drug to provide a "smooth" concentration–time profile to replicate basal insulin secretion and providing a prolonged duration of action.[9]

8.2.2 Distribution

Distribution is an important pharmacokinetic characteristic for therapeutic proteins, as their pharmacological target is often an extracellular protein on the surface of cells within tissues. The extent of distribution of therapeutic proteins depends on the rates of extravasation (i.e. the ability to transfer out of the blood) in tissue and partition into the interstitial space. The molecular charge and size have a major impact on distribution and tissue uptake of biologics leading to very small volumes of distribution, typically between volume of plasma

Table 8.2 Pharmaceutical, pharmacokinetic, and pharmacodynamic differences between drugs that are synthetic or naturally occurring small chemicals and biologic medicines (such as immunoglobulins).

Characteristics	Small synthetic chemicals	Biologic medicines
Size (kDa)	Approximately 0.5	About 150
Structure	Chemical moiety	Protein (e.g. immunoglobulin)
Production	Controlled chemical synthetic processes or isolation	Cell-based production, complex isolation and purification required, Glycosylation, PEGylation
Formulation	Multiple dose forms (e.g. solid, liquid)	Typically, sterile isotonic solutions
Molecular target	Intracellular, extracellular proteins/transporters/channel, binding in receptor pocket, variety of tissues	Extracellular or soluble (shed) target antigens
Target specificity	Low	Extremely high
Routes of administration	All (oral, parenteral, topical, transdermal, local delivery)	Parenteral (mainly intravenous, subcutaneous, intramuscular)
Absorption	Bioavailability can vary (impacted by solubility, lipophilicity, transporter substrate specificity, first pass metabolism)	Slow absorption after parenteral administration (via lymphatics)
Distribution mechanism	Diffusion (rapid), influenced by solute and efflux transporters	Receptor-mediated endocytosis (slow), convection (slower)
Volume of distribution	Varied (low to very high) based on physicochemical properties, protein binding, transporters	Very low limited by molecular size
Access to target tissues	Access all tissues (including CNS depending on molecular properties, substrate specificity)	Limited tissue access due to molecular size
Elimination pathway	Hepatic (metabolism, excretion), renal pathways (filtration, secretion and/or reabsorption)	Reticuloendothelial system (catabolism), target-mediated clearance, mAb protected from degradation by FcRn-mediated recycling; smaller therapeutic proteins (less than 60 kDa) undergo renal excretion and metabolism
Clearance	High; influenced by pathway and organ function	Very low; influenced by target-mediated processes
Determinants of clearance	Intrinsic (body size, age, organ function, genetics, disease); extrinsic (drug–drug interactions, diet)	Intrinsic (albumin concentration, anti-drug antibodies, burden of disease/target); extrinsic (concomitant immunosuppressants)
Half-life	Hours to days	Weeks to months
Dose–exposure relationship	Typically, linear (rarely nonlinear in therapeutic range)	Often nonlinear due to saturable target-mediated disposition
Dose frequency	Usually daily	Weekly or monthly

kDa, kilodaltons; FcRn, neonatal Fc-receptor.

(0.04 L/kg) and the extracellular space (0.23 L/kg).[5,11] The extravasation of biologics may occur through passive diffusion (mainly for "smaller" proteins), convective transport, and transcytosis (transcellular transport in which macromolecules are transported across the interior of a cell). Unlike small molecule drugs, passive diffusion plays a minor role in the extravasation of most therapeutic proteins including monoclonal antibodies, especially due to the large molecular size of many biologics. Convection, the flux of fluid from the vascular space to the tissue driven by the blood-tissue hydrostatic gradient (liquid pressure gradient) with sieving by paracellular (transfer of solutes across an epithelium passing through the intercellular space between the cells) pores in the vascular epithelium, provides the major pathway by which monoclonal antibodies and other protein therapeutics enter the extracellular space in the tissue.[12] This is followed by a transfer into the lymphatic system, particularly for therapeutic proteins larger than 16 kDa, which slowly drains back into the systemic

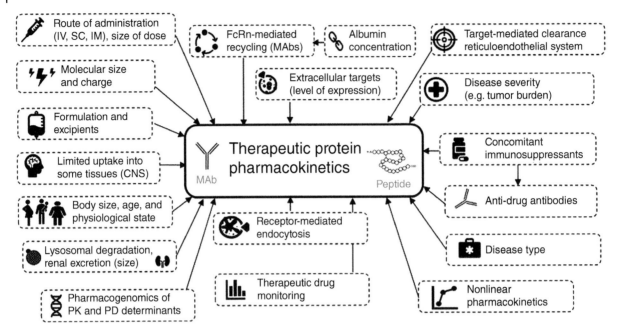

Figure 8.2 Schematic summarizing the many unique factors that influence the pharmacokinetics of therapeutic proteins.

circulation (15 mL/day) typically at the thoracic duct.[3] This slow transfer from the lymphatic to blood circulation also explains the delayed uptake (relatively long t_{max}) following SC and IM administration. Another less prominent pathway of distribution of mAbs is transcytosis which can occur via epithelial cells lining the vasculature, mediated via the neonatal Fc receptor (FcRn, also known as Brambell receptor). FcRn is expressed in vascular endothelial cells and the reticuloendothelial system (RES), with a lower expression on monocytes and macrophages.[2] Once in the extracellular tissue compartment, mAbs may bind to their target and enter the cell via receptor-mediated endocytosis or be recycled by the FcRn (Figure 8.3).[13] Interactions between nonspecific Fc region of mAbs and FcRn are pH-dependent with binding occurring at slightly acidic environment of endosomes (pH 6.0–6.5). The mAbs are then transported back to the cell surface and released from the complex with FcRn into the circulation, while unbound mAbs are catabolized to amino acids by lysosomes.

The plasma concentration–time (pharmacokinetic) profiles of mAbs following IV administration (Figure 8.1) typically follow a biexponential decline except in the nonlinear disposition of mAbs.[14,15] This can be best described by a two-compartment pharmacokinetic model which is basically a mathematical model that divides the body into hypothetical central and peripheral compartments. The latter represents a cluster of tissues

and interstitial spaces into which protein drugs are slowly distributed. The central compartment (vascular space and the interstitial space of well-perfused organs, e.g. liver and kidney) typically has a volume of distribution equal or slightly larger than volume of plasma (range: 3–5 L or 0.04–0.07 L/kg). The volume of distribution at steady state (V_{ss}, the sum of central and peripheral compartments) of approximately 5–10 L (0.07–0.14 L/kg) suggests a limited distribution outside vascular space, which is consistent with the behavior of endogenous IgG.[12] The fraction of the interstitial space that is not available for distribution (excluded volume) depends on the molecular size and charge of protein therapeutics.[16] The estimated V_{ss} of mAbs are relatively smaller and more homogenous compared to that of non-antibody protein drugs (Table 8.3). The latter may have V_{ss} close to total volume of extracellular space (range 8–20 L or 0.11–0.29 L/kg) with few exceptions, e.g. cyclosporine (cyclic polypeptide, 1200 Da) and interferon β-1b (18.5 kDa) with V_{ss} of 3–5 and 0.25–2.8 L/kg, respectively. These unusually large volumes of distribution are attributed to highly lipophilic and neutral properties of cyclosporine and high binding of interferon β-1b to intra- and extravascular proteins.[16]

Biodistribution of antibody–drug conjugates are more complex as they comprise a potent cytotoxic small molecule chemically linked to a large antibody molecule. This conjugation process leads to distribution of the cytotoxic payloads that is confined to plasma space and

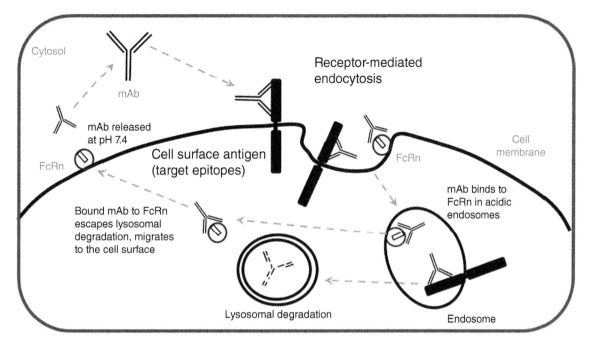

Figure 8.3 The process of receptor-mediated endocytosis and FcRn recycling of monoclonal antibodies – major impacts on their distribution and elimination.

target-expressing cells.[19] Once cleaved, the cytotoxic small molecule may diffuse to neighboring cells that do not express the target antigen (so-called, "bystander effect"), increasing the tumor killing activity and the risk of off-target toxicity.[20]

8.2.3 Elimination

Therapeutic proteins tend to be eliminated via different mechanisms when compared to small synthetic or naturally occurring drug molecules (Table 8.2). For example, proteolytic catabolism of mAbs within the RES is the primary route of drug elimination.[2,4] Monoclonal antibodies binding to the target antigen undergo phagocytosis in the RES and then their elimination occurs via intracellular catabolism, after receptor-mediated endocytosis (Figure 8.3). This is facilitated by the interaction of the binding domains of the monoclonal antibody with target epitopes found on cell surfaces. This type of endocytosis and elimination is a form of target-mediated disposition where the interaction of the drug and its pharmacological target (e.g. a target receptor or endogenous chemical/antigen) serves as a significant contributor to the pharmacokinetics of antibody distribution and elimination.[5]

Taken together, the rate of uptake into cells and the elimination of mAbs by target-mediated pathways is significantly influenced by the dose size and the level of expression of the target antigen or endogenous chemi-

cal and are influenced by the kinetics of receptor internalization (endocytosis) and intracellular (lysosomal) catabolism (depicted in Figure 8.3).

The rate of elimination in tissues is governed by the convective elimination clearance which is a function of the lymph flow rate (approximately 15 mL/day) and by the rate of protein catabolism within the tissue.[21] These relatively slow processes of elimination (endocytosis and catabolism) along with the recycling of mAbs via FcRn pathway leads to most mAbs having a relatively long half-life (in the order of weeks to months), despite the very small volume of distribution.[2,4,5] The long half-life for mAbs has implications for the frequency of dosing (typically weeks to months) and the time taken to achieve steady state during multiple dosing. Unlike mAbs, the elimination of non-antibody therapeutic proteins is restricted to nonspecific proteolytic catabolism pathways within plasma and tissues due to the lack of fragment of antigen binding (Fab) which can specifically bind to their targets and Fc region which may form a complex with Fc-γ-receptors (FcγR). These two saturable binding processes may trigger endocytosis (internalization) and catabolism.[3,21] Therefore, nonlinearity in pharmacokinetics of non-antibody biologics is less likely to be observed over the therapeutic dosing range as the elimination of these proteins relies on nonspecific catabolism pathways. The absence of FcRn-mediated salvage pathway also leads to most non-antibody biologics having substantially shorter

Table 8.3 Pharmacokinetic parameters for selected therapeutic proteins including Fc-fusion proteins and monoclonal antibodies.

Active molecule	Pharmacological indication	Route of administration	Volume of distribution (V_{ss}, L/kg)[a]	Clearance (L/d/kg)[a,b]	Half-life (units shown)
Peptide and protein (non-antibody) drugs					
Agalsidase β	Fabry disease	IV infusion	N.R.	Nonlinear PK	45–102 min
Alteplase	Acute ischemic stroke, myocardial infarction	IV	0.04	7.9–11.8	0.5 h
Anakinra	Rheumatoid arthritis	SC	0.13	2.9	4–6 h
Asparaginase (from *Escherichia coli*)[c]	Acute lymphoblastic leukemia	IV, IM	0.02	0.02	19 h
Asparaginase (recombinant)[c]	Acute lymphoblastic leukemia	IV, IM	0.08	0.05	18 h
Asparaginase (from *Erwinia chrysanthemi*)[c]	Acute lymphoblastic leukemia	IV, IM	N.R.	N.R.	16 h
PEGylated asparaginase[c]	Acute lymphoblastic leukemia	IV, IM	1.27 L/m^2	0.1–0.7 L/d/m^2	2.5–12 d
Cyclosporine	Autoimmune conditions, organ transplantation	Oral	3–5	7.2–10.1	8.4 h (range: 5–18 h)
Degarelix	Advanced prostate cancer	SC	0.65–0.82	1.0–1.2	43 d (range: 27–73 d)
Ecallantide	Hereditary angioedema	SC	0.38	3.1	2 h
Epoetin α	Anemia	IV, SC	N.R.	1.2	15–19 h
Eptifibatide	Acute coronary syndrome	IV	N.R.	1.4	1.5–2 h
Exenatide	Diabetes mellitus type 2	SC	0.40	3.1	2.4 h
Filgrastim	Low neutrophil counts	SC, IV	0.15	0.7–1.0	3.5 h
PEGylated filgrastim (pegfilgrastim)	Low neutrophil counts	SC	N.R.	N.R.	15–80 h
Follitropin α	Infertility	SC	0.14	0.2	32 h
Glucagon	Severe hypoglycemia	IV, IM, SC	0.25	19.4	18 min (IV), 45 min (IM)
Insulin (regular)	Diabetes mellitus	SC, IV infusion	0.26–0.36	34.3	1.5 h
Insulin aspart	Diabetes mellitus	SC, IV infusion	N.R.	38.4	1.5 h
Insulin lispro	Diabetes mellitus	SC, IV infusion	0.26–0.36	N.R.	1 h
Insulin degludec	Diabetes mellitus	SC	N.R.	N.R.	25 h
Insulin glargine	Diabetes mellitus	SC	N.R.	N.R.	12 h
Interferon β-1b	Multiple sclerosis	SC	0.25–2.88	13.4–41.5	8 min–4.3 h
PEGylated interferon α-2a (peginterferon α-2a)	Chronic hepatitis B and C infections	SC	N.R.	0.03	80 h (range: 50–140 h)
Leuprolide acetate	Endometriosis, prostate cancer	SC, IM	0.39	2.9	3 h
Liraglutide	Diabetes mellitus type 2	SC	0.16	0.2–0.5	13 h
Nafarelin acetate	Endometriosis, early puberty	Intranasal	N.R.	N.R.	3 h
Peginesatide	Anemia	SC, IV	0.05	0.01	53 h
Tenecteplase	Acute myocardial infarction	IV	0.04	1.9–2.4	1.5–2 h
Fc-fusion proteins					
Abatacept	Rheumatoid arthritis, psoriatic arthritis	IV infusion, SC	0.07	0.005	14 d

Table 8.3 (Continued)

Active molecule	Pharmacological indication	Route of administration	Volume of distribution (V_{ss}, L/kg)^a	Clearance (L/d/kg)^{a,b}	Half-life (units shown)
Alefacept	Psoriasis	IV, IM	0.09	0.008	11–14 d
Antihemophilic factor-Fc fusion protein (Eloctate®)	Hemophilia A	IV	0.05	0.05	20 h
Eftrenonacog alfa (Factor IX-Fc fusion protein)	Hemophilia B	IV	0.33	0.08	3.6 d
Etanercept	Rheumatoid arthritis, psoriatic arthritis	SC	0.11	0.02	4.3 d
Rilonacept	Cryopyrin-associated periodic syndrome	SC	0.13	0.01	8 d
Monoclonal antibodies					
Abciximab	Percutaneous coronary intervention	IV infusion	0.12	N.R.	0.5 h
Adalimumab	Rheumatoid arthritis, psoriatic arthritis	SC	0.09	0.003	14 d (range: 10–20 d)
Alemtuzumab	Chronic lymphocytic leukemia	SC, IV infusion	0.16	N.R.	6 d
Atezolizumab^d	Urothelial carcinoma, NSCLC, small cell lung cancers, triple-negative breast cancer	IV infusion	0.10	0.003	27 d
Avelumab^d	Merkel cell carcinoma, urothelial carcinoma, renal cell carcinoma	IV infusion	0.06	0.008	6.1 d
Basiliximab	Kidney transplantation	IV infusion	0.11	0.01	7.2 d
Bevacizumab	Metastatic colorectal cancer, NSCLC	IV infusion	0.08	0.003	20 d
Canakinumab	Cryopyrin-associated periodic syndrome	SC	0.09	0.002	26 d
Certolizumab pegol	Rheumatoid arthritis, psoriatic arthritis, Crohn's disease	SC	0.09	0.006	14 d
Cetuximab	Metastatic colorectal cancer, head and neck cancer	IV infusion	0.08	0.004	3–4 d
Denosumab	Osteoporosis, skeletal-related events due to bone metastases from solid tumors	SC	0.06	0.001	28 d
Durvalumab^d	Urothelial carcinoma, NSCLC	IV infusion	0.10	0.003	21 d
Eculizumab	Paroxysmal nocturnal hemoglobinuria	IV infusion	0.11	0.008	11.3 d
Golimumab	Rheumatoid arthritis, psoriatic arthritis, ankylosing spondylitis	SC, IV infusion	0.10	0.006	12.5 d
Ibritumomab tiuxetan	Non-Hodgkin's lymphoma	IV	N.R.	N.R.	28 h

(Continued)

Table 8.3 (Continued)

Active molecule	Pharmacological indication	Route of administration	Volume of distribution (V_{ss}, L/kg)[a]	Clearance (L/d/kg)[a,b]	Half-life (units shown)
Infliximab	Crohn's disease, ulcerative colitis, rheumatoid arthritis	IV infusion	0.09	0.004	8–9.5 d
Ipilimumab[d]	Melanoma, renal cell carcinoma, metastatic colorectal cancer	IV infusion	0.10	0.005	15 d
Natalizumab	Multiple sclerosis, Crohn's disease	IV infusion	N.R.	0.004	16 d
Nivolumab[d]	Melanoma, renal cell carcinoma, NSCLC, urothelial carcinoma, head and neck cancer	IV infusion	0.09	0.003	25 d
Ofatumumab	Chronic lymphocytic leukemia	IV	0.05	0.07	14 d
Omalizumab	Moderate to severe allergic asthma	SC	0.08	0.003	26 d
Palivizumab	Respiratory syncytial virus infection	IM, IV	0.06	0.005	18 d
Panitumumab	Metastatic colorectal cancer	IV infusion	0.09	0.004	7.5 d
Pembrolizumab[d]	Melanoma, microsatellite instability-high or mismatch repair deficient solid tumors	IV infusion	0.11	0.003	27 d
Ranibizumab	Age-related macular degeneration	Intravitreal	0.04	0.41	9 d
Rituximab	Non-Hodgkin's lymphoma, chronic lymphocytic leukemia, rheumatoid arthritis	IV	0.09	0.004	8.6 d
Tocilizumab	Rheumatoid arthritis, cytokine release syndrome	IV infusion, SC	0.09	0.004	8–14 d
Trastuzumab	Breast and gastric cancers overexpressing HER2	IV infusion	0.09	0.004	28.5 d
Ustekinumab	Plaque psoriasis, psoriatic arthritis, Crohn's disease, ulcerative colitis	SC, IV infusion	0.22	0.007	15–32 d
Antibody–drug conjugates					
Brentuximab vedotin	Hodgkin's lymphoma, anaplastic large-cell lymphoma	IV infusion	0.09–0.14	0.02–0.03	4–6 d
Gemtuzumab ozogamicin	Acute myeloid leukemia	IV	0.30	0.12	2.8 d
Inotuzumab ozogamicin	Acute lymphoblastic leukemia	IV	0.17	0.01	12.3 d
Trastuzumab emtansine	HER2-positive breast cancer	IV	0.06	0.01	4.2 d

Table 8.3 (Continued)

Active molecule	Pharmacological indication	Route of administration	Volume of distribution (V_{ss}, L/kg)[a]	Clearance (L/d/kg)[a,b]	Half-life (units shown)
Bispecific antibodies					
Blinatumomab	Acute lymphoblastic leukemia	IV	0.06	0.04	2.1 h
Emicizumab	Hemophilia A	SC	0.15	0.004	27 d

Source: Adapted from Refs. 3, 15, 17, 18.

[a] Mean clearance and volume of distribution reported as absolute values were normalized to a 70-kg body weight.

[b] Clearance values for routes of administration other than IV injection and infusion were reported as the apparent clearance (CL/F).

[c] Clinical pharmacokinetic data were collected from children with acute lymphoblastic leukemia.

[d] Belong to immune checkpoint inhibitors.

IM, intramuscular; IV, intravenous; N.R., not reported; NSCLC, non-small cell lung cancer; PK, pharmacokinetics; SC, subcutaneous; V_{ss}, volume of distribution at steady state.

half-lives compared to that of mAbs (Table 8.3). There are few exceptions though, for example, degarelix, a gonadotropin-releasing hormone (GnRH) antagonist used for hormonal therapy of prostate cancer has an unexpectedly long half-life (43 days) as a result of a very slow release of the drug from the depot formed at the SC injection site.[22] As discussed in Section 8.2.1, this phenomenon is known as absorption rate-limited elimination (also called "flip-flop pharmacokinetics") where the absorption rate is much slower than the elimination rate). Despite being an antibody, abciximab has a very short half-life (0.5 hour) due to a lack of Fc portion in its structure. Blinatumomab, a bispecific antibody that binds to CD19 and CD3 receptors present on B- and T-cells, respectively, also has an unusually short half-life (two hours) due to being composed of two short-chain variable regions (scFvs) joined together without an Fc portion.[23] Conversely, etanercept, abatacept, and alefacept are Fc-fusion proteins (relatively unstable proteins attached to the Fc domain of IgG) with half-lives in the order of days to weeks (Table 8.3). This highlights the importance of Fc regions in stabilizing protein therapeutics and prolonging their half-lives. PEGylation (covalent and non-covalent attachment of polyethylene glycol) offers another strategy to increase the stability and half-life of non-antibody protein drugs, albeit less effective than the attachment of Fc portion. For example, half-life of PEGylated asparaginase in pediatrics is significantly longer than that of native asparaginase from *Escherichia coli* (2.5–12 days vs 19 hours) with comparable volumes of distribution.[24,25] PEGylation involves conjugation of polyethylene glycol (PEG) chains with therapeutic proteins, which in turn increases the molecular mass of the protein protecting

the molecule from degradation by proteolytic enzymes. PEGylation is widely used for biologics to improve their pharmacokinetic profiles (typically reduced clearances and extended half-lives).

Once broken down, elimination of the antibody component in antibody–drug conjugates follows the pathways as described above, while the small molecule payloads are typically eliminated via hepatic and renal clearance pathways. The conjugation process significantly extends the half-life of small molecule payload from hours to days. However, the half-lives of antibody–drug conjugates are relatively shorter (only two to five days) than that of the corresponding unconjugated (naked) mAbs which may be related to the stability of the immunoconjugates.[19]

Little is known about pharmacokinetic properties of some biologics such as vaccines. Vaccines contain attenuated (weakened or killed) microorganisms or viruses or proteins from these organisms that stimulate the production of antibodies against one or several diseases and hence are essentially antigens. Concentrations of antigens (vaccines) in the lymphatic system, in which the immune systems are developed, are more likely to better correlate with the corresponding antibody titer (immunoprotective effect) compared to that in blood (plasma).[26] However, measurement of antigen concentrations in lymphatic circulation in the clinical setting is by no means a straightforward procedure. Moreover, pharmacokinetic evaluations of vaccines are not currently required for regulatory approval. This might explain the paucity of published clinical pharmacokinetic data of these biologic products. A phase I clinical trial of a pasteurized human tetanus immunoglobulin vaccine in healthy individuals hinted at a relatively slow elimination process (plasma half-life of the order of weeks)[27] which is comparable to that for many therapeutic mAbs.

8.2.4 Renal Elimination

Most therapeutic proteins generally have low renal clearances (when evaluated as urinary excretion rates relative to their plasma concentrations) due to their large molecular size and extensive metabolism in the proximal tubule for proteins that undergo glomerular filtration. The molecular weight cutoff for glomerular filtration is approximately 60 kDa, above which proteins are excluded from the glomerulus.[16] Peptides and proteins below 5 kDa undergo extensive renal filtration at a rate that can approach the glomerular filtration rate (GFR = 125 mL/min). Filtration rates of therapeutic proteins with molecular weights exceeding 30 kDa fall off sharply.[16] However, the efficiency of renal filtration rates of therapeutic proteins is not solely based on molecular size. Anionic proteins (e.g. albumin) are filtered much less readily than cationic and neutral proteins at similar molecular sizes due to the overall negative charge of glomerular capillary membrane. Once proteins undergo glomerular filtration, they can be catabolized predominantly by enzymes (exopeptidases) localized on the brush border membrane of proximal tubule. Proteins in the glomerular filtrate may also undergo reabsorption preventing loss into urine. FcRn plays a significant role in this reabsorption process as has been observed with endogenous IgG and albumin (despite lacking an Fc domain).[28] Endocytic receptors, e.g. megalin and cubilin, have also been reported to play a major role.[29,30] Tubular secretion has not been reported to any significant extent in the renal excretion of therapeutic proteins.

Unlike small molecule drugs, there is a paucity of clinical data of the effect of renal impairment on the pharmacokinetics of therapeutic proteins.[30] For protein drugs that undergo substantial renal elimination, e.g. anakinra (17 kDa, an interleukin-1 receptor antagonist), the total clearance may decrease in patients with renal impairment.[31] Additionally, total clearance of the antitopotecan antibody 8C2 (a 14 kDa murine antibody) was almost double in a murine model of diabetic nephropathy compared to that in control animals, with a strong correlation with urinary albumin excretion rate.[32]

8.2.5 Nonlinear Pharmacokinetics of Monoclonal Antibodies

Target-mediated elimination of monoclonal antibodies and FcRn-mediated recycling is capacity limited (saturable) because of limited expression of the target and of course this varies with the expression of the target (reflected in the burden of disease) influenced by whether this is a membrane-bound epitope, soluble ligand, or shed antigen (or a combination) (Figures 8.1 and 8.2). The abundance (level of expression) of target antigens may dictate the nonlinear relationship between dose and systemic exposure. For example, in individuals with a high burden of disease, target-mediated elimination is high and this form of clearance changes as the burden of disease is reduced during successful treatment. Interestingly, the nonlinear behavior is observed at lower doses (concentrations) in the body (yet to saturate membrane-bound or circulating targets). This nonlinear phenomenon in which the pharmacokinetics of drugs are affected by their binding to drug receptors (pharmacodynamics) is termed *target-mediated drug disposition* (TMDD).[33] This usually involves a high affinity binding of drugs to low capacity receptors and is more frequently observed with mAbs compared to non-antibody therapeutic proteins and small molecule drugs due to the presence of Fab in the former that specifically and (relatively) strongly binds to its target.[34] At low systemic concentrations of mAbs, TMDD accounts for a significant fraction of total mAb clearance, whereas at high mAb concentrations, target-mediated elimination becomes saturated and total mAb clearance decreases approaching a first-order (linear) process.

TMDD generally occurs for antibodies that target surface (membrane-bound) antigens rather than soluble antigens, the latter typically exhibits linear pharmacokinetics.[13,34] Examples of monoclonal antibodies with different membrane-bound, soluble, and shed targets are presented in Table 8.4. Shed extracellular domain (ECD) of the membrane-bound targets may affect the pharmacokinetics of mAbs as observed for trastuzumab.[35–37] There is a mutual relationship between concentrations of trastuzumab and shed antigen. The first-order rate of antigen shedding was driven by trastuzumab concentration and conversely, concentration of shed antigen in plasma and interstitial space may affect the target-mediated clearance of trastuzumab.[38] In a population pharmacokinetic analysis, plasma concentration of shed ECD of HER2 (target antigen for trastuzumab) was one of the significant covariates (influential factors) for trastuzumab clearance. However, its effect was modest compared to a high interindividual variability (coefficient of variation of clearance of 43%). The steady-state concentration of trastuzumab in women with shed HER2 concentrations of 200 ng/mL or above was lower by only 12% compared to those with a baseline level of 8 ng/mL.[36] The relatively small effect of antigen

Table 8.4 Examples of monoclonal antibodies with membrane-bound and shed targets and those with soluble ligands. Impact on linearity of pharmacokinetics.

Monoclonal antibody (product name)	Isotype	Target	Nature of target	Linearity of pharmacokinetics (therapeutic dosing range)
Adalimumab (HUMIRA®)	Humanized IgG1	TNFα	Soluble ligand[a]	Linear
Bevacizumab (AVASTIN®)	Humanized IgG1	VEGF	Soluble ligand	Linear
Ranibizumab (LUCENTIS®)	Humanized IgG1κ mAb fragment	VEGF	Soluble ligand	Linear
Infliximab (REMICADE®)	Chimeric IgG1	TNFα	Soluble ligand[a]	Linear
Trastuzumab (HERCEPTIN®)	Humanized IgG1κ	HER2	Membrane-bound and shed HER2	Nonlinear
Rituximab (RITUXAN®)	Human/Mouse chimeric IgG1κ	CD20	Membrane-bound and circulating CD20	Nonlinear/linear
Panitumumab (STELARA®)	Human IgG2/κ	EGFR	Membrane-bound and circulating EGFR	Nonlinear

Source: Adapted from Samineni et al.[35]
[a] Also exists in membrane-bound form.

shedding on pharmacokinetics of trastuzumab was in agreement with an evaluation using a more mechanistic modeling approach of a physiologically based pharmacokinetic (PBPK) model and simulation.[37]

Taken together, nonlinearity in the elimination of mAbs has implications for interpreting and predicting pharmacokinetic behavior between and within individuals prompting the increasing use of PBPK models[14,17,39] by researchers to inform mAb pharmacokinetics and individualized dosing (see Section 8.5). The formation of anti-drug antibodies (ADAs) may also alter pharmacokinetics of mAbs and lead to nonlinearity characterized by a much steeper decline of plasma concentrations due to increased clearance of the immune complexes (see Section 8.3.4).[34]

8.3 Understanding Pharmacokinetic Variability for Monoclonal Antibodies

Figure 8.2 captures the many factors that can influence the pharmacokinetics of monoclonal antibodies. There is an increasing understanding of intrinsic and extrinsic factors that influence monoclonal antibody pharmacokinetics. Body weight (especially obesity), disease severity, target antigen or endogenous chemical expression and turnover, and concomitant medicines can all play a role.[40]

8.3.1 Body Size

Body size (body weight) is the most commonly identified significant covariate (variables or factors that are specific to an individual and may explain pharmacoki-

netic variability) in population pharmacokinetic analyses of mAbs.[14] Body weight- and body surface area (BSA)-based dosing is often implemented to minimize interindividual variability in systemic exposure to protein therapeutics. This approach may be justifiable for most of non-antibody therapeutic protein drugs as volume of endosomal space and thus, the capacity and rate of nonspecific proteolytic catabolism is closely related to body size.[41] Conversely, in most cases, there is a lack of correlation between body size and target-mediated clearance for mAbs.[40] The extent of binding of mAbs to membrane-bound antigen is not likely to be determined by body weight, but rather by disease state (e.g. tumor load), target expression level, and the affinity of the mAbs for the antigen.[41] Based on an analysis of published pharmacokinetic models of 12 mAbs in adult patients, a fixed-dosing approach resulted in interindividual variabilities in systemic exposure that were comparable to that of body size-based dosing.[42] Similarly, a larger population pharmacokinetic-based analysis suggested that for most mAbs, the pharmacokinetic variability introduced by either dosing strategy is moderate relative to the overall variability. Additionally, the preference of dosing strategy may be evaluated on the basis of the exponential functions for the body size covariate in the population pharmacokinetic models.[43] However, it is noteworthy that the effect of body size on pharmacokinetic variability of mAbs in children is likely to be greater than that in adults, as was observed for blinatumomab (a bispecific antibody for the treatment of adult and childhood acute lymphoblastic leukemia)[44] and hence, body size-based dosing may be preferred over fixed-dosing regimen in this special patient population.

8.3.2 Albumin, FcRn, and Inflammation

Antibodies and albumin are recycled by FcRn receptors, protecting these proteins from catabolism leading to a long half-life in the body.[2] When IgG concentrations and/or albumin concentrations fluctuate, they can alter mAb clearance by interacting with FcRn receptor. Low albumin concentrations are strongly associated with severe inflammation in inflammatory bowel disease.[2] A close correlation between low serum albumin concentration and increased clearance (hence subtherapeutic concentrations) of infliximab has been observed.[45] Severe inflammation and diffuse ulcerative disease, characterized by elevated C-reactive protein and hypoalbuminemia, has been associated with decreased serum biologic drug concentrations.[2,45] Changes in albumin concentration and FcRn expression are unlikely to have an effect on the pharmacokinetics of non-antibody therapeutic proteins but inflammation certainly has the potential to influence the clearance and distribution of smaller therapeutic proteins via its effects on the activity and expression of catabolic enzymes and transporters.

8.3.3 Pharmacogenomics

Pharmacokinetic behavior and efficacy of monoclonal antibodies is significantly influenced by pharmacogenomic endpoints. This relates to allelic variation in genes that regulate pathways that are responsible for antibody recycling, target expression, and catabolism. All of these have the potential to influence mAb dosing requirements and treatment outcomes.[46] Chapter 9 of this book has comprehensively covered pharmacogenomic considerations of biologic drugs, including monoclonal antibodies.

8.3.4 Impact of ADAs on Pharmacokinetics and Pharmacodynamics

Many therapeutic proteins have the potential to generate an immunogenic response leading to the formation of ADAs. Monoclonal antibodies are exogenous immunoglobulins and as such all mAbs can generate ADAs. Partly (-xumab) or fully (-umab) humanized monoclonal antibodies have a significantly lower potential to form ADAs compared to murine (-omab) or chimeric (-ximab) antibodies (Table 8.1).[47] Characteristics of the mAb (or biosimilar) products (including glycosylation and/or impurities), formulation (aggregate formation, excipients), and the route of administration can all have an impact on the extent to which ADAs can form.[2,4] Interestingly, ADAs seem more likely to form at lower doses of mAbs when compared to higher doses and less likely when administered intravenously compared to SC administration.[2] The formation of ADAs seems more prevalent in combination therapies of mAbs compared to the corresponding monotherapies. The incidence of ADA formation against nivolumab (a mAb targeting PD-1) was significantly higher in patients being treated with a combination with ipilimumab (a CTLA-4 ligand) compared to that in monotherapy (22 vs 10%).[47] ADAs fall into two broad types, neutralizing and non-neutralizing depending on where they bind to the monoclonal antibody (Figure 8.4). All ADAs have the ability to form an immune complex which facilitates removal by the RES, leading to increased systemic clearance and lower concentrations. Neutralizing ADAs interact with the Fab region of the monoclonal antibody and significantly affect the ability to bind to the target and hence substantially reduced efficacy of the mAb. Whereas non-neutralizing antibodies bind to epitopes not essential for activity of mAbs (e.g. the Fc region), altering clearance, but the mAb–ADA complex retains some ability to bind the target molecule via the Fab region.[40]

Coadministration of immunosuppressants reduced the formation of ADAs reversing the effects on clearance. For example, concomitant treatment with methotrexate in people who have developed ADAs to infliximab can restore its therapeutic trough concentrations by decreasing the clearance of this mAb.[48]

8.3.5 Drug Interactions Involving Biologics

Biologic medicines do not share many elimination or clearance pathways with small molecule drugs that influence pharmacokinetics reducing the likelihood of direct drug–drug interactions between biologics and small molecule drugs (Table 8.2). Yet there are some examples of these interactions with mAbs explored in Table 8.5.[49] Many of these interactions are indirect, with small molecules or monoclonal antibodies impacting on pathways that regulate pharmacokinetic processes and primarily influencing the clearance of the affected ("victim") drugs. For example, anti-TNF and anti-IL 6 mAbs (e.g. infliximab and tocilizumab, respectively) and Fc-fusion proteins (e.g. etanercept) that alter the concentrations of active cytokines potentially reverse the inhibitory effect of these cytokines on the activity of drug-metabolizing enzymes (cytochrome P450 enzymes) and transporters (e.g. P-glycoprotein or ABCB1).[6,50] Conversely, interactions involving the small molecule portion (cytotoxic payload) of antibody–drug conjugates may be direct in

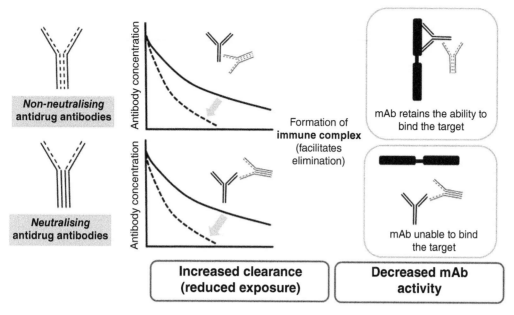

Figure 8.4 The impact of antidrug antibodies on the pharmacokinetics and pharmacodynamics of monoclonal antibodies.

Table 8.5 Drug interactions involving monoclonal antibodies.[2,49]

Class	Interaction	Comment
mAb as perpetrator mAb–small molecule DDIs	Regulation of drug metabolism (CYP) and transporters by mAbs (e.g. infliximab) alter the clearance of small molecules (e.g. verapamil)	mAbs potentially alter pharmacokinetics of concomitantly administered drugs. Inflammation (cytokines, TNFα) influences enzyme expression. Infliximab can alter drug metabolizing enzyme and transporter activity
mAbs as victim mAb–small molecule DDIs	Immunosuppressants (e.g. methotrexate) and anticancer drugs (e.g. paclitaxel) reduced the clearance of mAbs (infliximab, trastuzumab)	Coadministered drugs potentially affect mAbs pharmacokinetics. Mechanism is unclear but likely to relate to reduced immunogenicity (ADAs) or expression of extracellular ligands
		Immunosuppressants and cytotoxic agents have the potential to impact on disease severity, altering the expression of the mAb target, with implications for clinical efficacy and target-mediated clearance
mAb–mAb DDI	FcRn-mediated recycling is capacity limited which provides the possibility that coadministration of mAbs (at sufficient doses) may lead to changes in pharmacokinetics	Few examples of direct antibody–antibody pharmacokinetic interactions have been identified. Anti-FcRn mAbs and high-dose of IVIg increased the systemic clearance of pathogenic auto-antibody in patients with immune thrombocytopenia

The "perpetrator" drug causes the interaction with the "victim" drug whose pharmacokinetics are altered.
ADA, anti-drug antibodies; DDI, drug–drug interaction; IVIg, intravenous immunoglobulin; mAbs, monoclonal antibodies.

nature. The concentrations of the small molecule component are generally not high enough to cause an interaction ("perpetrators"); however, interactions can be found as "victim" drugs. For example, ketoconazole and rifampicin (potent CYP3A modulators) increased and lowered the systemic exposures of monomethyl aurista-

tin E (MMAE) in brentuximab vedotin by 34 and 46%, respectively.[6]

Direct antibody–antibody interactions involving a saturation of FcRn-mediated salvage pathway have been observed clinically. Although FcRn recycling is capacity limited, its saturation is not typically achieved with therapeutic doses of mAbs. Conversely, saturation may be observed at a massive dose of "perpetrator" antibody which leads to an increase in clearance of the affected antibody. This strategy has been implemented in the treatment of immune thrombocytopenic purpura (an autoimmune disorder characterized by low platelet counts due to increased platelet destruction) by a high-dose intravenous immunoglobulin (IVIg, 2 g/kg) which is sufficient to saturate FcRn recycling pathway and increase the clearance of endogenous, pathogenic autoantibodies.[21,51] This can also be achieved by high affinity mAbs against human FcRn, e.g. rozanolixizumab at a much lower dose.[52]

8.4 Pharmacokinetics of Biosimilar and Biobetter Biologics

A biosimilar is a biological medicine which is highly similar or comparable (and therefore theoretically interchangeable) to another therapeutic protein product already approved (reference/innovator product).[53] Biosimilars have the same amino acid sequence as the reference product but may be produced from different clones and manufacturing processes.[54] Unlike generic medicines for small molecule drugs, evaluation of biosimilarity of protein drugs requires more extensive data, including preclinical and clinical studies, e.g. structural and analytical studies, toxicological tests in animal model(s), and clinical pharmacokinetic (and pharmacodynamic) and immunogenicity evaluations. The natural variability and the complexity of the manufacturing process makes it impossible to produce exact copies of therapeutic proteins.[53]

Pharmacokinetic and pharmacodynamic evaluations are the cornerstone to investigating the comparability of biosimilar products with their innovator product. It is noteworthy that immunogenicity (development of ADAs) may affect pharmacokinetic profiles of biologics and this may vary between biosimilars and their reference products. Unfortunately, even small differences in manufacturing may lead to differences in incidence of ADA formation.[55] The general acceptable criteria for key pharmacokinetic parameters, e.g. peak plasma concentration (C_{max}) and area under the plasma concentration-time curve (AUC), are that the 90% confidence intervals for the ratio of the geometric mean value of the biosimilar to the reference products lie within the internationally accepted limits of 0.80–1.25.[53] Interestingly, most of the reference biological medicines available in

the United States and European countries have comparable pharmacokinetic and pharmacodynamic properties, thus, clinical bridging studies between reference products for global marketing may not be necessary.[56] There remains a strong focus on pharmacovigilance to assess safety and immunogenicity of biosimilar products once they enter the market (see Chapter 11).

Biobetter biologics have a similar pharmacological activity (i.e., target the same receptor) as a marketed biologic product but have been engineered to have improved pharmacokinetic (and pharmacodynamic) properties.[54] These modifications may include amino acid substitution (e.g. insulin lispro and aspart that are absorbed more rapidly from the injection site compared to regular insulin and insulin glargine that forms a depot at the injection site due to an isoelectric point close to neutral pH to give sustained activity),[9] optimizing glycosylation profile to enhance effector function (antibody-dependent cellular cytotoxicity), attachment to polyethylene glycol (PEGylation, e.g. PEGylated asparaginase, filgrastim, and interferons), and engineering the Fc domain of antibodies to reduce elimination and therefore, extending their half-lives.[54] Each of these modifications is designed to produce a "better" biologic than the innovator therapeutic protein and much of the focus is on improved pharmacokinetic properties. This chapter has discussed the factors that influence the pharmacokinetic of therapeutic proteins; it is these factors which are manipulated to produce biobetters. Pharmacokinetic studies are a vital aspect of the evaluation and regulation for biobetters to ensure that these re-engineered therapeutic proteins demonstrate improved patient therapeutic outcomes.

8.5 Modeling and Simulation of Monoclonal Antibody Pharmacokinetics and Pharmacodynamics

Given the complexity of the pharmacokinetic behavior of therapeutic proteins (especially mAbs) and the factors that may influence the disposition of these drugs (Figure 8.2), modeling and simulation has been valuable to quantify variability in biologics pharmacokinetics and pharmacodynamics. The number of published reports of validated population pharmacokinetic (also called "pharmacometric") models[14,15] to describe mAb PK/PD continues to expand to support the understanding of the impact of influential covariates, mechanistic insights, and Bayesian forecasting to guide dosing of mAbs. Of all the modeling approaches, PBPK modeling of mAb has provided novel insights and understanding of the mechanistic drivers for mAb pharmacokinetic behavior.[17] This PBPK approach provides a platform for decision-making in regulatory science and drug development.[57,58] PBPK models consist of multiple compartments representing different organs (tissues) connected by blood (and lymphatic) flow.[17] By incorporating anatomical and demographical data and physiological processes underlying disposition of mAbs, a PBPK model provides a mechanistic quantitative platform to explore potential sources of interindividual variability in pharmacokinetics of mAbs. Use of PBPK models for mAbs is useful in limiting the need for further clinical trials in some settings such as change of indication and dose regimen justification.[17]

For example, while most mAb target extracellular epitopes are membrane bound, there are a few mAb that bind to targets that are shed into the extracellular space and systemic circulation. Understanding how to dose these mAbs and describe their pharmacokinetic and pharmacodynamic behavior has been achieved with PK/PD modeling which uses mechanistic models to describe and predict mAbs concentrations but also biomarkers of drug response when they are integrated into the model.[35] Another example extends to the application of modeling to characterize the time-dependent nature of mAb pharmacokinetics, especially when the mAb clearance changes over time due to target-mediated disposition.[39]

8.6 Individualizing Therapy and the Role of TDM

Understanding the factors that influence the pharmacokinetics of biologics is essential to guiding optimal dosing and TDM has been identified as a key tool to support dose individualization for biologics. TDM has an important role in identifying patients at risk of treatment failure or for exploring poor response to mAb therapy. For most therapeutic proteins, dose adjustment is made based on a biomarker of response rather than drug concentration. For example, insulin doses are adjusted based on glycemic control and erythropoietin doses are moderated based on hematocrit. There are few clinically available assays for many therapeutic proteins and lack of data (evidence) on the concentration–effect relationship to guide dosing using a target concentration approach to TDM.

There is a general consensus of the utility of TDM to guide treatment in areas such as anti-TNF therapies

(e.g. infliximab) in individuals with inflammatory bowel disease[59]; however, it is clear that more research is needed to address some of the challenges such as optimal analytical methods, consideration of ADAs, appropriate pharmacokinetic parameter/variable or metric (typically trough concentration), optimal target concentration ranges associated with clinical response in different diseases, role of concomitant drugs, and the cost-effectiveness[60] of TDM of biologics.[61]

The unique properties of therapeutic proteins (including mAbs) present different challenges and kinds of analytical tools (mass, activity, or immunogenicity assays) for application in TDM. Enzyme-linked immunosorbent assays (ELISA), homogenous mobility shift assay (HMSA), and electrochemiluminescence immunoassay (ECLIA) are common for many mAbs. TDM requires a clear understanding of what is being measured, recognizing whether an assay is reporting free (or unbound) mAb or total mAb which may include what is bound to soluble or shed target and circulating ADAs. The usual challenges of specificity, cross-reactivity, and sensitivity can be additional challenges to TDM of biologics including for mAbs.[61]

The most common area where TDM has been applied is for anti-TNF biologic agents in the therapeutic area of inflammatory bowel disease.[62] Trough concentrations of infliximab and adalimumab have been associated with clinical remission in Crohn's disease and ulcerative colitis.[61] For biologics used in cancer treatment, there is growing evidence to support an association between trough concentration and clinical efficacy (typically, overall survival) for cetuximab, rituximab, trastuzumab, and ipilimumab.[61] As discussed above, implementation of TDM for non-antibody therapeutic proteins is more limited. One notable example is the monitoring of enzyme activity (a more sensitive metric than its concentration) of asparaginase in children with acute lymphoblastic leukemia.[63] The generally acceptable threshold is 100 U/L, above which asparaginase activity should be maintained throughout the course of treatment.[64] Interestingly, different preparations of asparaginase are available in the market (recombinant and native asparaginase from *E. coli*, asparaginase derived from *Erwinia chrysanthemi*, and PEGylated asparaginase/Oncaspar®) with nonidentical pharmacokinetic profiles and hence requiring different dosing regimens (2500–10 000 U/m² every two to three days, 25 000 U/m² every two to three days and 1000 U/m² every other week, respectively).[63] Patients may develop antibodies against the protein and/or PEG moieties (for Oncaspar®)[65] which leads to inferior treatment outcomes. Unfortunately, this is often not accompanied by observable allergic reactions (silent inactivation) and thus rendering TDM a potential integral part of treatment with asparaginase.[66]

Further refinement of pharmacodynamic targets associated with benefit and harm are needed to guide dose adjustment of therapeutic proteins of interest. The availability of point-of-care assays for selected therapeutic proteins and the use of Bayesian pharmacokinetic modeling integrated into dosing dashboards has the potential to see the innovation in analytical methods and rigorous clinical pharmacokinetic models translated into the clinical/practice setting.[67,68]

8.7 Conclusions

Biologic medicines display more complex pharmacokinetic and pharmacodynamic behaviors than small molecule drugs. The pharmacokinetic behavior of biological medicines is significantly influenced by their molecular size and charge. The method of production of the therapeutic proteins can have a major impact on tertiary structure, glycosylation, and folding which in turn influences their pharmacokinetic behavior. There is growing understanding of how aspects of the formulation can influence the disposition and immunogenicity of therapeutic proteins. Owing to large molecular sizes, the biodistribution of therapeutic proteins and many biologics are typically limited to vascular and interstitial spaces, proteins also undergo minimal renal excretion and are typically absorbed from extravascular administration sites via the lymphatic circulation. Potential for drug interactions are generally low since biologics are eliminated through a nonspecific pathway (proteolytic catabolism).

Monoclonal antibodies constitute a sizable portion of biologics available in the market. Their half-lives (in the order of weeks to months) are relatively longer than that of smaller (non-antibody) therapeutic proteins due to the presence of FcRn-mediated recycling pathway that gives protection from lysosomal degradation. Target-mediated clearance of several mAbs and formation of antibody against protein drugs (ADAs) may lead to nonlinear pharmacokinetic behavior. The disposition and effects of mAb as therapeutic agents can vary between and within patients. A greater understanding of the factors that influence mAb pharmacokinetics and pharmacodynamics is required, based on mechanistic insights into molecule-related factors, physiological processes, disease factors as well as other intrinsic and extrinsic factors.

Modeling and simulation methods, especially PBPK models, provide mechanistic insights into mAb pharmacokinetic behaviors and guide dosing in scenarios that have not been explored directly in clinical trials. There is growing evidence to support targeting specific concentration-effect relationship leading to improved health outcomes, which has paved the way for the use of TDM strategies and Bayesian informed dashboards to guide and individualize dosing of therapeutic mAbs.

Acknowledgements

The authors thank Professor Iqbal Ramzan and Dr. Stephanie Reuter Lange for helpful discussions and comments on this chapter.

References

1 Kinch, M.S. (2015). An overview of FDA-approved biologics medicines. *Drug Discov. Today* 20 (4): 393–398.

2 Mould, D.R. (2015). The pharmacokinetics of biologics: a primer. *Dig. Dis.* 33 (suppl. 1): 61–69.

3 Rathi, C. and Meibohm, B. (2015). Pharmacokinetics of peptides and proteins. *Rev. Cell Biol. Mol. Med.* 1 (2): 300–326.

4 Lobo, E.D., Hansen, R.J., and Balthasar, J.P. (2004). Antibody pharmacokinetics and pharmacodynamics. *J. Pharm. Sci.* 93 (11): 2645–2668.

5 Datta-Mannan, A. (2019). Mechanisms influencing the pharmacokinetics and disposition of monoclonal antibodies and peptides. *Drug Metab. Dispos.* 47 (10): 1100–1110.

6 Jing, X., Ji, P., Schrieber, S.J. et al. (2020). Update on therapeutic protein-drug interaction: information in labeling. *Clin. Pharmacokinet.* 59 (1): 25–36. https://doi.org/10.1007/s40262-019-00810-z.

7 Klein, K., De Bruin, M.L., Broekmans, A.W., and Stolk, P. (2015). Classification of recombinant biologics in the EU: divergence between national pharmacovigilance centers. *BioDrugs* 29 (6): 373–379.

8 Urquhart, L. (2019). Top drugs and companies by sales in 2018. *Nat. Rev. Drug Discov.* published online 12 March 2019). doi: https://doi.org/10.1038/d41573-019-00049-0.

9 Morello, C.M. (2011). Pharmacokinetics and pharmacodynamics of insulin analogs in special populations with type 2 diabetes mellitus. *Int. J. Gen. Med.* 4: 827–835.

10 Osterberg, O., Erichsen, L., Ingwersen, S.H. et al. (2003). Pharmacokinetic and pharmacodynamic properties of insulin aspart and human insulin. *J. Pharmacokinet. Pharmacodyn.* 30 (3): 221–235.

11 Glassman, P.M., Abuqayyas, L., and Balthasar, J.P. (2015). Assessments of antibody biodistribution. *J. Clin. Pharmacol.* 55 (suppl. 3): S29–S38.

12 Ryman, J.T. and Meibohm, B. (2017). Pharmacokinetics of monoclonal antibodies. *CPT Pharmacometrics Syst. Pharmacol.* 6 (9): 576–588.

13 Liu, L. (2018). Pharmacokinetics of monoclonal antibodies and Fc-fusion proteins. *Protein Cell* 9 (1): 15–32.

14 Dirks, N.L. and Meibohm, B. (2010). Population pharmacokinetics of therapeutic monoclonal antibodies. *Clin. Pharmacokinet.* 49 (10): 633–659.

15 Keizer, R.J., Huitema, A.D., Schellens, J.H., and Beijnen, J.H. (2010). Clinical pharmacokinetics of therapeutic monoclonal antibodies. *Clin. Pharmacokinet.* 49 (8): 493–507.

16 Meibohm, B. (2019). Pharmacokinetics and pharmacodynamics of therapeutic peptides and proteins. In: *Pharmaceutical Biotechnology: Fundamentals and Applications*, 5e (eds. D.J.A. Crommelin, R. Sindelar and B. Meibohm). New York: Springer.

17 Dostalek, M., Gardner, I., Gurbaxani, B.M. et al. (2013). Pharmacokinetics, pharmacodynamics and physiologically-based pharmacokinetic modelling of monoclonal antibodies. *Clin. Pharmacokinet.* 52 (2): 83–124.

18 Shah, D.K. (2015). Pharmacokinetic and pharmacodynamic considerations for the next generation protein therapeutics. *J. Pharmacokinet. Pharmacodyn.* 42 (5): 553–571.

19 Hinrichs, M.J. and Dixit, R. (2015). Antibody drug conjugates: nonclinical safety considerations. *AAPS J.* 17 (5): 1055–1064.

20 Malik, P., Phipps, C., Edginton, A., and Blay, J. (2017). Pharmacokinetic considerations for antibody-drug conjugates against cancer. *Pharm. Res.* 34 (12): 2579–2595.

21 Wang, W., Wang, E.Q., and Balthasar, J.P. (2008). Monoclonal antibody pharmacokinetics and pharmacodynamics. *Clin. Pharmacol. Ther.* 84 (5): 548–558.

22 Shore, N.D. (2013). Experience with degarelix in the treatment of prostate cancer. *Ther. Adv. Urol.* 5 (1): 11–24.

23 Duell, J., Lammers, P.E., Djuretic, I. et al. (2019). Bispecific antibodies in the treatment of hematologic malignancies. *Clin. Pharmacol. Ther.* 106 (4): 781–791.

24 Hempel, G., Muller, H.J., Lanvers-Kaminsky, C. et al. (2010). A population pharmacokinetic model for pegylated-asparaginase in children. *Br. J. Haematol.* 148 (1): 119–125.

25 Borghorst, S., Pieters, R., Kuehnel, H.J. et al. (2012). Population pharmacokinetics of native *Escherichia coli* asparaginase. *Pediatr. Hematol. Oncol.* 29 (2): 154–165.

26 Gomez-Mantilla, J.D., Troconiz, I.F., and Garrido, M.J. (2016). ADME processes in vaccines and PK/PD approaches for vaccination optimization. In: *ADME and Translational Pharmacokinetics/ Pharmacodynamics of Therapeutic Proteins* (eds. H. Zhou and F.P. Theil), 347–368. Hoboken, NJ: Wiley.

27 Forrat, R., Dumas, R., Seiberling, M. et al. (1998). Evaluation of the safety and pharmacokinetic profile of a new, pasteurized, human tetanus immunoglobulin administered as sham, postexposure prophylaxis of tetanus. *Antimicrob. Agents Chemother.* 42 (2): 298–305.

28 Sand, K.M., Bern, M., Nilsen, J. et al. (2014). Unraveling the interaction between FcRn and albumin: opportunities for design of albumin-based therapeutics. *Front. Immunol.* 5: 682.

29 Christensen, E.I., Birn, H., Storm, T. et al. (2012). Endocytic receptors in the renal proximal tubule. *Physiology (Bethesda)* 27 (4): 223–236.

30 Chadha, G.S. and Morris, M.E. (2016). Monoclonal antibody pharmacokinetics in type 2 diabetes mellitus and diabetic nephropathy. *Curr. Pharmacol. Rep.* 2: 45–56.

31 Yang, B.B., Baughman, S., and Sullivan, J.T. (2003). Pharmacokinetics of anakinra in subjects with different levels of renal function. *Clin. Pharmacol. Ther.* 74 (1): 85–94.

32 Engler, F.A., Zheng, B., and Balthasar, J.P. (2014). Investigation of the influence of nephropathy on monoclonal antibody disposition: a pharmacokinetic study in a mouse model of diabetic nephropathy. *Pharm. Res.* 31 (5): 1185–1193.

33 Mager, D.E. and Jusko, W.J. (2001). General pharmacokinetic model for drugs exhibiting target-mediated drug disposition. *J. Pharmacokinet. Pharmacodyn.* 28 (6): 507–532.

34 Kamath, A.V. (2016). Translational pharmacokinetics and pharmacodynamics of monoclonal antibodies. *Drug Discov. Today Technol.* 21–22: 75–83.

35 Samineni, D., Girish, S., and Li, C. (2016). Impact of shed/soluble targets on the PK/PD of approved therapeutic monoclonal antibodies. *Expert Rev. Clin. Pharmacol.* 9 (12): 1557–1569.

36 Bruno, R., Washington, C.B., Lu, J.F. et al. (2005). Population pharmacokinetics of trastuzumab in patients with HER2+ metastatic breast cancer. *Cancer Chemother. Pharmacol.* 56 (4): 361–369.

37 Malik, P.R.V., Hamadeh, A., Phipps, C., and Edginton, A.N. (2017). Population PBPK modelling of trastuzumab: a framework for quantifying and predicting inter-individual variability. *J. Pharmacokinet. Pharmacodyn.* 44 (3): 277–290.

38 Li, L., Gardner, I., Rose, R., and Jamei, M. (2014). Incorporating target shedding into a minimal PBPK-TMDD model for monoclonal antibodies. *CPT Pharmacometrics Syst. Pharmacol.* 3: e96.

39 Petitcollin, A., Bensalem, A., Verdier, M.C. et al. (2020). Modelling of the time-varying pharmacokinetics of therapeutic monoclonal antibodies: a literature review. *Clin. Pharmacokinet.* 59 (1): 37–49. https://doi.org/10.1007/ s40262-019-00816-7.

40 Thomas, V.A. and Balthasar, J.P. (2019). Understanding inter-individual variability in monoclonal antibody disposition. *Antibodies (Basel)* 8 (4): 56. https://doi.org/10.3390/antib8040056.

41 Hendrikx, J., Haanen, J., Voest, E.E. et al. (2017). Fixed dosing of monoclonal antibodies in oncology. *Oncologist* 22 (10): 1212–1221.

42 Wang, D.D., Zhang, S., Zhao, H. et al. (2009). Fixed dosing versus body size-based dosing of monoclonal antibodies in adult clinical trials. *J. Clin. Pharmacol.* 49 (9): 1012–1024.

43 Bai, S., Jorga, K., Xin, Y. et al. (2012). A guide to rational dosing of monoclonal antibodies. *Clin. Pharmacokinet.* 51 (2): 119–135.

44 Clements, J.D., Zhu, M., Kuchimanchi, M. et al. (2020). Population pharmacokinetics of blinatumomab in pediatric and adult patients with hematological malignancies. *Clin. Pharmacokinet.* 59 (4): 463–474. https://doi.org/10.1007/ s40262-019-00823-8.

45 Fasanmade, A.A., Adedokun, O.J., Olson, A. et al. (2010). Serum albumin concentration: a predictive factor of infliximab pharmacokinetics and clinical response in patients with ulcerative colitis. *Int. J. Clin. Pharmacol. Ther.* 48 (5): 297–308.

46 Shek, D., Read, S.A., Ahlenstiel, G., and Piatkov, I. (2019). Pharmacogenetics of anticancer monoclonal antibodies. *Cancer Drug Resist.* 2: 69–81.

47 van Brummelen, E.M., Ros, W., Wolbink, G. et al. (2016). Antidrug antibody formation in oncology: clinical relevance and challenges. *Oncologist* 21 (10): 1260–1268.

48 Klotz, U., Teml, A., and Schwab, M. (2007). Clinical pharmacokinetics and use of infliximab. *Clin. Pharmacokinet.* 46 (8): 645–660.

49 Zhou, H. and Mascelli, M.A. (2011). Mechanisms of monoclonal antibody-drug interactions. *Annu. Rev. Pharmacol. Toxicol.* 51: 359–372.

50 Ferri, N., Bellosta, S., Baldessin, L. et al. (2016). Pharmacokinetics interactions of monoclonal antibodies. *Pharmacol. Res.* 111: 592–599.

51 Jin, F. and Balthasar, J.P. (2005). Mechanisms of intravenous immunoglobulin action in immune thrombocytopenic purpura. *Hum. Immunol.* 66 (4): 403–410.

52 Kiessling, P., Lledo-Garcia, R., Watanabe, S. et al. (2017). The FcRn inhibitor rozanolixizumab reduces human serum IgG concentration: a randomized phase I study. *Sci. Transl. Med.* 9 (414): eaan1208. https://doi. org/10.1126/scitranslmed.aan1208.

53 Ishii-Watabe, A. and Kuwabara, T. (2019). Biosimilarity assessment of biosimilar therapeutic monoclonal antibodies. *Drug Metab. Pharmacokinet.* 34 (1): 64–70.

54 Beck, A. (2011). Biosimilar, biobetter and next generation therapeutic antibodies. *MAbs* 3 (2): 107–110.

55 Schreitmuller, T., Barton, B., Zharkov, A., and Bakalos, G. (2019). Comparative immunogenicity assessment of biosimilars. *Future Oncol.* 15 (3): 319–329.

56 Tu, C.L., Wang, Y.L., Hu, T.M., and Hsu, L.F. (2019). Analysis of pharmacokinetic and pharmacodynamic parameters in EU- versus US-licensed reference biological products: are *in vivo* bridging studies justified for biosimilar development? *BioDrugs* 33 (4): 437–446.

57 Younis, I.R., Robert Powell, J., Rostami-Hodjegan, A. et al. (2017). Utility of model-based approaches for informing dosing recommendations in specific populations: report from the public AAPS workshop. *J. Clin. Pharmacol.* 57 (1): 105–109.

58 Rowland, M., Lesko, L.J., and Rostami-Hodjegan, A. (2015). Physiologically based pharmacokinetics is impacting drug development and regulatory decision making. *CPT Pharmacometrics Syst. Pharmacol.* 4 (6): 313–315.

59 Papamichael, K., Cheifetz, A.S., Melmed, G.Y. et al. (2019). Appropriate therapeutic drug monitoring of biologic agents for patients with inflammatory bowel diseases. *Clin. Gastroenterol. Hepatol.* 17 (9): 1655–1668.

60 Scharnhorst, V., Schmitz, E.M.H., van de Kerkhof, D. et al. (2019). A value proposition for trough level-based anti-TNFalpha drug dosing. *Clin. Chim. Acta* 489: 89–95.

61 Imamura, C.K. (2019). Therapeutic drug monitoring of monoclonal antibodies: applicability based on their pharmacokinetic properties. *Drug Metab. Pharmacokinet.* 34 (1): 14–18.

62 Lin, K. and Mahadevan, U. (2014). Pharmacokinetics of biologics and the role of therapeutic monitoring. *Gastroenterol. Clin. North Am.* 43 (3): 565–579.

63 Schrey, D., Borghorst, S., Lanvers-Kaminsky, C. et al. (2010). Therapeutic drug monitoring of asparaginase in the ALL-BFM 2000 protocol between 2000 and 2007. *Pediatr. Blood Cancer* 54 (7): 952–958.

64 Voller, S., Pichlmeier, U., Zens, A., and Hempel, G. (2018). Pharmacokinetics of recombinant asparaginase in children with acute lymphoblastic leukemia. *Cancer Chemother. Pharmacol.* 81 (2): 305–314.

65 Armstrong, J.K., Hempel, G., Koling, S. et al. (2007). Antibody against poly(ethylene glycol) adversely affects PEG-asparaginase therapy in acute lymphoblastic leukemia patients. *Cancer* 110 (1): 103–111.

66 van der Sluis, I.M., Vrooman, L.M., Pieters, R. et al. (2016). Consensus expert recommendations for identification and management of asparaginase hypersensitivity and silent inactivation. *Haematologica* 101 (3): 279–285.

67 Strik, A.S., Wang, Y.C., Ruff, L.E. et al. (2018). Individualized dosing of therapeutic monoclonal antibodies-a changing treatment paradigm? *AAPS J.* 20 (6): 99.

68 Mould, D.R., Upton, R.N., Wojciechowski, J. et al. (2018). Dashboards for therapeutic monoclonal antibodies: learning and confirming. *AAPS J.* 20 (4): 76.

9

Pharmacogenomics of Biologics
Michael Ward

Clinical and Health Sciences, University of South Australia, Adelaide, South Australia, Australia

---

KEY POINTS

- Pharmacogenomics seeks to explain the interindividual variability in response to drugs due to genetic variation.
- Biologic medicines have transformed the management of many conditions but there remain patients who do not respond adequately.
- Despite extensive investigation, with the exception of mutation associated with cancer, the impact of pharmacogenomics on biologics is limited at this time, possibly in part due to the complex, multigenic conditions that many of these medicines are used to treat.

Abbreviations

Abbreviation	Full name	Abbreviation	Full name
ACR	American College of Rheumatology	GWAS	Genome-Wide Association Studies
ADAs	Anti-Drug Antibodies	HLA	Human Leukocyte Antigen
AMD	Age-related Macular Degeneration	HTRA1	High Temperature Requirement Factor A1
ARMS2	Age-Related Maculopathy Susceptibility 2	IBD	Inflammatory Bowel Disease
bDMARDs	biological Disease Modifying Anti-Rheumatic Drugs	IL	Interleukin
CD	Crohn's Disease	NCBI	National Center for Biotechnology Information
CFH	Complement Factor H	PASI	Psoriasis Area and Severity Index
COPD	Chronic Obstructive Pulmonary Disease	RA	Rheumatoid Arthritis
DAS28	Disease Activity Score-28	SNP(s)	Single Nucleotide Polymorphism(s)
DMARDs	Disease Modifying Anti-Rheumatic Drugs	STR	Short Tandem Repeat
DNA	Deoxyribonucleic Acid	TLR	Toll-Like Receptors
EULAR	European League Against Rheumatism	TNF	Tumor Necrosis Factor
		UC	Ulcerative Colitis
FcRn	neonatal Fc Receptor	VEGF	Vascular Endothelial Growth Factor
		VNTR	Variable Number Tandem Repeat

Biologics, Biosimilars, and Biobetters: An Introduction for Pharmacists, Physicians, and Other Health Practitioners,
First Edition. Edited by Iqbal Ramzan.

9.1 Introduction

Pharmacogenetics seeks to explain the interindividual variability in response to drugs due to genetic variation. With the sequencing of the entire human genome, the term pharmacogenomics has been introduced. Today, the terms pharmacogenetics and pharmacogenomics are often used interchangeably. In the context of predicting drug response, pharmacogenetics/pharmacogenomics is primarily concerned with changes in the deoxyribonucleic acid (DNA) sequence that are passed from parents to offspring via the sperm and egg, known as germline mutations. The closely related term precision medicine includes not only information related to germline mutations but also a range of other factors that might influence drug response. For example, in the setting of cancer, additional factors such as genetic mutations that develop within the cancer itself, but in no other cells in the body, can significantly impact on response to treatment, including biologic medicines such as trastuzumab and cetuximab.[1] Precision medicine may also include the impact of other genetic factors such as gene expression profiles, which provide an overall picture of biochemical pathways that are active within a particular cell or tissue at a given time through the simultaneous measurement of the expression level of thousands of genes, and epigenetic changes, which are changes to the structure of DNA but not the genetic sequence itself that can influence gene expression. This chapter will specifically explore the pharmacogenomics of biologics with regard to germline mutations.

The predominant focus of pharmacogenomics is on genetic variants known as single nucleotide polymorphisms (SNPs, pronounced "snips"), in which one nucleotide in the DNA sequence is substituted for another (e.g. at a specific location the nucleotide adenine [A] may be changed to the nucleotide cytosine [C]; described using the notation A > C). SNPs may be located within a gene or in intergenic regions. SNPs located within a gene may be further located either within a region that encodes for the protein product of the gene (known as an exon) or in a noncoding region, either an intron which is located between exons or in a regulatory region of the gene. SNPs located within coding region of a gene may change the amino acid sequence of the protein produced, referred to as a nonsynonymous SNP, or may not, referred to as synonymous, depending upon the specific change to the genetic sequence. SNPs located in noncoding or intergenic regions may increase or decrease the level of gene expression. In addition to SNPs, there are several other simple genetic variants that are often investigated in pharmacogenomic studies including small-scale multi-nucleotide insertions or deletions in which a small number of nucleotides are either added to or absent from the genetic sequence (e.g. the genetic sequence ...ACGA... with a deletion variant could be ...ACA... or with an insertion could be ...ACTGA...) and tandem repeat variations in which a sequence of DNA is repeated multiple times arranged in a head to tail manner (e.g. the sequence ACC might be repeated three times ... ACC ACC ACC ... or four times ... ACC ACC ACC ACC ...). Tandem repeats may be referred to as either a short tandem repeat (STR) in which the repeating unit consists of 2–13 nucleotides or a variable number tandem repeat (VNTR) in which the unit consists of up to 100 nucleotides. Given the large number of SNPs and other simple genetic variants that exist within the human genome, in order to identify each SNP, a reference system has been developed known as the Reference SNP cluster Identification number, otherwise referred to as the "rs number." The rs number is a unique accession number assigned by the dbSNP, a public-domain archive hosted at the National Center for Biotechnology Information (NCBI).[2] For example, rs1800629 refers to a G > A SNP within tumor necrosis factor (TNF) alpha.[2]

The process of assessing an individual for the presence of one or more genetic variants is known as genotyping. Each version of a genetic variant is referred to as an allele and the most common allele is referred to as the wild-type allele. An individual who has two of the same versions of an allele is referred to as a homozygote while an individual who has two different versions is referred to as a heterozygote. Accordingly, an individual who possesses two copies of most common allele is referred to as being homozygous wild type, an individual who has two copies of a variant form is referred to being homozygous variant, and an individual who has one wild-type allele and one variant allele is described as a heterozygote.

While the advent of biologic medicines has transformed the management of many conditions, there remain patients that do not respond adequately to these medicines. For many conditions currently treated with biologics, the number of medicines continues to expand with new agents becoming available with different pharmacological targets. This creates the challenge of selecting the most appropriate agent for an individual patient. Additionally, biologic medicines are typically expensive therapies creating an economic imperative to optimize their therapeutic use. Together these factors have prompted the search for factors, such as genetic variants, that might explain this inter-patient variation in drug response to help guide drug selection and dosing to improve treatment efficacy, minimize the risk of side effects, and to improve cost effectiveness.

9.2 Approaches to the Identification of Genetic Variants Influencing Response to Biologic Medicines

There are two main approaches to the identification of genetic variants that influence response to medicines: candidate gene studies and genome-wide association studies (GWAS). Candidate gene studies represent a more traditional approach to the identification of genetic variants that predict drug response. These studies typically investigate only a relatively limited number of genetic variants within a limited number of genes that have been specifically selected by researchers for inclusion in the study. In contrast, a more recent approach to conducting pharmacogenomic studies is to utilize the enormous developments in genetic technologies to investigate many thousands of SNPs throughout the whole genome, with the aim of identifying genetic variants that modify drug response but without first specifically selecting these SNPs for investigation; hence the term GWAS.

9.2.1 Candidate Gene Studies

Candidate gene studies investigate genetic variants located within specific genes that have been identified by researchers on the basis of a biological plausibility with regard to influencing response to the medicine under investigation. Typical candidate genes investigated in these studies include those known to be important in determining the pharmacokinetic properties of the medicine, such as drug metabolizing enzymes and drug transporters, those involved in pharmacodynamic pathways, such as drug targets, or those known to be involved in disease susceptibility. Such a targeted approach in selecting genetic variants to be investigated was necessary when genotyping was a technically challenging task, limiting the number of genetic variants that could be feasibly genotyped in a study. As genetic technologies have developed, the number of SNPs that can be investigated in a candidate gene approach has expanded greatly, although the premise of the study design remains the same.

The outcome of interest of most pharmacogenetic studies is to predict an overall clinical response to treatment or to predict the risk of adverse events such as drug toxicity. The assessment of clinical response varies according to the therapeutic indication, particularly given the relatively complex nature of many of the conditions for which biologics are used. While the prediction of an overall clinical response is of clear clinical importance, it might also prove more challenging to identify genetic variants that predict these outcomes given the variety of factors that might influence outcomes of this nature. On this basis it might be possible to identify genetic variants that predict more specific aspects of response to these medicines. For example, many biologic medicines have the potential to stimulate the formation of anti-drug antibodies (ADAs), which can significantly impact on the pharmacokinetics, pharmacological activity, and the risk of side effects such as allergic reactions (see Chapter 8). For this reason, some candidate gene studies focus specifically on SNPs located within genes that might be associated with this specific risk such as genes that are involved in immune pathways.

When comparing candidate gene studies of small molecule drugs with those of biologics, a major difference exists with regard to the selection of candidate genes related to drug metabolism. While for many small molecule drugs, the drug metabolizing enzymes such as the Cytochromes P450 have been frequently studied, these are not relevant for biologics as elimination of biologics occurs via other pathways. For example, elimination of the monoclonal antibodies occurs mostly via endocytosis and pinocytosis followed by proteolytic catabolism. This distinct difference in elimination pathways means that other candidate genes are often considered such as *FCGRT*, which is the gene encoding for the neonatal Fc receptor (FcRn). It has been identified that within the promoter region of *FCGRT,* genetic variation exists in the form of VNTR with variants including between one and five to five repeats of the repeating unit; two (VNTR2) and three (VNTR3) repeats are most common in Caucasians (7.5 and 92.0%, respectively).[3] The remaining alleles are relatively rare in the population with frequencies in the order of 0.1–2%. Individuals who are VNTR3 homozygous express more FcRn (1.66-fold) than individuals who are VNTR2/VNTR3 heterozygous.[3] Given the importance of FcRn in the recycling and elimination of monoclonal antibodies (see Chapter 8), this forms a biological basis for a potential difference between patients with regard to the pharmacokinetics of monoclonal antibodies.

9.2.2 Genome-Wide Association Studies (GWAS)

In contrast to the hypothesis-based approach of candidate gene studies, GWAS are driven by large amounts of genetic data enabled by the major developments in genetic technologies. By assessing genetic variants throughout the genome, the GWAS approach enables the identification of genes that would not be considered in a candidate gene approach because of lack of prior association with relevant disease or pharmacological pathways. In the context of pharmacogenomic studies, GWAS present a powerful tool to investigate complex conditions in which many genetic variations located within many different genes may each contribute to determining drug response. However, there are a number of challenges associated with this approach. The large number of genetic variants included in these analyses poses a statistical challenge with regard to defining statistically significant associations.[4] The statistical analysis of the large data sets produced in a GWAS must minimize the likelihood of false positive associations between genetic variants and drug response while also minimizing false negative results, which would limit the ability to identify genetic variants with a small contribution to determining drug response. In addition, it is important to recognize that the genetic variants identified by a GWAS may not be the biologically functional genetic variants influencing drug response. Instead, the identified genetic variants may simply be markers of another co-inherited genetic variant that is the biological basis for the observed effects and as such the outcomes of GWAS may require further investigation to identify the genetic variant that is truly influencing drug response.

9.2.3 Limitations of Pharmacogenomic Studies

This chapter will explore the pharmacogenomic associations of biologic medicines. In general, the impact of genetics in predicting response to biologics has been limited. For a number of biologic medicines, some of which have multiple therapeutic indications, there are many studies that have been conducted. However, the results of these studies are often contradictory. The interpretation of the results of these studies is often complicated by factors such as relatively small sample sizes and differences in study design, including the clinical endpoint used to define drug response. Many of the pharmacogenomic studies of biological medicines are relatively small, with sample sizes in the order of 100–300 patients not uncommon, and only a limited number of studies with sample sizes in the thousands of patients. The relatively small size of these studies limits statistical power to detect associations between genetic variants and drug response, particularly given the complex nature of the conditions for which these medicines are often used.[5] Given these limitations, the best evidence is often found in meta-analyses, which combine findings of each of the smaller studies. When a specific genetic variant has been found to be associated with drug response in multiple studies, the meta-analysis can help to improve the understanding of the strength of the association. Where studies have conflicting results, with some studies suggesting an association and other studies suggesting no association, meta-analysis can increase the statistical power of the analysis providing greater insight as to whether an association does or does not exist.

9.3 Pharmacogenomics of Biologics in Rheumatoid Arthritis

Rheumatoid arthritis (RA) is an autoimmune disease that causes pain and swelling of the joints. If left untreated, RA can cause the destruction of bone and joint tissues resulting in irreversible joint damage and permanent immobility. There is no cure for RA but the early diagnosis and initiation of medicines to control the condition is key to obtaining good patient outcomes.[6] A range of small molecule medicines such as methotrexate, often referred to as the disease modifying anti-rheumatic drugs (DMARDs) are considered first-line in the management of RA. Patients that do not respond to DMARDs may then be treated with a biologic medicine,[6] often referred to as biological DMARDS (bDMARDs).

The introduction of biologic medicines has transformed the management of RA and other inflammatory arthropathies providing treatment options for patients that have failed to respond to traditional small molecule medicines. A number of different biologic medicines are now available for RA including those which target TNFα (including infliximab, adalimumab, and etanercept), interleukin 1 (anakinra), and CD20 (rituximab). While these agents provide an important advance in management, not all patients respond. A meta-analysis of 62 studies representing 27 280 patients found that a good response was achieved in approximately 25% of patients, a moderate response in a little under 50% of patients with remainder failing to achieve either of these outcomes.[7] For this reason, a large number of studies have sought to investigate genetic variants using both candidate gene and GWAS approaches for association with drug response in the hope that genetic variants might be help inform decisions about treatment selection.

9.3.1 Predictors of Clinical Response

In 2017, a systematic review and meta-analysis of studies investigating genetic predictors of clinical response to biologics in RA was conducted.[8] In total, the authors identified 43 candidate gene studies and five GWAS that reported an association between one or more genetic variants and response to anti-TNFα treatments in patients with RA. Clinical response in these studies was defined by a number of different measures such as the European League Against Rheumatism (EULAR) response criteria, the American College of Rheumatology (ACR) criteria, and the Disease Activity Score-28 (DAS28). These studies most commonly considered the anti-TNFα agents as a group, ignoring potential differences between drugs within this class, and included patients treated with any of infliximab, adalimumab, or etanercept, while other studies were limited to patients treated with only a single specific agent (e.g. only patients treated with adalimumab).

Consistent with the hypothesis-driven approach of candidate gene studies in which investigators select genetic variants that they consider likely to be of importance in determining drug response, the systematic review identified hundreds of different polymorphisms located in a range of different biochemical pathways that have been investigated.[8] Overall, across the numerous studies, a large number of different SNPs were found to be associated with response to anti-TNFα treatment. Frequently though this association was limited to a single study and only a small number of SNPs were identified that had been included in more than one study. A total of 23 SNPs located within 21 different genes were identified as having been investigated in a minimum of two studies, at least one of which identified an association with response, and accordingly these SNPs were subjected to meta-analysis. After meta-analysis of these 23 SNPs, only six were found to be associated with treatment response (rs3761847, rs1801274, rs10919563, rs3136645, and rs3136645).[8] This result highlights the challenge associated with the interpretation of the importance of many of the published studies in isolation.

The SNPs identified in the meta-analysis as being associated with response to anti-TNF treatments are located in genes such as TNF receptor-associated factor 1 (*TRAF1*), Fc fragment of IgG receptor IIa (*FCGR2A*), and protein tyrosine phosphatase receptor type C (PTPRC).[8] Consistent with the candidate gene approach of the studies from which they were identified, each of these genes has a clear biological plausibility with regard to determining response to anti-TNF treatment. For example, *TRAF1* is involved in mediating the signal transduction from multiple receptors within the TNFR superfamily and has been identified as a risk locus for the development of RA.[9] *FCGR2A* is a member of the immunoglobulin Fc receptor located on the surface of phagocytic cells such as macrophages and neutrophils.

PTPRC is a regulator of T- and B-cell antigen receptor signaling and acts as a negative regulator of the Janus kinase (JAK), which is independently a therapeutic target for a number of small molecules such as tofacitinib and baricitinib, which are more recent therapeutic options in the management of RA.

As demonstrated by the systematic review described above, there has been limited replication of SNPs among the numerous studies conducted. On this basis, a subsequent study has sought to replicate 12 SNPs identified in the three largest published GWAS, using a cohort of 755 patients treated with infliximab, adalimumab, or etanercept.[10] In this replication study, no significant association between any of the 12 SNPs studied and treatment response was identified. A sub-analysis investigating individual medicines was conducted. This identified a single SNP (rs2378945) that may be associated with response to etanercept, with the less common allele associated with an inferior response to treatment. Given the structural difference between etanercept, an Fc-fusion protein, and the monoclonal antibodies infliximab and adalimumab, it is possible that there could be different genetic factors influencing drug response. However, the SNP identified is located within a gene (*NUBPL*) which encodes for a protein involved in the mitochondrial membrane respiratory chain NADH dehydrogenase (complex I) but has no known functions relevant to autoimmune conditions and as such the biological plausibility of this association is uncertain.

As described in Section 9.2.2, while GWAS are a potentially powerful technique to enable the identification of genetic predictors of drug response, the analysis of the results can be challenging. In the context of predicting response to anti-TNF treatments in RA, a unique approach to this challenge has been employed. Using GWAS data from 2706 patients with RA, a competition was established in which 73 international research groups applied a range of state-of-the-art modeling methodologies in an attempt to develop models of combinations of genetic variants to predict clinical response (DAS28 and EULAR) to anti-TNF agents.[11] Despite the large amount of highly sophisticated analysis conducted by the participants in this competition, the conclusion of the project was that, based upon the data available at that time, SNPs do not provide a meaningful contribution to predicting treatment response anti-TNF agents.

9.3.2 Predictors of Anti-drug Antibody Formation

Human leukocyte antigen (HLA) alleles have been identified as strong pharmacogenetic predictors of allergic reactions to small molecule drugs such as abacavir. On this basis, HLA alleles have been assessed as potential predictors for the risk of the development of ADAs toward biologic medicines used in the treatment of RA. HLA alleles have been investigated in 248 patients with RA, ankylosing spondylitis, and psoriatic arthritis who had received at least six months of treatment with an anti-TNF agent.[12] Three HLA alleles (HLA-DRβ-11, HLA-DQ-03, and HLA-DQ-05) were identified as being associated with an increased risk of the development of ADAs. It should be noted that this analysis is based on small numbers of patients with ADAs; including only two patients with ADAs to infliximab, four with ADAs to adalimumab, four with ADAs to certolizumab pegol, and a single patient with ADAs to golimumab.

The influence of HLA alleles on the risk of ADA formation has also been assessed in participants in adalimumab clinical trials in RA and hidradenitis suppurativa.[13] The authors sought to identify participants with the strongest immunogenicity phenotype, which was defined on the basis of multiple measured ADA levels and adalimumab concentrations. Of 634 participants, only 37 were defined as having a strong immunogenicity phenotype. Assessments of HLA genetic variants identified two HLA alleles that were associated with the strong immunogenicity phenotype (HLA-DRB1*03 and HLA-DRB1*011) and three HLA alleles that were protective against the strong immunogenicity phenotype (HLA-DQB1*05, HLA-DRB1*01 and HLA-DRB1*07).

9.4 Pharmacogenomics of Biologics in Inflammatory Bowel Disease (IBD)

The term inflammatory bowel disease (IBD) covers a group of disorders which are characterized by inflammation of the gastrointestinal tract. The two major types of IBD are Crohn's disease (CD), in which inflammation most commonly affects the small intestine and start of the colon, and ulcerative colitis (UC), which is associated with inflammation of the colon and rectum. The symptoms vary between patients depending upon the site and severity of inflammation but typically include diarrhea, bloody stools, and unintended weight loss. The treatment of IBD involves drugs such as corticosteroids, 5-aminosalicylates (e.g. sulfasalazine), and azathioprine. Patients that do not respond to these treatments may then be treated with a biologic medicine.[14]

As has occurred for RA, the introduction of biologics has transformed the management of IBD. The first biologics available for the treatment were the anti-TNF agents (e.g. infliximab and adaliumumab). More recently, additional agents including vedolizumab, targeting human α4β7 integrin, and ustekinumab, targeting IL12/23 have provided further therapeutic options for these patients. With a number of different biologic agents available, with different pharmacological targets, genetics may assist with the selection of the most appropriate agent. A large number of studies have sought to identify genetic predictors of response to biologics in patients with IBD using both candidate gene and GWAS approaches.

9.4.1 Predictors of Clinical Response

In 2015, a systematic review and meta-analysis of previously published studies was conducted.[15] In total, the authors identified 15 candidate gene studies that reported an association between one or more genetic markers and clinical response or biological response to anti-TNF agents in patients with IBD including UC and CD. Clinical response in these studies was defined by standardized clinical response measures including the Harvey–Bradshaw index in patients with UC and CD activity index in patients with CD, and biological response was assessed on the basis of changes in C-reactive protein level, a marker of inflammation. A total of 23 SNPs in 18 genes were identified as having been investigated in a minimum of two studies, at least one of which identified an association with clinical response, and accordingly these SNPs were subjected to meta-analysis. After meta-analysis of these 23 SNPs, only eight were found to be associated with treatment response. The SNPs identified as being associated with response after meta-analysis are located in genes including Toll-like receptors (*TLR*) 2, 4, and 9, TNF receptor superfamily member 1A (*TNFRSF1A*), interferon gamma (*IFNG*), and interleukins (*IL*) 6 and 1B. These seven genes can be clustered into two broad groups; those involved in cytokine pathways (*TNFRSF1A*, *IFNG*, *IL6, and IL1B*) and those involved in the innate immune response including recognition of bacteria (*TLR2, TLR4, TLR9*). This latter group of genes is noteworthy given the increasing appreciation that the interaction between the host and gut microbiome may influence response to treatment in IBD.[16] However, it must be noted that whilst the authors of the meta-analysis required SNPs to have been investigated in more than one study, all eight SNPs were actually identified from a single study that investigated patients with UC and CD separately.[17]

9.4.2 Predictors of Pharmacokinetics

The impact of genetic variability in *FCGRT*, the gene encoding for the FcRn on infliximab and adalimumab pharmacokinetics has been investigated.[18] Reflecting the complexity associated with biologics, investigating the impact of FCGRT on pharmacokinetics requires consideration of ADAs given that they can impact upon drug clearance (see Chapter 8). When patients who developed ADAs to infliximab were excluded from the analysis, patients heterozygous for VNTR2/ VNTR3 had 14% lower drug exposure, as assessed by area-under the curve (AUC) than individuals who were homozygous for VNTR3. For patients treated with adalimumab, drug exposure was 24% lower in VNTR2/VNTR3 heterozygote patients as compared with VNTR3 homozygous patients. The difference in the magnitude of the effect of FCGRT genetic variants between infliximab and adalimumab may be due to the difference in route of administration of the two agents. Infliximab is administered by intravenous infusion while adalimumab is administered subcutaneously. The increased magnitude of effect observed for adalimumab relative to infliximab may be due to an effect of FcRn in reducing pre-systemic antibody catabolism at the site of injection and a resultant impact on bioavailability following subcutaneous administration.[19]

9.4.3 Predictors of Anti-drug Antibody Formation

Genetic predictors for the risk of ADA formation have been assessed in patients with IBD. Using a candidate gene approach, HLA-DRB1 alleles have been assessed on the basis that they had previously been demonstrated to impact upon immunogenicity to interferon-β therapy in patients with multiple sclerosis.[20] Among 192 IBD patients who received infliximab, 76 of whom developed ADAs, carriage of HLA-DRB1*03, was associated with an increased risk of the development of ADAs. More recently, a GWAS conducted in an initial discovery cohort of 1240 biologic-naïve patients with CD starting treatment with either infliximab or adalimumab identified HLA-DQA1*05 as a risk factor for the development of ADAs.[21] The finding was confirmed in a replication cohort. In order to select the patients with the strongest immune response to treatment, study participants were only considered ADA positive if they returned an ADA titer above a defined threshold. Carriage of the HLA allele HLA-DQA1*05 approximately doubled the rate of ADA positivity. Patients treated with infliximab and who were HLA-DQA1*05 positive experienced the highest rates of ADA positivity,

92% at 1 year, while those who were HLA-DQA1*05 negative and treated with adalimumab displayed the lowest rates of ADA positivity, 10% at 1 year. The concomitant administration of immunosuppressants such as azathioprine or methotrexate is generally associated with a reduced risk of ADA formation. In this study, the effect of HLA-DQA1*05 was independent of concomitant immunosuppression. Subsequently, a single center, retrospective cohort study was carried out in 262 patients with IBD who were treated with infliximab.[22] HLA-DQA1*05 carriers had a significantly increased risk of ADA positivity. The authors also identified that HLA-DQA1*05 carriers were at increased risk of loss of response to treatment and treatment discontinuation, which is consistent with the known effects of the development of ADAs to infliximab.

9.5 Pharmacogenomics of Biologics in Psoriasis

Psoriasis is a chronic immune-mediated skin condition. Development in our understanding of the pathogenesis of this condition have identified a number of pathways that can be manipulated through the use of biologics including those that target TNFα (e.g. adalimumab, etanercept, infliximab), IL12/23 (ustekinumab), or IL17 (secukinumab, ixekizumab). While providing new treatment options, response to these agents is subject to inter-patient variation. A number of authors have sought to identify genetic basis underlying this variability in response. As a multifactorial condition, genetics has been investigated as a contributing factor for development of psoriasis with a number of susceptibility alleles identified. Many of these alleles have subsequently been investigated as potential predictors of response to biologics used in the management of psoriasis.

Early studies investigating genetic risk factors for the development of psoriasis identified a prominent disease marker at chromosome 6p21 (PSORS1), which is within the major histocompatibility complex.[23–25] The human leukocyte antigen, HLA-C*06:02, was identified as the major risk allele located within this region. HLA-C*06:02 positivity is associated with earlier onset in the development of psoriasis, differences in lesion severity and distribution, and a higher incidence of the Koebner phenomenon.[26,27] Each copy of the HLA-C*06:02 allele carried by an individual increases their risk of developing psoriasis fivefold.[28]

Owing to the importance of HLA-C*06:02 in the susceptibility to psoriasis, a number of authors have subsequently investigated HLA-C*06:02 as a potential predictor of response to biologic treatment in psoriasis, particularly ustekinumab. However, these have typically been small cohort studies with limited statistical power. A meta-analysis of these studies has been conducted,[29] the aim of which was to assess whether HLAC*06:02 status influences the response to ustekinumab at six months of treatment. The meta-analysis of eight studies, ranging in size from 30 to 332 individuals, indicated that the pooled risk difference between HLA-C*06:02-positive and HLA-C*06:02-negative individuals for attainment of treatment response of a 75% improvement in Psoriasis Area and Severity Index (PASI75) at six months was 0.24 (95% confidence interval: 0.14–0.35; $P < 0.001$). In a clinical setting, this would mean that for every 100 patients of each HLA-C*6:02 genotype treated, 24 more individuals in the HLA-C*06:02-positive group would achieve a PASI75 response than in the HLA-C*06:02-negative group. The results suggest that HLA-C*06:02-positive individuals may have a greater probability of attaining a PASI75 response. However, the response rate for this outcome in the HLA-C*06:02-negative individuals was also high. Within the studies included in the meta-analysis, the PASI75 response rate in the HLA-C*06:02-positive group ranged from 62 to 98% (median = 92%) and from 40 to 84% (median = 67%) in the HLA-C*06:02-negative group. On this basis, the findings do not support the use of HLA-C*06:02 as a marker of whether or not an individual might benefit from treatment with ustekinumab.

More recently, HLA-C*06:02 has been investigated as a biomarker to assist in the selection between the two potential biologic treatments for psoriasis with different pharmacological targets; adalimumab, targeting TNFα, and ustekinumab, targeting interleukins 12 and 23.[30] This analysis included 1326 individuals, of which 704 were HLA-C*06:02 positive (defined as having one or two copies of the allele). While previous studies of HLA-C*06:02 utilized PASI75 as the response outcome, this study utilized the more stringent 90% improvement in psoriasis symptoms (PASI90) on the basis that this corresponds to patients being clear or nearly clear of psoriasis indicating that these patients could be considered to have had a very good response to treatment. In this analysis, in those that were HLA-C*06:02 positive, there was no difference in response between adalimumab and ustekinumab at any time point. In contrast, HLA-C*06:02-negative individuals were significantly more likely to respond to adalimumab than to ustekinumab at 3, 6, and 12 months.[30] Given the apparent importance of HLA-C*06:02 in the development of psoriasis, this observation suggests that the signaling pathways

downstream of TNFα may be of relatively greater importance in disease pathogenesis in those who are HLA-C*06:02 negative as compared with those who are HLA-C*06:02 positive. For a clinician choosing between ustekinumab and adalimumab for an individual patient knowing the patient's HLA-C*06:02 status may assist in the choice of the biologic.

Secukinumab and ixekizumab are monoclonal antibodies that bind to IL-17A. Following a candidate gene approach, genetic variants within IL-17A have been investigated for potential associations with response to secukinumab and ixekizumab in patients with psoriasis. Treatment response was assessed according to the change in PASI. A number of genetic variants in both the protein-coding region and noncoding regions of *IL17-A* have been identified. However, no associations were identified between five variants with known or suspected effects on the regulation of IL-17A expression and response to treatment.[31]

While candidate gene studies have typically focused on a relatively small number of SNPs, improvements in genotyping technologies have enabled studies using this approach to explore pharmacological pathways more broadly. This approach has been taken to examine the TLRs and NOD-like receptor pathways, which can activate nuclear factor-κB. Nuclear factor-κB, which can modulate inflammation through the regulation of the expression of cytokines such as IL-1β, is an important mediator in the initiation and maintenance of psoriatic plaques.[32] A total of 62 SNPs located within 44 different genes were investigated in 376 patients who received an anti-TNF agent and 230 patients who received ustekinumab. Patients were classified as either responders if they remained on treatment for at least 225 days or as primary or secondary non-responders if they were treated for less than this period of time. Overall, the results suggest that SNPs that possibly increase IL-1β levels might be associated with poorer response to both anti-TNFα agents and ustekinumab and that variants associated with increased interferon-γ might be associated with a response to ustekinumab.

9.6 Pharmacogenomics of Biologics in Age-Related Macular Degeneration (AMD)

Age-related macular degeneration (AMD) is the most common cause of blindness in industrialized countries and the third leading cause worldwide.[33] AMD can be classified as either non-exudative AMD, or "dry" AMD, or neovascular AMD, also known as "wet" AMD.

Treatment of the neovascular form of AMD consists of intravitreal injections of anti-vascular endothelial growth factor (VEGF) agents, ranibizumab or aflibercept, which have been demonstrated to be effective at maintaining visual acuity. A number of genes related to the mechanism of action of anti-VEGF treatments or genes that are associated with an increased risk of developing AMD have been investigated as potential predictors of response to treatment.

Following a traditional candidate gene approach, polymorphisms within the VEGF signaling pathways have been explored as potential predictors of response to the agents targeting these pathways in the management of neovascular AMD. The VEGF gene family consists of five members (VEGF-A, VEGF-B, VEGF-C, VEGF-D, and placental growth factor) which exert their effects through three different VEGF receptors (VEGFR-1, VEGFR-2, and VEGFR-3) of which VEGF-A binding to VEGFR-2 is thought to be the major mediator of angiogenesis and vascular leakage in neovascular AMD. A meta-analysis of eight small studies investigating SNPs within VEGF-A and VEGFR-2 has been conducted.[34] The SNPs subjected to meta-analysis included those located within the promoter regions of VEGF-A and VEGFR-2, which are believed to influence expression levels of these genes. Of these SNPs, the meta-analysis identified only one SNP within VEGF-A (rs833061) that was associated with treatment response, with anti-VEGF treatment being more effective in those with the CC genotype. These results are consistent with previous observations that the C allele of this polymorphism is associated with increased promoter activity of the gene.[35] However, in an allele model of C versus T and in a dominant model of homozygous CC genotype or heterozygous CT versus homozygous (TT), no association with treatment response was identified.

In the mid-2000s, GWAS and subsequent work identified the complement factor H gene as being strongly associated with the development of AMD.[36] This observation is consistent with prior observations of the presence of complement proteins within AMD drusen (deposits comprised of protein and lipid that are observed in the macula of patients with AMD). The SNP rs1061170, a non-synonymous SNP in which a T-to-C substitution leads to a change from tyrosine to histidine within the protein, and as such is also known as the Y402H polymorphism, is the most studied and has been estimated to account for a population attributable risk fraction of 10% for the development of early AMD and 53% for the development of late AMD. The specific mechanism by which this variant increases the risk of

development and progression of AMD has not been determined.[36]

On the basis of the strength of this association with the development of and progression of AMD, numerous studies have investigated the Y402H polymorphism for association with treatment response. The most recent meta-analysis of 14 published studies including a total of 2963 patients with AMD assessed the impact of the Y402H polymorphism on the visual acuity or observed anatomical changes in the retina following treatment with anti-VEGF agents.[37] This analysis found that over a follow-up period between 3 and 12 months in the included studies, the homozygous TT genotype was associated with an increased probability of achieving a "good" response outcome compared with the CC genotype. Individuals that were CC homozygous were more likely to experience poorer visual acuity outcomes, greater anatomical changes, and require a greater number of intravitreal anti-VEGF injections.

More recently, a longer duration study has been conducted which examined the impact of the Y402H polymorphism on response to intravitreal ranibizumab, as assessed by best visual acuity, over a five-year period.[38] Consistent with the results of the meta-analysis of studies over a shorter treatment duration,[37] the average best corrected visual acuity scores in the second, third, fourth, and fifth years of treatment were significantly higher in the TT homozygous genotype group than in the TC and CC genotypes.

In addition to the CFH locus, a second risk locus related to AMD has been identified. This locus contains a number of tightly linked genes including age-related maculopathy susceptibility 2 (ARMS2) and the high temperature requirement factor A1 (HTRA1). Polymorphisms located within both of these genes have been investigated as potential predictors of response to anti-VEGF treatment. For ARMS2, a meta-analysis of 12 studies comprising a total of 2389 patients investigated a non-synonymous SNP (rs10490924) which leads to an alanine to serine substitution. Results suggested that the G allele was associated with a better response to anti-VEGF treatment, with a subgroup analysis suggesting an association in East Asian patients but not in Caucasians.[39] For HTRA1, a meta-analysis of five studies comprising a total of 1570 patients has investigated a SNP (rs11200638) which is located within the promoter region.[40] However, in this analysis no association with anti-VEGF treatment response was identified.

Given the limited impact of candidate genes in predicting response to anti-VEGF treatment, a GWAS has been conducted with the aim of identifying genetic variants associated with the change in visual acuity after three anti-VEGF intravitreal injections.[41] This analysis identified variants within the two genes (*C10ORF88* and *UNC93B1*) which were associated with a poorer response to treatment. However, these variants were rare (variant allele frequencies less than 1%) and were only identified within a specific cohort of patients from Jerusalem and as such the broader significance is uncertain.

9.7 Pharmacogenomics of Biologics in Asthma and Chronic Obstructive Pulmonary Disease

One of the more recent therapeutic areas to benefit from the emergence of biologics are the respiratory conditions including asthma and chronic obstructive pulmonary disease (COPD). Given the recency of biologics in the management of these conditions, only limited pharmacogenomic investigations have taken place thus far.

The potential impact of genetics on response to mepolizumab, an anti-IL5 monoclonal antibody, in patients with severe eosinophilic asthma has been assessed using clinical samples and data obtained during clinical trials.[42] Using a candidate gene approach, a total of 105 SNPs in 53 gene regions previously linked to asthma, asthma severity, IL-5, and eosinophil pathway genes were investigated for association with clinical response, as assessed by the rate of clinically significant exacerbations. An exploratory GWAS was also conducted. None of the SNPs included in the candidate gene analysis were significantly associated with rate of clinically significant asthma exacerbations and no variants in the GWAS reached statistical significance. In the

GWAS, six SNPs were identified just as being below the statistical significance threshold for association and as such warrant further investigation. Four of these SNPs map to genes that are of no clear biological relevance to asthma. However, two of the variants (rs10811516 and rs10811517) are located in close proximity to interferon alpha 14, a type I interferon. While no specific role of interferon alpha 14 in infection-mediated immune responses has been identified, it is notable that other type I interferons regulate immune cells that produce IL5, the target of mepolizumab.[43]

The potential impact of genetics on response to mepolizumab in COPD has also been assessed in a GWAS using clinical samples and data obtained during clinical trials.[44] No genetic variants were significantly associated with moderate and/or severe COPD exacerbations or any of the other clinical endpoints tested.

9.8 Conclusions

Overall, despite significant investigations, the clinical impact of pharmacogenomics to individualize treatment with biologics is limited. For many of the biologics that are currently used, it appears likely that drug response might be influenced by combinations of genes all contributing small effects, which is perhaps not surprising given the impact that genetics has in the susceptibility to those same disease conditions. As the range and class of biologics continues to expand, with new pharmacological targets in new therapeutic indications, it will be important to continue to investigate pharmacogenomics as a way of guiding biologic drug and dose selection to improve patient outcomes and cost effectiveness of these highly effective but expensive therapies.

References

1 Hyman, D.M., Taylor, B.S., and Baselga, J. (2017). Implementing genome-driven oncology. *Cell* 168 (4): 584–599.

2 NCBI (2013). *The NCBI Handbook* [Internet]. Bethesda, MD: National Center for Biotechnology Information (US) https://www.ncbi.nlm.nih.gov/books/NBK143764 (accessed 19 June 2020).

3 Sachs, U.J., Socher, I., Braeunlich, C.G. et al. (2006). A variable number of tandem repeats polymorphism influences the transcriptional activity of the neonatal Fc receptor alpha-chain promoter. *Immunology* 119 (1): 83–89.

4 Peters, B.J., Rodin, A.S., de Boer, A., and Maitland-van der Zee, A.H. (2010). Methodological and statistical issues in pharmacogenomics. *J. Pharm. Pharmacol.* 62 (2): 161–166.

5 Kelly, P.J., Stallard, N., and Whittaker, J.C. (2005). Statistical design and analysis of pharmacogenetic trials. *Stat. Med.* 24 (10): 1495–1508.

6 Smolen, J.S., Aletaha, D., and McInnes, I.B. (2016). Rheumatoid arthritis. *Lancet* 388 (10055): 2023–2038.

7 Roberts, L., Tymms, K., de Jager, J. et al. (2014). Efficacy of biologic medications in active rheumatoid arthritis: a systematic review. *Arthritis Rheum.* 66: S1055.

8 Bek, S., Bojesen, A.B., Nielsen, J.V. et al. (2017). Systematic review and meta-analysis: pharmacogenetics of anti-TNF treatment response in rheumatoid arthritis. *Pharmacogenomics J.* 17 (5): 403–411.

9 Kurreeman, F.A., Padyukov, L., Marques, R.B. et al. (2007). A candidate gene approach identifies the TRAF1/C5 region as a risk factor for rheumatoid arthritis. *PLoS Med.* 4 (9): e278.

10 Ferreiro-Iglesias, A., Montes, A., Perez-Pampin, E. et al. (2019). Evaluation of 12 GWAS-drawn SNPs as biomarkers of rheumatoid arthritis response to TNF inhibitors: a potential SNP association with response to etanercept. *PLoS One* 14 (2): e0213073.

11 Sieberts, S.K., Zhu, F., Garcia-Garcia, J. et al. (2016). Crowdsourced assessment of common genetic contribution to predicting anti-TNF treatment response in rheumatoid arthritis. *Nat. Commun.* 7: 12460.

12 Benucci, M., Damiani, A., Li Gobbi, F. et al. (2018). Correlation between HLA haplotypes and the development of antidrug antibodies in a cohort of patients with rheumatic diseases. *Biologics* 12: 37–41.

13 Liu, M., Degner, J., Davis, J.W. et al. (2018). Identification of HLA-DRB1 association to adalimumab immunogenicity. *PLoS One* 13 (4): e0195325.

14 Hendrickson, B.A., Gokhale, R., and Cho, J.H. (2002). Clinical aspects and pathophysiology of inflammatory bowel disease. *Clin. Microbiol. Rev.* 15 (1): 79–94.

15 Bek, S., Nielsen, J.V., Bojesen, A.B. et al. (2016). Systematic review: genetic biomarkers associated with anti-TNF treatment response in inflammatory bowel diseases. *Aliment. Pharmacol. Ther.* 44 (6): 554–567.

16 Aden, K., Rehman, A., Waschina, S. et al. (2019). Metabolic functions of gut microbes associated with efficacy of tumor necrosis factor antagonists in patients with inflammatory bowel diseases. *Gastroenterology* 157 (5): 1279–1292.e11.

17 Bank, S., Andersen, P.S., Burisch, J. et al. (2014). Associations between functional polymorphisms in the NFkappaB signaling pathway and response to anti-TNF treatment in Danish patients with inflammatory bowel disease. *Pharmacogenomics J.* 14 (6): 526–534.

18 Billiet, T., Dreesen, E., Cleynen, I. et al. (2016). A genetic variation in the neonatal Fc-receptor affects anti-TNF drug concentrations in inflammatory bowel disease. *Am. J. Gastroenterol.* 111 (10): 1438–1445.

19 Datta-Mannan, A., Witcher, D.R., Lu, J., and Wroblewski, V.J. (2012). Influence of improved FcRn binding on the subcutaneous bioavailability of monoclonal antibodies in cynomolgus monkeys. *MAbs* 4 (2): 267–273.

20 Billiet, T., Vande Casteele, N., Van Stappen, T. et al. (2015). Immunogenicity to infliximab is associated with HLA-DRB1. *Gut* 64 (8): 1344–1345.

21 Sazonovs, A., Kennedy, N.A., Moutsianas, L. et al. (2020). HLA-DQA1*05 carriage associated with development of anti-drug antibodies to infliximab and adalimumab in patients with Crohn's disease. *Gastroenterology* 158: 189–199.

22 Wilson, A., Peel, C., Wang, Q. et al. (2020). HLADQA1*05 genotype predicts anti-drug antibody formation and loss of response during infliximab therapy for inflammatory bowel disease. *Aliment. Pharmacol. Ther.* 51: 356–363.

23 Nair, R.P., Stuart, P., Henseler, T. et al. (2000). Localization of psoriasis-susceptibility locus PSORS1 to a 60-kb interval telomeric to HLA-C. *Am. J. Hum. Genet.* 66 (6): 1833–1844.

24 Nair, R.P., Stuart, P.E., Nistor, I. et al. (2006). Sequence and haplotype analysis supports HLA-C as the psoriasis susceptibility 1 gene. *Am. J. Hum. Genet.* 78 (5): 827–851.

25 Trembath, R.C., Clough, R.L., Rosbotham, J.L. et al. (1997). Identification of a major susceptibility locus on chromosome 6p and evidence for further disease loci revealed by a two stage genome-wide search in psoriasis. *Hum. Mol. Genet.* 6 (5): 813–820.

26 Chen, L. and Tsai, T.F. (2018). HLA-Cw6 and psoriasis. *Br. J. Dermatol.* 178 (4): 854–862.

27 Gudjonsson, J.E., Karason, A., Runarsdottir, E.H. et al. (2006). Distinct clinical differences between HLA-Cw*0602 positive and negative psoriasis patients – an analysis of 1019 HLA-C- and HLA-B-typed patients. *J. Invest. Dermatol.* 126 (4): 740–745.

28 Strange, A., Capon, F., Spencer, C.C.A. et al. (2010). A genome-wide association study identifies new psoriasis susceptibility loci and an interaction between HLA-C and ERAP1. *Nat. Genet.* 42 (11): 985–990.

29 van Vugt, L.J., van den Reek, J., Hannink, G. et al. (2019). Association of HLA-C*06:02 status with differential response to ustekinumab in patients with psoriasis: a systematic review and meta-analysis. *JAMA Dermatol.* 155 (6): 708–715.

30 Dand, N., Duckworth, M., Baudry, D. et al. (2019). HLA-C*06:02 genotype is a predictive biomarker of biologic treatment response in psoriasis. *J. Allergy Clin. Immunol.* 143 (6): 2120–2130.

31 van Vugt, L.J., van den Reek, J.M.P.A., Meulewaeter, E. et al. (2020). Response to IL-17A inhibitors secukinumab and ixekizumab cannot be explained by

genetic variation in the protein-coding and untranslated regions of the IL-17A gene: results from a multicentre study of four European psoriasis cohorts. *J. Eur. Acad. Dermatol. Venereol.* 34 (1): 112–118.

32 Loft, N.D., Skov, L., Iversen, L. et al. (2018). Associations between functional polymorphisms and response to biological treatment in Danish patients with psoriasis. *Pharmacogenomics J.* 18 (3): 494–500.

33 Pennington, K.L. and DeAngelis, M.M. (2016). Epidemiology of age-related macular degeneration (AMD): associations with cardiovascular disease phenotypes and lipid factors. *Eye Vis. (Lond).* 3: 34.

34 Wu, M., Xiong, H., Xu, Y. et al. (2017). Association between VEGF-A and VEGFR-2 polymorphisms and response to treatment of neovascular AMD with anti-VEGF agents: a meta-analysis. *Br. J. Ophthalmol.* 101 (7): 976–984.

35 Stevens, A., Soden, J., Brenchley, P.E. et al. (2003). Haplotype analysis of the polymorphic human vascular endothelial growth factor gene promoter. *Cancer Res.* 63 (4): 812–816.

36 Toomey, C.B., Johnson, L.V., and Bowes Rickman, C. (2018). Complement factor H in AMD: bridging genetic associations and pathobiology. *Prog. Retin. Eye Res.* 62: 38–57.

37 Hong, N., Shen, Y., Yu, C.Y. et al. (2016). Association of the polymorphism Y402H in the CFH gene with response to anti-VEGF treatment in age-related macular degeneration: a systematic review and meta-analysis. *Acta Ophthalmol.* 94 (4): 334–345.

38 Sengul, E.A., Artunay, O., Rasier, R. et al. (2018). Pharmacogenetic aspect of intravitreal ranibizumab treatment in neovascular age-related macular degeneration: a five-year follow-up. *Ocul. Immunol. Inflamm.* 26 (6): 971–977.

39 Hu, Z., Xie, P., Ding, Y. et al. (2015). Association between variants A69S in ARMS2 gene and response to treatment of exudative AMD: a meta-analysis. *Br. J. Ophthalmol.* 99 (5): 593–598.

40 Zhou, Y.L., Chen, C.L., Wang, Y.X. et al. (2017). Association between polymorphism rs11200638 in the HTRA1 gene and the response to anti-VEGF treatment of exudative AMD: a meta-analysis. *BMC Ophthalmol.* 17 (1): 97.

41 Lores-Motta, L., Riaz, M., Grunin, M. et al. (2018). Association of genetic variants with response to anti-vascular endothelial growth factor therapy in age-related macular degeneration. *JAMA Ophthalmol.* 136 (8): 875–884.

42 Condreay, L., Chiano, M., Ortega, H. et al. (2017). No genetic association detected with mepolizumab efficacy in severe asthma. *Respir. Med.* 132: 178–188.

43 Duerr, C.U., McCarthy, C.D., Mindt, B.C. et al. (2016). Type I interferon restricts type 2 immunopathology through the regulation of group 2 innate lymphoid cells. *Nat. Immunol.* 17 (1): 65–75.

44 Condreay, L.D., Gao, C., Bradford, E. et al. (2019). No genetic associations with mepolizumab efficacy in COPD with peripheral blood eosinophilia. *Respir. Med.* 155: 26–28.

10

International Regulatory Processes and Policies for Innovator Biologics, Biosimilars, and Biobetters

Ankur Punia and Hemant Malhotra

Department of Medical Oncology, Sri Ram Cancer Center, Mahatma Gandhi Medical College Hospital, Jaipur, India

---

KEY POINTS

- Biologics are manufactured in living cells or tissues of human, animal, bacterial, or viral origin using biotechnology methods.
- Biosimilars are highly similar versions of innovator biologics and provide more affordable treatment options.
- World Health Organization, WHO, has played a key role in establishing the WHO Biological Reference Materials for standardizing biologics and for developing guidelines on the production and control of biological technologies and products.
- A dedicated pathway for approval of biosimilars was introduced in the European Union (EU) in 2004 which has pioneered biosimilar regulation. EU has approved the highest number of biosimilars and has the most experience with their efficacy and safety.
- Biobetters are considered as new products under existing regulatory frameworks.

Abbreviations

Abbreviation	Full name
ANVISA	Agência Nacional de Vigilância Sanitária
ARTG	Australian Register of Therapeutic Goods
BGTD	The Biologics and Genetic Therapies Directorate
BLA	Biologics License Application
CBER	Center for Biologics Evaluation and Research
CDE	Center for Drug Evaluation
CDER	Center for Drug Evaluation and Research
CDSCO	Central Drugs Standard Control Organization
CFDA	China Food and Drug Administration
CHMP	Committee for Medicinal Products for Human Use
CMC	Chemistry, Manufacturing, and Controls
DCGI	Drugs Controller General of India
EMA	European Medicines Agency
FD&C Act	Federal, Food & Cosmetic Act
FDA	Food & Drug Administration

Abbreviation	Full name
GLP	Good Laboratory Practice
GMP	Good Manufacturing Practice
GPSP	Good Practice Systems and Programs
IND	Investigational New Drug
MFDS	Ministry of Food and Drug Safety
MHLW	Ministry for Health Labour and Welfare
NDA	New Drug Application
NIFDC	National Institute for Food and Drug Control
NIFDS	National Institute of Food and Drug Safety Evaluation
NMPA	National Medical Products Administration
PD	Pharmacodynamic
PK	Pharmacokinetic
PMDA	Pharmaceuticals and Medical Devices Agency
TGA	Therapeutic Goods Administration
TPD	Therapeutic Products Directorate
WHO	World Health Organization

Biologics, Biosimilars, and Biobetters: An Introduction for Pharmacists, Physicians, and Other Health Practitioners, First Edition. Edited by Iqbal Ramzan.

10.1 Introduction

Biologicals or biologics replicate endogenous substances such as enzymes, hormones, or antibodies. They are derived from living cells or tissues of human, animal, and/or bacterial or viral origin using biotechnology.

The inherent variations that may result during their manufacture present challenges in characterizing these products. Minor differences between different batches of the same biologic product are routine. Various regulatory frameworks have been put in place to assess their manufacture and such product variations; these are designed to ensure that biologics are able to deliver consistent clinical outcomes.

Biosimilars are highly similar versions of innovator biologics and provide more affordable treatment options. Biosimilars are approved via abbreviated pathways that avoid duplicating costly clinical trials.

10.2 Major International Regulatory Agencies

Most countries (or umbrella organizations) have embedded specific regulations for biologics approval. Summary of the various regulatory frameworks is now provided.

10.2.1 World Health Organization (WHO)

WHO has played a key role for over 50 years in establishing WHO Biological Reference Materials necessary for standardizing biologics and developing guidelines on the production and control of biological technologies and products. These are based on scientific consensus through international consultations and assist WHO member states to ensure quality and safety of biologics and related biological diagnostic tests.[1]

10.2.2 United States of America (USA) – Food & Drug Administration (FDA)

FDA's Center for Drug Evaluation and Research (CDER) and Center for Biologics Evaluation and Research (CBER) regulate pharmaceuticals and biological products. CBER regulates a wide range of biologics, including:

- Allergenic extracts (e.g. for allergy shots and tests)
- Blood and blood components
- Gene therapy products
- Devices and test kits

- Human tissue and cellular products used in transplantation
- Vaccines

CDER regulates other biologics, mostly produced by biotechnology including:

- Monoclonal antibodies designed as targeted therapies
- Cytokines involved in immune responses
- Growth factors that control cell growth
- Enzymes that catalyze biochemical reactions
- Immunomodulators that affect immune responses

10.2.3 Europe (EU) – European Medicines Agency (EMA)

EMA plays a key role in centralized procedure for development and approval of biological products. Approval of medicinal products has three stages: (i) research and development, (ii) marketing authorization, and (iii) post authorization. The first stage provides an opportunity for early dialogue before submission of a marketing application. Stage two involves scientific evaluation of the application; once approval is granted, the centralized marketing authorization is valid in EU Member States, Iceland, Norway, and Liechtenstein. Stage three primarily deals with responsibilities of authorization holders like pharmacovigilance.

The national competent authorities in each EU country are responsible for authorization of medicines available in the EU that have not undergone the centralized EU procedure.

10.2.4 Australia – Therapeutic Goods Administration (TGA)

TGA has regulatory authority for therapeutic goods and provides the regulatory framework for biologics. It regulates supply, import, export, manufacturing, and advertising of therapeutic goods, prescription medicines, vaccines, sunscreens, vitamins and minerals, medical devices, blood, and blood products. TGA manages the Australian Register of Therapeutic Goods (ARTG) for all therapeutic goods for human use and grants exemptions from the ARTG for unapproved therapeutic goods to be used in clinical trials.

10.2.5 Canada – The Biologics and Genetic Therapies Directorate (BGTD)

The Therapeutic Products Directorate (TPD) and the Biologics and Genetic Therapies Directorate (BGTD), respectively, regulate pharmaceutical drugs and medical devices, and biological drugs and radiopharmaceuticals for human use. TPD's Office of Clinical Trials and the BGTD's Office of Regulatory Affairs, among others, are directly involved with clinical trial review and approval of pharmaceutical, biological, and radiopharmaceutical drugs.

BGTD ensures that drugs in Schedule C (Radiopharmaceuticals and their Kits) and D (Biologicals) in Canada are safe, effective, and of high quality.

BGTD develops new regulations, policies, and guidelines under four acts[2] and reviews and provides market authorization for all biologics for human use.

10.2.6 China – National Medical Products Administration (NMPA)

The NMPA, previously known as the China Food and Drug Administration (CFDA), implements guidelines, policies, and decision-making for the supervision and administration of drugs, medical devices, and cosmetics. It is responsible for safety supervision, standards management, drug registration, quality management, risk management, pharmacist licensing, inspection systems, international cooperation, guiding provincial and municipal drug administration, and other tasks assigned by the State Council and Party Central Committee. NMPA is charged with accelerating approval of innovative drugs, establishing a system of listing license holders, promoting electronic review, and introducing overall efficiencies.

Drug Registration Management Department of NMPA formulates, supervises, and implements drug standards while the Drug Administration Department formulates, supervises, and implements pharmaceutical production quality management standards for drugs, Chinese medicines, biological products, and radioactive and toxic materials. The NMPA includes the National Institutes for Food and Drug Control (NIFDC) and the Center for Drug Evaluation (CDE).[3]

The establishment of NMPA is an important reform for foreign access, eliminating the conflicting standards that prevailed among provincial government agencies; it has centralized the Chinese health regulatory system and increased transparency.

10.2.7 India – Central Drugs Standard Control Organization (CDSCO)

The Drugs Controller General of India (DCGI) heads CDSCO and is referred to as the Central Licensing Authority in the Indian regulatory framework.

CDSCO is responsible for approving new drugs, conducting clinical trials, establishing drug standards, overseeing the quality of imported drugs, providing expert advice, and coordinating the state licensing authorities that regulate the manufacture, sale, and distribution of drugs.

10.2.8 Brazil – Agência Nacional de Vigilância Sanitária (ANVISA)

ANVISA is an independent administrative agency responsible, under the authority of the Ministry of Health – *Ministério de Saúde* – of the Brazilian Government, for drug registration and license of pharmaceutical laboratories and other companies within the pharmaceutical production area. It also establishes regulations for clinical trials and drug pricing.

The goods and products under the agency's purview include medicines for human use and their active ingredients, immune biologicals and their active substances, and blood and blood derivatives.

10.2.9 South Korea – Ministry of Food and Drug Safety (MFDS)

MFDS was formerly the Korean Food and Drug Administration. The agency, through its Biopharmaceuticals and Herbal Medicines Bureau, is responsible for the scientific evaluation of medicines developed by pharmaceutical companies for use in South Korea.

National Institute of Food and Drug Safety Evaluation (NIFDS), affiliated with MDFS, conducts research and development (R&D) to secure scientific evidence as the basis for safety management decisions and policies. The NIFDS consists of six departments, three research departments, and three review departments.[4]

South Korea has detailed procedures for the review of safety/efficacy, specifications/analytical procedures, criteria/requirements for granting product approval or approval changes for manufacture/marketing/importing of biologics, recombinant DNA or cell culture-derived products, and cell/gene therapy or other similar products.[5]

10.2.10 Japan – Pharmaceuticals and Medical Devices Agency (PMDA)

The Ministry for Health Labour and Welfare (MHLW) has overall responsibility for the use of medicines and medical devices in Japan.

PMDA is the Japanese regulatory agency, working on behalf of MHLW. It reviews applications for marketing authorizations of pharmaceuticals and medical devices and monitors their post-marketing safety. It also provides relief compensation for individuals with adverse drug reactions and infections from pharmaceuticals or biological products.

PMDA evaluates the quality, efficacy, and safety of drugs, medical devices, and cellular and tissue-based products in line with current scientific and technological standards. In addition, PMDA's reviews and related services consist of various activities, such as "consultations" providing advice in relation to regulatory submission, GLP/GCP/GPSP inspections to ensure the submitted data comply with ethical and scientific standards, and GMP/QMS/GCTP inspections to ensure quality of the manufacturing facility.[6]

PMDA's Office of Biologicals handles biotechnology medicine applications including reference biologics and biosimilars.

10.3 General Requirements for Biologics Approval

10.3.1 Innovator/Reference Biologics

For licensing of a biological product, national regulatory authorities and manufacturers provide data on quality, nonclinical, and clinical aspects of rDNA-derived biotherapeutic protein products. WHO provides the basic framework and standardized approach to be followed but the legal status of investigational products varies in each country.

Basic requirements for these products are: (i) manufacturing and quality control (control of source materials, manufacturing processes, and drug substance/drug product), (ii) nonclinical evaluation (identify safe dose, potential toxicity target, safety parameters including pharmacokinetics), and (iii) clinical evaluation (Phase I and II studies and biomarker identification; Phase IV studies may be required if the biologic has been approved via an abridged process).

10.3.2 Biosimilars

A biosimilar is highly similar to and has no clinically meaningful differences from an existing approved reference biologic. A reference product is the biologic against which a proposed biosimilar is compared and whose approval is based on full safety and efficacy data.

A biosimilar sponsor must demonstrate that its product is highly similar to the reference product by extensively characterizing the structure (chemical identity, purity) and function (bioactivity) of both the reference product and the proposed biosimilar. This involves a comprehensive head-to-head comparison of the biosimilar and the reference medicine. Comparability is a stepwise process that is tailor-made for each product; knowledge from the initial quality comparability studies (step 1) is used to determine the extent and type of nonclinical (step 2) and clinical (step 3) studies required in the next phase of development, with the ultimate aim of ruling out differences in clinical performance between the biosimilar and the reference biologic.

A detailed discussion on what is required from an analytical perspective to demonstrate "highly similar" is provided in Chapter 6.

10.4 Specific Requirements for Innovator/Reference Biologics and Biosimilars

10.4.1 Innovator/Reference Biologics

10.4.1.1 USA (FDA)

FDA's regulatory authority for the approval of biologics resides in the Public Health Service Act (PHS). However, biologics are also subject to regulation under the Federal Food, Drug, and Cosmetic Act (FD&C Act) because most biological products also meet the definition of "drugs" within this Act.

The 2009 amendment to the PHS Act introduced the term "protein" (except any chemically synthesized polypeptide) in the definition of a biological product stating that chemically synthesized polypeptides are chemicals and not biologics, as long as the synthesized polypeptide is less than 100 amino acids.

Hormone protein biologics are regulated exclusively under the FD&C Act. FDA regulates most other protein biologics exclusively under the PHS Act. However, protein enzyme biologics are regulated by FDA under both Acts.

The antibody drug conjugate Mylotarg® (gemtuzumab ozogamicin) was approved in 2000 by the FDA under the FD&C Act as a New Drug Application; following withdrawal from the market and resubmission, it was approved in 2017 by the FDA via the PHS Act using a Biologic License Application.

According to the "deemed to be a license" provision of the Biologics Price Competition and Innovation Act of 2009, all biologic proteins must be regulated under the

PHS Act from 23 March 2020. Biologic proteins already approved under the FD&C Act will have to be redesignated under the PHS Act.

Some medical devices used to produce biologics are regulated by CBER under the FD&C Act's Medical Device Amendments of 1976. FDA also:

- approves new biologic products and new indications for already approved products
- protects against threats of emerging infectious diseases
- provides public information for safe and appropriate use of biologics
- inspects biologics manufacturing facilities prior to approval and regularly thereafter
- monitors safety of biologics post-marketing.

CBER regulates the Investigational New Drug (IND) or Device Exemption (IDE) Process (CBER), Expanded Access to Experimental Biologics; BLA Process (CBER); NDA Process (CBER).[7]

10.4.1.2 Europe (EMA)

The EMA's scientific guidelines on biologics provide detailed guidelines under two categories:[8]

1) Biologicals: active substance
   - Manufacture, characterization and control of the active substance
   - Specifications
   - Comparability and biosimilarity
   - Plasma-derived medicinal products
   - Plasma master file
   - Vaccines
   - Stability
2) Biologicals: finished product
   - Product information
   - Pharmaceutical development
   - Adventitious agents – safety evaluation
   - Transmissible spongiform encephalopathies, TSEs (animal and human)
   - Investigational medicinal products
   - Genetically modified organisms (GMOs)
   - Specifications
   - Lifecycle management

10.4.1.3 Australia (TGA)

Two basic guidelines are provided on the legislative requirements and how to apply for biologics approval: (i) pathways for supply of biologicals and (ii) autologous human cells and tissues regulation.[9]

The Australian Regulatory Guidelines for Biologicals provide sponsors with guidance on the regulation of biologicals. Not everything that meets TGA's definition of a biological is regulated as a biological as it needs to be verified whether it is excluded from TGA regulation or regulated as therapeutic goods. The pathway for supply of biologic products in Australia varies depending on whether the product is (i) exempt from some TGA regulation, (ii) an "unapproved" biological authorized for supply, and (iii) included on the ARTG.

GMP clearance and other certification requirements apply to: (i) biologicals that comprise, contain, or are derived from human cells and tissues must comply with GMP code for human blood and blood components, human tissues, and human cellular therapy products and (ii) biologicals that comprise/contain live animal cells, tissues, or organs must comply with the PIC/S guide to GMP, except for the same manufacturing principles that apply to medicines.

Certain autologous human cell and tissue products may be eligible for exemption from some regulatory requirements:

- samples collected from a patient under clinical care
- manufactured and used by the practitioner for clinical care OR by a person under their supervision
- for a single indication in a single clinical procedure
- minimally manipulated products for homologous use

Where one or more criteria are not met, full regulation by TGA are applied.

Biologics that are on the ARTG have been evaluated for quality, safety, and efficacy and their supply is approved. However, if a biologic that is not on the ARTG (an "unapproved" biological), is to be supplied, the following pathways are available: (i) as part of a clinical trial (clinical trial scheme), (ii) for an individual patient (special access scheme), or (iii) by an individual practitioner for multiple patients (authorized prescriber scheme).

10.4.1.4 Canada (BGTD)

In Canada, biologics are listed in Schedule D of the Food and Drugs Act.

A sponsor must collect enough scientific evidence about a biologic before BGTD can consider approving it. The evidence must show that the biologic is safe, effective, and of suitable quality.

Along with the information needed for approving other drugs, more detailed chemistry and manufacturing information is required for biologics to ensure the purity and quality of the product. For example, specific processes are required to ensure that the product is not contaminated by an undesired microorganism or another biologic. For these reasons, the BGTD has unique programs for biologic drugs including on-site evaluation (OSE) and lot release program.[10]

OSE is unique to biologics in Canada. It is a product-specific assessment that the BGTD may conduct at the manufacturing site and it also confirms the ability of the manufacturer to consistently produce a safe biologic. Lot release covers both pre- and post-market stages and each lot is subject to this program before distribution/sale.

Canada's level of regulatory oversight (testing and/or protocol review) is based on the degree of risk linked to the product. BGTD assigns the product to 1 of 4 different risk-based evaluation groups (Group 1: preapproval stage; Group 2: sample testing and protocol review; Group 3: protocol review and periodic testing; Group 4: notification and periodic testing). BGTD also provides guidance, information, and support on meeting Health Canada's regulatory requirements for biologics; these documents also assist its staff in implementing Health Canada's mandate in a fair, effective, and consistent manner. Canada has adopted a number of WHO guidance documents on quality, safety, and efficacy of biologics.

10.4.1.5 China (NMPA)

The Drug Registration Regulation (Revised 2007) pathway classifies therapeutic biologicals into 15 categories[11]:

1) Products that have not been marketed in China and other countries.
2) Monoclonal antibodies.
3) Gene therapy, somatic cell therapy, and related products.
4) Allergens.
5) Multicomponent bioactive products extracted from human/animal tissue/body fluid, or produced by fermentation.
6) New combination products made from marketed biologics.
7) Products which have been marketed in other countries but not in China.
8) Microbiological products containing components made from strains that have not been approved for use in China.
9) Products that do not have the exact same structure as marketed products and have not been marketed in China or overseas (including locus mutation or absence of amino acid, changes in post-translational mutation, or absence of amino acid, changes in post-translational modification caused by using different expression systems, and chemical modification of the product).

10) Biologicals produced by different methods compared with the marketed products, such as different expression systems or host cells.
11) The first product produced by recombinant DNA method (for example, replacement of synthesis, tissue extraction, or fermentation technologies by recombinant DNA technology).
12) Products changed from non-injection route to injection route or from topical use to systemic use, which have not been marketed in China or other countries.
13) Marketed products with a new formulation but same route of administration.
14) Marketed products with a new route of administration (excluding Category 12).
15) Products with national standard.

To conduct an investigational clinical trial in China, a clinical trial application is filed which requires very comprehensive information including highly detailed manufacturing protocols, which raises concern as this information is considered proprietary by companies. Also, at the time of submission, sample analysis (by the NIFDC) and inspection of the biologic manufacturing site is required. Overall, China's drug registration and regulation is complicated, time-consuming, and an ever-changing process, involving several regulatory bodies at various levels of central and local governments.

10.4.1.6 India (CDSCO)

CDSCO is the regulatory authority for approving manufacturing and import of biologic products in India. For biologics, additional approvals are also required by other agencies, including the Genetic Engineering Approval Council (GEAC), Recombinant DNA Advisory Committee (RDAC), Review Committee on Genetic Manipulation (RCGM), Institutional Biosafety Committees (IBSC), State Biosafety Coordination Committees (SBCC), and the District Level Committees (DLC).

Various guidelines also need to be followed for biologicals including (i) Recombinant DNA Safety Guidelines, (ii) Generating preclinical and clinical data for rDNA vaccines, (iii) Diagnostics and other biological guidelines, (iv) Guidelines and Handbook for IBCs, and (v) CDSCO guidance for industry.

Biologicals approval requires the following data:

- Clinical trial application for evaluating safety and efficacy
- Conditions for permission of new drugs approval

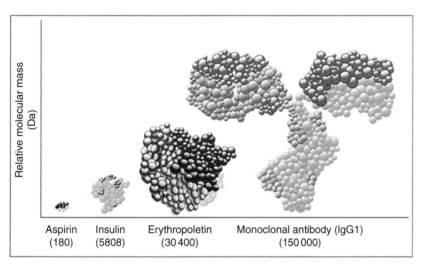

Figure 3.1 Relative molecular mass of small molecule and biologic drugs. *Source:* Mellstedt.

Biologics, Biosimilars, and Biobetters: An Introduction for Pharmacists, Physicians, and Other Health Practitioners,
First Edition. Edited by Iqbal Ramzan.

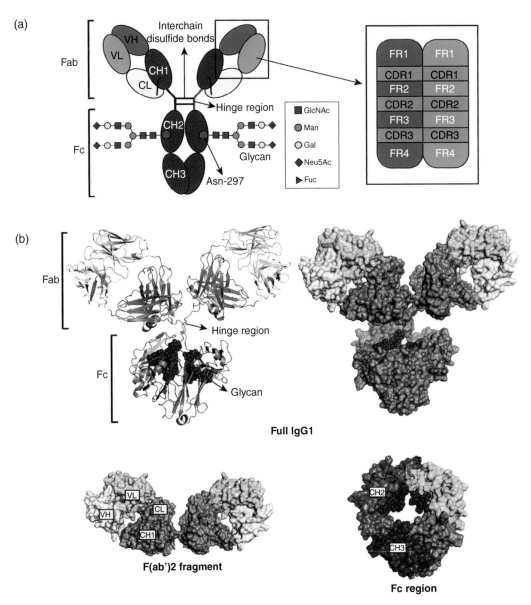

Figure 4.1 Structure of IgG1. (a) Schematic representation of an IgG1 displaying the various domains that make up the heavy chain (VH, CH1, CH2, and CH3) and the light chain (VL and CL). A complex glycan represented as per the standardized Symbol Nomenclature for Glycans (SNFG) is shown bound to the asparagine 297 (Asn-297) residue on the CH2 of the heavy chain. The expanded region provides a closer look to the variable regions of the heavy and light chain, displaying three complementarity-determining regions (CDRs and four framework regions (FR) on each domain. (b) Crystal structure of a murine IgG1 protein (PDB 1IGY) displayed as both cartoon and surface representations. The Fc-glycan is shown in red. The F(ab')2 and Fc region are shown isolated to indicate the position of the domains in the surface representation. Note the orientation of the glycan toward the interior of the protein.

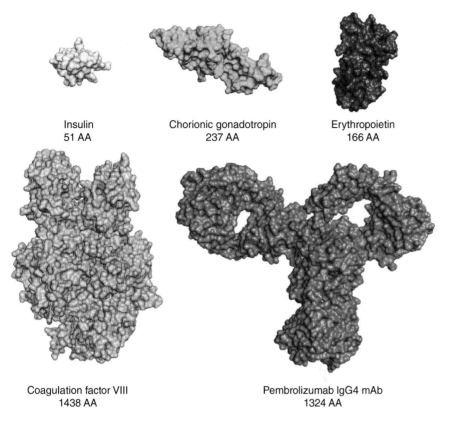

Insulin
51 AA

Chorionic gonadotropin
237 AA

Erythropoietin
166 AA

Coagulation factor VIII
1438 AA

Pembrolizumab IgG4 mAb
1324 AA

Figure 4.2 Structural features of various classes of therapeutic proteins.

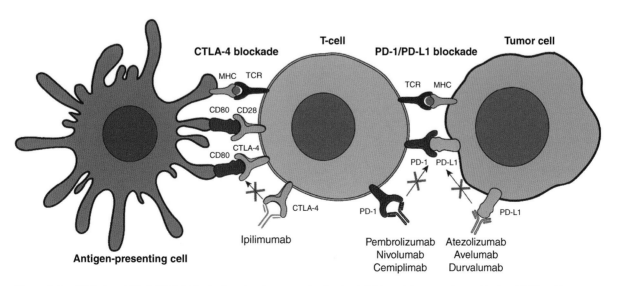

Figure 5.1 CTLA-4 and PD-1/PD-L1 blockade using immune checkpoint inhibitors. Tumor antigens bind to MHC and interact with TCR on T cells. CTLA-4 binds to CD80 to suppress T cell activation. Ipilimumab prevents CTLA-4 from binding to CD80 and activates immune response to tumor cells. Similarly, the immune-suppressing signal sent by PD-1/PD-L1 binding is blocked by anti-PD-1 and anti-PD-L1 antibodies. *Source:* Adapted from Cruz and Kayser.

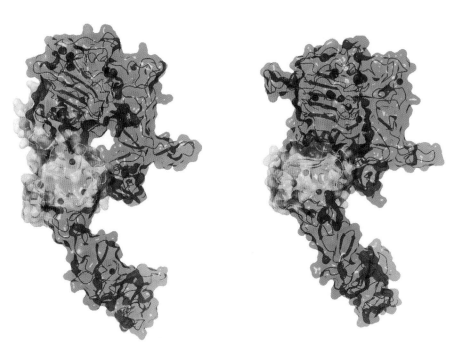

Figure 5.2 Extracellular domain structures of EGFR bound to EGF (PDB ID: 3NJP) (left) and HER2 (PDB ID: 6J71). Red – domain I; blue – domain II; green – domain III; purple – domain IV; yellow – EGF. *Source:* Adapted from Lu et al. and Wang et al.

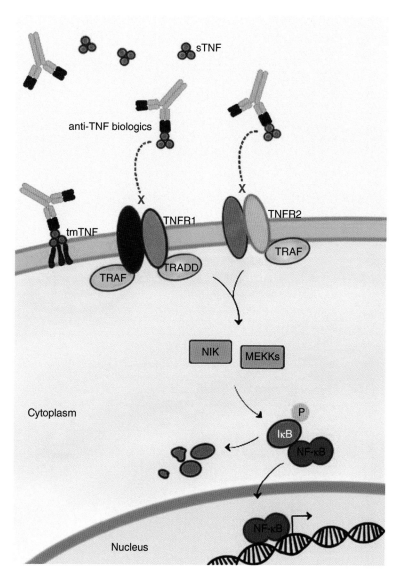

Figure 5.3 Mechanism of action of TNF-α biologics; infliximab is shown binding to tmTNF-α and sTNF-α. *Source:* Adapted from Pedersen et al.

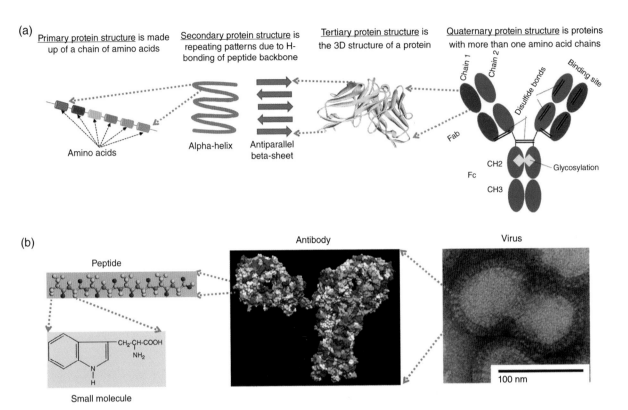

Figure 6.1 (a) Primary, secondary, tertiary, and quaternary protein structures. The quaternary structure depicts different domains of an IgG antibody and shows some of the disulfide bonds and its glycosylation site. It has two light chains (Chain 1) and two heavy chains (Chain 2). (b) Size comparison of small molecule, peptide, mAb, and influenza virus.

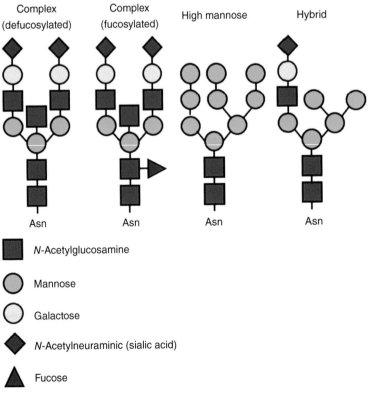

■ *N*-Acetylglucosamine

● Mannose

○ Galactose

◆ *N*-Acetylneuraminic (sialic acid)

▲ Fucose

Figure 6.2 Types of N-glycans observed in therapeutic proteins. All N-glycans are composed of the core structure highlighted in red.

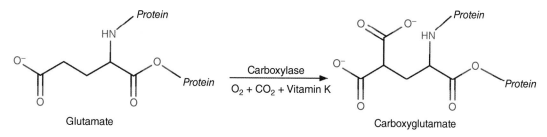

Figure 6.3 Post-translational γ-carboxylation of glutamate residues in proteins.

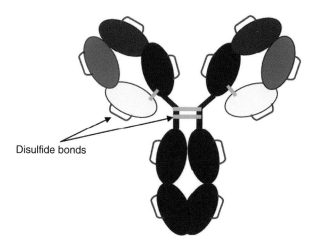

Figure 6.4 An IgG1 antibody with 4 interchain (green) and 12 intra-chain disulfide bonds (red).

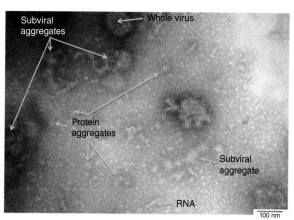

Figure 6.6 Different size particles present in an influenza vaccine sample as visualized by a negatively stained TEM image.

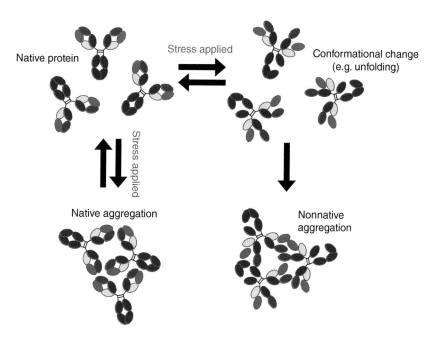

Figure 6.5 Native and non-native protein aggregation.

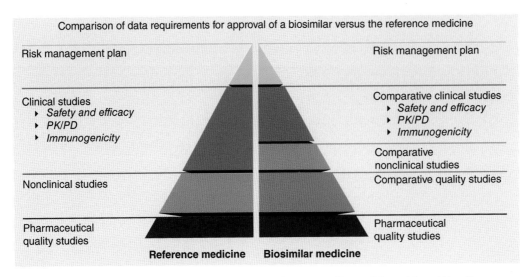

Figure 7.1 Development paradigm of biosimilars versus innovators. *Source:* Adapted from https://ec.europa.eu/docsroom/documents/22924/attachments/1/translations/en/renditions/native.

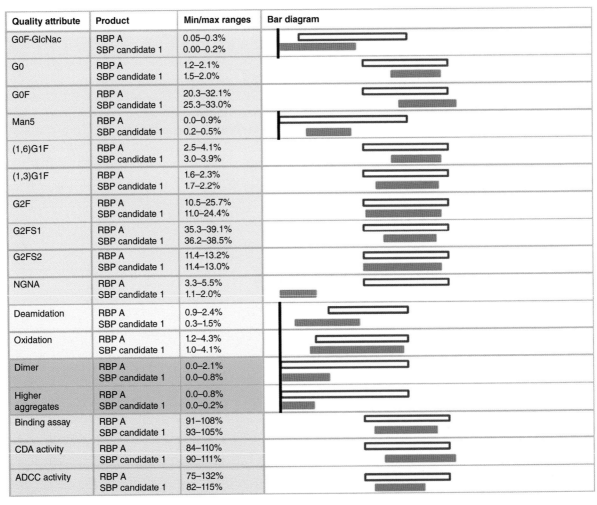

Quality attribute	Product	Min/max ranges	Bar diagram
G0F-GlcNac	RBP A	0.05–0.3%	
	SBP candidate 1	0.00–0.2%	
G0	RBP A	1.2–2.1%	
	SBP candidate 1	1.5–2.0%	
G0F	RBP A	20.3–32.1%	
	SBP candidate 1	25.3–33.0%	
Man5	RBP A	0.0–0.9%	
	SBP candidate 1	0.2–0.5%	
(1,6)G1F	RBP A	2.5–4.1%	
	SBP candidate 1	3.0–3.9%	
(1,3)G1F	RBP A	1.6–2.3%	
	SBP candidate 1	1.7–2.2%	
G2F	RBP A	10.5–25.7%	
	SBP candidate 1	11.0–24.4%	
G2FS1	RBP A	35.3–39.1%	
	SBP candidate 1	36.2–38.5%	
G2FS2	RBP A	11.4–13.2%	
	SBP candidate 1	11.4–13.0%	
NGNA	RBP A	3.3–5.5%	
	SBP candidate 1	1.1–2.0%	
Deamidation	RBP A	0.9–2.4%	
	SBP candidate 1	0.3–1.5%	
Oxidation	RBP A	1.2–4.3%	
	SBP candidate 1	1.0–4.1%	
Dimer	RBP A	0.0–2.1%	
	SBP candidate 1	0.0–0.8%	
Higher aggregates	RBP A	0.0–0.8%	
	SBP candidate 1	0.0–0.2%	
Binding assay	RBP A	91–108%	
	SBP candidate 1	93–105%	
CDA activity	RBP A	84–110%	
	SBP candidate 1	90–111%	
ADCC activity	RBP A	75–132%	
	SBP candidate 1	82–115%	

Figure 7.2 Quality attribute ranges of a biosimilar candidate and its reference product. A, targeting a cell membrane bound target and where Fc functionality is an important part of the clinical mode of action. The lengths of the bars show the relative boundaries of the quality attribute ranges. Attributes are shown as a black line on the left; this black line represents the point of origin (i.e. 0%). *Source:* From Schiestl et al.

- Biological products post approval changes: quality, safety, and efficacy document
- Quality information for drug submission for new drug approval: biotechnological or biological products

Guidance on various application forms for Indian regulatory agencies is available.[12]

10.4.1.7 Brazil (ANVISA)

Prior to 2002, Brazil had no specific guidance for biological products. In 2002, guidelines for biological products were published (RDC 80/2002), which need to be followed by sponsors of both originator biological and biosimilars ("follow-on biological products" as they are known in Brazil).

Brazil introduced new regulations in December 2020 to specifically address and establish specific pathways to license follow-on biological products. This Brazilian regulation (Resolution 55/2010) is based on different regulations and guidelines from around the world. They follow the same scientific principles as the WHO guidelines but contain differences due to the specific needs of Brazil.[13]

The guideline for development of biologics requires a complete dossier of information on quality, safety, and efficacy. Resolution 49/2011 establishes requirements and procedures for post-approval changes of Biological Products; Resolution 50/2011 establishes requirements and procedures for stability studies of Biological Products; Resolution 204/2017 establishes prioritization criteria for marketing authorization, post-approval changes; and IND submission and Resolution 205/2017 establishes a special procedure for clinical trials.

10.4.1.8 South Korea (MFDS)

Biopharmaceuticals and Herbal Medicines Bureau is a part of MFDS which establishes policy making whereas Biopharmaceuticals and Herbal Medicines Evaluation Department approves biologic products. These departments have five subdivisions concerned with biologicals, first two making policies and the remaining three implementing these policies.

Biopharmaceutical Policy Division establishes and revises policies for biologics and supports exports and international cooperation. **Biopharmaceutical Quality Management Division** coordinates administrative work: PMS, GMP inspection, and coordinates all compliance actions: product recalls, regulatory letters. **The Biologics Division** coordinates administrative work: market authorization for biologics, evaluates chemistry, manufacturing and control section, evaluates pharmacology, toxicology, and clinical data

section, and supports GMP, GLP, GCP compliance. **National Center for Lot Release** performs official lot release tests of biologics and conducts laboratory research relating to testing method and national STD. **Biologics Research Division** conducts research of vaccine effectiveness and collaborative activities with WHO.

Regulation on approval and review of biological products ensures safety and efficacy, specifications and analytical procedures, criteria and requirements for granting the product approval with regard to the approval or approval changes of the manufacture and marketing (importing) of the biologics, recombinant DNA products, cell culture-derived products, cell therapy products, gene therapy products, and other similar products.[14]

10.4.1.9 Japan (PMDA)

PMDA evaluates biologics via its:

- Office of Biologics I: Blood Products; CMC for Biologics (Pharmaceuticals); Advanced therapy products (Gene therapy products).
- Office of Biologics II: Biological Products (Vaccines, etc.); CMC for Biologics (Medical devices); Advanced therapy products (Cell/Tissue-based products, Regenerative medicine).

Development of biologics is based on the cooperative interaction between academia, industry, and regulatory agency. PMDA provides consultation from early stages of biologic development through to clinical trials and biomarker identification and for establishment of guidelines to ensure quality, efficacy, and safety. Various guidelines for biologics are available.[15]

10.4.2 Biosimilars

10.4.2.1 WHO

Clinical experience and established safety profile of the original biologic has contributed to biosimilar development. Data required for licensing of biosimilars is thus likely to be less than that required for the originator. With respect to quality, comparability data of a new biosimilar with an established innovator, showing that they are "highly similar," can be considered as additional data, over and above what is normally required for the originator product. This is the basis for reducing the requirements for nonclinical and clinical data in the context of the WHO. The WHO Q&A document serves to clarify guiding principles in the Guidelines on biosimilars and on similar mAbs.

WHO Guidelines have been instrumental in raising awareness of the complex scientific issues related to the licensing of similar biologics. Implementation of these guidelines is important in achieving international regulatory convergence. The first workshop on implementing "WHO Guidelines on Evaluating Similar Biotherapeutic Products" held in 2010 featured speakers from various regulatory agencies, international clinical/scientific experts, industry, and WHO representatives and resulted in 26 peer-reviewed publications.[16]

WHO Prequalification of Medicines Programme's mission is to work closely with national regulatory agencies/partner organizations to ensure rapid availability of quality priority medicines. The WHO has prequalified its first biosimilar, trastuzumab, on 18 December 2019, to make this expensive, life-saving treatment more affordable globally. It has also launched the prequalification project for rituximab.[17]

10.4.2.2 USA (FDA)

The Biologics Price Competition and Innovation (BPCI) Act, created an abbreviated licensure pathway for biological products that are demonstrated to be biosimilar to or interchangeable with an FDA-approved biologic. This pathway provides more treatment options, increases access to lifesaving drugs, and potentially lower health care costs through greater competition.

This abbreviated pathway does not mean a lower approval standard for a biosimilar or an interchangeable product; the data package is extensive. If a biosimilar manufacturer can demonstrate that its product is biosimilar to the reference product, then it is scientifically justified to rely on existing scientific knowledge on the safety and effectiveness of the reference product to support approval. This allows for a potentially shorter and less costly biosimilar development program.

A biosimilar application must include data demonstrating biosimilarity to the reference product, in the form of analytical studies demonstrating that the biosimilar is highly similar to the reference biologic, notwithstanding minor differences in clinically inactive components; animal studies with toxicity assessment and a clinical study/studies demonstrating safety, purity, and potency of the proposed biosimilar in one or more indication for which the reference product is licensed.[18] The totality of data is taken into account in making a decision.

An application for an interchangeable biosimilar must also include data demonstrating that: (i) the proposed interchangeable product is expected to produce the same clinical result as the reference product in any given patient, (ii) for a product administered more than once to an individual, switching between the proposed

interchangeable product and the reference product does not increase safety risks or decrease effectiveness compared to using the reference product without such switching.

FDA evaluates each biosimilar product on a case-specific basis to determine what data are needed and which data elements may be exempt if scientifically justified. This determination may be informed by what is already publicly known about the reference biologic.

Biosimilars may be approved for an indication without direct studies in that indication. This concept, "indication extrapolation," is critical to the goals of an abbreviated pathway and is scientifically justified and accepted but the biosimilar manufacturer must provide scientific justification to support such indication extrapolation.

FDA first approved a biosimilar in March 2015, a filgrastim biosimilar; there are now 26 approved biosimilars up to December 2019.

10.4.2.3 Europe (EMA)

In 2004, a dedicated biosimilar approval pathway was introduced which has pioneered the regulation of biosimilars since the approval of the first agent in 2006 (the growth hormone, somatropin). Since then, the EU has approved the highest number of biosimilars worldwide and consequently has the most extensive experience of their clinical efficacy and safety.

Over the years, EMA has issued scientific guidelines to assist biosimilar developers meet the regulatory requirements for biosimilars. The guidelines have evolved to keep pace with rapid advances in biotechnology and analytical sciences and take on board increasing experience of their clinical use.[19]

When a sponsor applies for marketing authorization with EMA, data are evaluated by EMA's expert panels. The review by EMA results in a scientific opinion, which is then sent to the European Commission, which ultimately makes the authorization decision.

10.4.2.4 Australia (TGA)

Biosimilar medicines are evaluated via the standard prescription medicines registration process[20] including mandatory general dossier requirements and Common Technical Document (CTD) data module.

CTD sets the specifications for the registration dossier of medicines; it was developed by the International Conference on Harmonization of Technical Requirements for Registration of Pharmaceuticals for Human Use (ICH) and adopted by the TGA in 2004. A number of laboratory and clinical studies are needed to demonstrate comparability (biosimilarity) of the new biosimilar to the reference biologic on the ARTG.

The TGA has adopted a number of European guidelines that outline the quality, nonclinical, and clinical data requirements specific to biosimilar medicines; and the ICH guideline on the assessment of comparability. For biosimilar registration, the reference biologic must be registered based on full data and have been marketed in Australia for a substantial period so that a sizeable body of efficacy and safety data for the approved indications is available.

If a reference biologic used for comparability studies is not on the ARTG, then the reference medicine must be approved by a regulatory authority with similar scientific/regulatory standards (e.g. EMA or US FDA) and a bridging study must be provided to demonstrate that the comparability studies are relevant to the Australian reference medicine (this bridging study may be abridged or omitted if evidence is provided that the medicine is manufactured at a single site for global distribution).

If the biosimilar manufacturing process changes significantly between clinical trial and commercial stages, then either (i) a second comparability study together with the clinical trial and commercial medicines is required (preferred) or (ii) a linked comparability study is needed showing the clinical trial and commercial batches are similar.

If direct comparison of the biosimilar and reference material is not possible (the concentration of the active substance in the reference product is too low or there are interfering excipients such as human serum albumin), then extraction or concentration validated techniques may be used. Reproduction of pharmacopoeia monograph methodologies is not sufficient.

If indication extrapolation is sought, then EMEA/CHMP guidelines for assessing nonclinical and clinical issues related to extrapolation are warranted.[21]

TGA usually requires a risk management plan (RMP) for a biosimilar medicine. Information on RMP requirements is available.[22]

10.4.2.5 Canada (BGTD)

Biosimilars are regulated as new drugs under the Food and Drugs Act and the Food and Drug Regulations. The manufacturer must provide information to show that the biosimilar and the reference biologic are highly similar, and that there are no clinically meaningful differences in terms of efficacy and safety between them. A decision to authorize a biosimilar is based upon a benefit/risk assessment after reviewing all the submitted data.[23]

Health Canada has contributed to two international statements on biosimilars; BGTD is also undertaking a three-year pilot to explore a stepwise review of biosimilars that would be complementary to the biosimilar development process.

10.4.2.6 China (NMPA)

The CDE, the technical review body under the China Food and Drug Administration (CFDA), released the Technical Guideline for the Research, Development and Evaluation of Biosimilars (Tentative) (Guideline) 2015.[3] This long-awaited Guideline outlines the regulatory framework for biosimilars, aiming to address clinical needs for biologics in China by improving accessibility and affordability of innovative products.

The Chinese copy biologicals guideline is based on four principles of comparison, stepwise, consistency, and similarity. It sets forth the definition of biosimilars and their reference products, basic principles for the technical review, criteria for comparability, and the conditions under which extrapolations of indications would be permissible. A biosimilar should, in principle, have the same amino acid sequence as the reference product. Similar to the regulatory authorities in the United States and EU, the CDE expects a detailed structural and functional comparison of the biosimilar to the reference product and also takes a stepwise approach to determining comparability.

Despite sharing the same principles for technical reviews, the Guideline has notable differences from the United States and EU. First, biosimilars do not have a separate abbreviated approval pathway in China; they are subject to the same approval pathway as innovator biologics. Secondly, the CDE does not accept an innovator biologic approved by foreign regulatory authorities as a reference product. The reference product may be awaiting NMPA approval during early stages of the biosimilar development process but must be approved by the NMPA when comparative clinical studies are conducted (i.e. the reference biologic must be an originator biologic approved in China). Importantly, first-to-market biosimilars are not entitled to any regulatory exclusivity in China but indication extrapolation is allowed and considered on a case-by-case basis.

The technical guidance does not address interchangeability with the reference product, the naming rules and labeling requirements for biosimilars. In addition, the CDE has not provided guidance specific to the stages of biosimilar development or product categories.

10.4.2.7 India (CDSCO)

CDSCO grants import/export licenses, clinical trial approvals, and permission for marketing and manufacturing of similar biologics. Each State Food and Drug Administration works with CDSCO and is responsible for issuance of license to manufacture of similar biologicals in each state.

The CDSCO "Guidelines on Similar Biologics" provide the regulatory pathway for a similar biologic claiming to be similar to an already approved reference biologic.[24] Various applicable guidelines are available. The competent authorities involved in the approval process include: (i) Review Committee on Genetic Manipulation (authorizes import/export for R&D/ review of data up to preclinical evaluation) and (ii) Genetic Engineering Appraisal Committee (statutory body for review and approval of activities involving large-scale use of genetically engineered organisms).

10.4.2.8 Brazil (ANVISA)

ANVISA follows key WHO biosimilarity principles but allows for some variations for specific needs of Brazil. ANVISA introduced new regulations (Resolution no. 55/2010)[13] to establish specific pathways to license follow-on biological products in 2010. The guideline provides two regulatory pathways: a comparative pathway and an individual development pathway. The individual development pathway introduces a more progressive approach in which the "copy" product does not require a full comparison with the original one and a reduced dossier may be submitted. The applicant needs to provide complete data on quality issues but it does not have to be comparative in nature. Therefore, this pathway might lead to approval of products with an unknown degree of dissimilarity.

ANVISA has not adopted the WHO recommendation or the FDA rule on naming. This differentiation, "is needed to provide the prescribing physician the option to designate the product to be dispensed and, perhaps more importantly, to ensure the necessary traceability for safety assessments."[25]

Interchangeability of indications for follow-on biologics is also not regulated in Brazil; this places pressures on physicians and conflicts of interest in the decision-making on the use of follow-on biologicals.[25]

10.4.2.9 South Korea (MFDS)

MFDS issued a guideline regarding the regulation of biosimilar products in July 2009 which was later revised in 2015 (Guidelines on Evaluation of Biosimilar Products, MFDS Guideline 2015).[26] This is based on the EU and WHO guidelines and is thus similar to these in its scope and data requirements.

A biosimilar is defined as "a biological product that is comparable to an already marketed reference product in terms of quality, safety and efficacy"; the reference biologic is used in demonstrating the "comparability" of a biosimilar in terms of quality, nonclinical, and clinical issues.

The quality needs to be evaluated in two ways: the evaluation of the quality of the biosimilar itself and

comparability to a reference product. These two evaluations should meet the Regulations on Review and Authorization of Biological Products.[26]

There is no market exclusivity for biologics in Korea. Since there is no regulation in Korea that prohibits an application for approval of a biosimilar from being filed, an application for biosimilar approval may be filed any time after the approval of the reference biologic. Practically, however, it would not be easy to file a biosimilar application before the expiration of the re-examination period of a reference drug, i.e. six years, in Korea. The re-examination is carried out to fully assess the originator drug's safety and efficacy through post-marketing surveillance.

In Korea, if similar efficacy/safety of the biosimilar and the reference product have been demonstrated for a clinical indication, then the biosimilar may receive authorization for other indications of the reference product. Extrapolation of clinical indications of a biosimilar is allowed for indications where the post-marketing surveillance period of the reference product has expired and if all of the following conditions are met: (i) a sensitive clinical model is used to detect potential differences between the biosimilar and the reference drug, (ii) clinically relevant mechanism(s) of action and the target(s) is the same for the different indications, and (iii) safety and immunogenicity of the biosimilar have been sufficiently characterized.

For biosimilars, MFDS has developed the Template for Assessment Summary Information for Biosimilars as Chair of the International Pharmaceutical Regulators Programme Biosimilars Working Group.[27]

MFDS has approved 13 biosimilar products to date (21 marketed products).

10.4.2.10 Japan (PMDA)

PMDA's Office of Biologicals provides consultations and handles biotechnology medicines including biosimilars. Guidelines, based on the EU's existing processes, were published in 2009. These guidelines consider biosimilar drugs that are equivalent and homogenous to the reference biological product in terms of efficacy, quality, and safety. The applications must be submitted with data from clinical trials, details of manufacturing methods, long-term stability data, and information on overseas use.[15]

In June 2009, Sandoz received approval for its recombinant human growth hormone somatropin, the first biosimilar in Japan.

Comparison of biosimilar guidelines and regulations in different regulatory jurisdictions is summarized in Table 10.1.

Table 10.1 Comparison of biosimilar regulations and guidelines internationally.

Country/organization	WHO	USA	EU	Australia	Canada	China	India	Brazil	South Korea	Japan
Regulatory authority	WHO	FDA	EMA	TGA	BGTD	NMPA	CDSCO	ANVISA	MFDS	PMDA
Laws and regulation	N/A	Biologics Price Competition and Innovation (BPCI) Act.	Committee for Medicinal Products for Human Use (CHMP).	Therapeutic Goods Act 1989.	Food and Drugs Act; Food and Drug Regulations.	CDE Guidelines 2015.	Drugs and Cosmetics Act, 1940.	Resolution no. 55/2010.	Guidelines on Evaluation of Biosimilar Products.	Guidelines for Quality, Safety and Efficacy Assurance of Follow-on Biologics.
Terminology	Similar biotherapeutic products.	Biosimilar	Biosimilar	Biosimilar	Biosimilar	Copy biologicals	Biosimilar	Follow-on biologicals.	Biosimilar	Follow-on biological.
Scope	Well-established, well-characterized biotherapeutic products like DNA-derived therapeutic proteins. Excludes vaccines, plasma-derived products/their recombinant analogues.	Any virus, therapeutic serum, toxin, antitoxin, vaccine, blood component or derivative, allergenic product/protein), or analogous product. No mention of "well established" or "characterized".	Omission of "cell culture/DNA-derived process" is a significant deficiency. No mention of "well-established and well-characterized." WHO standards specifically exclude vaccines, plasma-derived products/their recombinant analogues. EU standard allows their consideration.	Made from/contains human cells/tissues or comprises/contains live animal cells, tissues, or organs.	Biologics that contain well-characterized proteins derived via modern biotechnological methods.	15 categories	Biological products/biologics-medical products of human, animal, or microorganism origin.	Vaccines, hyper-immune serums, hemoderivatives, bio-drugs, monoclonal antibodies, and drugs containing live/attenuated, or killed microorganisms.	Biological products containing well-characterized proteins; for which comparability is demonstrated by nonclinical and clinical studies.	Recombinant proteins produced using microorganisms or cultured cells, have been highly purified, and have undergone characterization by appropriate analytical procedures. Polypeptides/their derivatives and drugs containing these constituents.

(*Continued*)

Table 10.1 (Continued)

Country/organization	WHO	USA	EU	Australia	Canada	China	India	Brazil	South Korea	Japan
Reference product (RP)	The same RP to be used throughout the entire comparability exercise. RP must be approved in the country/region. BSmP should be expressed/produced in the same host cell type as the RP.	RP is the single BLP under Section 351(a) of the Public Health Service Act against which a biosimilar is evaluated. No mention of any host cell requirement.	There may be only one RP which is approved on basis of a complete dossier; the same RP should be used throughout the comparability program. Data from products authorized outside the Community may provide only supportive information. No mention of any host cell requirement.	A biological medicine that has been registered in Australia based on full quality, safety, and efficacy data and has been marketed in Australia for a substantial period/ have a volume of marketed use so there is likely to be a substantial body of acceptable efficacy and safety data for approved indications.	The same RP product should be used in studies supporting safety, quality, and efficacy of the BSm. The RP product should: (1) be authorized for sale/marketing in Canada; (2) be originally authorized based on a complete data package; and (3) have significant safety and efficacy data accumulated to demonstrate similarity. No mention of any host cell requirement.	The RP used in analytical/ preclinical studies may be approved in China or elsewhere. It must be approved in China when comparative clinical studies are initiated. It should be usually/but not always, the OP. An approved BSmP cannot act as RP.	RP to be licensed in India and should be the IP. The RB should be licensed based on full safety/ efficacy/ quality data. A BSmP cannot be the RP.	RP needs to be registered with ANVISA based on a complete dossier and sold in Brazil. No mention of any host cell requirement.	RP to be authorized by a regulator based on a full dossier. No mention of any host cell requirement.	It must be approved in Japan and be the same product throughout the development period of the biosimilar where the host cell of the RP has been disclosed, it is preferable to proceed with the development using the same host cell.
Stability studies	Head-to-head accelerated stability studies to reveal hidden properties that might warrant additional evaluation.	An appropriate physicochemical/ functional comparison of BSm stability with RP under multiple stress conditions. Accelerated and stress stability studies/forced degradation studies, to establish degradation profiles of BSm and RP.	Justification of BSm shelf life with full stability data. Comparative real-time/ real-condition stability studies between BSm and RP not required.	Same as EU.	Stability data, from accelerated and stress conditions, to provide potential differences in degradation pathways and differences in product-related substances/ impurities.	Comparative stability studies under accelerated and stress conditions.	Side-by-side accelerated and stress studies showing comparable degradation profiles for BSm and RP.	Comparative stability studies conducted under accelerated and stress conditions.	Comparative stability study with RP is not required. Accelerated/ stress stability studies to establish impurity profiles are useful in demonstrating comparability.	Long-term storage testing should be conducted.

Preclinical studies	Receptor-binding studies/ cell-based assays should be conducted to establish comparability in PD. Any animal study should be in relevant species. In vivo studies not mandated if highly reliable in vitro assays that reflect clinically relevant PD activity of the RP are available.	EU standard specifies that in vitro assays be predictive of in vivo behavior. In vitro studies need to establish comparability in reactivity and identify causative factors if comparability cannot be established. Comparative in vivo studies need to be in relevant animal species.	Same as EU.	Receptor-binding studies/cell-based assays are needed when appropriate. In vivo studies should include animal PD relevant to the clinical application. At least one repeat-dose toxicity study conducted in a relevant species in addition to other relevant safety observations.	In-vitro studies, if minor differences are found, some subsequent comparability testing may be exempt. Adequate in vivo studies may be required.	Comparability of test and RB should be established by in vitro cell-based assays. In vivo evaluation of biological/ PD activity may not be required if in vitro assays are known to reliably reflect clinically relevant PD of the RP.	Relevant PD studies for the intended therapeutic indications and cumulative toxicity studies (repeated doses), in relevant species to quantify TK parameters.	In vivo and in vitro studies should be tailored to the specific product, on a case-by-case basis.	Does not give details of in vivo and in vitro studies needed.
Pharmacokinetic (PK) and pharmacodynamic (PD) studies	The comparability exercise is a stepwise procedure that should begin with PK and PD studies followed by pivotal clinical trials.	Comparative human PK and PD (where there is a relevant PD measure) studies "generally" will be expected, unless an applicant can justify that these are unnecessary.	More specific than WHO standards. Choice of PK study design should be justified; cross-over design is not appropriate for proteins with long half-lives or for those with potential for formation of anti-drug antibodies. PD effect should be compared in a population where possible differences are able to be observed.	Same as EU.	Comparative PK studies should be conducted. Comparative PD studies should be clinically relevant and surrogate markers should be validated.	Comparative PK and PD studies should be conducted.	Like PK studies in the BSm clinical development program, PD studies should also be comparative in nature.	PD studies may be performed in combination with PK studies, provided the PK/ PD relationship is characterized. PD parameters should be relevant to demonstrating clinical efficacy.	For PK, PD, or PK/PD studies, the study population should be selected based on the properties of the BSm, the RP, and the target disease.

(Continued)

Table 10.1 (Continued)

Country/ organization	WHO	USA	EU	Australia	Canada	China	India	Brazil	South Korea	Japan
Efficacy and safety studies	Similar efficacy to be demonstrated in a controlled double blind, adequately powered study. Potential differences between the products should be examined in a sensitive/ well-established clinical model.	Data derived from a clinical study/studies to demonstrate safety, purity, and potency in one/ more appropriate conditions of RP use/intended use.	Specificity, scope, and criteria exceed WHO's. Presumes that extensive data from efficacy trials will be augmented to collect safety data.	Same as EU.	Comparative trials for efficacy and safety are critical, and equivalence trials are generally preferred unless non-inferiority trials are justified.	Guideline is not clear on whether large-scale confirmatory clinical studies can be waived even if comparative PK/PD and immunogenicity studies are performed in patients and have demonstrated high comparability.	Comparative clinical trials are critical to demonstrate similarity in safety and efficacy profiles between the BSm and RB with few exceptions. Waivers are not common.	Must submit protocols/reports of pivotal clinical safety and efficacy studies.	An efficacy equivalence trial (adequately powered, randomized, parallel groups) should be conducted. If confirmatory PK/PD data demonstrate comparability, a clinical efficacy study may be omitted.	Studies designed to compare efficacy may be designed also to investigate safety (adverse event types and incidence).
Immunogenicity	Immunogenicity should always be investigated in humans prior to authorization. Animal data may not be predictive of human PK/PD or safety.	At least one clinical study that includes a comparison of the immunogenicity of the BSm and the RP is generally expected.	Criteria more specific/broader than WHO. For recombinant proteins, immunogenicity of a BSm must always be examined and should include data in enough patients to characterize antibody response variability.	Same as EU.	Immunogenicity should be evaluated using appropriate studies and methods, considering the potential impact on efficacy/safety.	Comparability study designed based on nonclinical immunogenicity.	Required but details not provided.	Must submit an immunogenicity study report. However, details are not provided.	Details of antibody testing are required. A screening assay with adequate sensitivity is needed for antibody detection and a neutralization assay should be available for further antibody characterization. Interference by circulating antigen with the antibody assay should be accounted for.	At an appropriate stage of clinical development, sponsors should conduct studies to evaluate immunogenicity. However, details are not provided.

Indication extrapolation	If prerequisites for extrapolation of efficacy/ safety of the BSm to other indication(s) of the RP are not fulfilled, the manufacturer needs to submit its own clinical data to support the desired indication(s).	Extrapolation to other approved indications of the RP may be possible if the applicant provides scientific justification to support each condition of use.	Lacks WHO criteria; interpretation of guidance will determine compliance level.	Same as EU.	Extrapolation should be justified based on pathophysiological mechanisms of disease, safety profile in the respective conditions and/or populations, and clinical experience with the RP. Detailed scientific rationale addressing the risks/benefits of the proposed extrapolation should be provided.	Indications with same mode of action and/or receptor can be extrapolated, provided safety and immunogenicity of extrapolated indications are fully evaluated.	Extrapolation of the safety and efficacy data of a clinical indication (for which clinical studies have been carried out) of BSm to other clinical indications may be possible if certain conditions are met.	Must have comparable safety and efficacy in a clinical model able to detect potential differences between the products, the safety, and immunogenicity of the BSm must be sufficiently characterized, and the mechanism and receptors involved must be the same.	Other indications "for which a post-marketing survey was completed" may be possible if certain conditions are met. If an applicant intends to extrapolate efficacy data to support approval of other indications of the RP, an equivalence clinical study design is more desirable than a non-inferiority design.	If the RP is approved for multiple indications, the BSm sponsor may be able to extrapolate study results to other indications if a similar pharmacological result can be expected for the relevant indications.
Pharmacovigilance	Further close monitoring of clinical safety of these products in all approved indications and continued benefit–risk assessment is necessary in the post-marketing phase.	Adequate mechanisms should be in place to differentiate adverse events associated with the BSm and the RP.	Cites monitoring efficacy and implies traceability.	Outlines mandatory reporting requirements and offers recommendations on pharmacovigilance best practice.	The RMP should monitor/detect known/potentially unknown safety signals that may result from the impurity profile/ other characteristics of the BSm. The PVP should include periodic safety update reports.	Same as WHO.	The RMP should consist of: PVP, Adverse Drug Reaction (ADR) Reporting, and Post Marketing Studies (PMS).	The applicant must present a PVP/RMP and a risk minimization strategy in accordance with (other) legal requirements.	Not addressed.	The design/ method of the post-marketing surveillance study and RMP for a BSm should be discussed with the regulator and submitted with the application. Post-marketing surveillance findings should be reported in a timely manner to the regulator.

(Continued)

Table 10.1 (Continued)

Country/organization	WHO	USA	EU	Australia	Canada	China	India	Brazil	South Korea	Japan
Interchangeability	Not addressed specifically; to be determined by national authority.	FDA may find a BSm to be interchangeable with the RP if the information submitted demonstrates that: the product is BSm to the RP; and the product can be expected to produce the same clinical result as the RP in any given patient.	In the EU, interchangeability and substitution practices are determined at the Member State level. This topic is not addressed in the EMA guidance.	Not addressed.	Biosimilars are not "generic biologics." Authorization of a BSm is not a declaration of pharmaceutical or therapeutic equivalence to the RP.	Not addressed.	Not addressed.	Not addressed.	Not addressed.	Notwithstanding any switch of the RP/BSm with similar indications to BSm, their substitution/ combined use should be avoided during the treatment period.

BLP, biologic product; BSm, biosimilar; BSmP, biosimilar product; PD, pharmacodynamic; PK, pharmacokinetic; PMS, post-marketing studies; PVP, pharmacovigilance plan; RB, reference biologic; RMP, risk management plan; RP, reference product.

10.5 Approval of Biobetters

Biosimilars have established "similarity" to an innovator biologic. Biobetters work on the same target but claim superiority in structure, function, or other properties and thus are expected to have enhanced clinical profile.

No firm separate regulatory pathway exists to demonstrate that an altered biologic is a "biobetter"; under existing regulatory pathways, biobetters are viewed as new drugs. Lack of such a specific biobetter approval pathway presents opportunities for their developers. Approval of a biobetter may lead to patent protection and market exclusivity and may also assist in defending a company's market share against biosimilars or originator biologics. Some development advantages may be available for biobetter developers. Knowledge of the target may reduce R&D costs, prior information on choice of biomarkers and safety monitoring to focus on known side effects of the already established target pathway. Further, if a biobetter gains marketing authorization, this may lead to market exclusivity, even if no patent protection is available. Sometimes, biobetter development is even used as a defensive strategy by innovator companies to protect their market niche against possible biosimilars via line extension, as was the case of a subcutaneous formulation of Roche's trastuzumab, which gained marketing authorization in 2013 shortly before Roche's Herceptin® (intravenous trastuzumab) patent expired in 2014.

10.6 Conclusions

With an increasing number of biologics in the development pipeline, there is need to harmonize their regulation across different countries.[28,29] Some collaborative initiatives exist to standardize such regulation:

The International Pharmaceutical Regulators Programme (IPRP)[30] was created in 2018 to promote regulatory convergence for pharmaceutical products for human use. IPRP creates an environment for its members/observers to exchange information and promote convergence of medicines regulation.

The U.S. FDA as part of its Oncology Center of Excellence (OCE) has announced Project Orbis.[31] This project provides a framework for concurrent submission and review of oncology drugs among international partners. The FDA, TGA, and Health Canada have collaboratively reviewed applications for two oncology drugs, allowing for simultaneous decisions in the three countries.

In conclusion, global initiatives are urgently required to harmonize biologics approvals to allow biotech industries to develop their products uniformly for all countries and provide timely market access.

References

1 World Health Organization. *WHO Expert Committee on Biological Standardization (ECBS)*. https://www.who.int/biologicals/WHO_ECBS/en (accessed 16 January 2020).

2 Government of Canada. *Biologics, Radiopharmaceuticals and Genetic Therapies*. https://www.canada.ca/en/health-canada/services/drugs-health-products/biologics-radiopharmaceuticals-genetic-therapies.html (accessed 16 January 2020).

3 Center for Drug Evaluation NMPA. www.cde.org.cn (accessed 15 January 2020).

4 National Institute of Food and Drug Safety Evaluation, Republic of Korea. *Vision and Mission*. http://www.nifds.go.kr/en/wpge/m_7/cont_01/cont_01_01.do (accessed 16 January 2020).

5 Ministry of Food and Drug Safety Evaluation, Republic of Korea. *Biological Products/Herbal Medicines/Quasi-Drugs*. https://www.mfds.go.kr/eng/brd/m_27/list.do (accessed 16 January 2020).

6 PMDA. *Pharmaceuticals and Medical Devices Agency*. https://www.pmda.go.jp/english/index.html (accessed 16 January 2020).

7 U.S. Food and Drug Administration. *Development & Approval Process (CBER)*. https://www.fda.gov/vaccines-blood-biologics/development-approval-process-cber (accessed 16 January 2020).

8 European Medicines Agency. *Biological Guidelines*. https://www.ema.europa.eu/en/human-regulatory/research-development/scientific-guidelines/biological-guidelines (accessed 16 January 2020).

9 Australian Government, Department of Health, Therapeutic Goods Administration. *Biologicals*. https://www.tga.gov.au/biologicals-0 (accessed 16 January 2020).

10 Government of Canada, *Regulatory Roadmap for Biologic (Schedule D) Drugs in Canada*. https://www.canada.ca/en/health-canada/services/drugs-health-products/biologics-radiopharmaceuticals-genetic-therapies/

regulatory-roadmap-for-biologic-drugs.html (accessed 16 January 2020).

11 China NMPA. www.sfdachina.com (accessed 16 January 2020).

12 *Regulation of Biologicals in India*. https://morulaa. com/cdsco/regulation-of-biologics-in-india-2 (accessed 10 January 2020).

13 ANVISA. *Drugs*. http://portal.anvisa.gov.br/drugs (accessed 15 January 2020).

14 Ministry of Food and Drug Safety. *Biological Drugs*. https://www.mfds.go.kr/eng/wpge/m_22/ de011012l001.do (accessed 16 January 2020).

15 Pharmaceuticals and Medical Devices Agency. Notifications and Administrative Notices. https:// www.pmda.go.jp/english/review-services/regulatory-info/0003.html (accessed 10 January 2020).

16 World Health Organization. *Biologicals*. https://www. who.int/biologicals/biotherapeutics/similar_ biotherapeutic_products/en (accessed 16 January 2020).

17 WHO – Prequalification of Medicines Programme. *WHO Medicines Prequalification Guidance*. https:// extranet.who.int/prequal/content/who-medicines-prequalification-guidance (accessed 20 January 2020).

18 U.S. Food and Drug Administration. *Biosimilars*. https://www.fda.gov/drugs/therapeutic-biologics-applications-bla/biosimilars (accessed 16 January 2020).

19 European Medicines Agency. *Regulatory Guidance*. https://www.ema.europa.eu/en/human-regulatory/ overview/biosimilar-medicines-overview#regulatory-guidance-section (accessed 16 January 2020).

20 Australian Government, Department of Health, Therapeutic Goods Administration. *Biosimilar Medicines Regulation*. https://www.tga.gov.au/ publication/biosimilar-medicines-regulation (accessed 16 January 2020).

21 Australian Government, Department of Health, Therapeutic Goods Administration. *EU and ICH Guidelines Adopted in Australia*. https://www.tga.gov. au/ws-sg-index?search_api_views_fulltext=BMWP/ 42832/2005 (accessed 16 January 2020).

22 Therapeutic Goods Administration (TGA). *Risk Management Plans for Medicines and Biologicals*. https://www.tga.gov.au/publication/risk-management-plans-medicines-and-biologicals (accessed 23 January 2020).

23 Government of Canada. *Biosimilar Biologic Drugs*. https://www.canada.ca/en/health-canada/services/ drugs-health-products/biologics-radiopharmaceuticals-genetic-therapies/biosimilar-biologic-drugs.html (accessed 16 January 2020).

24 *Guidelines on Similar Biologics: Regulatory Requirements for Marketing Authorization in India*. http://dbtbiosafety.nic.in/Files%5CCDSCO-DBTSimilarBiologicsfinal.pdf (accessed 30 December 2019).

25 de Assis, M.R. and Pinto, V. (2018). Strengths and weaknesses of the Brazilian regulation on biosimilars: a critical view of the regulatory requirements for biosimilars in Brazil. *Therapeutic Advances in Musculoskelet Disease* 10 (12): 253–259.

26 Ministry of Food and Drug Safety Evaluation, Republic of Korea. *Biosimilar*. https://mfds.go.kr/eng/ wpge/m_37/de011024l001.do (accessed 16 January 2020).

27 International Pharmaceuticals Regulators Programme (IPRP). *Public Assessment Summary Information for Biosimilar (PASIB)*. http://www.iprp.global/page/ public-assessment-summary-information-biosimilar-pasib (accessed 16 January 2020).

28 Malhotra, H. (2011). Biosimilars and non-innovator biotherapeutics in India: an overview of the current situation. *Biologicals* 39: 321–324.

29 Krishnan, A., Mody, R., and Malhotra, H. (2015). Global regulatory landscape of biosimilars: emerging and established market perspectives. *Biosimilars* 5: 19–32.

30 International Pharmaceutical Regulators Programme (IPRP). http://www.iprp.global/home (accessed 26 January 2020).

31 FDA. *Project Orbis*. https://www.fda.gov/about-fda/ oncology-center-excellence/project-orbis (accessed 16 January 2020).

11

Pharmacovigilance of Innovator Biologics and Biosimilars

Catherine A. Panozzo[1], Kevin Huang[2], and Christine Y. Lu[1]

[1] Department of Population Medicine, Harvard Medical School and Harvard Pilgrim Health Care Institute, Boston, MA, USA
[2] Department of Medicine, Harvard Medical School, Boston, MA, USA

KEY POINTS

- Pharmacovigilance is the science and activities related to the detection, assessment, understanding, and prevention of adverse effects or any other drug-related problem.
- Many of the approaches (e.g. signal refinement processes) and systems (e.g. existing passive and active surveillance programs) used for pharmacovigilance of small molecule drugs and medical devices apply to innovator biologics and biosimilars.
- The structural complexity of biosimilars, potential for post-translational modifications, and potential for rare, delayed immune reactions in combination with their accelerated regulatory approval cycles warrant careful premarketing (preclinical) and post-marketing (general population use) safety monitoring.

Abbreviations

Abbreviation	Full name
AsPEN	Asian Pharmacoepidemiology Network
BBCIC	Biologics and Biosimilars Collective Intelligence Consortium
COPD	Chronic Obstructive Pulmonary Disease
CMS	Centers for Medicare and Medicaid Services
CNODES	Canadian Network for Observational Drug Effect Studies
DRNs	Distributed Research Networks
EMA	European Medicines Agency
EU	European Union
EU-ADR	Exploring and Understanding Adverse Drug Reactions
FAERS	FDA's Adverse Event Reporting System
FDA	Food and Drug Administration
HCPCS	Healthcare Common Procedure Coding System
ICD-9-CM	International Classification of Diseases, 9th Revision, Clinical Modification

Abbreviation	Full name
ICD-10-CM	International Classification of Diseases, 10th Revision, Clinical Modification
IMI	Innovative Medicines Initiative
NDC	National Drug Code
PCORI	Patient-Centered Outcomes Research Institute
PCORnet	Patient-Centered Outcomes Research Network
RCTs	Randomized, Controlled Clinical Trials
RELIANCE	Roflumilast or Azithromycin to Prevent COPD Exacerbations
REMS	Risk Evaluation and Mitigation Strategy
SEER	Surveillance Epidemiology and End Results
UK	United Kingdom
UMC	Uppsala Monitoring Centre
UNESCO	United Nations Educational, Scientific and Cultural Organization
US	United States
WEB-RADR	Recognizing Adverse Drug Reactions
WHO	World Health Organization

Biologics, Biosimilars, and Biobetters: An Introduction for Pharmacists, Physicians, and Other Health Practitioners,
First Edition. Edited by Iqbal Ramzan.
© 2021 John Wiley & Sons, Inc. Published 2021 by John Wiley & Sons, Inc.

11.1 Introduction

Pharmacovigilance is defined by the World Health Organization (WHO) as being the "science and activities relating to the detection, assessment, understanding and prevention of adverse effects or any other drug-related problem."[1] Pharmacovigilance of medical products include herbal & traditional medicines, blood products, biologics, medical devices, and vaccines.[2] The goals of pharmacovigilance include the improvement of patient care and safety in relation to the use of medicines, the improvement of public health and safety in relation to the use of medicines, and the assessment of benefit, harm, effectiveness, and risk of medicines to encourage their safe, effective, and rational use.[2] Other issues that pharmacovigilance is concerned with include medication errors, lack of efficacy reports, use of medicines for indications that are not approved, case reports of poisoning, assessment of drug-related mortality, abuse and misuse of medicines, and adverse interactions of medicines with other products.[2]

Before medicines are approved for use in patients, they are subjected to safety and efficacy testing in clinical trials. However, patients in clinical trials are often carefully selected and monitored under closely controlled conditions over a relatively short period of time.[3] As a result, most medicines have only been used in limited populations for limited periods of time at time of their approval. After marketing approval, real-world use of medicines follows larger and diverse populations for longer or more variable periods of time and with concurrent use of other products, situations which may result in the emergence and detection of adverse drug reactions.[3] Pharmacovigilance ensures that the safety of medicines is monitored throughout the entire time they are used in humans, both during premarketing (preclinical) and post-marketing (general population use) phases.

To date, only 26 biosimilars have been approved in the United States, while over 50 have been approved in the EU.[4,5] The first biosimilar (filgrastim-sndz or Zarxio®) was approved in the United States in 2015, while the first biosimilar (recombinant human grown hormone or Omnitrope®) was approved in the EU in 2006. Therefore, while the European Medicines Agency (EMA) and the EU have more experience with pharmacovigilance than the Food and Drug Administration (FDA) and the United States, nevertheless the pharmacovigilance of biosimilars is a relatively nascent field. Compounding the importance of conducting pharmacovigilance studies on biosimilars is the fact that biosimilars are not identical in molecular/structural terms to their innovator biologic counterparts in the same way that generic drugs are identical to their innovator small molecule drugs. Innovator biologics and biosimilars are often hundreds of times larger and more structurally complex than small molecule drugs.[6,7] In addition, multiple post-translational modifications such as glycosylation may affect efficacy and/or safety.[7] Moreover, pharmacovigilance may be complicated by delayed immunologic reactions, complicating the link between the observed reaction and the specific biosimilar product.[7]

This chapter covers the emerging science of pharmacovigilance for biosimilar medicines. Pharmacovigilance considerations during both premarketing and post-marketing periods are discussed. A careful examination of both current and proposed rules for biosimilar pharmacovigilance, as well as the key challenges and innovative solutions that are evolving and being developed and tested in the field, is discussed in the context of the infrastructure that exists for the safety monitoring of small molecule drugs and innovator biologics.

11.2 Premarketing Period

11.2.1 The EU Experience

Pharmacovigilance considerations during the premarketing period have been set up with the goal of ensuring that biosimilars are sufficiently similar to their innovator biologic counterparts and safe for use in humans. For example, the EMA requires that biosimilars are "highly similar" to the reference medicine in terms of physical, chemical, and biological properties, have no clinically meaningful differences compared with the reference medicine, and have limited variability (minor variability is allowed when scientific evidence shows it does not affect safety and/or efficacy, and the range of variability allowed is the same as that for the reference biologic).[7] The EMA also requires that biosimilars contain the same amino acid sequence and 3D structure (protein folding characteristics) as the reference biologic, although minor variability in glycosylation is allowed. In addition, biosimilars must have the same dosage and administration route as the reference biologic, although there can be differences in formulation, presentation, or dose administration device.[7] To obtain approval from the EMA, biosimilars must demonstrate sufficient similarity to the reference biologic via comprehensive comparability studies and robust pharmaceutical quality data; by demonstrating similarity, they can, however, largely rely on the safety and efficacy experience of the reference medicine.[7] Comparability studies include head-to-head comparisons of the biosimilar and reference medicine, with knowledge from initial quality comparability studies used to determine the range of nonclinical and clinical studies required in further development.[7] Comparative quality studies are performed first and involve *in vitro* studies that compare protein structure and biological function (both chemical and biophysical properties as well as biological and pharmacological activity) of the biosimilar compared to the reference medicine. These studies inform the nature of comparative nonclinical studies, which include *in vitro* pharmacodynamic studies that investigate binding and modulation of physiological targets as well as direct physiological effects in cells and comparative clinical studies, which do not seek to demonstrate safety or efficacy but confirm similarity and may involve pharmacokinetic, pharmacodynamic, safety, efficacy, and immunogenicity (for checking if immune reactions are induced) studies.[7] Since biosimilars build on existing safety and efficacy knowledge of the reference biologic, less clinical data are required for marketing approval and the reference medicine's entire clinical development program (e.g. pivotal studies) does not need to be repeated. Comparative clinical trials are specifically designed to rule out clinically relevant differences in pharmacokinetics, safety or efficacy, and immunogenicity between the biosimilar and reference medicine.[7] Safety data are collected throughout the clinical studies, including pharmacokinetic and pharmacodynamic information. The amount and variety of data collected are dependent upon the type and severity of safety concerns for the reference medicine; in theory, adverse reactions should be expected to occur at similar rates for biosimilars and their reference biologics.[7]

Case Study: Epoetin alfa Biosimilar Approval in the EU

A case study of the EMA's approval of epoetin alfa biosimilars in Europe illustrates some of the principles associated with premarketing pharmacovigilance of biosimilars.[8] Epoetin alfa is indicated to treat anemia in patients with chronic kidney disease or cancer. A number of epoetin alfa biosimilars have been approved for marketing in the EU, all of which have undergone extensive nonclinical and clinical safety and efficacy studies. According to quality guidelines published by the Committee for Medicinal Products for Human Use (CHMP), applications for biosimilar erythropoietin should include information on nonclinical and clinical studies similar to those performed for the innovator biologic and comparative studies (biosimilar versus innovator biologic).[9] Nonclinical studies should include *in vitro* (comparative bioassays like receptor-binding and cell proliferation) studies and *in vivo* (comparative erythrogenic [inflammation]) pharmacodynamic studies as well as toxicology studies with data from at least one repeat dose toxicity study of at least four weeks duration.[9] Clinical studies should include pharmacokinetic studies (single-dose crossover studies for different routes [for example, intravenous or subcutaneous] of administration in healthy volunteers), pharmacodynamic studies (evaluated as part of the comparative pharmacokinetic studies), and clinical efficacy studies in randomized, controlled clinical trials (RCTs) for different routes of administration.[9] Comparative clinical safety studies to address concerns surrounding immunogenicity also need to be conducted (12-month minimum assessment) and adverse events of special interest like hypertension and thromboembolic events should be reviewed.[9] In addition, the use of validated, highly sensitive antibody assays to detect early and late immune responses to the biosimilar are required for approval.

The importance of premarketing pharmacovigilance is highlighted in a clinical trial evaluating the safety and efficacy of the epoetin alfa biosimilar, Binocrit®, relative to the reference product, Eprex®. The trial was suspended because 2 of the 174 patients receiving the biosimilar developed erythropoietin-neutralizing antibodies, one of whom developed pure red cell aplasia.[10,11] This rate of neutralizing antibodies was higher than expected and believed to be caused by tungsten-induced denaturation of epoetin alfa and formation of immunogenic complexes during the manufacturing of prefilled syringes.[10-13] As a result, subcutaneous administration of biosimilar epoetin alfa was contraindicated in many countries until coated stoppers were introduced; since then, no increases in pure red cell aplasia have been observed during routine pharmacovigilance of epoetin alfa biosimilars.[10,14]

11.2.2 The US Experience

Premarketing pharmacovigilance for biosimilars in the United States generally mirrors that of biosimilars in the EU. The US FDA requires that biosimilars are "highly similar" to their reference biologics in terms of structure and function, including characteristics like purity, chemical identity, and bioactivity.[15] Minor differences between the biosimilar and reference biologic in clinically inactive components are acceptable; other differences expected during the manufacturing process are carefully reviewed by the FDA.[15] The FDA assesses the biosimilarity between the biosimilar and reference biologic, not the independent safety and efficacy of the biosimilar.[15] In order to demonstrate biosimilarity, detailed analytical (structural and functional) characterization is required, as well as animal studies (including toxicity assessments), clinical pharmacology studies, and additional clinical studies as needed.[15] As long as no clinically meaningful differences are demonstrated, biosimilars do not need as many clinical trials and may rely in part on the safety and efficacy data generated for the reference biologic; this demonstration is achieved through clinical safety, purity, and potency studies that assess pharmacokinetics, pharmacodynamics, and immunogenicity.[15]

FDA mandates additional requirements for biosimilars that seek to be labeled as interchangeable products.[16] Interchangeability standards set by the FDA include that the biological product is biosimilar to the reference product,[17] can be expected to produce the same clinical results as the reference product in any patient,[17] and its risk in terms of safety or diminished efficacy of alternating or switching from the reference biologic to the biosimilar is not greater than the risk of using the reference product without the switch.[17] A product designated as interchangeable means that it may be substituted for the reference biologic without the intervention of the healthcare provider (physician) who prescribed the reference product.[18] State laws differ in how long pharmacists have to notify prescribers, whether pharmacists must notify patients, and how long pharmacists have to retain records for substitutions.[19] To date, no biosimilar has received an interchangeable designation by the FDA. Implications of interchangeability to pharmacovigilance include the (i) need to conduct switching studies during pre- and post-marketing phases[20] and (ii) need to accurately capture the detailed identifiable name of the biologic in pre- and post-marketing pharmacovigilance databases (preferably down to the manufacturer and batch level).[10]

Case Study: Filgrastim Biosimilar Approval in the United States

A case study of the approval of filgrastim-sndz (Zarxio®) in the United States (the first biosimilar to be approved by the FDA) illustrates some of the principles associated with premarketing pharmacovigilance of biosimilars. A biosimilar of filgrastim (Neupogen), Zarxio®, benefited from having a well-characterized structure and established mechanism of action as well as a smaller size and lack of glycosylation compared to other biologics seeking approval.[21] It also may have benefited from marketing approval by the EU in 2009.[22] In order to demonstrate biosimilarity of Zarxio®, analytical studies, pharmacokinetic and pharmacodynamic studies, immunogenicity results from five clinical studies, two safety and efficacy studies, and rationale for extrapolation to other indications were provided.[21] Analytical similarity of filgrastim-sndz to the reference product was assessed by measuring critical quality attributes like primary structure, bioactivity, receptor binding, protein content, higher order structures, sequence variants, and post-translational modifications.[21] Four clinical studies obtained pharmacokinetic and pharmacodynamic parameters and comparative safety and efficacy were assessed in a pivotal trial in 214 patients with breast cancer through primary endpoint of severe neutropenia duration and secondary endpoints of febrile neutropenia, days of fever, absolute neutrophil count, and time to recovery.[21] To demonstrate interchangeability and support approval as an interchangeable product, the pivotal trial incorporated three switches between the biosimilar and reference product and compared the results to those of patients who were not switched; the switching had no impact on clinical safety or efficacy.[21] However, the manufacturer did not ultimately seek a designation of interchangeability for Zarxio®.[23] When the biosimilar filgrastim was approved by the FDA in 2015, its approval was extrapolated to include all five licensed indications of the reference product and the label mirrored that of the reference biologic.[24,25] It is interesting to note that the biosimilar label refers to safety and efficacy data from the reference product's clinical trials and does not provide details of the biosimilar clinical trials.

11.2.3 Limitations

There are several key limitations to premarketing pharmacovigilance evidence for biosimilars. First and foremost, since clinical safety and efficacy studies of biosimilars still rely on RCTs (as do the studies of the innovator biologics upon which they are based for regulatory approval), the limitations of RCTs apply. These include a lack of external validity (i.e. an inability to generalize findings to patients outside the study population).[26] In order to demonstrate clinical benefit, study populations in RCTs are often carefully chosen and well-defined with specific inclusion and exclusion criteria and therefore may not be representative of the specific target patient population. RCTs also usually do not have sufficient study periods or sample size and may not be sufficiently statistically powered to assess duration of treatment effects or to identify rare but serious adverse events, which are often identified through post-marketing surveillance and pharmacovigilance plans.[26] Finally, the high costs and time constraints of RCTs can lead to reliance on surrogate endpoints that may not correlate well with patient outcomes clinically.[26]

A particular area of concern surrounds innovator biologics that have been approved through FDA-designated priority review, so-called breakthrough therapies, accelerated approvals, or fast track pathways. An increasing number of innovator biologics have been approved through these alternative, expedited pathways in recent years; however, concerns have been raised regarding the rigor of their safety and efficacy evaluations. Many of these pathways allow for modified clinical trials that rely on unvalidated surrogate endpoints (only reasonably likely to predict clinical response) for approval.[27] Drugs that are granted accelerated approvals must undergo confirmatory trials, but many of these confirmatory trials share similar design elements with preapproval trials, including reliance on surrogate measures rather than clinical outcomes.[28] Furthermore, important safety signals may be missed until post-marketing studies are performed.[27] These observations raise concerns for the approval of biosimilars based on innovator biologics that have undergone these expedited approval pathways, particularly in terms of safety and pharmacovigilance.

11.3 Post-marketing Period

11.3.1 Rationale and Challenges

Post-marketing pharmacovigilance refers to pharmacovigilance that occurs after a new drug has already been approved for marketing by regulators. In addition to premarketing pharmacovigilance, post-marketing pharmacovigilance plays an important role in ensuring the safety of new medical products. As noted in the previous section, there are several limitations to premarketing pharmacovigilance that post-marketing pharmacovigilance seeks to address. First, the size of the patient population studied in clinical trials is relatively small compared to the patient population that is treated with the drug in clinical practice. Second, the population and indications studied in clinical trials are narrow, often excluding or providing insufficient data on special patient groups who are more prevalent in the treated population (e.g. the elderly or those with comorbidities).[29] Finally, the duration of treatment or follow-up in clinical trials is generally shorter than the duration of treatment or follow-up once products are approved and used in the real world.[29] Post-marketing pharmacovigilance has the ability to address many of these limitations, by assessing rare and/or delayed adverse events not identified in clinical trials, capturing adverse events from the entire treated population (including high-risk groups and in various indications), examining and identifying any long-term adverse effects of a drug, detecting drug–drug and drug–food interactions and identifying increased severity or reporting frequency of known adverse reactions.[29]

As an increasing number of medical products are approved through accelerated or expedited approval processes, the importance of post-marketing pharmacovigilance will continue to grow and evolve. Post-marketing pharmacovigilance allows for the collection of real-world evidence and the study of drugs as they are prescribed in the general population and used in medical practice, free from the constraints of clinical trials. Post-marketing pharmacovigilance is especially important for biologics (including biosimilars), since they have the potential to stimulate immune reactions due to their large molecular size. In addition, innovator biologics and biosimilars are large, complex molecules made in living cells that are sensitive to changes in manufacturing processes, environmental conditions, container closure systems, and structural changes that can affect their immunogenicity.[30] Rare, adverse drug reactions like immunogenicity are difficult to predict since the sample size in clinical trials is small and they may only arise after the patient has taken the biologic for a long period of time. Particularly for biosimilars, since they are not identical to their reference biologics, minor structural differences may result in rare or delayed safety events like immunogenicity that are only detected in larger, diverse patient populations after marketing approval.[10] Finally, the relatively abbreviated safety and efficacy data necessary for regulatory approval of biosimilars, which emphasize pharmacokinetics, pharmacodynamics, and clinical immunogenicity as opposed to large comparative clinical trials, warrant more stringent post-marketing pharmacovigilance.[31] The potential for biosimilar approval to be extrapolated to indications (indication extrapolation) in which the biosimilar has not been tested in clinical trials similarly highlights the importance of post-marketing pharmacovigilance for biosimilars.

One challenge common across post-marketing surveillance systems concerns the ability to accurately identify the biosimilar of interest. In addition to the potential for missing details or errors during manual transfer of data across various reporting systems, naming conventions vary across jurisdictions.[32,33] For example, in the EU, biosimilars are licensed under an international non-proprietary name (INN) (e.g. infliximab), while in the United States, biosimilars are licensed under the INN with an additional four-letter suffix (e.g. infliximab-dyyb).[34] This may pose challenges to studies utilizing global spontaneous reporting systems (e.g. VibiBase) or electronic health records when seeking to identify the biosimilar at the manufacturer level.

Studies utilizing administrative claims data may also experience challenges identifying biosimilars due to coding limitations and practices. For example, a recently published study using administrative claims data from the United States found that as consistent with their use in the clinical setting, filgrastim and infliximab biosimilars were most commonly identified via Healthcare Common Procedure Coding System (HCPCS) codes, although a substantial proportion of use was identified via National Drug Code (NDC) dispensing, revealing the need to monitor both types of billing codes when conducting surveillance in pharmacoepidemiologic databases.[35] Although HCPCS may be useful in identifying filgrastim, infliximab, and other biosimilars in US administrative claims data, some limitations include that prior to 1 April 2018, the US Centers for Medicare and Medicaid Services (CMS) used the same HCPCS code for all biosimilars related to the innovator biologic.[36] For example, the same HCPCS code, Q5102, was used to identify infliximab biosimilars for all manufacturers with an FDA-approved infliximab biosimilar.[36] To distinguish between manufacturers, CMS required the addition of a two-digit modifier unique to each manufacturer (e.g. Q5102-ZB specifies the infliximab biosimi-

lar manufactured by Pfizer and Q5102-ZC specifies the infliximab biosimilar manufactured by Bioepis/Merck),[36] but the common data model of a major US post-marketing safety surveillance system was not designed to include this information, hampering monitoring efforts.[37] However, since 1 April 2018, CMS has been assigning unique HCPCS codes to every biosimilar so this limitation was relatively short-lived (e.g. infliximab previously coded as Q5102-ZB is now Q5103 and Q5102-ZC is now Q5104).[36] However, an additional limitation is that in the United States, new biologics and biosimilars that are not adequately described by an existing HCPCS code may be billed under a miscellaneous or "not otherwise classified" code such as J3590.[36] Thus, monitoring newly marketed biologic (including biosimilar) products using HCPCS may not be feasible, delaying the identification of potential safety and effectiveness signals in pharmacoepidemiologic databases.[38]

11.3.2 Processes

Post-marketing pharmacovigilance is a multistep process that involves signal detection, refinement, and confirmation. A signal is defined as "information that arises from one or multiple sources (including observation and experiments), which suggests a new potentially causal association, or a new aspect of a known association, between an intervention and an event or set of related events either adverse or beneficial, which would command regulatory, societal, or clinical attention, and is judged to be of sufficient likelihood to justify verificatory action, and when necessary, remedial actions" by the Council for International Organizations of Medical Sciences, an international, nongovernmental, nonprofit organization representing the biomedical scientific community established by WHO and United Nations Educational, Scientific and Cultural Organization (UNESCO).[39] Such signals can come from a wide variety of sources, including spontaneous reporting systems, clinical trial data, scientific literature, and pharmacoepidemiologic studies.[40] Each of these sources has different strengths and limitations. For example, clinical trials collect high-quality clinical data, but are generally underpowered to detect rare adverse events while many pharmacoepidemiologic studies are well-powered to detect rare events but may offer limited clinical information. A more detailed characterization of spontaneous reporting systems is discussed in Section 11.3.3.

Signal detection and monitoring is usually achieved through a combination of passive and active surveillance mechanisms which are further discussed in

Sections 11.3.3 and 11.3.4. Passive surveillance may be used to generate hypotheses while active surveillance may be more commonly used to provide more focused examination of a signal identified via passive surveillance. The mere presence of a safety signal does not mean that a drug has caused the reported adverse event or reaction. The signal must be refined and then undergo confirmation in order to establish whether there is a causal relationship between the drug and the adverse event.[40]

The steps of signal refinement in pharmacovigilance may include (i) checking the data, (ii) examining descriptive statistics, (iii) checking computer code, (iv) looking for patterns over time from exposure to outcome, (v) adjusting for confounders, (vi) using other comparison groups, (vii) conducting patient chart review, (viii) comparing results for similar outcomes, (ix) comparing results with other existing data, and (x) collecting more data and/or conducting a new study.[41] These are summarized in Table 11.1. If the signal remains a potential risk after refinement, it then undergoes confirmation through pharmacoepidemiologic or mechanistic studies, clinical trials, and other types of studies that may be able to establish causality.[42] Pharmacoepidemiologic studies and distributed research networks (DRNs) that support such studies are further described in Section 11.3.4.

Pharmacovigilance activities of signal detection, refinement, and confirmation can be encompassed under the umbrella of "signal management." Signal management activities may include passive (not actively solicited but self-reported) receipt of adverse event reports from healthcare providers and consumers, surveillance of medical literature for investigator-initiated study results and case reports, data-mining of large drug safety databases (e.g. Sentinel System [US], Drug Safety Research Unit [UK], Shanghai Drug Monitoring and Evaluative System [China]), patient registries (e.g. UK Renal Registry, EUROTRAPS – Autoinflammatory diseases [EU], Surveillance Epidemiology and End Results Program [SEER]-Oncology [US]), and regulatory agency-directed post-approval safety studies (e.g. ongoing observational study for the innovator biologic filgrastim [Neupogen®] led by the manufacturer [Amgen Inc.]), "Incidence of Hematologic and Non-Hematologic Malignancies, Thrombotic Events, and Autoimmune Disorders in Unrelated Normal Donors Undergoing Bone Marrow Harvest Versus Peripheral Blood Stem Cell Mobilization with Recombinant Human Granulocyte Colony-Stimulating Factor (20130209)."[43,44]

Table 11.1 Pharmacovigilance signal refinement.[41]

Action	Description
1. Check the data	Examine observed and expected counts and rates; compare with incidence and prevalence estimates from the scientific literature
2. Examine descriptive statistics	Tabulate descriptive statistics by age, sex, and study site; assess secular and seasonal trends
3. Check computer code	Review analytic code
4. Look for patterns over time from exposure to outcome	Assess time from exposure to adverse event using descriptive histograms; consider formal statistical inference which can be performed using the temporal scan statistic
5. Adjust for confounders	If the previous analysis adjusted for confounders, adjust for a different and larger set of potential confounders; consider refining their definitions
6. Use other comparison groups	Consider assessing different comparison groups (e.g. groups from different time periods, matched controls using different criteria, different time periods in self-controlled designs)
7. Conduct chart review	Perform chart review to exclude erroneously coded cases on a random sample of the exposed and/or unexposed, or a complete chart review
8. Compare results for similar outcomes	Compare the signal generated by 1 exposure–outcome pair with results for subdiagnostic groups and with results/outcomes for similar innovator biologics or biosimilars
9. Compare results with other existing data	Compare results from phase III clinical trials, phase IV post-marketing trials, and other observational data sets
10. Collect more data and/or conduct a new study	Continue prospective monitoring, or design and conduct a completely new study

Source: Adapted from Yih et al.[41]

These activities serve to ensure that potential safety signals and drug-associated adverse events are not missed.

11.3.3 Passive Surveillance

Passive surveillance involves regular reporting of disease or morbidity data by all institutions that see patients (or test patient specimens) and are part of a reporting network.[45, 46] No active measures are taken; that is, reporting is entirely dependent on the initiative and motivation of potential reporters.[47] Passive surveillance thus primarily involves spontaneous and voluntary reporting of adverse events. The Uppsala Monitoring Centre (UMC), an independent nonprofit foundation, was established in 1978 as the WHO Collaborating Centre for International Drug Monitoring.[48] UMC operates the technical and scientific aspects of the WHO's worldwide pharmacovigilance network which currently consists of 136 countries that are full members.[48] UMC has developed and maintains VigiBase®, the WHO global database containing >20 million individual safety case reports covering >90% of the world's population.[48] Member countries submit spontaneous reports of suspected adverse events which assist in identifying signals for rare events more rapidly than if such reports were not reported in a global database. Importantly, the database system is linked to medical and drug classifications (e.g. WHO-ART/MedDRA) and the medicinal products dictionary, WHODrug, to allow structured data entry, retrieval, and analysis.[49]

In the EU, the EMA supports various databases for the collection of spontaneous reports; the largest and most centralized of these is EudraVigilance, an electronic database operated by the EMA on behalf of the EU medicines regulatory network.[50] It is used to collect worldwide reports of suspected adverse drug reactions which have been authorized or being studied in clinical trials in the European Economic Area.[10] Drug manufacturers and EU member states are required to submit suspected adverse event reports to EudraVigilance, which allows for the detection of safety signals.[10] If a signal is detected, it is evaluated by the EMA's scientific committees, which determine whether any regulatory action is warranted.[7] All biologics (including biosimilars) approved by the EMA are subject to "additional monitoring" and a black triangle label encourages healthcare professionals and patients to report any suspected adverse drug reactions.[7] However, traceability issues have been a concern for adverse drug reaction reports submitted to EudraVigilance, with a low percentage of reports including details like batch numbers.

A study of 49 003 adverse drug reaction reports related to 10 classes of innovator biologics for which biosimilars or related biologic products have been approved were assessed by EudraVigilance between January 2011 and June 2016.[51] Adequate identifiers were reported for 96.7% of suspected biologicals, ranging from 89.5% for filgrastim to 99.8% for beta-1a, but batch traceability was low at 20.5%.[51] In addition, reports were often heavily weighted toward products that attract media attention, such as epoetins during the pure red cell aplasia outbreak (see Section 11.2.1).[10,51]

In the US, spontaneous reports are submitted to the FDA via Medwatch, of which less than 10% are submitted directly to the FDA and the vast majority are submitted by manufacturers (who are required to report adverse events). These reports are captured by the FDA's Adverse Event Reporting System (FAERS), a fully automated and computerized database that contains both human drug and therapeutic innovator biologic and biosimilar reports which are coded using a standardized medical dictionary and can be reviewed by safety analysts for potential signals or trends.[52] Advantages of the FAERS program include its ability to provide ongoing large-scale surveillance of real-world evidence relatively cheaply and to detect rare, short latency events for all marketed products.[52] In addition, reports can be submitted by multiple stakeholders, including patients, providers, and manufacturers, even when causality is uncertain or details are missing. In the case of missing data, there are generally opportunities to follow-up with the reporter to obtain more details. Compared to pharmacovigilance conducted during clinical trials, the inclusivity of FAERS allows more heterogeneous populations to be monitored and use of drugs at all stages of treated disease to be followed.[53] However, limitations of the program include underreporting due to its voluntary nature, reporting bias (e.g. underreporting of common, mild adverse events and stimulated reporting in response to media attention and increased public awareness), variable reporting quality that may result in lack of information (e.g. brand, manufacturer, and batch numbers may be missing), duplicate reporting (e.g. provider and patient/consumer may report the same adverse event for a given exposure), and an inability to estimate incidence or prevalence of adverse events due to the lack of a denominator (i.e. no unexposed comparison group).[53,54] These are summarized in Table 11.2.

Factors that may affect reporting and thus contribute to reporting bias include media attention, litigation, the nature of the adverse event, the type of drug product

Table 11.2 Advantages and limitations of spontaneous reporting systems.[54]

Advantages	Limitations
Large and diverse population able to report	Reporting bias, including underreporting and stimulated reporting
Can contain detailed information on the exposure, characteristics of individuals exposed, and adverse events	Quality and completeness of reports vary
Possible to obtain follow-up information	No unexposed comparison group makes it impossible to estimate and compare rates of adverse events in exposed versus unexposed individuals
Direct reporting can help rapid detection of safety signals, especially those with short latencies	Causality cannot generally be determined

Source: Reproduced with permission of Elsevier.

and indication, length of time on market, and extent and quality of the manufacturer's surveillance system.[53] It is thus important to interpret a safety signal in the context of the reporting environment. To overcome the lack of an available denominator, one method that is commonly used to analyze FAERS data is disproportionality analysis. This type of analysis involves statistical techniques like empirical Bayesian data mining and the proportional reporting ratio to assess disproportional reporting of specific innovator biologic or biosimilar–adverse event pair.[54] For example, the proportion of biosimilar epoetin alfa reports with thromboembolic events to innovator biologic epoetin alfa reports with thromboembolic events could be compared.[55,56] Other analysis techniques include performing descriptive analyses using historical comparisons and reporting trends over time.[54]

11.3.4 Active Surveillance

Active surveillance has been defined by the WHO as the collection of safety case information as a continuous pre-organized process. Active surveillance can be drug based (identifying adverse events in patients taking certain products), setting based (identifying adverse events in certain healthcare settings where patients are likely to present for treatment), or event based (identifying adverse events that are likely to be associated with medical products).[57] Active surveillance involves actively following patients after treatment and detecting events by asking patients directly or monitoring patient records.[47] This may involve the use of disease registries, medical records, and laboratory biochemical and other pathology results.[47]

One method of active surveillance engages disease registries or lists of patients presenting with the same disease characteristics. These registries may assist data collection on drug exposure and other factors associated with a clinical condition and may also be used as the basis for a case–control study comparing drug exposure of cases identified from the registry and controls selected from patients outside the registry.[58] One example of a disease registry in the United States is the SEER, a collection of population-based cancer registries that has existed since 1973.[59] Registries currently contributing cases include certain states (e.g. Connecticut), regions (e.g. Seattle-Puget Sound), and racial groups (e.g. Alaska Native Tumor Registry).[59] SEER and other population-based registries serve a wide-range of purposes, including (i) monitoring the distribution of (cancer) cases among certain occupations, communities, ethnicities, ages, and other demographic groups; (ii) improving patient care by linking follow-up services and cost-effectiveness information; and (iii) providing services to hospital cancer programs such as shared follow-up, death clearance, and pooled data on treatment, stage of disease, or survival.[59]

Registries can vary in their timeliness, completion, quality of data elements, and availability for research. A systematic review of worldwide renal registries found that among 48 registries meeting the inclusion criteria, only 17 had good public accessibility to annual reports, publications, or basic data, and only 13 made patient-level data available to external researchers.[60] Another limitation involved in using registries occurs when patients without the disease of interest are required for a study (e.g. case control study). In these instances, it may be necessary to seek controls outside the registry which may present both logistic and scientific challenges. In addition, while the development of a prospective registry can help maintain complete data on exposures and possibly outcomes for as long as the registry is maintained and curated appropriately, they are expensive to establish and maintain.[30] Thus, information gaps exist, particularly among populations residing in low- and middle-income countries that may not have the resources to maintain registries. Registries are also burdensome for healthcare providers to use as they are time-poor.[30] Furthermore, challenges facing regulators using existing registries or establishing new ones include a lack of coordination between ongoing initiatives, harmonized protocols and methods, data sharing and transparency (privacy issues), and sustainability, which have led to inefficiency and duplication of efforts.[61] In response, the EMA has set up an initiative to make better use of existing registries and facilitate the establishment of new registries, including disease-specific workshops for hemophilia, CAR-T cell therapies, multiple sclerosis, and cystic fibrosis.[61]

Another active surveillance tool is the use of prospective or retrospective cohort studies to examine the effect of a medication in real-world use conditions. In prospective cohort studies involving medication exposure, comparison cohorts are selected on the basis of medical product use and followed over time. The population at risk for an event is followed for occurrence of the event, and therefore, information on exposure status is known throughout the follow-up period for each patient, and incident event rates and rate ratios can be calculated from the at-risk population during follow-up.[58] In retrospective cohort studies, the data for both the exposures and outcomes are already collected, but investigators similarly analyze data by identifying the exposure and then looking forward in time for patient outcomes. While prospective and retrospective cohort studies are useful for determining the incidence rates and relative risks of multiple adverse events, it can be difficult to recruit sufficient number of patients in prospective cohort studies who are exposed to the drug of interest or to study very rare outcomes.[58] This challenge can be mitigated by identifying patients for cohort studies from large automated databases (e.g. from prescription payment claims databases).[58] In addition, cohort studies can be used to examine safety issues in special populations through over-sampling or stratification of those patients.[58] An example of a prospective cohort that have been used for pharmacovigilance is the Danish National Birth Cohort, which was established in 1996 and recruited a total of 60 000 pregnant women to study how prenatal and early childhood exposures affected health later in life.[62] This cohort has been used to investigate the effects of exposure to various drugs early in life, and thus has played a pivotal role in the pharmacovigilance of these medications.

In the US, where health care is decentralized, administrative health care claims databases consisting of single or multiple insurance plans may be particularly useful in identifying cohorts with sufficient exposure. Since persons exposed to a biosimilar may differ systematically in important ways from those who are unexposed (or receive the innovator biologic), safety studies utilizing these types of databases may take advantage of sophisticated pharmacoepidemiologic methods to control for these differences.[20] For example, controlling for confounders using traditional multivariable regression adjustment produces inaccurate estimates when the number of confounders is large, and number of outcomes is small[63]; therefore, it may be more appropriate to use confounder summary scores, such as propensity scores and disease risk scores.[20] Briefly, a propensity score is the probability of treatment assignment conditional on observed baseline characteristics.[64] It is a balancing score, meaning that conditional on the propensity score, the distribution of observed baseline characteristics is similar between treated and untreated patients (or those on comparison medications).[64] This mimics some of the features of a RCT; however, unlike a RCT, it does not necessarily balance unmeasured baseline characteristics.[64] The disease risk score is also a summary score, but represents the probability of disease conditional on observed baseline characteristics.[65]

If an exposure is particularly limited as may be the case during the marketing of a new biosimilar or the adverse event of interest is particularly rare, DRNs may further increase the size of a study cohort and may also increase the size of the population with sufficient longitudinal follow-up to address the study question.[66] Briefly, DRNs allow participating health systems to contribute data but maintain physical and operation control over electronic data at their site.[67] The data are thus "distributed" since there is no central data warehouse; the data remain behind the firewall of the data contributing institution.[68] This offers the advantage of allowing sites to maintain control of their data and all uses, mitigating the potential for privacy breaches, and allowing for secure remote analysis of separate data sets, each comprising a different medical organization's or health plan's records.[69] These advantages may allow for increased participation of various institutions owning healthcare data and thus larger sizes for post-marketing safety surveillance. A number of DRNs important to the current and future of biosimilar monitoring are described below and include but are not limited to: (i) the Patient-Centered Outcomes Research Network (PCORnet), (ii) Sentinel System (Sentinel), (iii) Biologics and Biosimilars Collective Intelligence Consortium

(BBCIC), (iv) Canadian Network for Observational Drug Effect Studies (CNODES), (v) Exploring and Understanding Adverse Drug Reactions by Integrative Mining of Clinical Records and Biomedical Knowledge Alliance (EU-ADR), and (vi) Asian Pharmacoepidemiology Network (AsPEN). A comprehensive review of adverse events surveillance systems worldwide is available.[70]

11.3.4.1 PCORnet

In 2014, the Patient-Centered Outcomes Research Institute (PCORI) developed PCORnet, whose mission is to "make informed healthcare decisions by efficiently conducting clinical research relevant to their needs."[71,72] One study funded by PCORI in 2018 is examining safety and effectiveness outcomes related to use of ustekinumab versus vedolizumab in adults with Crohn's disease, and use of tofacitinib versus vedolizumab in adults with ulcerative colitis, after lack of response to an anti-TNF therapy.[73]

11.3.4.2 Sentinel System

Sentinel, established in 2008, is the US FDA's national electronic system that monitors the safety of FDA-regulated medical products, including drugs, vaccines, biologics, and medical devices.[74–78] It has evolved to become a core component of FDA's medical product safety surveillance system, eliminating the need for post-marketing studies on nine potential safety issues associated with five products in the past two years.[78] Investigations focusing on innovator biologics and biosimilars included a comparative effectiveness safety study of infliximab and etanercept on the risk of serious infections according to patient characteristics,[79] and a study monitoring the post-marketing utilization of biosimilar filgrastim and infliximab.[35]

11.3.4.3 BBCIC

The BBCIC was established in 2015 as a nonprofit initiative under the auspices of the Academy of Managed Care Pharmacy in the US to address anticipated needs for post-marketing evidence generation for innovator biologics and biosimilars.[80–82] Specifically, its mission is to provide a range of research services that support questions about use, impact, safety, and effectiveness of innovator biologics and biosimilars; increase the rigor and credibility of real-world evidence; provide access to data from a large population for research; improve the efficiency and cost-effectiveness of observational studies; and develop standard approaches to common data needs, and address gaps in methods or data, like

improving capture of NDC on physician office claims, International Classification of Diseases, 9th Revision, Clinical Modification (ICD-9-CM) to International Classification of Diseases, 10th Revision, Clinical Modification (ICD-10-CM) diagnostic mapping, and comparative effectiveness statistical approach work group in tool and methods.[80] Examples of recently completed studies include a review of methodologic considerations for observational studies that involve switching from the innovator biologic to the biosimilar[20] and a descriptive analysis of key safety outcomes for intermediate and long-acting insulin in patients with Type 2 diabetes.[83] The descriptive analysis involving patients with Type 2 diabetes found that observed insulin patterns of use, and rates of severe hypoglycemic outcomes and major adverse cardiac events were consistent with other published literature, but that additional data sources would be required to incorporate A1c test results into studies.[83]

11.3.4.4 CNODES

Established in 2011, CNODES is a joint collaboration between Health Canada and the Canadian Institutes of Health Research, with an overall objective of providing high-quality and comprehensive information on the safety and effectiveness of pharmaceutical products marketed in Canada.[84] Peer-reviewed publications investigating innovator biologics include a population-based analysis of secular trends and prescribing patterns of antidiabetic medications.[85]

11.3.4.5 EU-ADR

The EU-ADR was initiated by the EMA, and includes the general population in Italy, United Kingdom, Denmark, Germany, and Spain.[86] Data sources include electronic health records, administrative claims, pediatrician records, pharmacy invoices, general practitioner databases, and registries.[86] Publications include a strategy to identify potential drug-induced acute myocardial infarction using the Eu-ADR network and a data mining study of potentially drug-induced acute liver injury among children.[87,88]

11.3.4.6 AsPEN

AsPEN is a collaboration between six countries (Australia, Japan, Korea, Sweden, Taiwan, United States) using eight databases.[89] The focus of AsPEN is drug safety among Asian populations and not solely from Asian countries.[89] It was first conceived in 2008 and was formed to provide a mechanism to support the conduct of pharmacoepidemiological research and to facilitate the prompt identification and validation of emerging safety issues among participating countries.[89] Published investigations to-date include a safety study on the risk of acute hyperglycaemia with antipsychotic use.[89]

11.3.4.7 REMS

Risk Evaluation and Mitigation Strategy (REMS) is a drug safety program in the US. This program remains an important component of pharmacovigilance in the US.[90] REMS is designed to help reduce the occurrence and/or severity of certain serious risks by informing and/or supporting the execution of the safe use conditions described in the medication's FDA-approved prescribing information.[91] The goal of REMS is to ensure that the benefits of the medication outweigh the risks. FDA may require use of REMS for certain medications with serious safety concerns; to-date, FDA has not required REMS for any approved biosimilar.

11.4 New Directions

New directions in the approach to pharmacovigilance of biosimilars hold promise for overcoming some current limitations by leveraging multiple data sources and/or research methods into future pharmacovigilance plans or studies. For example, in the US FDA's Sentinel System's five-year strategy from 2019 to 2023, new data expansion efforts strengthening post-marketing pharmacovigilance efforts include plans to expand mother–infant linkage to evaluate in-utero exposure, medical product usage during pregnancy, and postnatal outcomes, as well as integrating national (federal) and state registry linkages (e.g. the National Death Index, Surveillance Epidemiology, and Ends Results) into the existing system.[92] In the foreseeable future, the Sentinel System also seeks to determine optimal methods for incorporating unstructured electronic health record data and continue to develop new statistical methods for estimating causal risk in the distributed data setting.[92] In addition, another five-year goal is expanding and using data mining methods to detect safety signals that might otherwise go undetected through TreeScan, or other novel approaches. Briefly, TreeScan is designed to "simultaneously evaluate thousands of potential adverse events or disease outcomes to determine if any occur with higher probability among patients exposed to a specific pharmaceutical drug, device, or vaccine,

adjusting for the multiple tests inherent in the many adverse events evaluated."[93]

Other novel approaches involve leveraging new technologies to enhance pharmacovigilance. For example, in 2018, the FDA launched an open access MyStudies App, designed to facilitate the input of real-world data directly by patients which can be linked to electronic health data supporting traditional clinical trials, observational studies, registries, and pragmatic trials. (Pragmatic trials commonly take place in the setting where patients already receive their usual care, avoid the need for separately constructed infrastructure with specially trained research staff responsible for data collection, and minimize eligibility criteria in order to gain a more representative population).[94–96] To demonstrate platform viability, the FDA MyStudies App was used to examine medication use and healthcare outcomes of pregnant women in the Kaiser Permanente Washington Health Research Network.[94] Several enhancements to the mobile app are in development and the FDA hopes to utilize the enhanced app in comparative effectiveness studies leveraging pragmatic trials or registries.[97] In Europe, various Innovative Medicines Initiative (IMI) (PROTECT) projects are also leveraging technology to improve adverse event reporting. For example, Recognizing Adverse Drug Reactions (WEB-RADR) developed a mobile application which allows patients to directly report potential drug adverse events and also to receive reliable information on their drugs.[98]

The FDA has also instituted the FDA-Catalyst program to enhance post-marketing surveillance. Catalyst is being used now to define and promote the effective use of real-world data and real-world evidence approaches, including through demonstration projects like the FDA MyStudies App described above.[99] Catalyst studies involve direct interactions or interventions with patients and/or healthcare providers, relying on the Sentinel System's infrastructure to identify cohorts of interest and conducting analyses.[99] The program is also seeking to link data collected by other sources. For example, it is combining data from the Roflumilast or Azithromycin to Prevent COPD Exacerbations (RELIANCE), which is a pragmatic, randomized controlled trial funded by PCORI, with traditional fee-for-service Medicare administrative claims data for the purpose of examining long-term roflumilast versus long-term azithromycin use for the prevention of chronic obstructive pulmonary disease (COPD) exacerbations, death, patient-reported physical function, problems with sleep, fatigue, anxiety, and depression.[92,100] The RELIANCE study team is currently recruiting and randomizing up to 3200 adults who were hospitalized with COPD in the past year to receive either roflumilast or azithromycin for six months to three years.[100]

Other additional novel approaches include harnessing new sources of data such as patient networks, social media, and wearables in order to enhance post-marketing surveillance. For example, the FDA has explored the potential of Facebook and Twitter for safety signal detection in a manner that is analogous to passive surveillance reporting of spontaneous adverse events by checking signals based on these social network data against known signals.[101] However, after analyzing approximately 4.2 million Facebook posts and tweets, and >42 000 posts from >400 online patient fora, a 2019 report by the IMI Web-RADR team concluded that social media are not recommended for detecting potential safety issues except in certain niche areas, such as exposure to medicines during pregnancy and abuse (or misuse) of medicines.[102] In addition, dedicated social networking sites for patients like PatientsLikeMe® may be used to facilitate direct patient reports of adverse events to the FDA. Briefly, PatientsLikeMe has 350 000 members with 2500 conditions who report their real-world experiences online.[103] For example, patients who use infliximab to treat rheumatoid arthritis or Crohn's disease have reported 70 side effects, including fatigue,

headache, and nausea.[103] In 2008, PatientsLikeMe launched a pilot program allowing patients with multiple sclerosis to report adverse events directly to FDA.[103] In 2018, PatientsLikeMe formally started collaborating with the FDA to study whether data from its network can assist in the earlier identification of adverse events.[103] Finally, the advent of smartphones and wearable devices like smartwatches may allow for remote monitoring of patients for adverse drug-related events. These devices allow both active and passive collection of data from patients at all times, and offer new opportunities to collect additional information from consumers such as vital signs and biometric measurements.[104]

11.5 Conclusions

This chapter has provided an overview of pre-marketing, post-marketing, and new innovations in pharmacovigilance related to biosimilars. Although the pharmacovigilance of biosimilars is a relatively new field, many of the approaches used to monitor other medical products apply to biosimilars. However, the structural complexity and potential for post-translational modifications and delayed immune reactions in combination with accelerated regulatory approval cycles warrant additional robust monitoring throughout the pre- and post-marketing life cycle of biosimilars.

References

1 World Health Organization (2019). Pharmacovigilance. https://www.who.int/medicines/areas/quality_safety/safety_efficacy/pharmvigi/en (accessed 24 November 2019).

2 World Health Organization (2002). The importance of pharmacovigilance: safety monitoring of medicinal products. https://apps.who.int/iris/bitstream/handle/10665/42493/a75646.pdf?sequence=1&isAllowed=y (accessed 16 June 2020).

3 European Medicines Agency. Pharmacovigilance: overview. https://www.ema.europa.eu/en/human-regulatory/overview/pharmacovigilance-overview (accessed 24 November 2019).

4 U.S. Food and Drug Administration (2019). Biosimilar product information (15 November). https://www.fda.gov/drugs/biosimilars/biosimilar-product-information (accessed 24 November 2019).

5 Generics and Biosimilars Initiative (2019). Biosimilars approved in Europe (25 October). http://www.gabionline.net/Biosimilars/General/Biosimilars-approved-in-Europe (accessed 24 November 2019).

6 Casadevall, N., Edwards, I.R., Felix, T. et al. (2013). Pharmacovigilance and biosimilars: considerations, needs, and challenges. *Expert Opin. Biol. Ther.* 13: 1039–1047.

7 European Medicines Agency (2014). Biosimilars in the EU. Information guide for healthcare professionals. https://www.ema.europa.eu/en/documents/leaflet/biosimilars-eu-information-guide-healthcare-professionals_en.pdf (accessed 24 November 2019).

8 Aapro, M., Krendyukov, A., Hobel, N. et al. (2018). Development and 10-year history of a biosimilar: the example of Binocrit. *Ther. Adv. Med. Oncol.* 10: 1758835918768419.

9 Generics and Biosimilars Initiative (2017). Pharmacovigilance, traceability and building trust in biosimilar medicines. *GaBI J.* 6 (3): 135–140. http://gabi-journal.net/pharmacovigilance-traceability-and-building-trust-in-biosimilar-medicines.html (accessed 24 November 2019).

10 Felix, T., Jordan, J.B., Akers, C. et al. (2019). Current state of biologic pharmacovigilance in the European Union: improvements are needed. *Expert Opin. Drug Saf.* 18 (3): 231–240.

11 Haag-Weber, M., Eckardt, K.U., Horl, W.H. et al. (2012). Safety, immunogenicity and efficacy of subcutaneous biosimilar epoetin-alpha (HX575) in non-dialysis patients with renal anemia: a multi-center, randomized, double-blind study. *Clin. Nephrol.* 77: 8–17.

12 Bennett, C.L., Luminari, S., Nissenson, A.R. et al. (2004). Pure red-cell aplasia and epoetin therapy. *N. Engl. J. Med.* 351: 1403–1408.

13 Seidl, A., Hainzl, O., Richter, M. et al. (2012). Tungsten-induced denaturation and aggregation of epoetin alfa during primary packaging as a cause of immunogenicity. *Pharm. Res.* 29: 1454–1467.

14 Macdougall, I.C., Casadevall, N., Locatelli, F. et al. (2015). Incidence of erythropoietin antibody-mediated pure red cell aplasia: the prospective immunogenicity surveillance registry (PRIMS). *Nephrol. Dial. Transplant.* 30: 451–460.

15 U.S. Food and Drug Administration. Biological product definitions. https://www.fda.gov/media/108557/download (accessed 24 November 2019).

16 U.S. Food and Drug Administration (2019). Considerations in demonstrating interchangeability with a reference product: guidance for industry (May). https://www.fda.gov/media/124907/download (accessed 2 January 2020).

17 Section 351(k)(4)(A) of the PHS Act.

18 Section 351(i)(3) of the PHS Act.

19 Hung, A., Vu, Q., and Mostovoy, L. (2017). A systematic review of U.S. biosimilar approvals: what evidence does the FDA require and how are manufacturers responding? *J. Manag. Care Spec. Pharm.* 23 (12): 1234–1244.

20 Desai, R.J., Kim, S.C., Curtis, J.R. et al. (2019). Methodologic considerations for noninterventional studies of switching from reference biologic to biosimilars. *Pharmacoepidemiol. Drug Saf.* https://doi.org/10.1002/pds.4809.

21 Duke Margolis Center for Health Policy (2016). The future of U.S. biosimilars market: development, education, and utilization. https://www.focr.org/sites/default/files/pdf/The%20Future%20of%20U.S.%20Biosimilars_0.pdf (accessed 24 November 2019).

22 Harston, A. (2018). How the U.S. compares to Europe on biosimilar approvals and products in the pipeline. https://www.biosimilarsip.com/2018/10/29/how-the-u-s-compares-to-europe-on-biosimilar-approvals-and-products-in-the-pipeline-3 (accessed 24 November 2019).

23 Sandoz. Zarxio (filgrastim-sndz). https://www.zarxio.com/globalassets/zarxio5/assets/practical-information-for-zarxio-brochure.pdf (accessed 24 November 2019).

24 Highlights of prescribing information for Zarxio (2015). https://www.accessdata.fda.gov/drugsatfda_docs/label/2015/125553lbl.pdf (accessed 24 November 2019).

25 Highlights of prescribing information for Neupogen (1998). https://www.accessdata.fda.gov/drugsatfda_docs/label/1998/filgamg040298lb.pdf (accessed 24 November 2019).

26 Frieden, T.R. (2017). Evidence for health decision making-beyond randomized, controlled trials. *N. Engl. J. Med.* 377: 465–475.

27 Gyawali, B. and Kesselheim, A.S. (2018). Reinforcing the social compromise of accelerated approval. *Nat. Rev.* 15: 596–597.

28 Naci, H., Smalley, K.R., and Kesseheim, A.S. (2017). Characteristics of preapproval and postapproval studies for drugs granted accelerated approval by the US Food and Drug Administration. *JAMA* 318 (7): 626–636.

29 Munoz, M. (2016). Introduction to post-marketing drug safety surveillance: pharmacovigilance in FDA/CDER. https://www.fda.gov/media/96408/download (accessed 24 November 2019).

30 Amgen Biosimilars (2018). Clinical and scientific considerations for biosimilars. https://www.amgenbiosimilars.com/pdfs/Clinical%20and%20Scientific%20Considerations%20for%20Biosimilars.%20USA-BIO-058953.pdf (accessed 24 November 2019).

31 Harvey, D.R. (2017). Science of biosimilars. *J. Oncol. Pract.* 13 (suppl. 9): 17s–23s.

32 Vermeer, N.S., Spierings, I., Mantel-Teeuwisse, A.K. et al. (2015). Traceability of biologicals: present

challenges in pharmacovigilance. *Expert Opin. Drug Saf.* 14 (1): 63–72.

33 Vermeer, N.S., Ebbers, H.C., Straus, S.M.J.M. et al. (2016). The effect of exposure misclassification in spontaneous ADR reports on the time to detection of product-specific risks for biologicals: a simulation study. *Pharmacoepidemiol. Drug Saf.* 25 (3): 297–306.

34 Ramzan, I. (2020). Interchangeability of biosimilars: a global perspective for pharmacists. *Pharmaceutical Journal.* July 22 2020. Available at: https://www. pharmaceutical-journal.com/research/perspective-article/interchangeability-of-biosimilars-a-global-perspective-for-pharmacists/20208123.article.

35 Dutcher, S.K., Fazio-Eynullayeva, E., Eworuke, E. et al. (2019). Understanding utilization patterns of biologics and biosimilars in the United States to support postmarketing studies of safety and effectiveness. *Pharmacoepidemiol. Drug Saf.* https://doi.org/10.1002/pds.4908.

36 Centers for Medicare and Medicaid Services (2017). Part B biosimilar biological product payment and required modifiers. https://www.cms.gov/Medicare/Medicare-Fee-for-Service-Part-B-Drugs/McrPartBDrugAvgSalesPrice/Part-B-Biosimilar-Biological-Product-Payment (accessed 3 January 2020).

37 Sentinel System (2018). Sentinel system common data model, v6.0.2 (October). https://dev.sentinelsystem.org/projects/SCDM/repos/sentinel_common_data_model/browse?at=refs%2Fheads%2Fscdm (accessed 16 June 2020).

38 Blandizzi, C., Meroni, P.L., and Lapadula, G. (2017). Comparing originator biologics and biosimilars: a review of the relevant issues. *Clin. Ther.* 39 (5): 1026–1039.

39 Hauben and Aronson (2009). Defining "signal" and its subtypes in pharmacovigilance based on a systematic review of previous definitions. *Drug Saf.* 32 (2): 99–110.

40 European Medicines Agency. Signal management. https://www.ema.europa.eu/en/human-regulatory/post-authorisation/pharmacovigilance/signal-management (accessed 24 November 2019).

41 Yih, W.K., Kulldorff, M., Fireman, B.H. et al. (2011). Active surveillance for adverse events: the experience of the vaccine safety datalink project. *Pediatrics* 127: S54.

42 Shibata, A. (2011). Pharmacovigilance, signal detection and signal intelligence overview. *14th International Conference on Information Fusion*, Chicago, IL (5–8 July 2011). http://fusion.isif.org/proceedings/Fusion_2011/data/papers/200.pdf (accessed 24 November 2019).

43 Beninger, P. (2018). Pharmacovigilance: an overview. *Clin. Ther.* 40: 1991–2004.

44 ENEPP (2017). Incidence of hematologic and non-hematologic malignancies, thrombotic events, and autoimmune disorders in unrelated normal donors undergoing bone marrow harvest versus peripheral blood stem cell mobilization with recombinant human granulocyte colony-stimulating factor (20130209). http://www.encepp.eu/encepp/viewResource.htm?id=20582 (accessed 3 January 2020).

45 World Health Organization (2019). National passive surveillance. https://www.who.int/immunization/monitoring_surveillance/burden/vpd/surveillance_type/passive/en (accessed 24 November 2019).

46 CDC (2014). CDC Science Ambassador Workshop 2014 supplemental powerpoint: public health surveillance. https://www.cdc.gov/careerpaths/scienceambassador/documents/hs-i-have-a-gut-feeling-2014.pptx (accessed 24 November 2019).

47 World Health Organization. A practical handbook on the pharmacovigilance of antimalarial medicines. https://www.who.int/medicines/areas/quality_safety/safety_efficacy/handbook_antimalarialpharmvigilance.pdf (accessed 24 November 2019).

48 Uppsala Monitoring Centre. www.who-umc.org (accessed 3 January 2020).

49 Uppsala Monitoring Centre (2019). VigiBase: signaling harm and pointing to safer use (June). https://www.who-umc.org/vigibase/vigibase/vigibase-signalling-harm-and-pointing-to-safer-use (accessed 3 January 2020).

50 EMA (2011). EudraVigilance. https://www.ema.europa.eu/en/human-regulatory/research-development/pharmacovigilance/eudravigilance (accessed 3 January 2020).

51 Vermeer, N.S., Giezen, T.J., Zastavnik, S. et al. (2019). Identifiability of biologicals in adverse drug reaction reports received from European clinical practice. *Clin. Pharmacol. Ther.* 105 (4): 962–969.

52 Wyeth, J., Zornberg, G., and U.S. Food and Drug Administration. Postmarket safety surveillance of drugs and therapeutic biologics. https://www.fda.gov/media/88997/download (accessed 24 November 2019).

53 Tobenkin, A. and U.S. Food and Drug Administration (2018). An introduction to drug safety surveillance and the FDA adverse event reporting system. https://www.fda.gov/media/112445/download (accessed 24 November 2019).

54 Shimabukuro, T.T., Nguyen, M., Martin, D. et al. (2015). Safety monitoring in the Vaccine Adverse Event Reporting System (VAERS). *Vaccine* 33 (36): 4398–4405.

55 Harpaz, R., DuMouchel, W., LePendu, P. et al. (2013). Performance of pharmacovigilance signal detection algoriths for the FDA Adverse Event Reporting System. *Clin. Pharmacol. Ther.* 93 (6) https://doi.org/10.1038/clpt.2013.24.

56 Almenoff, J., Tonning, J.M., Gould, A.L. et al. (2005). Perspectives on the use of data mining in pharmacovigilance. *Drug Saf.* 28: 981–1007.

57 U.S Food and Drug Administration (2005). Guidance for industry: good pharmacovigilance practices and pharmacoepidemiology assessment. Rockville, MD.

58 European Medicines Agency (2005). Pharmacovigilance planning. https://www.ema.europa.eu/en/documents/scientific-guideline/international-conference-harmonisation-technical-requirements-registration-pharmaceuticals-human-use_en-25.pdf (accessed 24 November 2019).

59 NIH, NCI. About the SEER program. https://seer.cancer.gov/about (accessed 3 January 2020).

60 Liu, F.X., Rutherford, P., Smoyer-Tomic, K. et al. (2015). A global overview of renal registries: a systematic review. *BMC Nephrol.* 16 (31) https://doi.org/10.1186/s12882-015-0028-2.

61 European Medicines Agency. Patient registries. https://www.ema.europa.eu/en/human-regulatory/post-authorisation/patient-registries (accessed 24 November 2019).

62 Olsen, J., Melbye, M., Olsen, S.F. et al. (2001). The Danish National Birth Cohort – its background, structure and aim. *Scand. J. Public Health* 29 (4): 300–307.

63 Peduzzi, P., Concato, J., Kemper, E. et al. (1996). A simulation study of the number of events per variable in logistic regression analysis. *J. Clin. Epidemiol.* 49 (12): 1373–1379.

64 Austin, P.C. (2011). An introduction to propensity score methods for reducing the effects of confounding in observational studies. *Multivar. Behav. Res.* 46 (3): 399–424.

65 Arbogast, P.G. and Ray, W.A. (2009). Use of disease risk scores in pharmacoepidemiologic studies. *Stat. Methods Med. Res.* 18 (1): 67–80.

66 Panozzo, C.A. and Haynes, K.H. (2018). The challenges and opportunities of using large administrative claims databases for biosimilar monitoring and research in the United States. *Curr. Epidemiol. Rep.* 5 (1): 10–17.

67 Curtis, L.H., Weiner, M.G., Boudreau, D.M. et al. (2012). Design considerations, architecture, and use of the Mini-Sentinel distributed data system. *Pharmacoepidemiol. Drug Saf.* 21 (suppl. 1): 23–31.

68 Popovic, J.R. (2017). Distributed data networks: a blueprint for Big Data sharing and healthcare analytics. *Ann. N. Y. Acad. Sci.* 1387 (1): 105–111.

69 Maro, J.C., Platt, R., Holmes, J.H. et al. (2009). Design of a national distributed health data network. *Ann. Intern. Med.* 151 (5): 341–344.

70 Huang, Y.L., Moon, J., and Segal, J.B. (2014). A comparison of active adverse event surveillance systems worldwide. *Drug Saf.* 37 (8): 581–596.

71 Fleurence, R.L., Curtis, L.H., Califf, R.M. et al. (2014). Launching PCORnet, a national patient-centered clinical research network. *JAMIA* 21 (4): 578–582.

72 PCORnet (2019). The National Patient-Centered Clinical Research Network. www.pcornet.org (accessed 3 April 2019).

73 Kappelman, M. Comparing treatments for patients with inflammatory bowel disease who don't respond to anti-TNF therapy. https://www.pcori.org/research-results/2018/comparing-treatments-patients-inflammatory-bowel-disease-who-dont-respond-anti (accessed 24 November 2019).

74 Sentinal Coordinating Center (2019). Sentinel is a national medical product monitoring system. www.sentinelsystem.org (accessed 5 April 2019).

75 Platt, R., Wilson, M., Chan, K.A. et al. (2009). The new Sentinel Network – improving the evidence of medical-product safety. *N. Engl. J. Med.* 361 (7): 645–647.

76 Platt, R., Carnahan, R.M., Brown, J.S. et al. (2012). The U.S. Food and Drug Administration's Mini-Sentinel program: status and direction. *Pharmacoepidemiol. Drug Saf.* 21 (suppl. 1): 1–8.

77 Ball, R., Robb, M., Anderson, S.A. et al. (2016). The FDA's sentinel initiative – a comprehensive approach to medical product surveillance. *Clin. Pharmacol. Ther.* 99 (3): 265–268.

78 Platt, R., Brown, J.S., Robb, M. et al. (2018). The FDA Sentinel initiative – an evolving national resource. *N. Engl. J. Med.* 379 (22): 2091–2093.

79 Toh, S., Li, L., Harrold, L.R. et al. (2012). Comparative safety of infliximab and etanercept on the risk of serious infections: does the association vary by patient characteristics? *Pharmacoepidemiol. Drug Saf.* 21 (5): 524–534.

80 BBCIC (2019). Biologics & biosimilars collective intelligence consortium. www.bbcic.org (accessed 24 November 2019).

81 Baldziki, M., Brown, J., Chan, H. et al. (2015). Utilizing data consortia to monitor safety and effectiveness of biosimilars and their innovator products. *J. Manag. Care Spec. Pharm.* 21 (1): 23–34.

82 Lockhart, C.M., McDermott, C.L., Felix, T. et al. (2019). Barriers and facilitators to conduct high-quality, large-scale safety and comparative effectiveness research: the Biologics and Biosimilars Collective Intelligence Consortium experience. *Pharmacoepidemiol. Drug Saf.* https://doi.org/10.1002/pds.4885.

83 Kent, D.J., McMahill-Walraven, C.N., Panozzo, C.A. et al. (2019). Descriptive analysis of long- and intermediate-acting insulin and key safety outcomes in adults with type 2 diabetes meillitus. *J. Manag. Care Spec. Pharm.* 12: 1–10.

84 Platt, R.W., Henry, D.A., and Suissa, S. (2020). The Canadian Network for Observational Drug Effect Studies (CNODES): reflections on the fist eight years, and a look to the future. *Pharmacoepidemiol. Drug Saf.* 29 (S1): 103–107.

85 Secrest, M.H., Azoulay, L., Dahl, M. et al. (2020). A population-based analysis of antidiabetic medications in four Canadian provinces: secular trends and prescribing patterns. *Pharmacoepidemiol. Drug Saf.* 29 (S1): 86–92.

86 Patadia, V.K., Coloma, P., Schuemie, M.J. et al. (2015). Using real-world healthcare data for pharmacovigilance signal detection – the experience of the EU-ADR project. *Exp Rev Clin Pharm* 8 (1): 95–102.

87 Coloma, P.M., Schuemie, M.J., Trifiro, G. et al. (2013). Drug-induced acute myocardial infarction: identifying "prime suspects" from electronic healthcare records-based surveillance system. *PLoS One* 8 (8): e72148.

88 Ferrajolo, C., Coloma, P.M., Verhamme, K.M. et al. (2014). Signal detection of potentially drug-induced acute liver injury in children using a multi-country healthcare database network. *Drug Saf.* 37 (2): 99–108.

89 AsPEN collaborators, Andersen, M., Bergman, U. et al. (2013). The Asian Pharmacoepidemiology Network (AsPEN): promoting multi-national collaboration for pharmacoepidemiologic research in Asia. *Pharmacoepidemiol. Drug Saf.* 22: 700–704.

90 U.S. Food and Drug Administration. Approved risk evaluation and mitigation strategies. https://www.accessdata.fda.gov/scripts/cder/rems/index.cfm (accessed 24 November 2019).

91 U.S. Food and Drug Administration (2019). Risk evaluation and mitigation strategies: REMS (8 August). https://www.fda.gov/drugs/drug-safety-and-availability/risk-evaluation-and-mitigation-strategies-rems (accessed 24 November 2019).

92 U.S. Food and Drug Administration (2019). Sentinel system: five-year strategy, 2019–2023 (January). https://www.fda.gov/media/120333/download (accessed 24 November 2019).

93 TreeScan (2014). Software for the tree-based scan statistic. www.treescan.org (accessed 24 November 2019).

94 Sentinel System, Wyner, Z., Dublin, S., Reynolds, J., et al. (2018). Collection of patient-provided information through a mobile device application for use in comparative effectiveness and drug safety research (10 October). https://www.sentinelinitiative.org/sites/default/files/Final_Report-Collection_of_Patient-Provided_Information_Through_a_Mobile_Device_Application.pdf (accessed 24 November 2019).

95 U.S. Food and Drug Administration (2018). FDA in brief: FDA launches new digital tool to help capture real world data from patients to help inform regulatory decision-making (6 November). https://www.fda.gov/news-events/fda-brief/fda-brief-fda-launches-new-digital-tool-help-capture-real-world-data-patients-help-inform-regulatory-0 (accessed 24 November 2019).

96 Weinfurt, K. Definition of a pragmatic clinical trial. In: *NIH Collaboratory Living Textbook of Pragmatic Clinical Trials* (eds. K. Staman, J. McCall and L. Wing). https://rethinkingclinicaltrials.org/chapters/pragmatic-clinical-trial/what-is-a-pragmatic-clinical-trial-2 (accessed 3 January 2020).

97 Sentinel System (2018). FDA-Catalyst MyStudies App alignment with pragmatic trials and/or registries (16 October). https://www.sentinelinitiative.org/content/FDA-Catalyst-MyStudies-App-Alignment-with-Pragmatic-Trials-and-or-Registries (accessed 24 November 2019).

98 Innovative Medicines Initiative (IMI). WEB-RADR: recognising adverse drug reactions. https://www.imi.europa.eu/projects-results/project-factsheets/web-radr (accessed 3 January 2020).

99 Sentinel System. FDA-catalyst-about. https://www.sentinelinitiative.org/FDA-catalyst/about (accessed 24 November 2019).

100 Sentinel System. FDA-catalyst alignment with the CMS linkage to the PCORI RELIANCE trial. https://www.sentinelinitiative.org/content/fda-catalyst-alignment-cms-linkage-pcori-reliance-trial (accessed 24 November 2019).

101 Saragoussi, D. and Schaumberg, D.A. (2018). New trends in drug safety and the growing role of real-world evidence (Fall). https://www.evidera.com/wp-content/uploads/2018/10/09-New-Trends-in-Drug-Safety_Fall2018.pdf (accessed 24 November 2019).

102 Van Stekelenborg, J., Ellenius, J., Maskell, S. et al. (2019). Recommendations for the use of social media in pharmacovigilance: lessons from IMI WEB-RADR. *Drug Saf.* 42 (12): 1393–1407.

103 Sullivan, T. (2018). PatientsLikeMe teams with FDA to explore patient-reported adverse events (5 May). https://www.policymed.com/2015/07/patientslikeme-teams-with-fda-to-explore-patient-reported-adverse-events.html (accessed 3 January 2020).

104 Deloitte (2018). Harnessing safety data from wearable devices. https://www2.deloitte.com/content/dam/Deloitte/us/Documents/life-sciences-health-care/us-lshc-harnessing-safety-data-from-wearable-devices.pdf (accessed 3 January 2020).

12

Pharmacoeconomics of Biologic Medicines and Biosimilars

Gregory Reardon

School of Pharmacy and Health Sciences, Keck Graduate Institute, Claremont, CA, USA

KEY POINTS

- In the face of strong demand due to improvements in efficacy and safety over routine care and high prices, biologics are rapidly trending toward a dominant proportion of total pharmaceutical spending in many countries.
- Following approval, biologics in most countries have 10 or more years of effective regulatory protection from competition. In the United States, later patents for incremental innovations and improvements can discourage the introduction of biosimilar competition for many years following expiration of patent and regulatory protections.
- While a delayed and incomplete introduction of biosimilars in the United States has been a disappointment for payers, in Europe, a more government-mandated and mature biosimilars market is generating substantial savings when compared to the costs of innovator biologic therapies.
- Access to innovator biologics and biosimilars remains a concern, especially among populations of low- and middle-income countries, and for many patients in the United States and Canada.

Abbreviations

Abbreviation	Full name	Abbreviation	Full name
ACP	American College of Physicians	FDA	Food and Drug Administration
AHA	American Heart Association	FTC	Federal Trade Commission
ASCO	American Society of Clinical Oncology	GDP	Gross Domestic Product
ASP	Average Sales Price	HTA	Health Technology Assessment
BPCIA	Biologics Price Competition and Innovation Act	IBD	Inflammatory Bowel Disease
		ICER	Incremental Cost-Effectiveness Ratio
CBO	Congressional Budget Office	LMIC	Low- and Middle-Income
CFTR	Cystic Fibrosis Transmembrane Conductance Regulator	NHS	National Health Service
CMS	Centers for Medicare and Medicaid Services	NICE	National Institute for Health and Care Excellence
DMARD	Disease-Modifying Antirheumatic Drug	NMS	Nonmedical Switching
DRD	Drugs for Rare Diseases	OECD	Organisation for Economic Cooperation and Development
EEA	European Economic Area	PAN	Patient Access Network
EMA	European Medicines Agency	PBM	Pharmacy Benefit Manager
EU	European Union	PBS	Pharmaceutical Benefits Scheme
EULAR	European League Against Rheumatism		

Biologics, Biosimilars, and Biobetters: An Introduction for Pharmacists, Physicians, and Other Health Practitioners,
First Edition. Edited by Iqbal Ramzan.
© 2021 John Wiley & Sons, Inc. Published 2021 by John Wiley & Sons, Inc.

Abbreviation	Full name
PCORI	Patient-Centered Outcomes Research Institute
PCSK9	Proprotein Convertase Subtilisin/Kexin Type 9
PDP	Prescription Drug Plan
QALY	Quality-adjusted Life Year

Abbreviation	Full name
RA	Rheumatoid Arthritis
RD	Rheumatological Disease
TLV	*Tandvårds- och läkemedelsförmånsverket* (Swedish: Dental and Pharmaceutical Benefits Agency)

12.1 Pharmacoeconomics of Innovator Biologics

The global market share of modern biologics has steadily increased since regulatory approval of recombinant human insulin (Humulin®), in 1982, and monoclonal antibody muromonab-CD3 (Orthoclone OKT3®), in 1985. Since then, a series of newer biologics agents have been introduced to global markets.[1] Biologics today account for a growing share of total pharmaceutical spend. Total spending on specialty medications (high-cost or complex-use medications, dominated by biologics) has continued a long upward trend, growing from 41% of total pharmaceutical spend in the United States (US), in 2017, to 45% in 2018.[2] Due to high per-unit costs, specialty drugs actually accounted for only 2.2% of prescription volume in the United States in 2018.[3] In 2017, 11 of the best-selling pharmaceuticals worldwide were biologics.[4] Five biologics, alone (Remicade®, Neulasta®, Herceptin®, Rituxan®, Avastin®), accounted for 33% of all US commercial benefit drug spending in 2017.[5]

A recent, market disrupting shift in biologics spending is traceable to a more recent phenomenon. The period between 2012 and 2014 saw the introduction of breakthrough treatments for hepatitis C (Sovaldi® and Harvoni®), cystic fibrosis (Kalydeco®), and short-bowel syndrome (Gattex®). In the United States, Medicare expenditures for hepatitis C drugs alone increased from US$300 million in 2013 to US$4.5 billion in 2014.[6,7] Medicare Part D (drug coverage) catastrophic spending soared to nearly US$28 billion in 2014, a 363% increase over the previous eight years.[7] By 2015, specialty drugs had accounted for 89% of Medicare Part D catastrophic expenses.[8] Payers scrambled to fund demand. Some public programs, including the US Veterans Administration and the state Medicaid plan in California, required supplemental appropriations to cover the increased cost of treating hepatitis C patients.[7] Between 2010 and 2016, per-patient costs for privately insured patients with cystic fibrosis almost doubled, largely due to specialty drugs, especially the new genotype-targeted cystic fibrosis transmembrane conductance regulator (CFTR) modulator medications like Kalydeco®.[9]

Introduction and uptake of biologics has been cited, in the example of inflammatory bowel disease (IBD), as shifting the former dominance of indirect costs (e.g. sick-leave, disability/disablement pension, and early retirements) within the economic burden of illness toward a greater proportion of direct medical costs.[10] For instance, rheumatoid arthritis (RA) patients receiving biologic disease-modifying antirheumatic drugs (DMARDs) have 56% (US$20 262) of total health costs accounted for by RA-associated medical expenses vs. only 30% (US$3723) for those receiving any treatment, biologics, or not.[11]

How have expensive biologics come so quickly to dominate the pharmaceuticals market? As a general but simplistic principle, market demand can be expressed as a monetary aggregate of per-unit net price multiplied by unit purchases. For biologics, extremely high per-unit price (relative to traditional pharmaceuticals) and growing unit demand have combined to create this growing market dominance. Reasons for high unit demand for biologic agents, primarily related to improved targeting, efficacy, and safety over non-biologic routine care, have been cited throughout other chapters of this book.

Reasons for higher per-unit prices for biologics are more complex. These might include growing affluence in developed countries that allow a substantial proportion of total demand for biologics to meet or exceed negotiated price within these countries. Disease markets where the economic burden of illness is especially high, such as hepatitis C, cystic fibrosis, and RA, are attractive targets for biologics manufacturers. In these markets, premium prices are often justified by manufacturers based on the expected savings from high direct or indirect costs of routine care, for which such biologic agents are offered as desired substitutes. Other factors, such as the capital investments required for successful development and approval of biologics, higher regulatory requirements for these types of complex molecules, adjudication of intellectual property rights among competitors like innovator and biosimilar manufacturers, or industry marketing practices that prevent effective competition on pricing, might also strongly contribute to the new market dominance of biologics. Factors, such as these, are explored further in this chapter.

Some health markets can be described, as in the case of nonprescription drugs and food and nutritional supplements, as single sided: a firm sells a given product to a customer who, largely acting alone, or on behalf of a family, determines demand. Other health markets, as in the case of prescription pharmaceuticals, can be described as complex and multisided: a firm sells a product where demand is directed toward a traditional consumer or patient, but this demand is largely dependent on demand from or coordination with one or more other intermediaries, such as prescribers, payers, or regulators. For pharmaceuticals, a physician decides, the pharmacy provides, the private insurance company or government agency funds, and the patient consumes the product. Due to such complexities of demand among the various market "actors," financing is not directly aligned with product selection and consumption. The price that a pharmaceutical company offers for a biologic is not directly related to how end users, like patients, might assign value to the product.[12]

Complexity can be further illustrated by issues related to intellectual property rights. Biologics have, for pharmaceutical manufacturers, a distinct marketing advantage over traditional small-molecule drug products. Biologics maintain high patent monopolies and weak competition after expiration of patents claimed during product development.[13] Seemingly anticompetitive, patents represent a legal "right to exclude," producing a limited monopoly while the patents for a product remain enforced.[14] Many see the pricing practices for innovator or brand products, including biologics, as symptomatic of an underlying marketplace failure, where systemic obstructions impede ability to function appropriately and maintain therapy costs.[15] For instance, pricing in the pharmaceutical industry does not always follow economic principles. When a new competitor enters the market, it is common for a biologic currently on the market to actually raise its price.[7,16] The biologics industry defends practices like these by citing the need for high prices to offset a similarly obstructive development and approval process that is often characterized as expensive, time consuming, and susceptible to delays.[15,17]

12.2 Variation in Usage of Innovator Biologics

Usage rates of biologics vary between and within countries.[10] Reasons for variability are complex, but differences in national treatment guidelines and reimbursement rules are likely key drivers.[18] Many countries limit access by using reimbursement criteria that are stricter than

biologic usage recommended by consensus treatment guidelines, such as the European League Against Rheumatism (EULAR). In one earlier cross-sectional study of 46 European countries, biologics for RA were not reimbursed at all in 10 countries.[19] Within the other 36 countries, at least one biologic was reimbursed. However, across countries, significant differences in eligibility criteria for the initiation of biologics were reported in terms of RA disease duration and activity and the need to first demonstrate a failure to a number of synthetic DMARDs prior to receiving the biologic.[18,19]

In many countries, modern biologics remain unaffordable, even after pricing controls are applied. In 12 of 30 countries studied, the local price of hepatitis-C treatment Sovaldi®, after rebates, was generally equivalent to one or more years of the average annual wage of individuals in those countries. Across all 30 countries, the annualized number of work years required to balance the cost of Sovaldi® ranged from 0.27 years in Egypt to 5.28 years in Turkey.[20] Data from the RA study, above, showed similar inequities. Among the 46 studied European countries, 22% had no available biological reimbursement for RA while 59% had a one-year treatment cost exceeding the per capita gross domestic product (GDP) (up to a maximum of 11 times).[19,21] A study of biologics for Crohn's disease across 10 European countries showed a lack of discernible patterns regarding biologic usage when countries were plotted separately, then compared on prevalence of biologic availability, affordability, and per-capita GDP.[22]

Under reinsurance policy for high-cost enrollees, a common designation for those receiving modern biologics and other "specialty" drugs, traditional Medicare insurance in the United States shares 80% of the costs of these drugs. The remaining 20% of the cost is assigned, with risk-adjusted Medicare subsidization and patient cost sharing, to a prescription drug plan (PDP) selected by the patient under Medicare Part D coverage. This permits the PDP to negotiate its assigned share for purchase of the biologic. (Medicare is not permitted by legislation to negotiate specialty prices, thus the 80% is not subject to price controls.) In a complex, iterative scheme, the beneficiary (patient) is responsible for sharing a cost of the biologic product until a US$5000 out-of-pocket expense is realized. Thereafter, the beneficiary pays 5%, the PDP plan pays 15%, and the Medicare program pays 80%.[23] Under this complex insurance scheme, the final cost to the patient for a biologic can be substantial. One study reported median out-of-pocket patient costs for Medicare Part D enrollees of US$4413 to US$11 538, across 12 biologics.[24]

The impact of limited scope of payer negotiation for total biologics cost within Medicare Part D is telling.

For proprotein convertase subtilisin/kexin type 9 (PCSK9) inhibitors, agents that have a medical consensus role as third-line alternatives in refractory low-density lipoprotein levels in hypercholesterolemia, the price of marketed PCSK9 inhibitors is US$14 000 per year in the United States vs. US$5500 in the United Kingdom. In the United States, a substantial share of this cost burden extends to the patient. Although the Patient Access Network (PAN) Foundation, supported by the pharmaceutical industry, provides limited financial relief for patients,[25,26] this type of support has declined. Many patients who initially received a PCSK9 inhibitor at minimal or no cost can no longer afford these agents due to the substantial out-of-pocket costs.[26] Patients in Canada are facing similar cost burdens as cost transfers to patients are widely expected to increase with further uptake of specialty drugs in that country.[27]

"Out-of-pocket" patient cost sharing for traditional drug therapy, in the form of deductibles or co-payments, was designed, in part, to incentivize patients to avoid consumption of unnecessarily costly therapies, particularly those for which cheaper branded or generic products are available. However, for expensive therapies like biologics, the large out-of-pocket expenses associated with these biologics might be self-defeating. In their review, De Vera et al. found that economic factors, such as patient-borne expenses, were associated with nonadherence with biologics among RA patients. In some studies, the proportion of patients, deemed as adherent with these biologics, was as low as 11%.[28]

12.3 Payer Management of Innovator Biologic Costs

Several strategies are available in European countries to provide affordable access to medications. Tendering, a competitive bidding strategy, often resulting in a "winner take all" share of the local market, is commonly used by global hospital and outpatient sectors. This has proven successful but requires a robust framework and appropriate design with clear strategic goals to prevent shortages.[29] Tendering could be used to offer access to more affordable biosimilars. However, this strategy does not typically play a role in the purchase of innovator biologics, where often no equivalent therapeutic substitute to the innovator's product exists.

Reference pricing is commonly applied in many countries for certain drug products deemed to be therapeutically equivalent. Here, pricing is processed using three different methods: free pricing (in Germany, The Netherlands, and Denmark), direct price controls (in

France, Italy, Portugal, Greece, Spain, Turkey, and Belgium), and margin controls (in the United Kingdom, where profit is negotiated between the Department of Health and the pharmaceutical industry).[30] External reference pricing has been shown to generate savings but might contribute to delays in product launch where product prices are low.[29] The main advantage of reference pricing is that it is relatively easy to implement on a comprehensive basis, covering all new and existing drugs for which therapeutic equivalents exist.[31] Key methodological questions for setting up a reference-pricing scheme are how similar do different drugs have to be, to be clustered, and what method should be used to set the reference price for a new cluster?[31] As with tendering, reference pricing is problematic in cases where therapeutic equivalents to an innovator biologic do not exist. These would include disease markets without biosimilars, particularly if the innovator biologic offers substantial incremental effectiveness over other therapies used to treat a given condition.

Value-based pricing, with its emphasis on higher payments being negotiated for better outcomes produced in actual care settings, shows promise as strategy for payment of biologics in the future, yet will be challenging to execute. Managed entry agreements might be used to facilitate market entry in cases of limited evidence of value, while controlling the initial pricing of such medications.[29] In a systematic review of six biologic agents (brentuximab vedotin, bosutinib, ponatinib, idelalisib, vismodegib, ceritinib), evaluated from 2011 to 2016, the Swedish Dental and Pharmaceutical Benefits Agency (TLV), Scottish Medicines Consortium, and National Institute of Health and Care Excellence approved five of six, five of five, and four of five biologics, respectively, using prospective case series, despite initial payer concerns about uncertain cost-effectiveness.[32] The major approaches of these agencies, when applying value-based pricing to quantify drug treatment effect size, in the absence of comparative studies, were indirect comparison using results from another clinical trial or register data, comparison with retrospectively obtained information on results of prior treatment, and comparison with a subset of nonresponders in a pivotal study.[32]

Health technology assessment (HTA), where used, allows consideration of all costs and benefits of innovative drugs across a wide range of indications and patient subgroups. It is also more flexible than the strategies described above.[31] Approval rates of biologic reimbursement submissions vary between countries. In a review of 45 new oncology drugs in five countries, predominately biologic agents, the most restrictive, to the least restrictive coverage, was found in Australia, Canada,

France, the United Kingdom (UK), and the United States.[33] Australia, Canada, and the United Kingdom used cost-effectiveness analyses explicitly in coverage decisions. France considered cost implicitly, but did not compare health gains against an explicit cost-effectiveness threshold. Cost-effectiveness in coverage decisions remains limited in the United States.[33]

Using the results from HTA to restrict access to more cost-effective therapies can be illustrated by a comparison of payer policies within Australia and the United States. In the United States, a total of 34 new molecular entity drugs and biologics were approved for cancer, then covered for payment by US Medicare from 2000 to 2009. During that same period, just one-third of these agents were approved for payment under the Australian Pharmaceutical Benefits Scheme (PBS).[34] The stated goal of the PBS is to "provide timely access to the medicines that Australians need, at a cost individuals and the community can afford." The PBS is described as focused on "purchasing outcomes," with minimal out-of-pocket costs (e.g. a maximum of AU$33.30 per prescription) as a trade-off for restricted product availability and coverage and prolonged time between regulatory approval and payer coverage. In contrast, the US system appears to be less evidence-based, permitting a wider coverage of approved agents than Australia. Nonetheless, this approach comes at the expense of high maximum out-of-pocket costs (e.g. as noted, a patient in the United States might incur annual costs in thousands of US$ for certain high-cost drugs).[34]

The United States remains the only country among the 34-member Organisation for Economic Cooperation and Development (OECD) that lacks significant government oversight or regulation of prescription drug pricing.[35] For instance, in the United States, the government-supported Patient-Centered Outcomes Research Institute (PCORI) is restricted by regulation from the Affordable Care Act from reporting quality-adjusted life years (QALYs) in its sponsored research publications. The United States has the highest level of per-capita spending on prescription drugs and tends to introduce new drugs to market faster and with higher utilization rates.[35] One consequence of such uncontrolled spending has been calls, from key stakeholders, for changes to payer practices. For instance, the American College of Physicians (ACP),[35] the American Society of Clinical Oncology (ASCO),[36] and the American Heart Association (AHA) have each issued statements that support legislative or regulatory solutions to improve payer control of high drug costs.[15]

Both the United States and Canada remain exceptions in terms of continuing to use private drug insurance to cover a majority of the working population and dependents.[27] Canada might be especially sensitive to buffering the upward trend in the cost of biologics. Where cost pressures from specialty drugs in the United States have generated greater pressure on employers' total payroll, Canada has lagged behind in cost containment efforts. Efforts to mitigate the cost of specialty drugs in Canada have placed risk and cost transfers onto plan sponsors, patients, and provincial public programs.[27]

12.4 Value of Innovator Biologics

Marginal gains in expected QALYs, and potential savings from direct medical and indirect cost offsets, are drivers for health provider and payer acceptance of high-priced biologics versus routine care, in many treatment scenarios. For instance, one report found that, without the cure offered by recent biologics, 350 000 more patients in the United States would have been living with advanced stages of hepatitis C between 2015 and 2025, at a burden of illness cost of US$115 billion.[37] Still to be resolved is the estimated three million persons in the United States with hepatitis C who have not yet been treated. If all were treated now with Sovaldi®, at US$84 000 per course of therapy, the expected cost would be hundreds of billions of dollars.[35] The *status quo* is a problem, in the meantime, averted by a policy of payers to withhold this treatment for many with hepatitis C until, and if, this condition sufficiently advances to a later severity stage.

In a review of 154 specialty and 125 traditional (non-specialty) drugs approved by the Food and Drug Administration (FDA) from 1999 to 2011, Chambers et al. compared both differences in QALYs, and cost per QALY, between these two classes of therapy.[38] The mean (median) incremental gains across specialty drugs, versus routine care, was 0.25 (0.183) versus 0.08 (0.002) QALYs for drugs approved during this period. Mean (median) differences in costs was likewise US$72 917 ($12 238) for specialty drugs versus US$3237 ($784) for routine care. When incremental costs per QALY (also known as incremental cost-effectiveness ratios or ICERs) were calculated for each of the drugs, study authors found no significant differences between specialty and traditional drugs, leading authors to conclude that specialty drugs were associated with greater incremental costs, but also with larger QALY gains.[38] However, ICERs for biologics versus routine care often vary widely by product. For instance, a review of treatments for ulcerative colitis showed significant variance across monotherapies and combinations of biologics, with ICERs ranging from US$36 309 to US$456 979 per QALY gained. The lowest value shown was obtained for infliximab and the highest for a treatment scheme including infliximab 5 mg/kg and infliximab 10 mg/kg plus adalimumab.[39]

Direct medical cost offsets are those costs estimated to be reduced or avoided because of biologic use in place of routine care. These can be calculated for biologics in different ways. For instance, these cost savings can be subtracted, using a budget impact model, from the total product cost increases expected from introduction of the biologic, thereby increasing economic support for use of the biologic. Alternatively, in an HTA, where the ICER is commonly used to define value, these cost offsets can be subtracted from product cost of the innovator biologic shown in the numerator of the ICER equation ($cost_{innovator} - cost_{routine care}$)/($QALY_{innovator} - QALY_{routine care}$). Hence, the incremental cost of the innovator equals net cost of the biologic after consideration of these savings, minus the cost of routine care. This would effectively lower the cost per QALY for the biologic further than if the negotiated price of the biologic was considered alone, without factoring such savings.

Evidence supporting direct medical cost offsets from biologics is mixed. In the cases of hepatitis C or refractory elevated low-density lipoprotein levels in patients with hyperlipidemia at high risk of serious cardiovascular events, the cost-offset benefits are the substantial expected savings of avoided hospitalizations and liver transplants in the former disease, and in the latter, avoidable or delayed hospitalizations and medical management costs of treating cardiovascular events like myocardial infarction, stroke, or a myriad of downstream consequences (e.g. heart failure, hemiplegia, and long-term care). Other claims for medical cost offsets are less supported. For instance, Bansback et al. found, in their review of biologics for RA, no evidence to fully support the rationale that biologics have significantly reduced nondrug resource utilization such as joint surgeries and physician visits.[40]

12.5 Innovator Biologic Development Costs

Due to proprietary concerns and perennially limited efforts at economic research to answer this specific question, little is known about the average allocated costs by a pharmaceutical company to research, discover, develop, and obtain a successful regulatory launch of a biologic. The most commonly cited statistic is US$2.6 billion, including a $1.4 billion cash outlay

and $1.2 billion in opportunity costs of capital, estimated by the Tufts Center for the Study of Drug Development, published in 2016.[41] However, the utility of this statistic for estimating the cost of an approved biologic is dubious, since it was derived only to describe average costs of development of all pharmaceuticals, not specifically biologics. Higher development costs for a biologic, compared to small molecule agents, would add to this estimate, as would inflationary adjustments since Tufts study was published. Yet, given the persistent market dominance, often exceeding a decade, for approved biologics, the time-to-payback is likely to be quick. In US sales alone, the top 20 selling biologics generated median annual sales of $US3.73 billion, ranging from US$2.6 billion (Truvada®) to US$13.7 billion (Humira®).[42] Including ex-US sales would add another 50% in additional annual revenues to those obtained in US markets.

12.6 Coverage of Rare Disease Drugs

Due to the smaller populations covered, drugs for rare diseases (DRD), also known as orphan drugs in the United States, are typically much more expensive than pharmaceuticals approved for more highly prevalent conditions. In a DRD market dominated by biologics, in 2016, the median cost to the payer of a DRD was US$32 000 per year in the United States, ranging from several products costing less than $US5000 to 19 products exceeding costs of US$300 000 per year.[43] Across payers, coverage decisions for these drugs remain a challenge. While DRD typically represent the only active treatment option for rare progressive or life-threatening conditions, evidence of clinical benefit is often limited due to the small patient populations studied.[44] Standardized and special reimbursement techniques for biologics play different roles in different countries, with a large gap appearing between Eastern and Western Europe. Harmonization across countries appears to be lacking.[45] With few exceptions, countries have not created separate centralized review processes for DRD, rather choosing to modify components of existing mechanisms and add-on safety nets.[44]

12.7 Pharmacoeconomics of Biosimilars

A recent pharmaceutical market report suggests an upcoming period of rapid maturation and transformation from the *status quo* of biologics. Four major market trends are listed: (i) biologics will continue to enter nontraditional biologic disease areas, such as asthma, dyslipidemia, and allergy; (ii) confidence in the market success of biologics will encourage greater investment and acquisitions in this sector; (iii) new biologics will be launched to compete with and expand the current biologics market; and (iv) biosimilar usage will expand as the majority of top biologics lose exclusivity.[46] For the biopharmaceutical industry, the first two trends are encouraging, suggesting an even greater dominance of biologics in market revenue share, relative to less-expensive small molecules, and a greater investment directed toward biologics, perhaps relative to other sectors of healthcare spending. The third trend, launch of new biologic innovators, may not be especially beneficial to payers, given past pricing behavior where innovator prices actually rose following arrival of new competitors.[7] However, some opportunities for country or institutional negotiation to drive costs downward, such as through reference or value-based pricing, might improve this trend for payers, especially where multiple competitors with similar safety and efficacy profiles arrive to challenge biologic originators in certain indications. For countries and institutions specifically aiming to control spending, perhaps only the last trend, expansion of biosimilars, offers real hope for substantive change in the near term. Although biopharmaceuticals are an established sector and available globally, biosimilars have only began to establish themselves in the mainstream of worldwide markets, especially in Europe. The economics of these agents is discussed in detail in this section.

12.7.1 Biosimilars in Europe

A biosimilars pathway was fully enacted in Europe, in 2006. This included guidance for a variety of drug classes.[5,47] That same year the European Medicines Agency (EMA) approved Sandoz's somatropin (Omnitrope®), a biosimilar recombinant human growth hormone, as the first biosimilar in the European Union (EU).[48] As of October 2019, 54 biosimilars have been approved for use by the EMA.[48]

Market competition in many European Economic Area (EEA) countries has resulted in increased treatment utilization and significant price reductions for biosimilars, compared with their reference products. These price reductions appear to further depress both the pricing of the reference product itself as well as the price of the entire product class that contains the reference and biosimilar products.[49–51] Uptake of biosimilars appears across medical conditions. For instance, rapid penetration of infliximab and etanercept biosimilars in the UK IBD and the rheumatological disease (RD) markets indicates that these products are prescribed for both stabilized and biological naive patients.[52]

The number of biosimilars and date of first biosimilar market entry appear to be critical factors for competition.[21,53] The introduction of new infliximab, etanercept, and adalimumab biosimilars is expected to generate considerable cost savings and have a substantial favorable impact on the UK National Health Service (NHS) budget. Aladul et al. predicted that infliximab and etanercept biosimilars would replace their corresponding reference agents by 2020, with adalimumab biosimilars expected to achieve 19% of the rheumatology and gastroenterology market by 2020.[52]

Policies enforcing uptake of biosimilars in some European countries have been strong. For example, in the United Kingdom, National Institute for Health and Care Excellence (NICE) guidelines state, in one review, that with the availability of more than one suitable treatment option, the less expensive agent, including biosimilars, is to be chosen.[54] Savings from an NHS deal in 2018 with AbbVie and four biosimilar manufacturers was expected to cut US$378 million (£300 million) from its previous US$504 million (£400 million) annual expenditure for Humira® (adalimumab). All five adalimumabs will be available in varying volumes with preference given to the version with the lowest price.[55] NHS England estimates that biosimilars will save as much as £300 million a year by 2021; discounts of up to 50–60% off the cost of their originator counterparts have been reported.[56] Other countries have increased prescribing of biosimilars by implementing quotas for physicians.[57] For instance, Scandinavian countries, such as Denmark and Norway, have implemented strict reimbursement policies that favor the use of infliximab biosimilars, regardless of physician preferences. This is expected to grow market share and translate into significant savings.[21]

Decentralization of healthcare budgets and issuance of local guidelines means that, even within a given country, the use of biosimilars can vary substantially by region. For example, between the 21 counties of Sweden, market shares of biosimilar infliximab varied widely, ranging from 18 to 96% in 2017.[58] Such variations are largely explained by the discounted price difference between biosimilars and the originator product, with counties using different strategies to leverage competition. The presence of key opinion leaders, local guidelines, and gainsharing arrangements appeared to play a further role in local market dynamics.[58]

12.7.2 Biosimilars in the United States

In contrast to an established biosimilars market in Europe and Asia, the rise of biosimilars has been a disappointment for US payers, where this sector is currently in relative infancy.[59] In retrospect, projections of biosimilar penetration in the US market were overly optimistic. In 2010, the Congressional Budget Office (CBO) estimated that US$25 billion dollars would be saved within 10 years from the passage of the Biologics Price Competition and Innovation Act (BPCIA) that provided the regulatory pathway for biosimilars.[60] The Rand Corporation estimated that biosimilars would result in a US$44 billion reduction in biologic spending from 2014 to 2024, which equals approximately four percent of total projected biologic spending over that time frame.[61,62] In reality, the realized savings have been far less. By 2018, biosimilars had penetrated less than 0.8% of the US market for biologics, and even within those limited markets, were used less than one-third of the time versus innovator products.[3]

Compared to the 54 biosimilars approved and marketed in Europe,[48] with discounts of up to 70% from the innovator biologics, the FDA in 2019 had only approved 24 biosimilars, with only 11 actually launched to market, at discounts of only 15–35% below the originator's price.[5] In the United States, the much longer path to biosimilar regulatory approval and greater marketplace struggles are attributable to a number of factors, including patent disputes, a slow movement toward approval of first biosimilars due to earlier lack of clarity for the FDA 351[k] process, limited offered discounts for biosimilars when compared to generics, and the launch of second generation, follow-on, biologics by originator manufacturers.[47]

Due to different regulatory pathways for biologics in the United States, the FDA agency had lacked authority, prior to 2010, to approve biosimilars as patents for biologics expired.[63] The abbreviated 351[k] pathway for regulatory approval allowed for FDA review and approval with less-extensive testing, accepting clinical safety and efficacy of the biologic molecule as already demonstrated by the innovator.[63] In 2012, Teva's biologic tbo-filgrastim (Granix®) was approved under the traditional approval pathway for originator products, though many viewed this as an example of what the biosimilar pathway would look like in the United States.[64] The first true biosimilar developed under the BCPIA was the Sandoz drug filgrastim-sndz (Zarxio®), which was approved in 2015. By January 2018, more than 60 biosimilars had been enrolled in the FDA biosimilar development program. Additionally, the FDA commissioner reported that meeting requests had been received for biosimilars for 27 distinct reference biologics.[64] Despite an unimpressive first decade of biosimilar development and launch, biosimilars still offer potential for significant savings to the healthcare system, given that biologics accounting for US$100 billion in annual sales are set to lose exclusivity in 2020 in the United States alone.[15]

Assessing the impact of future biosimilar savings in the United States is a mix of highly variable pricing assumptions, marketing considerations, payer actions, patent litigation, Medicare reimbursement policy, and the FDA.[65] However, the United States remains, by far, the world's largest pharmaceutical market, including biologics, so there will be some market pressure to resolve legal issues.[66] Furthermore, biosimilar developers are not targeting small niche markets as entry points, but they are directly challenging originators having the largest market sales. Consequently, these biosimilar developers have a legitimate shot, like the originator, at also having a blockbuster product, exceeding US$1 billion in annual sales.[66]

12.7.3 Cost of Biosimilar Development

Little is known about the actual cost of developing a biosimilar. One informal estimate places the cost from US$200 to 600 million.[67] This is in contrast to small-molecule generics, which are relatively inexpensive to develop (between US$1 and 5 million) and can collectively quickly take 90% of the market away from an originator branded product due to the much lower price required.[67]

Compared to the originator, biosimilars have a shorter timeline for approval (approximately 8 years), compared to 12 years for innovator drugs, though biosimilars have a far longer approval time than generics. For biosimilars, development costs can be 10–20% of the innovator drug.[59] Time required to launch and earn revenues from earlier investments required for development of a new biosimilar is a key consideration for profitability. As described, the opportunity cost of capital (averaged cost for obtaining capital funds from investors and lenders over time) that is required for developing a new molecular entity can alone account for nearly one half of the total cost of development.[41] For large biopharmaceutical companies, corporate partnering or acquisitions are commonly used to develop biosimilars. Biosimilar development requires a specialized knowledge and skills to efficiently navigate the complexities of manufacturing to FDA/EMA mandated standards, costs of clinical trials to demonstrate similarity, and post-marketing surveillance or pharmacovigilance to report safety and immunogenicity outcomes and manage uncertainty regarding interchangeability.[68]

12.7.4 Data Exclusivity and Patent Battles

Approval of a biosimilar for the EU and United States is a complex and lengthy process, which can produce an uncertain economic outcome for a manufacturer. Even upon approval, a biosimilar manufacturer often finds itself blocked from release to market due to patent disputes, which might extend therapeutic market control by originator products for years. Chen et al. cite a recent example where, of the 12 biosimilars approved for marketing by the US FDA at one particular moment, 6 were then withheld from commercial launch due to "marginally inventive" patent disputes with the originator.[69]

Protection of the intellectual property of an originator product can take the form of standard patents and data exclusivity regulations. Patents that are filed in either the EU or United States provide protections for up to 20 years from the filing date in that region or country, respectively. However, patents are typically filed several years in advance of the lengthy process required for drug development and testing prior to approval by the EMA and FDA. The remaining patent life left on a product at time of regulatory approval is generally far less than 20 years, though originator manufacturers will typically file additional patent claims throughout the premarket and post-marketing periods, typically based on manufacturing process improvements. These might further extend the effective patent life beyond expiration of the earliest originator claims.

Although neither the EMA or FDA attempt to regulate or adjudicate patent protections, these agencies do have a clear role in overseeing a separate protection for the originator. Through "data exclusivity," these agencies enforce the period of time during which the manufacturer of a biosimilar (or generic drug for that matter) cannot rely on the clinical data submitted by an originator manufacturer in its approval applications.[59,70] As such, patent protection of originator intellectual property is largely independent of data exclusivity protection. Until the expiration of data exclusivity, any manufacturer attempting to obtain approval for a biosimilar would need to conduct the same lengthy process for clinical trials and approval as the originator had done earlier. After expiration of data exclusivity, biosimilar candidates can rely on the originator data, including preclinical tests and clinical trials. This greatly reduces the body of submitted evidence required to obtain regulatory approval.

Under EU guidelines, the EMA cannot accept applications for biosimilars until the end of the eight-year data exclusivity period after the reference product has been authorized to enter the market. Final approval cannot be granted until an additional 2 years after this initial authorization, yielding 10 years of effective protection.[70] In the United States, the BPCIA permits a four-year period of data exclusivity from reference product approval, followed by an additional eight years of protection for the reference product until final FDA approval is permitted for the biosimilar application. This yields 12 years of effective protection.

The BPCIA creates a two-step process for patent approval, termed the "patent dance." In the first step, an informational patent disclosure or "exchange" between a biosimilar candidate and the originator is triggered by acceptance of the biosimilar application by the FDA. This involves negotiations over claims of which patents may be asserted or challenged as patent infringement.[68] The second step is the subsequent litigation of disputed patents. Reluctance by the originator to fully comply with the exchange provision may contribute to slow biosimilar development.[71] Such delays and patent disputes benefit the originator. In the case of Abbvie, the composition of matter patent for Humira® expired in December 2016, yet Abbvie has asserted protection from 70 later patent filings regarding formulation, manufacturing, and methods of use, not due to expire until 2022, nearly 20 years after FDA approval of Humira®.[68]

A recent report detailed a number of concerns with originators asserting later patents for incremental innovations and improvements. These include threat of delays for biosimilar marketing launch resulting from pending patents for four originator products that have already been on the market for 20 years, extending patent exclusivity until 2033 for Herceptin®, 2030 for Rituxan®, 2029 for Enbrel®, and 2025 for Remicade®.[72] However, settlements between parties, in all or some of these cases, might change the timeline for biosimilar launches. As of this writing, Roche, which manufactures Herceptin®, has announced an agreement for biosimilar competitors to launch their products near the end of 2019.[73]

Another manner in which the introduction of biosimilars in the United States is delayed is termed "pay for delay," an arrangement previously observed between small molecule generic and branded pharmaceutical manufacturers. Here, the branded product drug maker offers patent settlements that pay generic companies not to bring lower-cost alternatives to market until an agreed-upon date. These contractual settlements effectively block all other generic (or biosimilar) drug competition for a growing number of branded drugs.[74]

Barlas describes a scenario where the Federal Trade Commission (FTC) could investigate deals between innovator biologic companies and biosimilar competitors, such as the one reached by AbbVie with Samsung Bioepis and Biogen. Here, the launch of the South Korean biosimilar to Humira® (adalimumab) to the US market had been agreed upon by these companies to be delayed until 2023.[75]

12.7.5 Provider Knowledge of and Attitudes Toward Biosimilars

Surveys of health provider attitudes and beliefs have produced a mix of opinions regarding acceptance of biosimilars. A systematic literature review of related articles published between 2014 and 2018 yielded 20 usable studies (17 from Europe and 3 from the United States).[76] Study authors found a general lack of familiarity with biosimilars across US and European healthcare settings. Both US and European healthcare providers cited several deterrents for biosimilar uptake, including limited knowledge of biosimilars, low prescribing comfort, and safety and efficacy concerns.[76] Most physicians perceived biosimilar medicines as second- or third-line treatment options, thus restricting provider-authorized usage to biologic treatment-naive patients only. Safety concerns, particularly related to immunogenicity and efficacy, deter most physicians from switching stabilized patients who are already tolerating originator therapies. Furthermore, the lack of long-term tolerability and extrapolated indications data evoke considerable provider concern, curtailing interest toward biosimilar prescribing and uptake.[76]

In contrast, a more recent study from the United Kingdom showed that healthcare professionals had good knowledge of biosimilars and were content to initiate them, particularly those whose departments shared part of the expected savings with health commissioners.[77] These shared savings could be used, for instance, to fund the perceived need for extra "biologic" nurse support. Reservations were mixed among respondents for other factors, like acceptance of switching stable patients, indication extrapolation, and specifics of cost savings sharing. Practitioners expressed a much more consistent concern about biosimilar substitution at the pharmacy level and multiple product switching.[77] Among US oncologists, Cook et al. found that understanding of biosimilars is low and educational needs are high. The information that oncologists deemed important to assess, before being comfortable to prescribe biosimilars, includes safety, efficacy, and cost.[78]

12.7.6 Biosimilar Savings

Greater controls over pricing will likely continue to yield substantial discounts in most nations. Biosimilars in European countries have reported discounts generally ranging from 5 to 35% over originator biologics, with discounts of up to 75% noted in some cases.[79] Biosimilar percentage discounts, still, are modest, however, when compared to those of small-molecule generics, which can be priced 80–90% lower than originator products.[79] Savings for biosimilars, when compared to the reference product, appears lower in the United States compared to other countries. These savings might increase as multiple biosimilars are introduced to markets formerly controlled by originator products. For instance, the second filgrastim biosimilar was priced 30% lower than the originator filgrastim and 20% lower than the first biosimilar approved by the FDA.[80]

A key concern for biosimilar manufacturers seeking market launch and success in the United States is the market advantage of originator manufacturers to assert control over pricing. Discounts, such as the 15–35% below the originator's price offered for recent biosimilars, might be easily matched by originator manufacturer rebates.[5] Another concern is the profit to be earned supplying and administering higher-priced originator products. This creates an incentive for physicians to use these higher-priced products since payment for outpatient drugs is often based on a product's average sales price (ASP) plus a 6% markup.[81]

For US payers looking for savings from biosimilar uptake, perhaps the most restrictive policy used by originator manufacturers is known as the "rebate trap." Here, the originator manufacturer can respond to a challenge by a biosimilar manufacturer for a payer's preferred formulary listing by contractually withdrawing the rebate on the reference biologic. For any patient continuing the originator biologic, the payer's costs for that patient could double once the rebate is withdrawn.[82] As a result, the payer may expend a greater total amount for all biologics in that indication, even with substantial, but not complete, biosimilar adoption.[65] In one scenario, the price of a biosimilar is assumed to be 60% less than the net price of the brand (following rebates and discounts). If, in this scenario, the payer is only able to convert 50% of its patients to the biosimilar, but the per-patient cost doubles for the remaining 50% of patients receiving the originator biologic (due to the withdrawn rebate for that product), the payer's total costs would actually increase relative to costs prior to biosimilar availability.[82] In a lawsuit alleging unfair marketing practices, Pfizer asserted that Johnson & Johnson's marketing plan for its originator,

Remicade®, included exclusive contracts that withheld rebates from insurers unless these excluded biosimilars from formularies or else imposed "fail-first" preconditions prior to biosimilar usage.[55]

Cost modeling can help to estimate projected biosimilar savings within countries, but models are imperfect. Using an example of rheumatic biosimilars, Araujo describes how the accuracy of budget impact models and true extent of economic gains from biosimilars will be difficult to predict for a given market. Many values for model variables are based on assumptions that might or not occur, or could vary among regions. These might include variables such as expected biosimilar market share in both newly treated and switching patients, actual (rather than official or "listed") acquisition prices of biosimilars and reference drugs, price discounts, treatment discontinuation, or switching rates to other biologics.[21]

A key social benefit from savings generated by biosimilars is the opportunity to expand access to biologics for a larger number of patients than would be possible if only using originator products. Abraham et al. cite how 100% conversion of erythropoiesis-stimulating agent users to biosimilars among a hypothetical population of 100 000 patients would allow, through shifting of generated savings to cancer budgets, an additional 12 913 rituximab, 5171 bevacizumab, or 4908 trastuzumab treatments under weight-based dosing.[83,84] Gyawali makes the case for biosimilar savings in low- and middle-income (LMIC) countries, citing a perceived double standard by the oncology community to raise its voice against the skyrocketing cost of cancer treatment and yet not prescribe biosimilars when available and backed up by safety and efficacy data.[84] For many LMIC countries, even the lowered cost of biosimilars remains unapproachable. The expected price of biosimilars would, even if priced at one-third the price of Herceptin® (trastuzumab), still exceed the per-capita gross national income by a factor of seven among countries like Rwanda and others within sub-Saharan Africa.[85]

12.7.7 Interchangeability, Switching, and Indication Extrapolation

In the United States, substitution laws for small-molecule oral generics allow pharmacists to automatically substitute approved generic equivalent drugs. These laws do not apply to biosimilars.[86] Interchangeability would permit, by the FDA, under the BCPIA, substitution of a biologic with a biosimilar specially designated as the statutory term "interchangeable." This is now possible with issuance of final guidance by this agency

in 2019.[87] To obtain an interchangeable designation, biosimilar manufacturers may submit, after market approval for a biosimilar is first granted, supplementary evidence as described in this guidance. This special designation would technically permit a pharmacist to substitute a biosimilar for its reference biologic, without the necessity of prescriber authorization. However, several obstacles are likely to block such automatic substitution in the near future. These disincentives include a number of states with enacted laws or rules that prohibit this unauthorized biosimilar substitution, evolving professional practice standards that could place the pharmacist at risk of civil liability if the prescribing physician is not consulted prior to substitution (for instance, for treating life-threatening conditions where medication adherence is threatened due to "nocebo" effects, described below), and expected restrictions by some payers to this practice. The latter are especially powerful disincentives in the United States. Health insurance intermediaries known as pharmacy benefit managers (PBMs) might put more expensive reference drugs, rather than biosimilars, in favored formulary positions since these are more expensive and therefore lead to higher rebate earnings for the PBM.[75] In this scenario, a biosimilar would thus not be eligible for payment by the PBM until failure of the originator product was demonstrated to the satisfaction of the PBM, even if the legal obstacles cited above were removed and the pharmacist was otherwise free to substitute the biosimilar for the reference product.

In contrast to the United States, the EMA does not assess or make recommendations on interchangeability. In Europe, interchangeability lacks meaning in terms of substitution authorized by regulation. It is, instead, a concept that is generally physician-led or driven by national policy.[88] In the United States, influence of the interchangeability designation will likely rest far less with an individual decision by a pharmacist to substitute without prescriber authorization, but much more with its comforting influence, with accompanying manufacturer evidence, on the pharmacy and therapeutics committees of hospitals and payers who would choose biosimilar products for favored formulary listings.

For conditions for which biosimilars are available in the market, a physician might initiate either an originator biologic or a biosimilar therapy as the first biologic that a patient receives to treat the condition (i.e. in treatment "naive" patients). Short of changing to another biologic or therapy altogether, switching of biologic products for these patients, from an originator to a biosimilar, biosimilar to originator, or biosimilar to another biosimilar, could occur at any further point in the treatment process. In nonmedical switching (NMS), patients are actively being switched for cost reasons.[52] In the United Kingdom, NHS has issued a guidance stating that nine of 10 new patients should be started on the best value biologic within three months of a biosimilar launch. Additionally, the guidance appears to strongly support NMS by stating that at least 80% of existing patients should be switched to the best value biologic within 12 months.[55]

NMS might be associated with additional costs for switched patients. In a systematic review, Liu et al. found that, among studies that reported the estimated cost impact associated with NMS, 33 reported net drug cost reduction, 12 reported additional healthcare costs post-NMS without a detailed breakdown, and 5 reported NMS setup and managing costs.[89] Three originator manufacturer-sponsored studies estimated significant expected short-term costs related to NMS.[90–92] Additional work required for the physician to verify adherence or patient outcomes on the switched therapy, or use of health services such as pharmacovigilance monitoring by a designated nurse, would all be potential marginal costs associated with a decision to switch a patient on any biologic therapy. The impact of these additional costs would need to be modeled or tracked to assess net savings. Use of a practical healthcare financial management technique, like payback period or break-even analysis, might be useful to determine the length of time required for NMS to a biosimilar to begin generating net savings.

The "nocebo" effect, in the context of biosimilars, is defined as the incitement or the worsening of symptoms induced by any negative attitude, for instance, as a consequence of negative provider or patient attitudes toward required biosimilar substitution in NMS.[93] Given the role that biosimilars can play in providing cost-effective alternatives to reference biologics, key strategies to help mitigate the nocebo effect in the NMS scenarios might include positive framing, such as discussion of the equality of the treatments as assessed by independent regulators, increasing understanding of biosimilars by patient and healthcare professionals, and utilizing an institution-wide managed switching program.[93] Other options might include reimbursing biosimilars for only newly diagnosed patients, using product-listing agreements for a biosimilar or the originator that compensate the payer for managing uncertainty regarding current or emerging biosimilar launches, or using tiered co-payments or other incentives to promote biosimilar use.[94]

Efforts that emphasize timely supportive evidence might be especially beneficial in cases where a payer encourages biosimilar usage for indication extrapola-

tion, where the approval of a biosimilar for use in an indication held by the reference product has not been studied in comparative clinical trials of the biosimilar. In such situations, payers and providers have reported concerns in instances where indications of the reference product are extrapolated to the biosimilar product, in the absence of clear clinical trial evidence for each indication. This would affect the ability to optimize the savings potential associated with biosimilar substitution.[62]

12.7.8 Incentives for Biosimilar Adoption

Several incentives that encourage and some disincentives that discourage biosimilar adoption have been described. A summary of these is listed in Table 12.1. Findings from these reports and other studies cited in this chapter suggest several incentives that might be critical "pressure points" for biosimilar uptake:

- Although national policies that encourage biosimilar uptake have been cited as being effective,[21,52,57,97] local policies have also demonstrated success.[58,88] Guidance from medical societies, pharmacist organizations, and other key opinion leaders might be especially productive.[58,88]

- Evidence suggests that a general willingness of payers to enforce biosimilar-favoring requirements might help to achieve high rates of biosimilar usage.[55,56,88] Policies with less emphasis on cost-containment are apparent among other payers, such as in the United States. These will likely have less success with enhancing biosimilar uptake.[53]

- Gainsharing with relevant stakeholders appears to be effective at encouraging biosimilar usage.[77,88,95,96] Reinforcing feedback from stakeholders, such as physicians,[64,81] and negative feedback from intermediaries like PBMs,[75] where each considers its own rewards or costs, have been cited. Additional costs related to switching have been noted.[89–92] Such costs would serve to inhibit biosimilar uptake by stakeholders not adequately compensated for the loss.

- Negative marketing campaigns by innovator companies might be discouraging biosimilar usage by physicians,[75] but such negative messaging will likely dissipate over time as innovator companies find it profitable to also launch biosimilars for indications where they currently have no market presence.

- Existing HTA data might help with biosimilar uptake, particularly if lower biosimilar costs would result in earlier restrictions from payment for originator biologics being lifted, such as when ICERs

Table 12.1 Examples of incentives and disincentives for biosimilar adoption.

Local policies encourage biosimilar uptake

Biosimilars are generally adopted following establishment of local policies concerning their use. Quotas for biosimilar prescriptions have resulted in high uptake of these agents in Germany. Gainsharing agreements enable healthcare professionals and patients to benefit directly from savings achieved through use of biosimilar medicines. Positive guidance from medical societies and pharmacist organizations may also result in healthcare professionals becoming more open to the use of biosimilars in their practice.[88]

Negative marketing information about biosimilars

Part of the failure of commercially available biosimilars to gain market share, aside from pricing, is attributable to confusion having to do with marketing campaigns by innovator companies to discourage physicians from prescribing biosimilars.[75] Such campaigns may be self-limiting over the long term, since many bio-originator manufacturers have biosimilars for other therapeutic indications approved or in the development pipeline.

A focus on outcomes and incentives

Governments and payers should take a strategic approach to obtain value for expenditures on biosimilars by (i) supporting generation of high-quality comprehensive outcomes data on the comparative effectiveness and safety of biosimilars and originator products, and (ii) ensuring that incentives exist for budget holders to benefit from price competition. This may create greater willingness on the part of budget holders and clinicians to use biosimilar and originator products with comparable outcomes interchangeably.[95]

Stakeholder communication and incentives

Most countries have specific supply-side policies for promoting access to biosimilars. Investments should be made to clearly communicate with and educate stakeholders about biosimilars. To create trust, physicians need to be especially informed on the entry and use of biosimilars. Further incentives should be offered to prescribe biosimilars. Gainsharing can be used as an incentive to prescribe, dispense, or use biosimilars. This approach, in combination with binding quotas, may support a sustainable biosimilar market.[96]

Supportive interchange policies and education of physicians and patients

Biosimilars may contribute to better patient access to biologics and provide savings to government programs. To increase acceptability, clinical evidence and real-world experiences are needed, as well as education of both physicians and patients. High biosimilar penetration rates in Norway, Denmark, and Poland suggest that policies which support interchange with the reference product may be important drivers of biosimilar uptake.[97]

Policies that encourage generic uptake might not adapt well to biosimilars

Per-capita pharmaceutical expenditure appears inversely correlated to biosimilar uptake. This is likely because countries with the highest drug expenditures are also prone to use more biologics and are exposed to overall less cost-containment pressure. Countries that are experiencing high uptake of generics are paradoxically exposed to less penetration of biosimilars. Such countries may be tempted to address biosimilars like generics. However, such policies have little effect on biosimilar uptake.[53] This implies that incentives like price discounting and those successfully used to encourage generic substitution are insufficient on their own to implement successful biosimilar uptake in these countries.

Avoiding policies that disadvantage and using HTA to permit biosimilar entry

Supply-side policies, such as internal or external reference pricing, price linkage, and tendering for biosimilars aim to push biologic reference medicine and biosimilar prices down to generate savings. However, compulsory price cuts on biosimilars may be disincentivizing for manufacturers. To compete with biosimilars, reference manufacturers might cut their prices to maintain uptake of branded biologics. This reduces competition and decreases the revenue and profitability of biosimilar producers, leading to price increases in the longer term. HTA may contribute to recognition of the value of biosimilars. In markets where the reference medicine is restricted from reimbursement, improvements in the cost-effectiveness ratios of biologics through biosimilar launches could alleviate the restrictions.[98]

Improve payment formulas to incentivize biosimilar prescribers

In the United States, Medicare typically reimburses for medication administered in a physician office or infusion clinic at average sales price (ASP) plus 6% as an administrative fee. To incentivize the prescribing of biosimilars, CMS has now set the administrative fee for the biosimilar based on the ASP of the reference product plus 6% of the reference product's ASP.[64] A corollary to this observation is the incentives to physicians to prescribe or not prescribe biosimilars must always be considered when developing policies at national or local levels.

---

found that these biosimilars meet acceptable thresholds.[98] For instance, the lower cost of the biosimilar would replace the higher cost of the innovator in an update of the original ICER developed when the innovator was first launched, where $(\text{cost}_{\text{innovator}} - \text{cost}_{\text{routine care}})/(\text{QALY}_{\text{innovator}} - \text{QALY}_{\text{routine care}})$. If this updated HTA assumed little or no difference in QALYs for the biosimilar versus the innovator, this would result in a lowered ICER. This, in turn, might be sufficiently low to meet a particular acceptable ICER threshold value for a given country. However, simple substitution of biosimilar cost inputs in older HTA models would not be sufficient. The HTA models used to evaluate originator biologics would need to be fully updated and modified to meet current treatment, economic, and market dynamics in place at the time the biosimilar is being reviewed by payers.

- Successful national or local policies for biosimilar uptake appear to be different from those successfully used for generic uptake.[53] An educational gap regarding the potential benefits and harms of biosimilar switching appears to exist among many physicians and patients.[96,97] There appears to be broad support from physicians for more outcomes data, specifically focused on real-world comparative effectiveness and safety of biosimilars when compared to originator products.[76,95,97] Issues related to nonmedical switching in stabilized patients, indication extrapolation, and the "nocebo" effect are particularly salient concerns.[93,94] Short of more concerted efforts by payers, these issues will likely inhibit efforts at greater biosimilar uptake until this knowledge gap is adequately addressed.

12.8 Conclusions

Due to demonstrated and often substantial gains in efficacy and safety, biologics have had tremendous market success since their introduction more than three decades ago. More biologics in the pipeline are anticipated to add to this success. The cost of biologics appears as outwardly large as their benefits. Accounting for only a small proportion of all pharmaceutical prescriptions, these agents account for a disproportionately massive share of prescription drug spending, across global markets. Countries that have adopted planned strategies to assess cost-effectiveness of biologics, and to encourage the uptake of "substitutes," like biosimilars, have demonstrated clear savings over those that have not. More work is still required, at all levels of healthcare policy and delivery. Affordability of, and access to, biologics remains a concern in all countries. Due to the high cost, some countries are effectively shut out of innovator biologic, and even biosimilar, usage altogether. Effective reinforcing (encouraging) and inhibiting (discouraging) policies and practices, locally adapted and carefully applied, are essential to balancing and fulfilling the needs of patients, providers, populations, nations, investors, and the biopharmaceutical industry itself, upon which all of these stakeholders rely.

References

1 Leber, M.B. (2018). Optimizing use and addressing challenges to uptake of biosimilars. *The American Journal of Managed Care* 24 (21 Suppl): S457–S461.

2 Express Scripts (2019). *2018 Drug Trend Report.* St. Louis, Missouri, USA: Express Scripts Report No.: 18-EME49136.

3 Aitken, M. and Kleinrock, M. (2019). *Medicine Use and Spending in the U.S. – A Review of 2018 and Outlook to 2023.* IQVIA Institute for Human Data Science: Parsippany, New Jersey, USA.

4 Philippidis, A. (2018). The top 15 best-selling drugs of 2017. *Genetic Engineering and Biotechnology News* (12 March) https://www.genengnews.com/a-lists/the-top-15-best-selling-drugs-of-2017 (accessed 16 June 2020).

5 Cohen, J. (2019). Biosimilars continue to exhibit market failure. *Forbes* (11 July) https://www.forbes.com/sites/joshuacohen/2019/07/11/biosimilars-continue-to-exhibit-market-failure/#1ec9c6e073ee (accessed 16 June 2020).

6 Ornstein, C. (2015). The cost of a cure: Medicare spent $4.5 billion on new hepatitis C drugs last year.

ProPublica (March 29) https://www.propublica.org/article/cost-of-a-cure-medicare-spent-4.5-billion-on-hepatitis-c-drugs-last-year (accessed 16 June 2020).

7 Alexander, G.C., Ballreich, J., Socal, M.P. et al. (2017). Reducing branded prescription drug prices: a review of policy options. *Pharmacotherapy* 37 (11): 1469–1478.

8 Trish, E., Xu, J., and Joyce, G. (2016). Medicare beneficiaries face growing out-of-pocket burden for specialty drugs while in catastrophic coverage phase. *Health Affairs* 35 (9): 1564–1571.

9 Grosse, S.D., Do, T.Q.N., Vu, M. et al. (2018). Healthcare expenditures for privately insured US patients with cystic fibrosis, 2010–2016. *Pediatric Pulmonology* 53 (12): 1611–1618.

10 Buer, L., Hoivik, M.L., Medhus, A.W., and Moum, B. (2017). Does the introduction of biosimilars change our understanding about treatment modalities for inflammatory bowel disease? *Digestive Diseases* 35 (1–2): 74–82.

11 Hresko, A., Lin, T.C., and Solomon, D.H. (2018). Medical care costs associated with rheumatoid

arthritis in the US: a systematic literature review and meta-analysis. *Arthritis Care & Research (Hoboken)* 70 (10): 1431–1438.

12 Gronde, T.V., Uyl-de Groot, C.A., and Pieters, T. (2017). Addressing the challenge of high-priced prescription drugs in the era of precision medicine: a systematic review of drug life cycles, therapeutic drug markets and regulatory frameworks. *PLoS One* 12 (8): e0182613.

13 Huggett, B. (2016). America's drug problem. *Nature Biotechnology* 34 (12): 1231–1241.

14 Levi, E.L. (2017). Using data exclusivity grants to incentivize cumulative innovation of biologics' manufacturing processes. *The American University Law Review* 66 (3): 911–970.

15 Antman, E.M., Creager, M.A., Houser, S.R. et al. (2017). American Heart Association principles on the accessibility and affordability of drugs and biologics: a presidential advisory from the American Heart Association. *Circulation* 136 (24): e441–e447.

16 Lu, J. (1998). Strategic pricing of new pharmaceuticals. *Review of Economics and Statistics* 80 (1): 108–118.

17 PhRMA (2016). *Policy Solutions: Delivering Innovative Treatments to Patients*. Washington, DC, USA: PhRMA.

18 Baumgart, D.C., Misery, L., Naeyaert, S., and Taylor, P.C. (2019). Biological therapies in immune-mediated inflammatory diseases: can biosimilars reduce access inequities? *Frontiers in Pharmacology* 10: 279.

19 Putrik, P., Ramiro, S., Kvien, T.K. et al. (2014). Inequities in access to biologic and synthetic DMARDs across 46 European countries. *Annals of the Rheumatic Diseases* 73 (1): 198–206.

20 Iyengar, S., Tay-Teo, K., Vogler, S. et al. (2016). Prices, costs, and affordability of new medicines for hepatitis C in 30 countries: an economic analysis. *PLoS Medicine* 13 (5): e1002032.

21 Araujo, F.C., Goncalves, J., and Fonseca, J.E. (2016). Pharmacoeconomics of biosimilars: what is there to gain from them? *Current Rheumatology Reports* 18 (8): 50.

22 Pentek, M., Lakatos, P.L., Oorsprong, T. et al. (2017). Access to biologicals in Crohn's disease in ten European countries. *World Journal of Gastroenterology* 23 (34): 6294–6305.

23 Padula, W.V., Ballreich, J., and Anderson, G.F. (2018). Paying for drugs after the Medicare part D beneficiary reaches the catastrophic limit: lessons on cost sharing from other US policy partnerships between government and commercial industry. *Applied Health Economics and Health Policy* 16 (6): 753–763.

24 Hoadley, J., Cubanski, J., and Nueman, T. (2015). *It Pays to Shop: Variation in Out-of-Pocket Costs for Medicare Part D Enrollees in 2016*. The Henry J. Kaiser Family Foundation: Washington, DC, USA.

25 Elgin, B. and Langreth, R. (2016). How big pharma uses charity programs to cover for drug price hikes. *Bloomberg* (19 May) https://www.bloomberg.com/news/articles/2016-05-19/the-real-reason-big-pharma-wants-to-help-pay-for-your-prescription (accessed 16 June 2020).

26 Whayne, T.F. (2018). Outcomes, access, and cost issues involving PCSK9 inhibitors to lower LDL-cholesterol. *Drugs* 78 (3): 287–291.

27 Charbonneau, M. and Gagnon, M.A. (2018). Surviving niche busters: main strategies employed by Canadian private insurers facing the arrival of high cost specialty drugs. *Health Policy* 122 (12): 1295–1301.

28 De Vera, M.A., Mailman, J., and Galo, J.S. (2014). Economics of non-adherence to biologic therapies in rheumatoid arthritis. *Current Rheumatology Reports* 16 (11): 460.

29 Vogler, S., Paris, V., Ferrario, A. et al. (2017). How can pricing and reimbursement policies improve affordable access to medicines? Lessons learned from European countries. *Applied Health Economics and Health Policy* 15 (3): 307–321.

30 Atikeler, E.K. and Ozcelikay, G. (2016). Comparison of pharmaceutical pricing and reimbursement systems in Turkey and certain EU countries. *Springerplus* 5 (1): 1876.

31 Drummond, M., Jonsson, B., Rutten, F., and Stargardt, T. (2011). Reimbursement of pharmaceuticals: reference pricing versus health technology assessment. *The European Journal of Health Economics: HEPAC: Health Economics in Prevention and Care* 12 (3): 263–271.

32 Wallerstedt, S.M. and Henriksson, M. (2018). Balancing early access with uncertainties in evidence for drugs authorized by prospective case series – systematic review of reimbursement decisions. *British Journal of Clinical Pharmacology* 84 (6): 1146–1155.

33 Zhang, Y., Hueser, H.C., and Hernandez, I. (2017). Comparing the approval and coverage decisions of new oncology drugs in the United States and other selected countries. *Journal of Managed Care & Specialty Pharmacy* 23 (2): 247–254.

34 Wilson, A. and Cohen, J. (2011). Patient access to new cancer drugs in the United States and Australia. *Value in Health: The Journal of the International Society for Pharmacoeconomics and Outcomes Research* 14 (6): 944–952.

35 Daniel, H. and Health, Public Policy Committee of the American College of Physicians (2016). Stemming the escalating cost of prescription drugs: a position paper of the American College of Physicians. *Annals of Internal Medicine* 165 (1): 50–52.

36 American Society of Clinical Oncology (2018). American Society of Clinical Oncology position statement on addressing the affordability of cancer drugs. *Journal of Oncology Practice* 14 (3): 187–192.

37 Pyenson, B., Bochner, A., and Cannon, R. (2015). *An Actuarial Approach to the Incremental Cost of Hepatitis C in the Absence of Curative Treatments*. New York, NY, USA: Milliman, Inc.

38 Chambers, J.D., Thorat, T., Pyo, J. et al. (2014). Despite high costs, specialty drugs may offer value for money comparable to that of traditional drugs. *Health Affairs* 33 (10): 1751–1760.

39 Stawowczyk, E. and Kawalec, P. (2018). A systematic review of the cost-effectiveness of biologics for ulcerative colitis. *PharmacoEconomics* 36 (4): 419–434.

40 Bansback, N., Fu, E., Sun, H. et al. (2017). Do biologic therapies for rheumatoid arthritis offset treatment-related resource utilization and cost? A review of the literature and an instrumental variable analysis. *Current Rheumatology Reports* 19 (9): 54.

41 DiMasi, J.A., Grabowski, H.G., and Hansen, R.W. (2016). Innovation in the pharmaceutical industry: new estimates of R&D costs. *Journal of Health Economics* 47: 20–33.

42 Blankenship, K. (2019). The top 20 drugs by 2018 U.S. sales. *FiercePharma* (17 June) https://www.fiercepharma.com/special-report/top-20-drugs-by-2018-u-s-sales (accessed 16 June 2020).

43 QuintilesIMS Institute (2017). *Orphan Drugs in the United States: Providing Context for Use and Cost*. Parsippany, NJ, USA: QuintilesIMS Institute Contract No.: 33.0003-1-10.17_QI.

44 Short, H., Stafinski, T., and Menon, D. (2015). A National Approach to Reimbursement Decision-Making on Drugs for Rare Diseases in Canada? Insights from Across the Ponds. *Health Policy* 10 (4): 24–46.

45 Szegedi, M., Zelei, T., Arickx, F. et al. (2018). The European challenges of funding orphan medicinal products. *Orphanet Journal of Rare Diseases* 13 (1): 184.

46 Kent, D., Rickwood, S., and Di Biase, S. (2017). *Disruption and Maturity: The Next Phase of Biologics*. QuintilesIMS: Durham, NC, USA.

47 Hirsch, B.R. and Lyman, G.H. (2014). Biosimilars: a cure to the U.S. health care cost conundrum? *Blood Reviews* 28 (6): 263–268.

48 Generics and Biosimilars Initiative (2011 (updated 25 October 2019). http://www.gabionline.net/

Biosimilars/General/Biosimilars-approved-in-Europe). Biosimilars approved in Europe. *Generics and Biosimilars Initiative* (7 August) (accessed 16 June 2020).

49 Aapro, M., Gascon, P., Patel, K. et al. (2018). Erythropoiesis-stimulating agents in the management of anemia in chronic kidney disease or cancer: a historical perspective. *Frontiers in Pharmacology* 9: 1498.

50 QuintilesIMS (2017). *The Impact of Biosimilar Competition in Europe*. London, UK: QuintilesIMS.

51 Kay, J. (2019). Are There Benefits and Risks to Biosimilars from a Patient Perspective? *Rheumatic Diseases Clinics of North America* 45 (3): 465–476.

52 Aladul, M.I., Fitzpatrick, R.W., and Chapman, S.R. (2019). The effect of new biosimilars in rheumatology and gastroenterology specialities on UK healthcare budgets: results of a budget impact analysis. *Research in Social & Administrative Pharmacy: RSAP* 15 (3): 310–317.

53 Remuzat, C., Dorey, J., Cristeau, O. et al. (2017). Key drivers for market penetration of biosimilars in Europe. *Journal of Market Access & Health Policy* 5 (1): 1272308.

54 National Institute for Health and Care Excellence (2018). *Review of TA329; Infliximab, adalimumab and golimumab for treating moderately to severely active ulcerative colitis after the failure of conventional therapy*. London, UK: National Institute for Health and Care Excellence.

55 Reinke, T. (2019). Show us (the U.S.) the savings. *Managed Care* 28 (1): 9–10.

56 The Lancet Gastroenterology & Hepatology (2018). Bullying biosimilars: cheaper drugs stymied in USA. *The Lancet Gastroenterology & Hepatology* 3 (6): 371.

57 Grabowski, H., Guha, R., and Salgado, M. (2014). Biosimilar competition: lessons from Europe. *Nature Reviews. Drug Discovery* 13 (2): 99–100.

58 Moorkens, E., Simoens, S., Troein, P. et al. (2019). Different policy measures and practices between Swedish counties influence market dynamics: part 1-biosimilar and originator infliximab in the hospital setting. *BioDrugs: Clinical Immunotherapeutics, Biopharmaceuticals and Gene Therapy* 33 (3): 285–297.

59 Agbogbo, F.K., Ecker, D.M., Farrand, A. et al. (2019). Current perspectives on biosimilars. *Journal of Industrial Microbiology & Biotechnology* 46: 1297–1311.

60 Congressional Budget Office (2008). *Congressional Budget Office Cost Estimate: S. 1695 – Biologics Price Competition and Innovation Act of 2007*. Washington, DC, USA: United States Congress.

61 Singh, S.C. and Bagnato, K.M. (2015). The economic implications of biosimilars. *The American Journal of Managed Care* 21 (16 Suppl): s331–s340.

62 Manolis, C.H., Rajasenan, K., Harwin, W. et al. (2016). Biosimilars: opportunities to promote optimization through payer and provider collaboration. *Journal of Managed Care & Specialty Pharmacy* 22 (9 Suppl): S3–S9.

63 Bhatt, V. (2018). Current market and regulatory landscape of biosimilars. *The American Journal of Managed Care* 24 (21 Suppl): S451–S456.

64 Dolan, C. (2018). Opportunities and challenges in biosimilar uptake in oncology. *The American Journal of Managed Care* 24 (11 Suppl): S237–S243.

65 Mehr, S.R. and Brook, R.A. (2017). Factors influencing the economics of biosimilars in the US. *Journal of Medical Economics* 20 (12): 1268–1271.

66 Reinke, T. (2017). The biosimilar pipeline seams seem to be bursting. *Managed Care* 26 (3): 24–25.

67 Reinke, T. (2016). Patent litigation could make 2017 no "dancing" matter. *Managed Care* 25 (12): 23–28.

68 Ha, C.Y. and Kornbluth, A. (2016). A critical review of biosimilars in IBD: The confluence of biologic drug development, regulatory requirements, clinical outcomes, and big business. *Inflammatory Bowel Diseases* 22 (10): 2513–2526.

69 Chen, B.K., Yang, Y.T., and Bennett, C.L. (2018). Why biologics and biosimilars remain so expensive: despite two wins for biosimilars, the supreme court's recent rulings do not solve fundamental barriers to competition. *Drugs* 78 (17): 1777–1781.

70 Diependaele, L., Cockbain, J., and Sterckx, S. (2018). Similar or the Same? Why Biosimilars are not the solution. *The Journal of Law, Medicine & Ethics: A Journal of the American Society of Law, Medicine & Ethics* 46 (3): 776–790.

71 Frank, R.G. (2018). Friction in the path to use of biosimilar drugs. *The New England Journal of Medicine* 378 (9): 791–793.

72 I-MAK (2018). *Overpatented, Overpriced: How Excessive Pharmaceutical Patenting Is Extending Monopolies and Driving Up Drug Prices*. I-MAK: New York, NY, USA.

73 Brennan, Z. (2019). FDA approves 20th biosimilar, 5th for Roche's Herceptin. *Regulatory Affairs Professionals Society (RAPS)* (13 June) https://www.raps.org/news-and-articles/news-articles/2019/6/fda-approves-20th-biosimilar-5th-for-roches-herc (accessed 16 June 2020).

74 Federal Trade Commission (2019). Pay for delay: when drug companies agree not to compete. Washington, DC, USA: Federal Trade Commission. https://www.ftc.gov/news-events/media-resources/mergers-competition/pay-delay (accessed 16 June 2020).

75 Barlas, S. (2019). Biosimilar roadblock removals seem tailor-made for Trump/Democrats agreement: development of new biosimilar drugs and marketing of few approved drugs stymied. *P & T: A Peer-Reviewed Journal for Formulary Management* 44 (2): 45–68.

76 Leonard, E., Wascovich, M., Oskouei, S. et al. (2019). Factors affecting health care provider knowledge and acceptance of biosimilar medicines: a systematic review. *Journal of Managed Care & Specialty Pharmacy* 25 (1): 102–112.

77 Aladul, M.I., Fitzpatrick, R.W., and Chapman, S.R. (2018). Healthcare professionals' perceptions and perspectives on biosimilar medicines and the barriers and facilitators to their prescribing in UK: a qualitative study. *BMJ Open* 8 (11): e023603.

78 Cook, J.W., McGrath, M.K., Dixon, M.D. et al. (2019). Academic oncology clinicians' understanding of biosimilars and information needed before prescribing. *Therapeutic Advances in Medical Oncology* 11: 1758835918818335.

79 Blackwell, K., Gligorov, J., Jacobs, I., and Twelves, C. (2018). The global need for a trastuzumab biosimilar for patients with HER2-positive breast cancer. *Clinical Breast Cancer* 18 (2): 95–113.

80 Chen, B., Nagai, S., Armitage, J.O. et al. (2019). Regulatory and clinical experiences with biosimilar filgrastim in the U.S., the European Union, Japan, and Canada. *The Oncologist* 24 (4): 537–548.

81 Johnson, S.R. (2016). What happened to the innovation and competition Zarxio was supposed to spark? *Modern Healthcare* 46 (13): 11.

82 Hakim, A. and Ross, J.S. (2017). Obstacles to the adoption of biosimilars for chronic diseases. *Journal of the American Medical Association* 317 (21): 2163–2164.

83 Abraham, I., Han, L., Sun, D. et al. (2014). Cost savings from anemia management with biosimilar epoetin alfa and increased access to targeted antineoplastic treatment: a simulation for the EU G5 countries. *Future Oncology* 10 (9): 1599–1609.

84 Gyawali, B. (2017). Biosimilars in oncology: everybody agrees but nobody uses? *Recenti Progressi in Medicina* 108 (4): 172–174.

85 Martei, Y.M., Binagwaho, A., and Shulman, L.N. (2017). Affordability of cancer drugs in Sub-Saharan Africa: effects of pricing on needless loss of life. *JAMA Oncology* 3 (10): 1301–1302.

86 Conti, R.M. (2017). Biosimilars: reimbursement issues in your oncology practice. *Journal of Oncology Practice* 13 (9_suppl): 12s–14s.

87 U.S. Food and Drug Administration (2019). *Considerations in Demonstrating Interchangeability with a Reference Product – Guidance for Industry*. Silver Spring, MD, USA: FDA.

88 O'Callaghan, J., Barry, S.P., Bermingham, M. et al. (2019). Regulation of biosimilar medicines and current perspectives on interchangeability and policy. *European Journal of Clinical Pharmacology* 75 (1): 1–11.

89 Liu, Y., Yang, M., Garg, V. et al. (2019). Economic impact of non-medical switching from originator biologics to biosimilars: a systematic literature review. *Advances in Therapy* 36 (8): 1851–1877.

90 Dolinar, R., Kohn, C.G., Lavernia, F., and Nguyen, E. (2019). The non-medical switching of prescription medications. *Postgraduate Medicine* 131 (5): 335–341.

91 Tarallo, M., Onishchenko, K., and Alexopoulos, S.T. (2019). Costs associated with non-medical switching from originator to biosimilar etanercept in patients with rheumatoid arthritis in the UK. *Journal of Medical Economics* 22 (11): 1162–1170.

92 Gibofsky, A., Skup, M., Yang, M. et al. (2019). Short-term costs associated with non-medical switching in autoimmune conditions. *Clinical and Experimental Rheumatology* 37 (1): 97–105.

93 Kristensen, L.E., Alten, R., Puig, L. et al. (2018). Non-pharmacological effects in switching medication: the nocebo effect in switching from originator to biosimilar agent. *BioDrugs: Clinical Immunotherapeutics, Biopharmaceuticals and Gene Therapy* 32 (5): 397–404.

94 Husereau, D., Feagan, B., and Selya-Hammer, C. (2018). Policy options for infliximab biosimilars in inflammatory bowel disease given emerging evidence for switching. *Applied Health Economics and Health Policy* 16 (3): 279–288.

95 Mestre-Ferrandiz, J., Towse, A., and Berdud, M. (2016). Biosimilars: how can payers get long-term savings? *PharmacoEconomics* 34 (6): 609–616.

96 Moorkens, E., Vulto, A.G., Huys, I. et al. (2017). Policies for biosimilar uptake in Europe: an overview. *PLoS One* 12 (12): e0190147.

97 Pentek, M., Zrubka, Z., and Gulacsi, L. (2017). The economic impact of biosimilars on chronic immune-mediated inflammatory diseases. *Current Pharmaceutical Design* 23 (44): 6770–6778.

98 Remuzat, C., Kapusniak, A., Caban, A. et al. (2017). Supply-side and demand-side policies for biosimilars: an overview in 10 European member states. *Journal of Market Access & Health Policy* 5 (1): 1307315.

13

New Emerging Biotherapies: Cutting-Edge Research to Experimental Therapies

Veysel Kayser[1] and Mehmet Sen[2]

[1] Sydney Pharmacy School, Faculty of Medicine and Health, The University of Sydney, Sydney, New South Wales, Australia
[2] Department of Biology and Biochemistry, University of Houston, Houston, TX, USA

KEY POINTS

- Biologics are the fastest growing therapies.
- Immune therapy is the most promising but at the same time most challenging to develop.
- Among biologics, ADCs, bispecifics, CAR T-cell and immune checkpoint therapies will most likely dominate research and development in the near future and perhaps the market more generally in the long run.

Abbreviations

Abbreviation	Full name
AcBUT	A Carbonyl-Containing Carboxylic Acid
ADC	Antibody–Drug Conjugate
ADCC	Antibody-Dependent Cell-Mediated Cytotoxicity
α4β7/αEβ7	Integrin β7-Receptors Heterodimers
ALL	Acute Lymphocytic Leukemia
AML	Acute Myelogenous Leukemia
Axi-cel	Axicabtagene Ciloleucel
B7	CD80 & CD86
BCMA/BCM	B-cell Maturation Antigen
BiTE	Bi-specific T-cell Engager
BsAb/bsAb	Bispecific Antibodies
CAR	chimeric antigen receptor
CCR4	C-C motif chemokine Receptor 4
C. difficile	Clostridium difficile, a gram-positive anaerobic bacterium
CD1d	MHC-I-Like Molecule
CD3	Cluster of Differentiation 3
CD3ζ	Cluster of Differentiation 3 Chain ζ
CD7	Cluster of Differentiation 7

Abbreviation	Full name
CD8+	Cluster of Differentiation 8-Carrying T-cells
CD19	Cluster of Differentiation 19
CD20	Cluster of Differentiation 20
CD22	Cluster of Differentiation 22
CD30	Cluster of Differentiation 30
CD33	Cluster of Differentiation 33
CD38	Cluster of Differentiation 38
CD40L	Cluster of Differentiation 40 Ligand
CD44	Cluster of Differentiation 44
CD56	Cluster of Differentiation 56
CD79b	Cluster of Differentiation 79b
CD80	Cluster of Differentiation 80
CD86	Cluster of Differentiation 86
CD135	Cluster of Differentiation 135
CD137	Cluster of Differentiation 137
CD138	Cluster of Differentiation 138
CD154	Cluster of Differentiation 154
CDC	Complement-Dependent Cytotoxicity
CHO	Chinese Hamster Ovary

Abbreviation	Full name	Abbreviation	Full name
Cll1	C-Type lectin-like Molecule-1	KTE-C19	Yescarta®
CQAs	Critical Quality Attributes	LMWH	Low-Molecular Weight Heparins
CRISPR	Clustered Regularly Interspaced Short Palindromic Repeats	mAb	Monoclonal Antibody
CSCC	Cutaneous Squamous Cell Carcinoma	MCC	Merkel Cell Carcinoma
CTL019	Tisagenlecleucel (Kymriah®)	MC-VC-PABC	Maleimidocaproylvaline-Citrulline-*p*-Aminobenzoyloxycarbonyl
CTLA-4	Cytotoxic T-Lymphocyte-Associated Protein 4	MERS	Middle East Respiratory Syndrome Coronavirus
DART	Dual-Affinity Re-Targeting	MHC	Major Histocompatibility Complex
DLBCL	Diffuse Large B-Cell Lymphoma	MMAE	Monomethylauristatin E
DNA	Deoxyribonucleic Acid	NACs	Nanoparticle–Antibody Complexes
DTT	Dithiothreitol	NCAM	Neural Cell Adhesion Molecule
EGFR	Epidermal Growth Factor Receptor	NK	Natural Killer
EMA	European Medicines Agency	NKT	Natural Killer T-cells
EpCAM	Epithelial Cell Adhesion Molecule	NME	New Molecular Entities
EU	European Union	NVPF	Nonviral Plasmid-free
Fab	Fragment antigen-Binding Region	P/G	Perforin and Granzyme
Fc	Fragment crystallizable Region	PCR	Polymerase Chain Reaction
FcRn	Fc Neonatal Receptor	PD-1	Programmed Cell Death Protein 1
FcγR	A Receptor for the Fc Portion of IgG	PD-L1	Programmed Cell Death Protein 1 ligand
FDA	United States Food and Drug Administration	PEG	Polyethylene Glycol
FLT-3	FMS like Tyrosine Kinase 3 (CD135)	PHSA	Public Health Service Act
FR β	Folate Receptor B	R&D	Research and Development
FUT8	A-1,6-Fucosyltransferase	RNA	Ribonucleic Acid
γδ	gamma-delta T-cells	sALCL	Systemic Anaplastic Large Cell Lymphoma
HER2	Human Epidermal Growth Factor Receptor 2	scFv	single-chain Variable Fragments
HIV	Human Immunodeficiency Virus	Siglec-3	Sialic acid-binding immunoglobulin-type lectin-3
HLA	Human Leukocyte Antigen	T-DM1	Trastuzumab Emtansine
hMSCs	Human Mesenchymal Stem and Stromal Cells	TAA	Tumor-Associated Antigens
HSCs	Hematopoietic Stem Cells	TB	Tuberculosis
ic9	Icaspase-9	TCR	T-Cell Receptor
IgE	Immunoglobulin E	TGF-β	Transforming Growth Factor
IgG	Immunoglobulin G	TIM-3	T-Cell Immunoglobulin and Mucin-Domain Containing-3
IL	Interleukin	TNF-α	Tumor Necrosis Factor Alpha
IL-2	Interleukin-2	tPA	tissue Plasminogen Activators
IL-4	Interleukin-4	Treg cell	T-regulatory Cells
IL-15	Interleukin-15	TRUCKs	T-cells Redirected for Universal Cytokine-Mediated Killings
IP	Intellectual property	US	United States of America
irAE	immune-related Adverse Event	USD	United States Dollar
IRs	Immune Checkpoint Receptors	VEGF	Vascular Endothelial Growth Factor
ITAM	Immunoreceptor Tyrosine-Based Activation Motifs	WHO	World Health Organization
IV	Intravenous	4-1BB	A Type-2 Transmembrane Glycoprotein Receptor Belonging to the TNF Superfamily
kDa	Kilodalton		
KIR	Killer-cell Immunoglobulin-like Receptors		

13.1 Introduction

The global biologics market is well over USD $250 billion.[1,2] Biologics are exceedingly successful but costly and used for prevention, diagnosis, and treatment of various disorders.[3,4] They include full-size monoclonal antibodies (mAbs),[5] other products based on different antibody formats such as bispecifics and antibody–drug conjugates (ADCs), various types of blood products, hormones, and very recent arrivals such as immune checkpoints and CAR T-cell therapies. More than one thousand biologics are already in clinical trials, but only about three hundred biologics are on the market currently or in regulatory reviews. In recent years, particular attention has been paid to biosimilars because of expiration of patent protections of highly successful biologics, and more complex products such as cell- and gene-therapies due to advancements in understanding the basics of immune response. As patent protections of existing biologics expire, it is expected that more biosimilars will come to market at a 10–50% reduced price. In addition, improved variants of existing biologics such as ADCs, bi- or multi-specifics, nanoparticle–antibody complexes (NACs), novel formulations for better half- and shelf-lives of products, and less invasive drug delivery methods aimed at greater patient compliance are also under heavy assessment at the moment across research institutes and the biopharma industry globally. Therefore, it is expected that next-generation biologics will include these new formats, as well as continue to include both biosimilars and improved versions of existing biologics (i.e. biobetters) and a new class of complex therapies such as cell- and gene-based products. In this chapter, we have focused on the most critical and widely studied products which have enormous potential to further develop, and thus, we discuss bispecifics, ADCs, CAR T-cell therapies, and immune checkpoints. Nevertheless, most of the discussion, including our suggested methods for addressing commonly observed issues, should be useful for a variety of different next-generation biologics.

Biobetters could be based on improvements of the existing products' properties such as: (i) addressing protein structural stability issues and hence reducing commonly observed protein aggregation, (ii) enhancing the drug's efficacy and cellular uptake, (iii) boosting its safety by reducing potential immunogenicity,[6,7] (iv) ensuring better formulation stability during long-term storage conditions, (v) improving patient convenience such as formulating for subcutaneous or even oral delivery rather than frequently used intravenous (IV) infusion, (vi) adding other quality attributes like better targeting using bi- or multi-specific mAb preparations, or (vii) preparing combination therapies or multiple treatment modalities. Currently, there are many products in clinical trials and in various developmental stages consisting of the abovementioned constructs. Most of the products that are in various stages of development are antibody-based drug candidates, and a majority of them are developed as cancer therapeutics. Consequently, it is expected that next-generation biologics will likely involve several of these concepts and emerge as full-size antibodies, antibody fragments, ADCs, hyper-glycosylated antibodies, bi- or multi-specifics, NACs among other formats; in addition to more complex preparations such as new vaccines, gene- and cell-based therapies as part of immunotherapies, targeted therapies, and personalized medicines.[8] Advancement of recent developments, such as clustered regularly interspaced short palindromic repeats (CRISPR) technology, will further enable us to prepare more advanced constructs for targeted and personalized therapies.

Other experimental approaches include bacteriophage-based therapies, particularly for antibiotic-resistant infections.[9–11] Although phage therapy fell out of favor in the Western world after the discovery of antibiotics, it again started to receive attention in recent years, the first clinical trial to test bacteriophage has been approved by the FDA.[12] However, phage therapy has been in use in some Asian countries for some time, particularly in ex-Soviet states such as Georgia. Bacteriophages are viruses that invade bacteria and cause lysis. They tend to be quite specific in their targets and hence each infection requires the generation of a new set of bacteriophages for that specific bacterium. Therefore, there will not be a broad-spectrum bacteriophage similar to that of broad-spectrum antibiotics. Understanding the mechanism of bacteriophage–bacteria interaction will result in novel bacteriophage therapies. There is substantial ongoing fundamental research on potential bacteriophage therapy for different infections such as tuberculosis (TB).[10]

Critical quality attributes (CQAs) of biologics and vaccines have been under scrutiny for several decades now for many different products. CQAs need to be elucidated mainly during research and development (R&D) phases of drug development, confirmed during production and long-term storage conditions because it is directly related to the efficacy and safety of the product. For example, protein aggregation needs to be controlled for many biologics, as a number of aggregates are naturally immunogenic due to their large size.[6,7] A small amount of protein aggregation may not be a major issue for vaccines; however, other biologics should be free of impurities such as aggregates and fragments in order to avoid any potential induced-immunogenicity. Emerging biologic products are anticipated to comply with this expectation and many novel formulations are investigated for both the enhancement of formulation stability in long-term storage conditions and discovery of improved preparations that progress through rapid regulatory approval and greater patient compliance. The importance of the latter was perhaps best demonstrated by Exubera®, which was approved by the FDA in 2006 for type-1 and type-2 diabetes, but was not a popular product among patients (the manufacturer took it off the market in 2007). Literature and market history are full of other examples where each aforementioned point created an issue for at least one product (e.g. formulation stability issues due to toxicity and visible particles in influenza vaccines for bioCSL and Novartis in 2010 and 2012, respectively). Hence, future biologics will likely be prepared to address such potential problems.

Moreover, in the new generation of biologics, additional improvements are also expected with side effects since serious cases of these have been observed with some current biologics. These side effects may result in permanent withdrawal of the therapy if the patient shows severe drug-related allergic reactions, toxicity, malignancy, heart failure, neurological issues, or lupus-like symptoms.[13] If the patient is pregnant, requires surgery, or shows infection, then a temporary withdrawal may be required. In fact, infection is quite common for many biologics because of the interference of biologics with the immune system either directly or indirectly, e.g. interactions with epidermal growth factor receptor (EGFR) or vascular endothelial growth factor (VEGF). Future developments are expected to address some of these side effects by providing more targeted and personalized medicine.

13.2 Methods to Enhance Stability of Biologics

Many biologics suffer from either physical or chemical instabilities. Arguably, the most concerning degradation of a biologic occurs via protein aggregation, which is generally overlooked in vaccine formulations but scrutinized in detail in recombinant proteins including antibody therapeutics. Other forms of degradation are also problematic and must be addressed in detail during R&D stages of a biotherapeutic. These include oxidation, deamidation, fragmentation, and other forms of physical and chemical degradation.

Major degradation pathways of biologics are discussed in Chapter 6 of this book. Potential instability problems may be addressed during early drug development stages by the following approaches (it must be noted that the recommendations below are generally for recombinant proteins, vaccines, and other biologics that can also benefit from these front-end investigation strategies):

1) Increasing structural stability and reducing the intrinsic aggregation propensity of a protein by point mutation or a set of mutations on or near aggregation-prone surface domains.[14] Finding such domains is no trivial task; however, recent computational methods have made this a distinct possibility.[15–18]

2) Another approach to obtain or derive stable and more effective biologics is glycan engineering. This could be carried out in two ways: (i) similar to Approach (1), where aggregation-prone regions are identified and stabilized, and protein–protein interaction is reduced by the addition of extra glycosylation on the protein surface, hence creating a "hyperglycosylated" protein; or (ii) where protein–receptor interactions are enhanced by afucosylation of naturally existing glycosylation of a mAb.[19–22] An example for the latter is mogamulizumab, an afucosylated mAb that targets CCR4,[23] is manufactured using a FUT8 knockout CHO cell line, and is used for the treatment of various cancers.[24] This product displays enhanced antibody–Fcγ receptor interactions because of afucosylation, and such interactions are desirable attributes for inducing greater cytotoxicity. Other types of glycan engineering techniques also exist, such as GlymaxX® technology from ProBioGen AG, and other systems such as the tobacco plant (*Pichia pastoris*) can be used to manufacture different types of fucosylated glycoforms, including aglycoforms.[25] In addition, if the protein aggregates exist due to the presence of a large hydrophobic surface exposure, the presence of a glycosylation site close to that region could enhance its structural stability, reducing protein–protein interactions sterically, and thus preventing protein aggregation.[26] Such hyperglycosylated antibodies have been shown to display reduced protein aggregation tendencies with enhanced solubility attributes.[21,22,27] Next-generation biologics, especially mAbs and antibody-based products, are expected to benefit from this approach.

3) The protein surface can also be modified by polyethylene glycol (PEG) or PEGylation near the aggregation-prone region, thereby reducing protein–protein

interaction, aggregation, and local hydrophobicity. Both the second and third approaches not only enhance conformational stability of the protein, but also increase protein solubility. Additional benefits of these latter methods (hyperglycosylation and PEGylation) include increased serum half-life, which is a major benefit for many biologics such as small-size bispecifics and nonhuman mAb-based ADCs which normally have short half-lives. However, neither glycosylation nor polymerization can be controlled externally, and both hyperglycosylation and PEGylation introduce additional heterogeneity to the system. This may be acceptable for many products, as long as extensive physicochemical characterization is carried out.

4) Another approach to stabilize biological products is formulation engineering. Existing formulation strategies are already aimed at addressing stability by maintaining the optimal level and long-term formulation of the product stability while retaining desired product attributes during long-term storage conditions. Formulation development is a multistep, empirical, laborious, and time-consuming process which involves screening the ideal pH range, identifying buffer composition, selecting the best additives that include various types of salts, sugars, surfactants, etc., performing accelerated studies and characterizing the product extensively. Future products are expected to benefit from both current practices as well as newly proposed formulation approaches, detection, and prediction methods.[28–30] Some of these include, but are not limited to, new formulations for less invasive delivery methods, such as microneedle delivery,[31–34] and novel additives such as ionic liquids and macrocycles.[35–39]

13.3 Bispecific Antibodies (bsAbs)

13.3.1 Current Perceptions, Clinical Trials, and Observed Issues

The concept of bispecifics goes back about 60 years[40]; however, only recent technological advances have made it possible to prepare successful bispecific biologics. Even though just two products (catumaxomab and blinatumomab) are commercially available, currently more than 200 different formats of bispecifics are being tested in 220 clinical trials.[41,42] It is expected that the bispecifics market will reach USD8 billion by 2025 with an annual growth rate between 10 and 20%.[43] Bispecifics

act as bridge molecules and bind to two targets simultaneously: mainly to a receptor on a tumor cell while at the same time engaging with an immune cell, such as a T-cell or natural killer (NK) cell. This simultaneous engagement has additional benefits; for example, selective and specific recruitment of a T-cell to the vicinity of a cancer cell would induce annihilation (death) of cancer cell more rapidly and efficiently.

Currently, there is no specific guideline to develop a bispecific product, but in April 2019, the FDA released a short draft guideline and indicated that developing bispecifics will be similar to the development of full-size mAb therapies.[44]

Bispecifics can generally be categorized into three major classes: protein fragment-based and symmetric- or asymmetric-protein-based. Almost all bispecific constructs utilize antibody platforms, but they all come in different flavors with varying size, shape, and specificity. More recently, a more advanced but complicated multiple target approach has been used to develop multispecifics, where three or more receptors are targeted. However, the development of both bi- and multi-specific antibodies suffers from great challenges, such as lack of adequate efficacy or too much cytotoxicity. This is despite promising antitumor outcomes, which were observed during *in vitro* studies and animal models; in addition, a short serum half-life, protein aggregation, and formulation instabilities were noted. Some of these challenges are expected to be addressed by providing a molecular level of understanding antibodies and protein engineering, better structural and formulation stable constructs, and developing improved animal models for more effective and targeted bsAbs therapies.[45]

Different classes of bispecifics have different advantages and limitations, which are mainly related to their pharmacokinetic profiles due to their different sizes and the presence of an Fc region. For example, one of the main limitations of smaller formats is having a relatively shorter serum half-life (similar to small molecules), requiring multiple or frequent dosing compared to full-size bispecific mAbs; however, smaller formats can have a better tissue penetration, which is usually a preferred attribute. If the fragment-crystallizable (Fc) domain is present on the construct, then not only is the bispecific's activity affected via triggering the Fcγ effector function, but the bispecific itself is also recycled due to neonatal Fcγ receptor recognition. Again, generally speaking, this is also a desired feature. Another potential issue is the immunogenicity of bispecifics, especially if the construct is based on a nonhuman antibody, but such a construct might demonstrate a better cytotoxicity as well.

13.3.2 Approved Bispecifics

13.3.2.1 Catumaxomab (Removab®)

Catumaxomab was developed by German companies, Fresenius Biotech and Trion Pharma, and is used to treat malignant ascites. It is a trifunctional, mouse–rat hybrid mAb comprising two fragment antigen-binding (Fab) arms. One of the arms binds to the epithelial cell adhesion molecule (EpCAM) of a tumor cell, while the other binds to CD3 of a CD8$^+$ T-cell and selectively engages with and activates Fcγ receptor I-, IIa-, or III positive accessory cells with its Fc domain. Catumaxomab has approximately 150 kDa molecular weight and 2.5 days half-life. The action of catumaxomab is quite complex, and it is believed that various different killing mechanisms such as T-cell-mediated cell lysis, cytokine-induced cytotoxicity, phagocytosis, and ADCC are involved. It was also shown that the immune system is activated by catumaxomab, which may induce a better and longer-lasting immunity against the tumor.[46] Catumaxomab was withdrawn from the market voluntarily in 2013 and 2017 from the US and EU markets, respectively.[47]

13.3.2.2 Blinatumomab (Blincyto®)

Blinatumomab developed by Amgen is a fusion protein and part of the bispecific T-cell engagers (BiTE) family. Blinatumomab is based on mAbs and is used for the treatment of acute lymphocytic leukemia (ALL). It binds to CD19 on the malignant B-cell (tumor cell) with one arm and to CD3 on the immune T-cell with the other arm. This enables T-cells to be activated, which in turn release perforin and granzyme (P/G) to mediate killing of the tumor cell via apoptosis.[48] The annual cost of blinatumomab is more than USD $150 000 per patient, although this varies between countries. It has about 55 kDa molecular weight, which is almost a third of the size of a full mAb, but has an extremely short half-life of ~2.1 hours. Two arms are joined with a linker and the whole construct is aglycosylated. One advantage of blinatumomab is that it engages with a variety of different T-cells, preventing resistance that is sometimes observed with downregulation of major histocompatibility complex (MHC) in T-cell-based treatments.

13.3.2.3 Emicizumab (Hemlibra®)

Emicizumab was developed for congenital factor VIII deficiency (hemophilia A) by Chugai Pharmaceuticals Co. Ltd. (a subsidiary of Roche) and approved by the FDA and EMA in 2017 and 2018, respectively, with accelerated assessments.[49] Patients missing coagulation factor VIII cannot mediate the function of factor X, which results in serious bleeding disorder. Emicizumab mimics the function of coagulation factor VIII by binding to activated factor X and mediating its function; it also binds to coagulation factor IX. Emicizumab is a humanized recombinant IgG4 antibody with 145 kDa molecular weight, formulated for subcutaneous administration at 30 or 150 mg/mL concentration, is well tolerated, and has about a 30 day of half-life.[50,51]

There are many different formats, platforms, and technologies to develop bispecifics, which includes, but is not limited to: DART® from MacroGenics, CrossMab from Roche, DVD-Ig™ from AbbVie, DuoBody® from Genmab, FcAb™ from F-Star, TandAb® from Affimed, Dock-and-lock™ from Immunomedics, and FIT-Ig® from EpimAb. Each of these constructs and formats are the propriety intellectual property (IP) of the company, with most platforms still in the developmental stages but highly likely to play an important role for the development of next-generation bispecifics. The core attributes of a mAb bispecific include working like a full-size mAb but behaving more like an improved modality rather than the full-size mAb because of its dual or more specific biological activity.[2]

13.4 Antibody–Drug Conjugates

ADCs are one of the most promising and prevalent emerging formats of the current antibody-based therapy. Although there are only six ADCs on the market (Table 13.1), which are mainly developed for different types of cancers, over 230 are in clinical trials[52] and thousands more are in the various stages of R&D. ADCs comprise of a mAb that is used for its specificity and targeting, a cytotoxic drug, a small molecule payload to kill the tumor cell, and a linker that connects the mAb with the toxin. The toxin and mAb are linked together covalently via surface-exposed amino acids (e.g. Lys or Cys) on the protein. Other moieties such as non-native amino acids or glycans can also be engineered specifically for the chemical linker to attach covalently both with the mAb and with the payload. Some of the chemical linkers can enable the release of the drug inside tumor cells via enzymatic cleavage, especially useful for agents that are otherwise too toxic to be administered systemically but only functional if they are free (i.e. not attached to any protein) in order to bind to DNA, tubulin, or any other part of the cell. In fact, chemical compounds that are highly toxic (for example, maytansinoids, calicheamicins, or auristatins), and are to be given at high concentrations systematically, can instead be delivered via ADCs, up to 1000-fold higher toxin concentrations can

Table 13.1 Some critical properties of approved antibody–drug conjugates (ADCs) in the market.

Drug/trade name®	Toxic agent/linker[a]	Target[b]	Manufacturer	Indication	Source	Cost (1K$US)	Approved
Gemtuzumab ozogamicin/ Mylotarg	Calicheamicins/AcBut	CD33	Pfizer/Wyeth	Relapsed acute myelogenous leukemia (AML)	Humanized	25–50	2000–2017
Brentuximab vedotin/ Adcetris	MMAE/valine–citrulline	CD30	Seattle Genetics, Millennium/Takeda	Relapsed HL and relapsed sALCL	Chimeric	100	2011
Trastuzumab emtansine/ Kadcyla	DM1/emtansine	HER2	Genentech, Roche	HER2+ metastatic breast cancer following treatment with trastuzumab and a taxane	Humanized	100	2013
Inotuzumab ozogamicin/ Besponsa	Calicheamicin/AcBut	CD22	Pfizer/Wyeth	Relapsed or refractory CD22-positive B-cell precursor acute lymphoblastic leukemia	Humanized	170	2017
Moxetumomab pasudotox/ Lumoxiti	Pasudotox (a toxic fragment of Pseudomonas exotoxin A)/a cleavable linker	CD22	Astra Zeneca	Adult patients with relapsed or refractory hairy cell leukemia who have received at least two prior systemic therapies	Mouse	N/A	2018
Polatuzumab vedotin-piiq/ Polivy	MMAE/valine–citrulline	CD79b	Genentech, Roche	Relapsed or refractory diffuse large B-cell lymphoma (DLBCL)	Humanized	90	2019

[a] AcBut, a carbonyl-containing carboxylic acid, 4-(4-acetylphenoxy) butanoic acid; Valine–citrulline, a protease-cleavable linker; maleimidocaproylvaline-citrulline-*p*-aminobenzoyloxycarbonyl or MC-VC-PABC; Emtansine, succinimidyl-*trans*-4-(maleimidylmethyl) cyclohexane-1-carboxylate.
[b] CD22, a lectin (sugar-binding transmembrane protein) found on the B cell surface; CD79b, B-cell antigen receptor complex-associated protein beta chain.

be delivered with ADCs safely.[53] This has numerous advantages, such as understanding the drug resistance mechanism, as well as overcoming resistance completely to the chemotherapeutic agent, especially if the tolerable dose is relatively low.[54] Even though ADCs show tremendous potential and are part of the targeted therapies due to being antibody-based products, their development is highly challenging and has been markedly hampered because of the structural complexity in their design and other difficulties involved in the development and manufacturing processes.[55,56] Like many other biologics, especially recombinant proteins, protein aggregation is one of the major issues with ADCs (mainly due to the hydrophobic payload on the antibody surface). It is known that if a protein surface has large aggregation-prone regions, it aggregates extensively.[4,14,29] Another issue is the low yield of the drug due to multiple processing steps, including purification after conjugation with the linker and payload. Another common problem with ADCs is the hepatotoxic side effect, which is generally attributed to suboptimal targeting efficiency and still remains a major concern for some products. This results in elimination of the drug via the hepatobiliary system prior to reaching its target tissue.[57] In fact, such high toxicity could be detrimental for the product, as was the case for the first approved ADC, Gemtuzumab ozogamicin (Mylotarg®, Wyeth). Gemtuzumab ozogamicin was approved in 2000 for the treatment of acute myeloid leukemia (AML),[58] but was voluntarily withdrawn by the manufacturer owing to concerns about its toxicity vs therapeutic benefits compared to other therapies.[59] Most of the toxic effects of Mylotarg® are ascribed to the stability of the linker in plasma, initiating the payload to be released into nontargeted tissues.[60] Therefore, the stability of the linker is clearly a matter of concern and needs to be optimized early-on in development for all ADCs. Nevertheless, this ADC was approved back in 2017 with an increased price tag.

Other products that have been approved earlier by the FDA include brentuximab vedotin (Adcetris®, Seattle Genetics) in 2011 for the treatment of several types of lymphomas,[61] and ado-trastuzumab emtansine (Kadcyla®, Genentech) in 2013 for the treatment of human epidermal growth factor receptor 2 (HER2) positive breast cancer,[62] as well as several other relatively recent approvals as shown in Table 13.1.

To overcome the aforementioned problems that are commonly observed in many ADCs, design optimization is required for component selection including the linker, payload, and conjugation chemistry and careful selection of the antigen target. The advent of new technologies and promising clinical trial outcomes so far are expected to boost a new generation of ADCs with enhanced clinical profiles in the near future.

13.4.1 Antigen Targeting, ADC Uptake, and the Linker

Many cancers overexpress certain surface proteins that can be used as biomarkers for both diagnosis and treatment; antibodies used for the development of ADCs are selected based on their abilities to recognize these tumor-specific marker antigens. High specificity of the antibody to the target is important in order to prevent side effects that are associated with off-target cytotoxicity so that the drug is safe to use.[63] Another important factor is the internalization of the drug efficiently after antibody–antigen complex formation. Many parameters can influence uptake including the antigen concentration level on the cell surface, which was shown to be critical for efficient uptake of the ADC.[64,65] Current ADC therapies utilize well known and common overexpressed cell surface receptors such as CD22 and HER2, among others (Table 13.1). After binding to the antigen, the ADC–antigen complex goes through a complicated endocytosis process and interacts with the Fc receptor.[66,67] Subsequently, the payload is released into the cytoplasm by: (i) hydrolysis, (ii) proteolysis during lysosomal trafficking, or (iii) reduction triggered in cytoplasm; depending on the type of the linker, enabling the drug to exert its cytotoxic effect.[53,68–71]

Various different types of linkers can be utilized to connect the mAb and payload. Some early linkers took advantage of the low pH environment of the cancerous cellular vicinity to release the payload, but this modality has very low serum stability. Currently, most commonly used linkers are AcBut and valine–citrulline, but there are many other linkers available and some marketed ADCs use other types (Table 13.1). The AcBut linker is a carbonyl-containing carboxylic acid, 4-(4-acetylphenoxy) butanoic acid, while the valine–citrulline is probably the most widely used protease cleavable dipeptide linker that is cleaved by cathepsin B, and some of the marketed products as well as a few ADCs that have been tested in clinical trials employ this linker.[72,73] Linker stability is very crucial, not only to avoid undesired off-target toxicity and thus efficient delivery of the payload to the tumor cells,[53,70] but also to have appropriate solubility and pharmacokinetics of the ADC. Disulfide linkers are alternative protease-susceptible linkers and can also be reduced to achieve enhanced ADC serum stability.

Other types of linkers such as non-cleavable moieties are also gaining traction due to their enhanced serum stabilities. In this system, the payload is released only after the antibody is internalized and digested inside the lysosome.[71,74] Some ADCs, including the popular Kadcyla®, are constructed with a non-reducible thioether linker (Table 13.1).[75,76] Future ADCs will probably use the last two types of linkers (protease cleavable and non-cleavable) due to their better serum stability compared to that of the very early linkers.

13.4.2 Conjugation of the Payload to the Antibody

One of the earliest conjugation methods was by employing existing amino acids, particularly partially reduced Cys and Lys residues.[77] This method is relatively straightforward and easy to implement; however, it creates a heterogenous ADC population because the number of conjugated amino acids is not controllable, i.e. one mAb could have between 0 and 8 conjugated Cys[78] or 0–8 conjugated Lys amino acids (even though there are about 80 Lys on a typical IgG1).[79–83] In addition, generally there are no free surface Cys residues on mAbs, although exceptions exist; and in order to achieve freely available Cys residues for payload conjugation, one or more of the four existing disulphide bonds have to be broken via a reducing agent such as dithiothreitol (DTT) or tris (2-carboxyethyl)phosphine (TCEP). This may result in conformational changes and negatively affect the structural integrity of the protein, leading to instabilities and aggregation. Nevertheless, for cancer biotherapeutics, some heterogeneity might be considered beneficial to tackle an already heterogenous cancer microenvironment. Still, if the drug-to-mAb ratio is too low, inadequate efficacy may be observed, while a high ratio might be too toxic. Hence, for optimal efficacy, potency, and safety, a homogenous ADC population with an appropriate amount of payload would be desirable in order to exert significant cytotoxicity to the cancer cells.[84,85] Despite such potential issues, conjugation of both Cys and Lys residues are widely used and marketed. Adcetris® and Kadcyla® were developed with these techniques, respectively,[62,86] and many products in clinical trials also employ these conjugation methods.

Site-specific conjugation is also possible for ADCs by point mutation of a surface amino acid to introduce, for example, a free Cys[87] or non-natural amino acid on the protein surface. This method would ensure both the number of payloads and their position and thus a homogenous preparation that might improve the therapeutic index of new ADCs.[81,88] An ADC prepared with this method could however be intrinsically aggregation-prone due to free surface Cys or dimer mAbs can be formed because of disulphide bonds.[89]

13.4.3 Payloads

The ADC format of drug delivery has paved the way to the delivery of payloads that were previously deemed to too toxic to be used as chemotherapy agents. Some of these agents include calicheamicins, maytansinoids, and auristatins as well as less commonly used compounds such as duocarymycins and pyrrolobenzodiazepine. For example, the first ADCs, Mylotarg® and inotuzumab ozogamicin, use a calicheamicin derivative (*N*-acetyl-γ calicheamicin 1,2-dimethyl hydrazine dichloride). Another example is DM, used as the payload for Kadcyla® and developed as an analog of maytansine, which was determined to be too toxic and could not be successfully developed as a chemotherapy compound in the 1970s. Many ADCs that are in clinical trials use both of these payloads as well as others for clinical development. Another widely employed class of payload is monomethyl auristatin E (MMAE); both Adcetris® and Polivy® use MMAE in addition to other ADCs in clinical trials. It is expected that several novel payloads will soon enter clinical tests.

13.4.4 Antibody Format

All marketed and developing ADCs are based on the IgG antibody as the targeting moiety. More specifically, most of the approved ADCs employ humanized IgG1 antibodies with the exception of Adcetris® and Lumoxiti™, which have a chimeric anti-CD30 and mouse anti-CD22 mAbs, respectively (Table 13.1). The majority of ADCs that are in clinical trials also avoid nonhuman mAbs and make use of either human or humanized antibodies because of the potential immunogenicity of chimeric or mouse antibodies. In addition, generating fully human or humanized antibodies is no more an onerous task than generating chimeric or mouse antibodies with currently available methods such as phage display or hybridoma platform technologies. Like other antibody-based therapeutics, IgG1 is the most common mAb format in the development of ADCs due not only to its higher structural stability and lower aggregation propensity compared to other types of antibodies but also probably because of its superior binding affinity toward Fcγ receptors to elicit a potential immune effector function.[90,91] However, other studies showed that the payload, pharmacokinetics, and ADC's uptake by the cancer tissue may contribute more than the cell response induced by the Fcγ immune effector.[92,93] Yet, some indications might display a better outcome if the mAb lacks effector cell response, in which case IgG2 or IgG4 can be used instead of IgG1.[94,95] Besponsa® is an example of such a modality; it is an anti-CD22 ADC

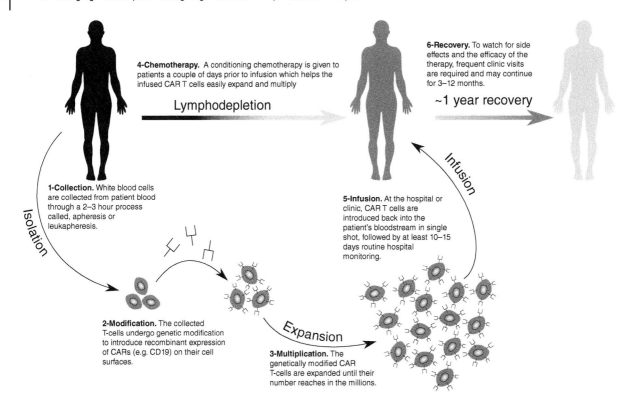

4-Chemotherapy. A conditioning chemotherapy is given to patients a couple of days prior to infusion which helps the infused CAR T cells easily expand and multiply

Lymphodepletion

6-Recovery. To watch for side effects and the efficacy of the therapy, frequent clinic visits are required and may continue for 3–12 months.

~1 year recovery

1-Collection. White blood cells are collected from patient blood through a 2–3 hour process called, apheresis or leukapheresis.

Isolation

5-Infusion. At the hospital or clinic, CAR T cells are introduced back into the patient's bloodstream in single shot, followed by at least 10–15 days routine hospital monitoring.

Infusion

2-Modification. The collected T-cells undergo genetic modification to introduce recombinant expression of CARs (e.g. CD19) on their cell surfaces.

Expansion

3-Multiplication. The genetically modified CAR T-cells are expanded until their number reaches in the millions.

Figure 13.1 CAR T-cell therapy treatment.

with a humanized IgG4 subtype mAb that was developed for the treatment of CD22$^+$ acute lymphoblastic leukemia (ALL) (Table 13.1).

Other forms of ADCs with antibody fragments such as svFv, single-domain antibodies, nanobodies, etc., are yet to be seen. However, smaller sizes of these formats might be beneficial for adequate payload concentration in the tumor vicinity,[96,97] better tissue penetration,[98] and even blood–brain barrier penetration. At the same time, however, the lack of the Fc domain means such smaller formats would undergo rapid clearance from the body, with a shorter serum half-life.

13.5 CAR T-cell Therapy

Chimeric antigen receptor-engineered T-cells have been highly promising in employing T-lymphocyte stimulation against blood malignancies and are perhaps the closest equivalent to "living drugs." Based on previously characterized high-affinity T-cell receptors (TCRs), which specifically recognize tumor-associated or tumor-specific antigens, the T-cells of a patient are modified using γ-retroviral or lentiviral delivery systems. During this process, T-lymphocytes acquire a new recognition receptor for cancer-specific targets. This special receptor

is known as the chimeric antigen receptor (CAR), with the eponymous CAR T-cell therapy referring to the scaling up of T-cells carrying specific CARs in a laboratory environment for adoptive T-cell immunotherapy (Figure 13.1). Despite being an incredibly recent development, CAR T-cell strategies have already been approved for clinical use and regulatory studies (for safety and efficacy evaluations) in the United States,[99,100] Europe, Australia, Japan, and Canada.

The concept of adoptive transfer of engineered T-lymphocytes against leukemia first appeared in the 1990s, and second-generation CARs have been generated to augment antitumor efficacy.[101] The US Food and Drug Administration (FDA) approved CAR T-cell therapy for (i) CTL019 or tisagenlecleucel (Kymriah™) to treat ALL in young patients aged up to 25 years or, in adults, for relapsed or refractory diffuse large B-cell lymphoma (DLBCL), and (ii) axicabtagene ciloleucel (Axi-cel), also known as KTE-C19 (Yescarta™), to treat large, mediastinal or high-grade B-cell lymphomas.

Aggressive chemotherapy treatment and stem cell transplantation together have provided an over 80% cure rate for children diagnosed with ALL. However, possible options in the case of cancer-return (either relapsed or refractory cases), even after chemo- and stem-cell therapies, are almost nonexistent, and relapsed and refractory cases are major contributors of childhood ALL death. In the initial trials of CAR T-cells in young patients and adults with relapsed ALL, 90% remission was reported (27/30 patients),[102] leading to approval in 2017 by the FDA. Basically, CAR T-cell therapy opened a new avenue for treatment of blood malignancies, especially for patients with aggressive lymphomas who have previously been considered essentially untreatable.

13.5.1 Treatment Steps

The making of a "living drug" such as CAR T-cell therapy is a multistep process and considerably expensive; therefore, the key elements of CAR T-cell therapy include: (i) a confirmed diagnosis of either DLBCL, primary mediastinal large B cell lymphoma, high-grade B-cell lymphoma, or transformed follicular lymphoma to DLBCL (all of which have not responded to the first and second lines of chemotherapeutic intervention or have relapsed after stem cell transplantation); and (ii) adequate organ function of the patient, including healthy cardiac and pulmonary systems.

CAR T-cell treatment is a multiphase process (Figure 13.2). Following the collection of a patient's T-cells from the blood, CARs are recombinantly introduced. To sharpen CAR T-cell strength, unmodified leukocytes are first depleted prior to their infusion (Figure 13.2).

13.5.2 Structure of CARs

The overall structure of CARs contains three regions: (i) an ectodomain that recognizes the malignant-specific molecule (e.g. CD19), (ii) a single helical transmembrane region, and (iii) a cytoplasmic intracellular tail that possesses immunoreceptor tyrosine-based activation motifs (ITAM) through which T-cell activation signaling is achieved upon ligand- or antigen-binding to CARs (Figure 13.2). Since ITAMs are activation gatekeepers, CAR-cytoplasmic tails containing different ITAMs have been tested and included in the design of CARs, leading to four major CAR definitions. The first-generation CARs consist of CD3ζ or FcRγ (without a co-stimulatory signal) and showed limited antitumor effects due to their sole dependence on interleukin-2 (IL-2) synthesis.[103] The second-generation CARs, which are currently used in clinics, include an additional co-stimulatory cytoplasmic receptor such as CD28 or 4-1BB, the addition of which stimulate *in vivo* T-cell proliferation and enhanced response (Figure 13.2).[104–106] The combination of multiple cytoplasmic co-stimulatory molecules, giving rise to third-generation CARs, have further augmented antitumor potency (Figure 13.2).[105,107] The fourth-generation CARs are equipped with a cytokine-expression stimulation domain, such as IL-12, as denoted by T-cells redirected for universal cytokine-mediated killings (Figure 13.2) (TRUCKs).[105] Newer designs have also quickly emerged

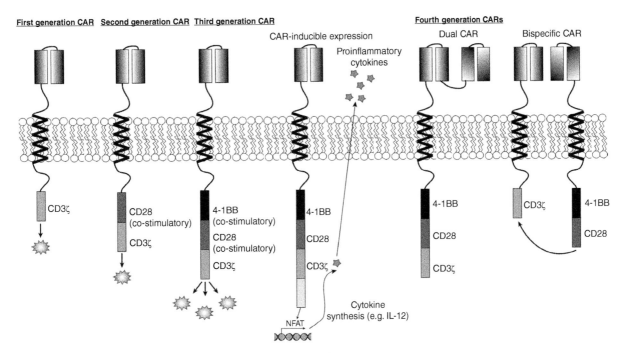

Figure 13.2 Cartoon illustration of the evolution of CAR-T therapies.

in the last two years, including dual or bispecific CAR molecules that have two unique single chain variable fragments (scFv) for two different tumor-specific antigens (Figure 13.2).

13.5.3 CAR T-Cell Clinical Uses and Trials in Blood Malignancies

CD19 is a cell surface molecule that is widely and ubiquitously expressed in all developing and mature B-cells, as well as in neoplastic B-cells, but are notably absent in plasma cells. Thus, targeting the malignant B-cell lineage via anti-CD19 CARs by both tisagenlecleucel and axicabtagene ciloleucel received FDA approval in 2017. CARs of tisagenlecleucel and axicabtagene ciloleucel contain CD3ζ and 4-1BB (CD137) or CD28 co-stimulatory domains, respectively, and are transfected using γ-retroviral or lentiviral vectors. Currently, lisocabtagene maraleucel, a CAR with a scFv anti-CD19, CD3ζ, and 4-1BB, and a truncated form of the human EGFR are in the late phase of clinical trial (NCT02631044).[108] On the other hand, the post-CAR T-cell therapy relapse patients have limited treatment options, mostly due to the short survival times of these patients. Therefore, third- and fourth-generation CARs, in addition to defining new adjunct treatments, are critical for creating new approaches for relapsed patients. One emerging strategy is to design bidentate receptor complexes that bind to a second target (e.g. CD20) together with CD19. Secondly, "armored" CAR T-cell lineages have been genetically engineered to secrete proinflammatory cytokines, such as IL-12 or IL-18, and constitutively express tumor

necrosis factor (TNF) superfamily molecules such as CD40L (CD154).[109,110] Additionally, the expression of immune checkpoint receptors (IRs) has been targeted, the loss of programmed cell death protein 1 (PD-1) improved *in vitro* proliferation and cytotoxicity of CAR T-cells.[111] IR antagonists such as durvalumabor, atezolizumab, and nivolumab are also being explored jointly with CD19-targeted CAR T-cell therapies[112–114] and are being tested in clinical trials (NCT02926833, NCT04134325, and NCT04003649).

CD19 expression is halted in plasma B-cells; therefore, malignant plasma cells, a pathological cause of multiple myeloma, cannot be targeted with anti-CD19 CAR T-cells. Thus, current efforts are solely focused on searching and identifying highly expressing myeloma-specific molecules. The primary outstanding antigen is B-cell maturation antigen (BCMA or BCM), member 17 of the TNF receptor superfamily, which is preferentially expressed in mature B-cells and plasma cells. Numerous phase 1 and phase 2 studies using CARs of BCMA, either alone, with several other CARs (e.g. CD19, CD38, CD56, CD138), or in combination with drugs (e.g. GSK3174998, REGN5458), are currently being assessed for relapsed and refractory multiple myeloma cases. The roles of different co-stimulatory signaling tails (e.g. CD3ζ and 4-1BB) and anti-EGFR therapy in BMCA CAR T-cell response have also been evaluated.[115] Although other alternatives to BCMA from cell adhesion molecules and receptors, such as CD44, NCAM (CD56), syndecan-1 (CD138), and integrin β7-receptors (α4β7 or αEβ7 heterodimers), have been investigated, these studies have still yet to mature.[115–118]

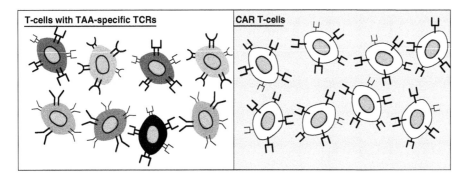

Figure 13.3 Comparison of MultiTAA and CAR T-cell therapy.

Furthermore, over 10 CAR T-cell phase trials are recruiting AML patients to test the CAR receptors engineered with either Siglec-3 (CD33) or IL3-receptor (CD123) or both. Other alternative antigen targets for designing CARs for AML are LewisY, FLT3 (CD135), Cll1, CD44-variant6, folate receptor β (FR β), CD38, and CD7,[119] but these are all in preclinical stages. AML remission and the successful response rate is high (around 70% for young and 50% for adult patients), but AML patients, especially those with a five-year survival rates less than 30%, represent a significant unmet health need.[120] Since expression of the aforementioned antigens are not exclusively restricted to myeloid cells but also shared with healthy hematopoietic stem cells (HSCs), severe hematotoxicity appears to be a major challenge in AML-directed CAR T-cell therapies. In short, new antigens expressed in AML, which deliver little to no toxicity when coupled to T-cells, are needed to further enrich the current armamentarium.

13.5.4 An Alternative Approach to CAR T-Cell Therapy

Rather than engineering CARs, another newly emerging approach is to select an ensemble of the patient's own T-cells that naturally have the TCRs against specific tumor-associated antigens (TAA). In the presence of cytokines, a patient's T-cell pool is exposed to peptides representing the blend of multiple TAAs. Then, the positively reacting or immune-response eliciting T-cell ensemble will be amplified in large quantities without any genetic modification (Figure 13.3). Because selected T-cells recognize a diverse set of TAAs, multiple types of tumors could be easily targeted. Also, since no *in vitro* CAR engineering is involved, the risk of exhausted T-cell response or rapid T-cell aging, and subsequent death is significantly mitigated. Toxicity in comparison to CAR T-cell therapy is also significantly less due to fewer numbers of the specific T-cells being infused back into the patients. Notably, the two-week hospitalization recovery period is not required and patients could immediately leave after treatment. As of late October 2019, MultiTAA-based clinical trials have been recruiting patients for four different phase 1 and phase 2 studies (Leukemia [NCT02475707], AML [NCT02494167], pancreatic cancer [NCT03192462], Hodgkin and non-Hodgkin lymphoma [NCT01333046], and one phase 2 breast cancer [NCT03093350] trial).

13.5.5 Nonviral Engineering of CAR T-Cells

The engineering process of T-cells with CARs is highly time-consuming, requiring up to four weeks; they are expensive and require a specialized facility. Additionally, both tisagenlecleucel and axicabtagene ciloleucel require a viral delivery of CARs and, thus, maintaining the integrity of CAR T-cells thereby poses another challenge. Unfortunately, only a limited number of patients have access to these therapies, due either to the financial burden or remote location of the patient. The price tags on tisagenlecleucel and axicabtagene ciloleucel treatments are exorbitant, around half a million dollars per patient in the United States.[121] Safety concerns for γ-retroviral or lentiviral modifications revolve around the potential for insertional mutagenesis and genotoxicity of these virus-dependent systems.[122]

Recently, small DNA vectors known as nonviral Sleeping Beauties, with minimal sets of genes that facilitate CARs expression in T-cells, were discovered to confer CD19–CAR T-cell potency.[123] Transposition of CAR constructs via electroporation provides a less costly therapeutic method than viral engineering of T-cells and reduces the amount of time needed to manufacture these virally engineered T-cells from four weeks to one week. Additionally, co-expression of the interleukin-15 (IL-15) cytokine as a membrane-bound molecule provides a greater T-cell proliferation signal without risk of T-cell exhaustion. The advantage of membrane-bound IL-15 also reduces the required number of CAR T-cells in the infusion step. Normally, unmodified T-cells in the patient's blood are deleted via lymphodepletion to avoid their competition for the soluble IL-15 prior to infusion. Lymphodepletion is time-consuming, expensive, and increases the toxicity of CAR T-cells. The nonviral TCR gene therapy, using the administration of autologous T-cells engineered with Sleeping Beauty in metastatic cancer patients, is in phase 2 clinical trial (NCT04102436).

The second nonviral approach using linear DNA without incorporation into the genome is only recently emerging and has promise of reducing the number of safety concerns to an almost nonexistent level. PCR-amplified linear DNA with protection tags at each end is transfected to T-cells, and transient expression of the nonviral, plasmid-free (NVPF) CARs provides a disruptive, inexpensive, and safer alternative. Despite its infancy, linear DNA CART19-based CAR-T clinical trials in China showed 100% remission of ALL patients (three out of three), and further US/Europe clinical trials are expected in the immediate future.

13.5.6 Is It Possible to Mass-Produce and Automate CAR T-Cells?

Engineering of CARs revolves around the genetic modification of the patient's own T-cells and represents a "true personalization" of medicine, the only disembodied modification is the CAR introduced during the process. Nonetheless, biopharmaceutical companies are more inclined to produce therapeutics in large quantities. Given the difficulty of finding an immediate HLA-match from bone-marrow donors, the current state of allogenic T-cell therapy has many challenges.

Manufacturing CAR T-cells for each patient, quality control, and reducing product variability, all in parallel, presents hurdles in current CAR T-cell production. It is also critical to reduce human error in engineering CARs as possible contamination during the manufacturing process poses a real risk. A large laboratory with controlled hygienic rooms and skilled personnel could deliver needed CAR T-cells for roughly hundreds of patients during clinical trials, the stage at which this technology stands today. However, if and when CAR T-cells are approved and move to clinical use, excessive demand in clinics simply would not be met by current production strategies. Indeed, what is needed is similar to the automated manufacturing of human mesenchymal stem and stromal cells (hMSCs) for a number of human pathologies including arthritis, diabetes, several malignancies, liver diseases, etc.[124] Small and closed automated systems for engineering CAR T-cells have recently emerged which control key parameters (e.g. temperature, cytokine concentrations), use defined media (e.g. serum-based uncontrolled media is avoided), and simply reduce batch-to-batch variations, despite being used for research purposes currently.[125] However, automation and even mass-production will require further biotechnological advancements.

13.5.7 Alternatives to T-Cells

Multiple lines of research and ongoing clinical trials aim to introduce further enhancement to CAR T-cell therapy, which include adjustments to dominant negative receptors, cytokines, and chemokine receptors; aforementioned checkpoint inhibitors; and expression of cytokines. To overcome a cancer-cell secreted suppressive cytokine like TGFβ, which induces metastasis, expression of the mutated TGFβ receptors on T-cells provides TGFβ resistance to its antiproliferative and anti-cytolytic effects; in brief, these mutated receptors promote antitumor immune functions.[126,127] Moreover, the negative or regulatory cytokine receptors pose other challenges in tuning the efficacy and persistence of the CAR T-cells. Chimeric receptors that combine the immunosuppressive extracellular domain and immune-activating cytoplasmic domains of two different cytokine receptors (e.g. IL-4 and IL-2, respectively) are

co-expressed with CARs, and have been shown to reduce T-cell exhaustion while also increasing the killing of malignant cells.[128] In parallel to tuning cytokine receptors, CAR T-cells that can oversecrete stimulatory cytokines, as seen in IL-15-modified CAR T-cells, have shown enhanced survival and improved *in vivo* antitumor efficacy,[129] all of which have led to a promising current phase 1 study (NCT03774654).

Other immune cell types apart from T-cells are also modified with CARs for research and clinical immunotherapy approaches. Gamma-Delta (γδ) T-cells, derived from lymphoid precursors, come in two major types: Vδ1, which is mostly found in mucosal surfaces and Vδ2 which circulates in the vascular system.[130] In preclinical studies, Vδ2 cells exhibited cytotoxic properties for malignancies ranging from lung carcinoma to breast cancer.[131,132] Since modification of the genetic engineered γδ T-cells using CARs attenuated neither tumor infiltration capacity nor cytotoxicity,[133] autologous γδ T-cells have been at the interface of both preclinical and clinical studies, with a new phase 1 study that will investigate the safety and tolerability of the allogeneic NKG2DL-targeting CAR-γδ T-cells in solid tumors (NCT04107142).

Another immune cell type that has a strong cytotoxic function, recognizing and killing their targets in an antigen-independent manner through apoptosis[134,135] is the NK cell. Alloreactivity of NK cells purified from the umbilical cord blood, in comparison to peripheral blood, is much lower (minimizing the risk of graft vs. host disease), which makes NK cells appealing for research and clinical immunotherapy studies. NK cells modified with CARs including iC9, anti-CD19, and IL-15, and constructs that include the anti-CD19 receptor and ectopically express IL-15 to directly sustain NK cell survival, proliferation,[129] and presentation of the iCaspase-9 suicide-inducing receptor[136] were shown to effectively kill malignant B-cells[137] and phase 2 studies are currently testing CAR NK cells (NCT03579927 and NCT03056339). Additionally, adjunct developments to increase the efficacy of NK therapies have been extensively ongoing; bi- or tri-specific killer receptor (KIR)-engaging molecules[138–140] and immunomodulatory molecules[141,142] are proven to induce KIR-dependent cytotoxicity.

The third cell type that is of particular interest for CAR engineering includes natural killer T-cells (NKT) cells. Antigen recognition of NKT cells is restricted to the antigen presentation by CD1d (an MHC-I-like molecule); therefore, CAR therapies theoretically have less off-target toxicity and increased applicability. Additionally, NKT cells, despite being limited in number, secrete a variety of cytokines and thus play critical roles in regulating the immune response of leukocytes as well. In fact, NKTs are a crucial link between innate and adaptive immunity and can be activated by antigenic and antigen-independent stimuli. For example, NKT cells can stimulate dendritic cells, cytotoxic CD8[+] T cells, and NK cells,[143–145] which could provide ancillary potent antitumor immunotherapy. CAR NKT cells are highly proliferative and have potent cytotoxic activity in tumor environments, combination of which mediates superb antitumor efficacy when used with other leukocytes.[146] Currently, two CAR NKT therapies have just recently started phase 1 studies for B-cell malignancies and neuroblastoma (NCT03774654 and NCT03294954) and more than 10 studies are about to start in the immediate future.

13.6 Immune Checkpoint Antagonism

Regulating immune response, which includes the maintenance of hemostasis and activation and termination of the response, occurs through immune checkpoint mechanisms, a process that is vital for host survival (Figure 13.4). Irregularities in this process lead to inflammation-directed tissue damage and autoimmune pathologies (Figure 13.4). T-cells are major effector adaptor cells that utilize a number of IRs, including cytotoxic T-lymphocyte-associated protein 4 (CTLA-4) and PD-1 (Figure 13.4). Normally, these receptors prevent the host

immune cells from attacking the host cells, a mechanism called tolerance. Several categories of cancer malignancies hijack this mechanism and antagonize immune cell attack by engaging the immune cell receptors on immune cell surfaces. Therefore, targeting of the inhibitory checkpoint molecules has recently been used in cancer immunotherapy in multiple cancer types. For their discoveries on how to harness the body's immunity to combat cancer, the Nobel Prize for Medicine was awarded to Tasaku Honjo and James Allison in 2018.

Given that tumors can evade immune stress by activating immune checkpoint mechanisms, the antitumor potential of antagonizing the suppressive roles of key molecules in immune checkpoints have proven to be highly successful in preclinical remissions in murine models.[147,148] CTLA-4 and then PD-1 and its ligand, PD-L1, were thus immediately targeted in clinical studies. Numerous clinical trials are testing immune checkpoint therapies, trials solely focusing on PD-1/PD-L1 already number in the thousands in 2018 and 2019 (ClinicalTrials.gov was searched for "PD-1" OR "PD-L1").

13.6.1 CTLA-4 Inhibitors

CTLA-4 plays a critical role in central tolerance and control, a mechanism that protects individuals against autoimmune pathologies, by removing autoreactive T-cells or activating suppressive T-regulatory (Treg) cells. The first CTLA-4 drug developed was an IgG$_1$ fully humanized antibody, ipilimumab, which recognized CTLA-4 and antagonized its ligand-binding[149] and enhanced the

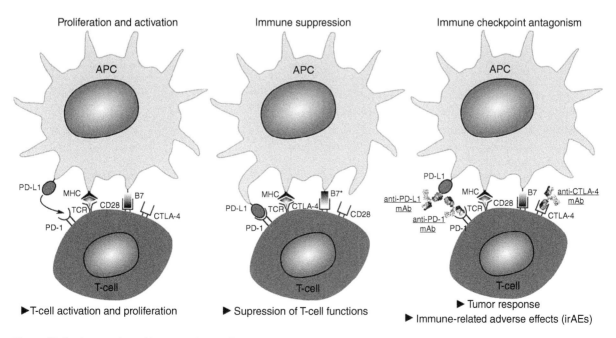

Figure 13.4 Antagonism of immune checkpoint receptors.

Table 13.2 Current immune checkpoint inhibitors in clinical use.

Antibody	Target	Clinical indication	Year of approval
Nivolumab	PD-1	• Melanoma	2014
Pembrolizumab	PD-1	• Melanoma • Non-small lung cancer • Head and neck squamous cell carcinoma	2014
Cemipliman	PD-1	• Cutaneous squamous cell carcinoma (CSCC)	2018
Atezolizumab	PD-L1	• Urothelial carcinoma • Non-small cell lung cancer	2016
Avelumab	PD-L1	• Merkel-cell carcinoma (MCC)	2017
Durvalumab	PD-L1	• Urothelial carcinoma • Non-small cell lung cancer	2017

survival of melanoma patients.[150,151] Currently, ipilimumab and its effective combination with other strategies such as the aforementioned CAR T-cell therapy have been evaluated in hundreds of clinical settings. The second antibody, tremelimumab, a complete human IgG$_2$ antibody, is an investigational antagonist of the interaction of CTLA-4 and its B7 molecule (CD80 and CD86) ligands. Tremelimumab has not achieved clinical approval alone, but its combination with other checkpoint inhibitors and CAR T-cell therapies are being extensively tested in numerous clinical trials.

13.6.2 PD-1/PD-L1 Inhibitors

In contrast to CTLA-4, the PD-1/PD-L1 interaction disables the functions of autoreactive T-cells after their departure from the thymus, a process known as peripheral tolerance. This mechanism simply eliminates the self-destruction of the body's own tissues by either converting the "incorrectly selected" T-cells into anergic or Treg cells, or by inducing T-cell death via a programmed cell death mechanism called apoptosis. PD-1, as reflected by its name, is critical in maintaining this peripheral tolerance mechanism. Currently, six different antibodies, three of which bind to PD-1 with the other three binding to PD-L1, have been approved for a number of malignancies (Table 13.2). PD-1/PD-L1 antagonists reported higher response rates in comparison to CTLA-4 inhibitors.[152] With the seven major pharmaceutical companies involved in the fray in developing more efficacious regimens of an antagonist or their combination thereof, extensive clinical studies of PD-1/PD-L1 inhibitors are assessing patient responses for a variety of classes of malignancies.

13.6.3 Adverse Effects of Immune Checkpoint Inhibitors

Immune checkpoint blockade enhances the level of activity of the immune system and therefore, therapy-induced inflammatory effects (Figure 13.4), called immune-related adverse events (irAEs), are naturally expected. Roughly, 1% of these irAEs are fatal.[153] irAEs or common side effects of immune checkpoint inhibition range from mild symptoms, including fatigue, appetite loss, skin discoloration, rash, and itching, to more serious problems in the dysfunction of the lungs, kidney, hormonal glands, and intestine.[154] Currently, there are not any well-established treatments for irAEs, but the common practice is to reduce dosage, delay the administration of immune checkpoint antagonist(s), or use immunosuppressive drugs such as glucocorticoids.[155] Given that clinical experience with these drugs has been accumulating, a multidisciplinary approach to assess and realize the full potential of immune checkpoint inhibitors should provide deeper insight into their applications in various malignancies.

13.7 Conclusions and Future Outlook

Biologics, especially immune therapies, are beginning to gain significant traction in recent years mainly due to being a game changer in the treatment of various pathologies, particularly in the treatment of several cancer types that previously had very low prognosis such as melanoma. New approaches such as CAR T-cells, immune checkpoint antagonists (and even their combination), and many different novel antibody-based constructs like ADCs go beyond what standard chemotherapy can achieve or offer. Such new approaches are progressing quite strongly and causing a paradigm shift. There are thousands of new constructs in clinical trials and probably thousands more at various stages of development that are expected to go into clinical trials in the near future. Having said that, some of these emerging new therapies are working very well in a small set of patients, e.g. only 10–20% of melanoma patients are cured with immune checkpoint therapy, while most other patients are not responsive at all to this new treatment. In addition, although some combination therapies are currently in use, we expect to see more of these in the near future in order to increase the response level observed with these new therapies. For instance, immune checkpoints along with chemotherapy or gut microbiota screening will be highly beneficial for some patients once we understand their relationships with the immune system further. Recent regulatory responses to such combination therapies is quite promising; the FDA recently granted accelerated approval for an ADC called polatuzumab vedotin-piiq (Polivy™) to be used in combination with bendamustine (a chemotherapy agent) and rituximab (a full mAb) for the treatment of a type of B-cell lymphoma for adult patients.[156] Such accelerated approvals in the United States for some biologics will probably be followed by other regulatory agencies around the globe, in fact, this is a must for many diseases that have no alternative treatment options other than simple palliative support. Some experimental treatments, such as new anti-Ebola virus

antibodies, first showed a high potential in Ebola treatment in a small set of infected patients and is now undergoing clinical trials to be used as a preventative treatment. Other treatments and new vaccines against different infectious diseases such as Zika virus are also being exhaustively studied and should be available clinically in the near future. Nevertheless, the lack of any viable treatment or vaccine for numerous potentially deadly diseases that are on the World Health Organization (WHO)'s Blueprint list of priority diseases,[157] including Zika virus, Middle East respiratory syndrome coronavirus (MERS), and Crimean–Congo hemorrhagic virus, as well as others that are not on the Blueprint list but are nonetheless widespread, such as malaria, influenza, HIV, and *C. Difficile*, *are* exceptionally concerning and there is an urgent need for drugs and vaccines for these infections.

Moreover, some of the new treatments mentioned earlier are already truly personalized, such as the case of CAR T-cell therapy but similar personalized medicines will be relatively common in the near future as new targeted therapies emerge.

In conclusion, the landscape for biologics in the recent decade is rapidly changing with the emergence of new therapeutics which are less toxic with fewer side effects, are highly efficacious, and importantly are far more effective. The future of biologics appears highly promising, considering their versatility. The number of variations and formats that may be designed are limitless, and they can also be combined with small molecules, other recombinant proteins and nanoparticles for the prevention, diagnosis, and treatment of many different types of diseases.

References

1 Company TBR (2018). Biologics global market opportunities and strategies to 2021 (July). https://www.reportlinker.com/p05482361/Biologics-Global-Market-Opportunities-And-Strategies-To.html (accessed 16 June 2020). Contract No.: ID: 5482361.

2 Global market for bioengineered protein drugs showing modest growth 2017. https://www.bccresearch.com/pressroom/bio/global-market-for-bioengineered-protein-drugs-showing-modest-growth (accessed 16 June 2020).

3 Ecker, D.M., Jones, S.D., and Levine, H.L. (2015). The therapeutic monoclonal antibody market. *MAbs* 7 (1): 9–14.

4 Elgundi, Z., Reslan, M., Cruz, E. et al. (2017). The state-of-play and future of antibody therapeutics. *Advanced Drug Delivery Reviews* 122 (Suppl. C): 2–19.

5 Antibody Society (2019). Antibody therapeutics approved or in regulatory review in the EU or US. https://www.antibodysociety.org/resources/approved-antibodies (accessed 16 June 2020).

6 Ahmadi, M., Bryson, C.J., Cloake, E.A. et al. (2015). Small amounts of sub-visible aggregates enhance the immunogenic potential of monoclonal antibody therapeutics. *Pharmaceutical Research* 32 (4): 1383–1394.

7 Bessa, J., Boeckle, S., Beck, H. et al. (2015). The immunogenicity of antibody aggregates in a novel transgenic mouse model. *Pharmaceutical Research* 32 (7): 2344–2359.

8 The Pharmaceutical Research and Manufacturers of America. Medicines in development for cancer – 2018 Report 2018. http://phrma-docs.phrma.org/files/dmfile/2018_MID_Cancer.pdf (accessed 16 June 2020)

9 Dedrick, R.M., Guerrero-Bustamante, C.A., Garlena, R.A. et al. (2019). Engineered bacteriophages for treatment of a patient with a disseminated drug-resistant *Mycobacterium abscessus*. *Nature Medicine* 25 (5): 730–733.

10 Furfaro, L.L., Payne, M.S., and Chang, B.J. (2018). Bacteriophage therapy: clinical trials and regulatory hurdles. *Frontiers in Cellular and Infection Microbiology* 8: 376.

11 Schmidt, C. (2019). Phage therapy's latest makeover. *Nature Biotechnology* 37 (6): 581–586.

12 Voelker, R. (2019). FDA approves bacteriophage trial FDA approves bacteriophage trial news from the food and drug administration. *JAMA* 321 (7): 638.

13 Mitropoulos, P. and Norman, R.A. (2010). Biologics. In: *Preventive Dermatology* (ed. R. Norman), 93–101. London: Springer.

14 Chennamsetty, N., Voynov, V., Kayser, V. et al. (2009). Design of therapeutic proteins with enhanced stability. *Proceedings of the National Academy of Sciences* 106 (29): 11937–11942.

15 Chennamsetty, N., Voynov, V., Kayser, V. et al. (2010). Prediction of aggregation prone regions of therapeutic proteins. *The Journal of Physical Chemistry B* 114 (19): 6614–6624.

16 Kuyucak, S. and Kayser, V. (2017). Biobetters from an integrated computational/experimental approach. *Computational and Structural Biotechnology Journal* 15: 138–145.

17 Agrawal, N.J., Kumar, S., Wang, X. et al. (2011). Aggregation in protein-based biotherapeutics: computational studies and tools to identify aggregation-prone regions. *Journal of Pharmaceutical Sciences* 100 (12): 5081–5095.

18 Zambrano, R., Jamroz, M., Szczasiuk, A. et al. (2015). AGGRESCAN3D (A3D): server for prediction of aggregation properties of protein structures. *Nucleic Acids Research* 43 (W1): W306–W313.

19 Jefferis, R. (2014). Glycosylation of antibody molecules. In: *Handbook of Therapeutic Antibodies* (ed. J.M. Reichert), 171–200. Wiley-VCH Verlag GmbH & Co. KGaA.

20 Wang, L.-X., Tong, X., Li, C. et al. (2019). Glycoengineering of antibodies for modulating functions. *Annual Review of Biochemistry* 88 (1): 433–459.

21 Nakamura, H., Oda-Ueda, N., Ueda, T., and Ohkuri, T. (2018). Introduction of a glycosylation site in the constant region decreases the aggregation of adalimumab Fab. *Biochemical and Biophysical Research Communications* 503 (2): 752–756.

22 Reslan, M., Sifniotis, V., Cruz, E. et al. (2020). Enhancing the stability of adalimumab by engineering additional glycosylation motifs. *International Journal of Biological Macromolecules* 158: 189–196. https://doi.org/10.1016/j.ijbiomac.2020.04.147.

23 Yu, X., Marshall, M.J.E., Cragg, M.S., and Crispin, M. (2017). Improving antibody-based cancer therapeutics through glycan engineering. *BioDrugs* 31 (3): 151–166.

24 Press Announcement. FDA approves treatment for two rare types of non-Hodgkin lymphoma 2018. https://www.fda.gov/news-events/press-announcements/fda-approves-treatment-two-rare-types-non-hodgkin-lymphoma (accessed 16 June 2020).

25 Li, H., Sethuraman, N., Stadheim, T.A. et al. (2006). Optimization of humanized IgGs in glycoengineered *Pichia pastoris*. *Nature Biotechnology* 24 (2): 210–215.

26 Kayser, V., Chennamsetty, N., Voynov, V. et al. (2011). Glycosylation influences on the aggregation propensity of therapeutic monoclonal antibodies. *Biotechnology Journal* 6 (1): 38–44.

27 Courtois, F., Agrawal, N.J., Lauer, T.M., and Trout, B.L. (2016). Rational design of therapeutic mAbs against aggregation through protein engineering and incorporation of glycosylation motifs applied to bevacizumab. *MAbs* 8 (1): 99–112.

28 Kayser, V., Chennamsetty, N., Voynov, V. et al. (2011). Evaluation of a non-arrhenius model for therapeutic monoclonal antibody aggregation. *Journal of Pharmaceutical Sciences* 100 (7): 2526–2542.

29 Roberts, C.J. (2014). Protein aggregation and its impact on product quality. *Current Opinion in Biotechnology* 30: 211–217.

30 The Pharmaceutical Research and Manufacturers of America (2013). Medicines in development: biologics: the pharmaceutical research and manufacturers of America. PhRMA. http://phrma-docs.phrma.org/sites/default/files/pdf/biologicsoverview2013.pdf (accessed 16 June 2020).

31 Kim, Y.-C., Park, J.-H., and Prausnitz, M.R. (2012). Microneedles for drug and vaccine delivery. *Advanced Drug Delivery Reviews* 64 (14): 1547–1568.

32 Demir, Y.K., Metin, A.Ü., Şatıroğlu, B. et al. (2017). Poly (methyl vinyl ether-co-maleic acid) – pectin based hydrogel-forming systems: gel, film, and microneedles. *European Journal of Pharmaceutics and Biopharmaceutics* 117: 182–194.

33 Henry, S., McAllister, D.V., Allen, M.G., and Prausnitz, M.R. (1998). Microfabricated microneedles: a novel approach to transdermal drug delivery. *Journal of Pharmaceutical Sciences* 87 (8): 922–925.

34 Raphael, A.P., Crichton, M.L., Falconer, R.J. et al. (2016). Formulations for microprojection/microneedle

vaccine delivery: structure, strength and release profiles. *Journal of Controlled Release* 225: 40–52.

35 Reslan, M., Demir, Y.K., Trout, B.L. et al. (2017). Lack of a synergistic effect of arginine–glutamic acid on the physical stability of spray-dried bovine serum albumin. *Pharmaceutical Development and Technology* 22 (6): 785–791. https://doi.org/10.1080/10837450. 2016.1185116.

36 Reslan, M. and Kayser, V. (2018). The effect of deuterium oxide on the conformational stability and aggregation of bovine serum albumin. *Pharmaceutical Development and Technology* 23 (10): 1030–1036. https://doi.org/10.1080/10837450.2016.1268157.

37 Reslan, M. and Kayser, V. (2018). Ionic liquids as biocompatible stabilizers of proteins. *Biophysical Reviews* 10 (3): 781–793.

38 Zhang, J., Frey, V., Corcoran, M. et al. (2016). Influence of arginine salts on the thermal stability and aggregation kinetics of monoclonal antibody: dominant role of anions. *Molecular Pharmaceutics* 13 (10): 3362–3369.

39 Martinez Morales, M., Zalar, M., Sonzini, S. et al. (2019). Interaction of a macrocycle with an aggregation-prone region of a monoclonal antibody. *Molecular Pharmaceutics* 16 (7): 3100–3108.

40 Nisonoff, A. and Rivers, M.M. (1961). Recombination of a mixture of univalent antibody fragments of different specificity. *Archives of Biochemistry and Biophysics* 93 (2): 460–462.

41 Sheridan, C. (2015). Amgen's bispecific antibody puffs across finish line. *Nature Biotechnology* 33 (3): 219–221.

42 NIH (ed.) (2019). *Bispecific*. NIH U.S. National Library of Medicine.

43 Global bispecific antibody market, drug sales & clinical pipeline insight 2025 (2019). https://www. researchandmarkets.com/research/xhhw54/global_ bispecific?w=5 (accessed 16 June 2020).

44 FDA. Bispecific antibody development programs – guidance for industry 2019 [updated 11 February 2019]. https://www.fda.gov/media/123313/ download (accessed 16 June 2020).

45 Thakur, A. and Lum, L.G. (2010). Cancer therapy with bispecific antibodies: clinical experience. *Current Opinion in Molecular Therapeutics* 12 (3): 340–349.

46 Linke, R., Klein, A., and Seimetz, D. (2010). Catumaxomab: clinical development and future directions. *mAbs* 2 (2): 129–136.

47 Neovii completes marketing authorisation withdrawal of Removab® in the European Union 2017. https:// neovii.com/neovii-completes-marketing-authorisation-withdrawal-of-removab-in-the-european-union (accessed 16 June 2020).

48 Rogala, B., Freyer, C.W., Ontiveros, E.P. et al. (2015). Blinatumomab: enlisting serial killer T-cells in the war against hematologic malignancies. *Expert Opinion on Biological Therapy* 15 (6): 895–908.

49 CHMP recommends EU approval of Roche's Hemlibra for people with severe haemophilia A without factor VIII inhibitors 2019. https://www.roche.com/media/ releases/med-cor-2019-02-01.htm (accessed 16 June 2020).

50 Uchida, N., Sambe, T., Yoneyama, K. et al. (2016). A first-in-human phase 1 study of ACE910, a novel factor VIII–mimetic bispecific antibody, in healthy subjects. *Blood* 127 (13): 1633–1641.

51 Shima, M., Hanabusa, H., Taki, M. et al. (2016). Factor VIII–mimetic function of humanized bispecific antibody in hemophilia A. *The New England Journal of Medicine* 374 (21): 2044–2053.

52 NIH (ed.) (2019). *Antibody Drug Conjugates*. NIH U.S. National Library of Medicine.

53 Peters, C. and Brown, S. (2015). Antibody–drug conjugates as novel anti-cancer chemotherapeutics. *Bioscience Reports* 35 (4): e00225. https://doi. org/10.1042/BSR20150089.

54 Kovtun, Y.V., Audette, C.A., Mayo, M.F. et al. (2010). Antibody-maytansinoid conjugates designed to bypass multidrug resistance. *Cancer Research* 70 (6): 2528–2537.

55 Ducry, L. (2012). Challenges in the development and manufacturing of antibody-drug conjugates. *Methods in Molecular Biology (Clifton, NJ)* 899: 489–497.

56 Beck, A., Goetsch, L., Dumontet, C., and Corvaïa, N. (2017). Strategies and challenges for the next generation of antibody–drug conjugates. *Nature Reviews. Drug Discovery* 16: 315.

57 Ducry, L. and Stump, B. (2010). Antibody-drug conjugates: linking cytotoxic payloads to monoclonal antibodies. *Bioconjugate Chemistry* 21 (1): 5–13.

58 Sorokin, P. (2000). Mylotarg approved for patients with CD33+ acute myeloid leukemia. *Clinical Journal of Oncology Nursing* 4 (6): 279–280.

59 Zaro, J.L. (2015). Mylotarg: revisiting its clinical potential post-withdrawal. In: *Antibody-Drug Conjugates*, AAPS Advances in the Pharmaceutical Sciences Series. 17 (eds. J. Wang, W.-C. Shen and J.L. Zaro), 179–190. Springer International Publishing.

60 Tuma, R.S. (2011). Enthusiasm for antibody–drug conjugates. *Journal of the National Cancer Institute* 103 (20): 1493–1494.

61 de Claro, R.A., McGinn, K., Kwitkowski, V. et al. (2012). U.S. Food and Drug Administration approval summary: brentuximab vedotin for the treatment of relapsed hodgkin lymphoma or relapsed systemic anaplastic large-cell lymphoma. *Clinical Cancer Research* 18 (21): 5845–5849.

62 Guerin, M., Sabatier, R., and Goncalves, A. (2015). Trastuzumab emtansine (Kadcyla®) approval in HER2-positive metastatic breast cancers. *Bulletin du Cancer* 102 (4): 390–397.

63 Gromek, S.M. and Balunas, M.J. (2015). Natural products as exquisitely potent cytotoxic payloads for antibody-drug conjugates. *Current Topics in Medicinal Chemistry* 14 (24): 2822–2834.

64 Bornstein, G.G. (2015). Antibody drug conjugates: preclinical considerations. *The AAPS Journal* 17 (3): 525–534.

65 Polakis, P. (2016). Antibody drug conjugates for cancer therapy. *Pharmacological Reviews* 68 (1): 3–19.

66 Schrama, D., Reisfeld, R.A., and Becker, J.C. (2006). Antibody targeted drugs as cancer therapeutics. *Nature Reviews. Drug Discovery* 5 (2): 147–159.

67 Senter, P.D. and Sievers, E.L. (2012). The discovery and development of brentuximab vedotin for use in relapsed Hodgkin lymphoma and systemic anaplastic large cell lymphoma. *Nature Biotechnology* 30 (7): 631–637.

68 Drachman, J.G. and Senter, P.D. (2013). Antibody-drug conjugates: the chemistry behind empowering antibodies to fight cancer. *Hematology/The Education Program of the American Society of Hematology American Society of Hematology Education Program* 2013: 306–310.

69 McCombs, J.R. and Owen, S.C. (2015). Antibody drug conjugates: design and selection of linker, payload and conjugation chemistry. *The AAPS Journal* 17 (2): 339–351.

70 Doronina, S.O., Mendelsohn, B.A., Bovee, T.D. et al. (2006). Enhanced activity of monomethylauristatin F through monoclonal antibody delivery: effects of linker technology on efficacy and toxicity. *Bioconjugate Chemistry* 17 (1): 114–124.

71 Erickson, H.K., Park, P.U., Widdison, W.C. et al. (2006). Antibody-maytansinoid conjugates are activated in targeted cancer cells by lysosomal degradation and linker-dependent intracellular processing. *Cancer Research* 66 (8): 4426–4433.

72 Vaklavas, C. and Forero-Torres, A. (2012). Safety and efficacy of brentuximab vedotin in patients with Hodgkin lymphoma or systemic anaplastic large cell lymphoma. *Therapeutic Advances in Hematology* 3 (4): 209–225.

73 Bendell, J., Saleh, M., Rose, A.A. et al. (2014). Phase I/II study of the antibody-drug conjugate glembatumumab vedotin in patients with locally advanced or metastatic breast cancer. *Journal of Clinical Oncology: Official Journal of the American Society of Clinical Oncology* 32 (32): 3619–3625.

74 Alley, S., Zhang, X., Okeley, N. et al. (2007). Effects of linker chemistry on tumor targeting by anti-CD70 antibody-drug conjugates. *Cancer Research* 67 (Suppl. 9): 916.

75 Girish, S., Gupta, M., Wang, B. et al. (2012). Clinical pharmacology of trastuzumab emtansine (T-DM1): an antibody-drug conjugate in development for the treatment of HER2-positive cancer. *Cancer Chemotherapy and Pharmacology* 69 (5): 1229–1240.

76 Erickson, H.K., Lewis Phillips, G.D., Leipold, D.D. et al. (2012). The effect of different linkers on target cell catabolism and pharmacokinetics/pharmacodynamics of trastuzumab maytansinoid conjugates. *Molecular Cancer Therapeutics* 11 (5): 1133–1142.

77 Lyon, R.P., Meyer, D.L., Setter, J.R., and Senter, P.D. (2012). Conjugation of anticancer drugs through endogenous monoclonal antibody cysteine residues. *Methods in Enzymology* 502: 123–138.

78 Hamblett, K.J., Senter, P.D., Chace, D.F. et al. (2004). Effects of drug loading on the antitumor activity of a monoclonal antibody drug conjugate. *Clinical Cancer Research: An Official Journal of the American Association for Cancer Research* 10 (20): 7063–7070.

79 Zimmerman, E.S., Heibeck, T.H., Gill, A. et al. (2014). Production of site-specific antibody-drug conjugates using optimized non-natural amino acids in a cell-free expression system. *Bioconjugate Chemistry* 25 (2): 351–361.

80 Sochaj, A.M., Swiderska, K.W., and Otlewski, J. (2015). Current methods for the synthesis of homogeneous antibody-drug conjugates. *Biotechnology Advances* 33 (6 Pt 1): 775–784.

81 Panowksi, S., Bhakta, S., Raab, H. et al. (2014). Site-specific antibody drug conjugates for cancer therapy. *mAbs* 6 (1): 34–45.

82 Gautier, V., Boumeester, A.J., Lossl, P., and Heck, A.J. (2015). Lysine conjugation properties in human IgGs studied by integrating high-resolution native mass spectrometry and bottom-up proteomics. *Proteomics* 15 (16): 2756–2765.

83 Francisco, J.A., Cerveny, C.G., Meyer, D.L. et al. (2003). cAC10-vcMMAE, an anti-CD30-monomethyl auristatin E conjugate with potent and selective antitumor activity. *Blood* 102 (4): 1458–1465.

84 Dennler, P., Fischer, E., and Schibli, R. (2015). Antibody conjugates: from heterogeneous populations to defined reagents. *Antibodies* 4 (3): 197.

85 Behrens, C.R., Ha, E.H., Chinn, L.L. et al. (2015). Antibody-drug conjugates (ADCs) derived from interchain cysteine cross-linking demonstrate improved homogeneity and other pharmacological properties over conventional heterogeneous ADCs. *Molecular Pharmaceutics* 12 (11): 3986–3998.

86 Okeley, N.M., Miyamoto, J.B., Zhang, X. et al. (2010). Intracellular activation of SGN-35, a potent anti-CD30 antibody-drug conjugate. *Clinical Cancer Research: An Official Journal of the American Association for Cancer Research* 16 (3): 888–897.

87 Junutula, J.R., Raab, H., Clark, S. et al. (2008). Site-specific conjugation of a cytotoxic drug to an antibody improves the therapeutic index. *Nature Biotechnology* 26 (8): 925–932.

88 Zhou, Q. and Kim, J. (2015). Advances in the development of site-specific antibody-drug conjugation. *Anti-Cancer Agents in Medicinal Chemistry* 15 (7): 828–836.

89 Gomez, N., Vinson, A.R., Ouyang, J. et al. (2010). Triple light chain antibodies: factors that influence its formation in cell culture. *Biotechnology and Bioengineering* 105 (4): 748–760.

90 Junttila, T.T., Li, G., Parsons, K. et al. (2011). Trastuzumab-DM1 (T-DM1) retains all the mechanisms of action of trastuzumab and efficiently inhibits growth of lapatinib insensitive breast cancer. *Breast Cancer Research and Treatment* 128 (2): 347–356. https://doi.org/10.1007/s10549-010-1090-x.

91 Brüggemann, M., Williams, G.T., Bindon, C.I. et al. (1987). Comparison of the effector functions of human immunoglobulins using a matched set of chimeric antibodies. *The Journal of Experimental Medicine* 166 (5): 1351–1361.

92 Perez, H.L., Cardarelli, P.M., Deshpande, S. et al. (2014). Antibody-drug conjugates: current status and future directions. *Drug Discovery Today* 19 (7): 869–881.

93 McDonagh, C.F., Kim, K.M., Turcott, E. et al. (2008). Engineered anti-CD70 antibody-drug conjugate with increased therapeutic index. *Molecular Cancer Therapeutics* 7 (9): 2913–2923.

94 Tse, K.F., Jeffers, M., Pollack, V.A. et al. (2006). CR011, a fully human monoclonal antibody-auristatin E conjugate, for the treatment of melanoma. *Clinical Cancer Research* 12 (4): 1373–1382.

95 Takeshita, A., Yamakage, N., Shinjo, K. et al. (2009). CMC-544 (inotuzumab ozogamicin), an anti-CD22 immuno-conjugate of calicheamicin, alters the levels of target molecules of malignant B-cells. *Leukemia* 23 (7): 1329–1336.

96 Nelson, A.L. (2010). Antibody fragments: hope and hype. *MAbs* 2 (1): 77–83.

97 Jain, R.K. (1990). Physiological barriers to delivery of monoclonal antibodies and other macromolecules in tumors. *Cancer Research* 50 (Suppl. 3): 814s–819s.

98 Adams, G.P., Schier, R., Marshall, K. et al. (1998). Increased affinity leads to improved selective tumor delivery of single-chain Fv antibodies. *Cancer Research* 58 (3): 485–490.

99 McConaghie, A. (2017). Novartis and FDA hail "historic" Kymriah, first ever approved CAR-T. *Pharmaphorum* https://pharmaphorum.com/news/novartis-fda-hail-historic-kymriah-first-ever-approved-car-t (accessed 16 June 2020).

100 van der Stegen, S.J., Hamieh, M., and Sadelain, M. (2015). The pharmacology of second-generation chimeric antigen receptors. *Nature Reviews. Drug Discovery* 14 (7): 499–509.

101 Eshhar, Z., Waks, T., Gross, G., and Schindler, D.G. (1993). Specific activation and targeting of cytotoxic lymphocytes through chimeric single chains consisting of antibody-binding domains and the gamma or zeta subunits of the immunoglobulin and T-cell receptors. *Proceedings of the National Academy of Sciences of the United States of America* 90 (2): 720–724.

102 Maude, S.L., Teachey, D.T., Porter, D.L., and Grupp, S.A. (2015). CD19-targeted chimeric antigen receptor T-cell therapy for acute lymphoblastic leukemia. *Blood* 125 (26): 4017–4023.

103 Brocker, T. (2000). Chimeric Fv-zeta or Fv-epsilon receptors are not sufficient to induce activation or cytokine production in peripheral T cells. *Blood* 96 (5): 1999–2001.

104 Savoldo, B., Ramos, C.A., Liu, E. et al. (2011). CD28 costimulation improves expansion and persistence of chimeric antigen receptor-modified T cells in lymphoma patients. *The Journal of Clinical Investigation* 121 (5): 1822–1826.

105 Zhang, C., Liu, J., Zhong, J.F., and Zhang, X. (2017). Engineering CAR-T cells. *Biomarker Research* 5: 22.

106 Milone, M.C., Fish, J.D., Carpenito, C. et al. (2009). Chimeric receptors containing CD137 signal transduction domains mediate enhanced survival of T cells and increased antileukemic efficacy in vivo. *Molecular Therapy* 17 (8): 1453–1464.

107 Haso, W., Lee, D.W., Shah, N.N. et al. (2013). Anti-CD22-chimeric antigen receptors targeting B-cell precursor acute lymphoblastic leukemia. *Blood* 121 (7): 1165–1174.

108 Abramson, J.S., Gordon, L.I., Palomba, M.L. et al. (2018). Updated safety and long term clinical outcomes in TRANSCEND NHL 001, pivotal trial of lisocabtagene maraleucel (JCAR017) in R/R aggressive NHL. *Journal of Clinical Oncology* 36 (Suppl. 15): 7505.

109 Yeku, O.O., Purdon, T.J., Koneru, M. et al. (2017). Armored CAR T cells enhance antitumor efficacy and overcome the tumor microenvironment. *Scientific Reports* 7 (1): 10541.

110 Kuhn, N.F., Purdon, T.J., van Leeuwen, D.G. et al. (2019). CD40 ligand-modified chimeric antigen receptor T cells enhance antitumor function by eliciting an endogenous antitumor response. *Cancer Cell* 35 (3): 473–488.e6.

111 Hu, B., Zou, Y., Zhang, L. et al. (2019). Nucleofection with plasmid DNA for CRISPR/Cas9-mediated inactivation of programmed cell death protein 1 in CD133-specific CAR T cells. *Human Gene Therapy* 30 (4): 446–458.

112 Locke, F.L., Neelapu, S.S., Bartlett, N.L. et al. (2017). Phase 1 results of ZUMA-1: a multicenter study of KTE-C19 anti-CD19 CAR T cell therapy in refractory aggressive lymphoma. *Molecular Therapy* 25 (1): 285–295.

113 Siddiqi, T., Abramson, J.S., Lee, H.J. et al. (2019). Safety of lisocabtagene maraleucel given with durvalumab in patients with relapsed/refractory aggressive b-cell non Hodgkin lymphoma: first results from the platform study. *Hematological Oncology* 37 (S2): 171–172.

114 Cao, Y., Lu, W., Sun, R. et al. (2019). Anti-CD19 chimeric antigen receptor T cells in combination with nivolumab are safe and effective against relapsed/refractory B-cell non-Hodgkin lymphoma. *Frontiers in Oncology* 9: 767.

115 Timmers, M., Roex, G., Wang, Y. et al. (2019). Chimeric antigen receptor-modified T cell therapy in multiple myeloma: beyond B cell maturation antigen. *Frontiers in Immunology* 10: 1613.

116 Hosen, N., Matsunaga, Y., Hasegawa, K. et al. (2017). The activated conformation of integrin beta7 is a novel multiple myeloma-specific target for CAR T cell therapy. *Nature Medicine* 23 (12): 1436–1443.

117 Carrabba, M.G., Casucci, M., Hudecek, M. et al. (2018). Phase I-IIa clinical trial to assess safety and efficacy of MLM-CAR44.1, a CD44v6 directed CAR-T in relapsed/refractory acute myeloid leukemia (AML) and multiple myeloma (MM). *Blood* 132 (Suppl. 1): 5790.

118 Sun, C., Mahendravada, A., Ballard, B. et al. (2019). Safety and efficacy of targeting CD138 with a chimeric antigen receptor for the treatment of multiple myeloma. *Oncotarget* 10 (24): 2369–2383.

119 Hofmann, S., Schubert, M.L., Wang, L. et al. (2019). Chimeric antigen receptor (CAR) T cell therapy in acute myeloid leukemia (AML). *Journal of Clinical Medicine* 8 (2): 200. https://doi.org/10.3390/jcm8020200.

120 Almeida, A.M. and Ramos, F. (2016). Acute myeloid leukemia in the older adults. *Leukemia Research Reports* 6: 1–7. https://doi.org/10.1016/j.lrr.2016.06.001.

121 Beasley, D. (2018). U.S. Medicare sets outpatient rate for Yescarta reimbursement.

122 Marcucci, K.T., Jadlowsky, J.K., Hwang, W.T. et al. (2018). Retroviral and lentiviral safety analysis of gene-modified T cell products and infused HIV and oncology patients. *Molecular Therapy* 26 (1): 269–279.

123 Monjezi, R., Miskey, C., Gogishvili, T. et al. (2017). Enhanced CAR T-cell engineering using non-viral Sleeping Beauty transposition from minicircle vectors. *Leukemia* 31 (1): 186–194.

124 Rafiq, Q.A., Twomey, K., Kulik, M. et al. (2016). Developing an automated robotic factory for novel stem cell therapy production. *Regenerative Medicine* 11 (4): 351–354.

125 Nicholas, S. (2017). T cell engineering for breakthrough immunotherapy research. https://www.selectscience.net/editorial-articles/t-cell-engineering-for-breakthrough-immunotherapy-research/?artID=44154 (accessed 16 June 2020).

126 Hou, A.J., Chang, Z.L., Lorenzini, M.H. et al. (2018). TGF-beta-responsive CAR-T cells promote anti-tumor immune function. *Bioengineering & Translational Medicine* 3 (2): 75–86.

127 Foster, A.E., Dotti, G., Lu, A. et al. (2008). Antitumor activity of EBV-specific T lymphocytes transduced with a dominant negative TGF-beta receptor. *Journal of Immunotherapy* 31 (5): 500–505.

128 Wilkie, S., Burbridge, S.E., Chiapero-Stanke, L. et al. (2010). Selective expansion of chimeric antigen receptor-targeted T-cells with potent effector function using interleukin-4. *The Journal of Biological Chemistry* 285 (33): 25538–25544.

129 Hoyos, V., Savoldo, B., Quintarelli, C. et al. (2010). Engineering CD19-specific T lymphocytes with interleukin-15 and a suicide gene to enhance their anti-lymphoma/leukemia effects and safety. *Leukemia* 24 (6): 1160–1170.

130 Kabelitz, D., Marischen, L., Oberg, H.H. et al. (2005). Epithelial defence by gamma delta T cells. *International Archives of Allergy and Immunology* 137 (1): 73–81.

131 Lamb, L.S. Jr. and Lopez, R.D. (2005). Gammadelta T cells: a new frontier for immunotherapy? *Biology of Blood and Marrow Transplantation* 11 (3): 161–168.

132 Liu, Z., Guo, B.L., Gehrs, B.C. et al. (2005). Ex vivo expanded human Vgamma9Vdelta2+ gammadelta-T cells mediate innate antitumor activity against human prostate cancer cells in vitro. *The Journal of Urology* 173 (5): 1552–1556.

133 Capsomidis, A., Benthall, G., Van Acker, H.H. et al. (2018). Chimeric antigen receptor-engineered human gamma delta T cells: enhanced cytotoxicity with retention of cross presentation. *Molecular Therapy* 26 (2): 354–365.

134 Burga, R.A., Nguyen, T., Zulovich, J. et al. (2016). Improving efficacy of cancer immunotherapy by genetic modification of natural killer cells. *Cytotherapy* 18 (11): 1410–1421.

135 Parham, P. (2006). Taking license with natural killer cell maturation and repertoire development. *Immunological Reviews* 214: 155–160.

136 Zhou, X., Di Stasi, A., and Brenner, M.K. (2015). iCaspase 9 suicide gene system. *Methods in Molecular Biology (Clifton, NJ)* 1317: 87–105.

137 Liu, E., Tong, Y., Dotti, G. et al. (2018). Cord blood NK cells engineered to express IL-15 and a CD19-targeted CAR show long-term persistence and potent antitumor activity. *Leukemia* 32 (2): 520–531.

138 Wiernik, A., Foley, B., Zhang, B. et al. (2013). Targeting natural killer cells to acute myeloid leukemia in vitro with a CD16 x 33 bispecific killer cell engager and ADAM17 inhibition. *Clinical Cancer Research: An Official Journal of the American Association for Cancer Research* 19 (14): 3844–3855.

139 Gleason, M.K., Ross, J.A., Warlick, E.D. et al. (2014). CD16xCD33 bispecific killer cell engager (BiKE) activates NK cells against primary MDS and MDSC CD33+ targets. *Blood* 123 (19): 3016–3026.

140 Gleason, M.K., Verneris, M.R., Todhunter, D.A. et al. (2012). Bispecific and trispecific killer cell engagers directly activate human NK cells through CD16 signaling and induce cytotoxicity and cytokine production. *Molecular Cancer Therapeutics* 11 (12): 2674–2684.

141 Jungkunz-Stier, I., Zekl, M., Stuhmer, T. et al. (2014). Modulation of natural killer cell effector functions through lenalidomide/dasatinib and their combined effects against multiple myeloma cells. *Leukemia & Lymphoma* 55 (1): 168–176.

142 Shortt, J., Hsu, A.K., and Johnstone, R.W. (2013). Thalidomide-analogue biology: immunological, molecular and epigenetic targets in cancer therapy. *Oncogene* 32 (36): 4191–4202.

143 Hermans, I.F., Silk, J.D., Gileadi, U. et al. (2003). NKT cells enhance CD4+ and CD8+ T cell responses to soluble antigen in vivo through direct interaction with dendritic cells. *Journal of Immunology* 171 (10): 5140–5147.

144 Crowe, N.Y., Coquet, J.M., Berzins, S.P. et al. (2005). Differential antitumor immunity mediated by NKT cell subsets in vivo. *The Journal of Experimental Medicine* 202 (9): 1279–1288.

145 Carnaud, C., Lee, D., Donnars, O. et al. (1999). Cutting edge: cross-talk between cells of the innate immune system: NKT cells rapidly activate NK cells. *Journal of Immunology* 163 (9): 4647–4650.

146 Heczey, A., Liu, D., Tian, G. et al. (2014). Invariant NKT cells with chimeric antigen receptor provide a novel platform for safe and effective cancer immunotherapy. *Blood* 124 (18): 2824–2833.

147 Leach, D.R., Krummel, M.F., and Allison, J.P. (1996). Enhancement of antitumor immunity by CTLA-4 blockade. *Science* 271 (5256): 1734–1736.

148 van Elsas, A., Hurwitz, A.A., and Allison, J.P. (1999). Combination immunotherapy of B16 melanoma using anti-cytotoxic T lymphocyte-associated antigen 4 (CTLA-4) and granulocyte/macrophage colony-stimulating factor (GM-CSF)-producing vaccines induces rejection of subcutaneous and metastatic tumors accompanied by autoimmune depigmentation. *The Journal of Experimental Medicine* 190 (3): 355–366.

149 Morse, M.A. (2005). Technology evaluation: ipilimumab, Medarex/Bristol-Myers Squibb. *Current Opinion in Molecular Therapeutics* 7 (6): 588–597.

150 Hodi, F.S., O'Day, S.J., McDermott, D.F. et al. (2010). Improved survival with ipilimumab in patients with metastatic melanoma. *The New England Journal of Medicine* 363 (8): 711–723.

151 Robert, C., Thomas, L., Bondarenko, I. et al. (2011). Ipilimumab plus dacarbazine for previously untreated metastatic melanoma. *The New England Journal of Medicine* 364 (26): 2517–2526.

152 Sharon, E., Streicher, H., Goncalves, P., and Chen, H.X. (2014). Immune checkpoint inhibitors in clinical trials. *Chinese Journal of Cancer* 33 (9): 434–444.

153 Wang, D.Y., Salem, J.E., Cohen, J.V. et al. (2018). Fatal toxic effects associated with immune checkpoint inhibitors: a systematic review and meta-analysis. *JAMA Oncology* 4 (12): 1721–1728.

154 Winer, A., Bodor, J.N., and Borghaei, H. (2018). Identifying and managing the adverse effects of immune checkpoint blockade. *Journal of Thoracic Disease* 10 (Suppl. 3): S480–S489.

155 Haanen, J., Carbonnel, F., Robert, C. et al. (2017). Management of toxicities from immunotherapy: ESMO Clinical Practice Guidelines for diagnosis, treatment and follow-up. *Annals of Oncology* 28 (Suppl. 4): iv119–iv142.

156 FDA News Release. FDA approves first chemoimmunotherapy regimen for patients with relapsed or refractory diffuse large B-cell lymphoma 2019. https://www.fda.gov/news-events/press-announcements/fda-approves-first-chemoimmunotherapy-regimen-patients-relapsed-or-refractory-diffuse-large-b-cell (accessed 16 June 2020).

157 WHO (2019). List of Blueprint priority diseases: WHO. https://www.who.int/blueprint/priority-diseases/en (accessed 16 June 2020).

14

Optimizing Use of Biologic Medicines Using a Quality Use of Medicines Approach

Lynn Weekes

Health Strategy and Sciences, Sydney, New South Wales, Australia
School of Pharmacy, University of Queensland, St. Lucia, Queensland, Australia

---

KEY POINTS

- The quality use of medicines (QUM) framework is a useful tool for ensuring biologic medicines are used to create greatest benefit and least harms for patients and greatest cost-effectiveness for health systems.
- The knowledge, attitudes, and beliefs of both patients and clinicians can act as barriers or enablers to uptake of biologic innovator medicines and interchange with biosimilars. Persistence with treatment has been an issue in some studies of biosimilar use and there is some evidence that the nocebo effect contributes to this.
- Implementation of guidelines, formulary management, and cost-effectiveness analysis are common ways of improving use of medicines at an institutional level.
- Objective information, education, and training should be available to support shared decision-making and promote persistence with therapy. This may require upskilling clinicians in understanding the technical or scientific analyses that replace usual clinical studies. Pharmacy and Therapeutic Committees can play an important role.

Abbreviations

Abbreviation	Full name
ARA	Australian Rheumatology Association
BCG	Bacille Calmette-Guérin
CATAG	Council of Australian Therapeutic Advisory Groups
DTC	Drugs and Therapeutics Committees
EMA	European Medicines Agency
FDA	Food and Drug Administration
FDMs	Formulary Decision Makers
GRADE	The Grading of Recommendations, Assessment, Development and Evaluation
HIV	Human Immunodeficiency Virus
IQVIA Institute	Formerly Quintiles and IMS Health
JIA	Juvenile Idiopathic Arthritis
PBS	Pharmaceutical Benefits Scheme

Abbreviation	Full name
PTCs	Pharmacy and Therapeutics Committees
TNF	Tumor Necrosis Factor
WHO	World Health Organization

14.1 Introduction

Biologic medicines can be very effective treatments for a variety of illnesses, most notably several cancers and chronic autoimmune diseases such as rheumatoid arthritis, ulcerative colitis, Crohn's disease, and psoriasis. The early biologics included recombinant human insulins, colony stimulating factors, growth hormone, and erythropoietin. A range of proteins, monoclonal antibodies, gene-based and cell-based therapies have

Biologics, Biosimilars, and Biobetters: An Introduction for Pharmacists, Physicians, and Other Health Practitioners,
First Edition. Edited by Iqbal Ramzan.
© 2021 John Wiley & Sons, Inc. Published 2021 by John Wiley & Sons, Inc.

followed and as the innovator products come off patent, we are seeing a burgeoning biosimilars industry. In 2017, the global biologics and biosimilars market size was USD209,400 million and the forecast for 2025 was almost USD300 billion.[1]

Until recently the cost of pharmaceuticals in health systems was driven by the high volume of prescriptions of relatively low-cost small molecule drugs. Use of generic medicines has been a major means for containing costs. Today high income countries and increasingly middle income countries are grappling with low-volume, high-cost biologic medicines for conditions not well managed previously and in small, niche groups of patients. Unlike small molecule generic medicines, biosimilars are not identical to the originator drug and so the concept of bioequivalence is not relevant. There may be divergence in cell lines and purification methods used in manufacturing but ultimately biosimilars must be shown to be "highly similar" and as effective and safe as the innovator biologic. Given the cost of innovator biologic medicines, it is not surprising that health systems and payers have seen biosimilars as a means of reducing pressure on the pharmaceuticals budget while maintaining access to effective treatment for patients.

A report from IQVia projected that between 2018 and 2022, competition from biosimilars could bring down spending on biologics by 10–30%, or by USD50–78 billion.[2] However, uptake of biosimilars versus innovator/reference products will depend on many factors such as the number of competitors; the speed with which competition enters the market, and the extent to which biosimilars compete on price, policy, and regulatory frameworks; acceptance by physicians and patients; and promotional/marketing activities. The patent period for several biologics has expired and regulators such as the European Medicines Agency and Australia's Therapeutic Goods Administration have registered biosimilars for these.

Industry predictions are that from 2018 to 2022, some 40–45 new active substances will be launched each year. Specialty medicines, led by oncology drugs, will drive growth in the innovator biologic market in high income countries. Indeed, specialty drugs are expected to reach 48% of spending in developed markets by 2022 compared with 39% of spending in 2017.[2]

The high costs associated with biologics and the growth of this market are anticipated to place an increasing burden on public healthcare spending globally. For example, in Australia, for the 2017–2018 financial year, 7 of the top 10 drugs by cost to the Australian Government were biologic medicines and the remaining three were new treatments for Hepatitis C.[3] Moreover, these medicines were used for a relatively small cohort – the list of top 10 drugs by volume continued to be dominated by small molecule medicines for cardiovascular disease and diabetes.

The challenge for health authorities then is to ensure that biologic medicines, whether innovator product or biosimilar, are chosen where they lead to maximum benefits and least harms and when they are cost-effective compared with alternatives. The quality use of medicines approach is helpful in this context as a means of both guiding appropriate therapy and providing the education and supports to see such guidance implemented.

14.2 Quality Use of Medicines Approach

Quality use of medicines has been defined as selecting management options wisely, choosing the most appropriate medicine when one is deemed necessary, and using medicines safely and effectively. This is a consumer-centric paradigm that includes a range of perspectives, from the individual, to health system, to the wider community. It involves all stakeholders, includes all stages of learning, and addresses all relevant settings.[4] Key concepts for quality use of medicines are: judicious use of medicines, appropriate selection, and safe and effective use of medicines.

14.2.1 Judicious Use

Judicious use refers to selecting management options wisely. This means that consideration should be given to non-pharmacological treatments as well as the range of medicines available that have been shown to be effective for the condition. The underlying philosophy is that the place of medicines in treating illness and maintaining health while critical must also recognize that there may be better options than a medicine to manage a given diagnosis in an individual patient.

14.2.2 Appropriate Selection

When a medicine is considered necessary, the most suitable medicine should be selected from those available to ensure the right medicine, for the right person, for the right condition, at the right time, and at the right dose. This means taking into account clinical, social, and cultural traits of the person taking the medicine as well as their preferences and beliefs. The comparative harms and benefits of specific medicines, coexisting conditions or therapies, the burden of monitoring treatment, and the quality of the evidence to support it should all be weighed up to tailor the best treatment for a person. The affordability of the medicine for the person, the community, and the health system will be an additional consideration in many situations. For biologic medicines, it is also important to consider the staging of medicines as many autoimmune diseases may only respond to therapy for a limited period – loss of efficacy can occur over time with all biologic medicines.

14.2.3 Safe and Effective Use

Medicines should be used safely and effectively to get the best possible results. This requires being clear about the therapeutic goal, monitoring progress, and making the necessary alterations, including stopping the medicine if the medicine does not achieve this outcome. Misuse, overuse, and underuse of medicines can all lead to suboptimal outcomes and these should be minimized. The person taking the medicine should be empowered to manage their medicines, problem solving when appropriate, and referring to their health professional as needed. Health literacy is important to support people taking medicines to access and use information and take part in decision-making with or without their health professional.

To support quality prescribing and use of medicines, six building blocks have been identified based on evidence and expert opinion about interventions, regulatory efforts, and programs to improve medication use. These are:

- Policy development and implementation.
- Facilitation and coordination of quality use of medicine initiatives.
- Provision of objective information and assurance of ethical promotion of medicines.
- Education and training.
- Provision of services and appropriate interventions to support health professionals and consumers.
- Strategic research, evaluation, and routine data collection.

14.3 Influences on Prescribing and Use of Biologic Medicines

An underlying premise of quality use of medicines is that in order to affect improvements, it is necessary not only to provide unbiased information but to do so in ways that will be influential for health professionals and consumers. This means that a rationalist approach is not enough if we expect to support the best use of medi-

cines and that we need to have a deep understanding of the influences on a given behavior, as well as, barriers, and enablers for change.

Many factors have been shown to influence clinical decision-making: the doctor's and the patient's demographic characteristics, perceived and actual patient expectations, specialist physician and hospital practices and policies, the healthcare system and financing, and pharmaceutical industry promotion, in addition to the knowledge, attitudes, and beliefs of both doctor and patient.

A range of biases come into play in the decision-making of all human beings, clinicians included. Decision-making is influenced not only by hard facts and rational arguments but by choices about risk, reward, time, and trade-offs.[5] The concept of cognitive bias shows systematic patterns of bias in how people think and reason – a field called heuristics. Research shows that people use heuristics – "quick natural assessments," "mental shortcuts," or "rules of thumb" – to efficiently make decisions especially about routine tasks. Often such cognitive strategies work well – they are fast and commonly lead individuals to closely approximate accurate probability assessments.

Generally, clinicians are disproportionally influenced by the most salient and digestible information and more concerned about harms than by achieving or missing out on benefits.[5] Programs that implement new guidelines or updates on therapy are usually asking clinicians to adopt new practices. While some clinical behaviors may be adopted relatively easily, many involve a complex interplay between social systems, communication style, and the decision-making process.[6]

For consumers or patients, their decisions to self-manage, play a role as an active decision maker, follow the doctor's advice, and continue with a therapy are similarly complex and subject to many influences. Previous experience; views of family and friends; trust in the clinician, education, and social factors; values and life view; access to objective information; health literacy; and expectations regarding their condition and its treatment are all relevant to how they will experience and benefit from a medicine. Adherence to and persistence with treatment is frequently noted to be an issue for management of chronic conditions and biologic medicines come with a raft of promises and challenges that are relevant in this context.

14.3.1 Health Professional Knowledge, Attitudes, and Beliefs

Studies looking at health professional knowledge and attitudes fall into two broad categories: those that consider biologic medicines as a group of new therapies and those that focus on acceptability of biosimilars.

A study of dermatologists in the United States in 2013 found that when respondents were asked to compare therapies, they perceived infliximab, ustekinumab, cyclosporine, and adalimumab to be the most effective treatments and etanercept, adalimumab, ultraviolet B, and ustekinumab to be the best tolerated.[7] However, 49% did not know what the data showed with respect to likelihood of efficacy or adverse effects. Moreover, their prescribing behavior could not be predicted by their stated views on the most effective or well-tolerated treatments.

A second study of US-based dermatologists found that they had concerns about the cost, long-term safety and tolerability of biologic medicines, and only two thirds said that they would initiate a biologic.[8] They also had concerns about loss of efficacy over time.

An interview study of senior rheumatologists in Sweden explored factors that influenced their decision to prescribe biologic medicines.[9] The individual clinician's experience and perceptions of the evidence, the health system (including if clinicians had budgetary responsibilities), peer pressure, political and administrative influences, and participation in clinical trials were all important. In addition, the patient and their views and preferences were critical factors in determining if a biologic would be prescribed.

Many studies on knowledge and attitudes to biosimilars come from Europe where these medicines have been used for longer. A study among French rheumatologists found that 55.2% reported limited knowledge of biosimilars and about half felt they had insufficient knowledge about efficacy and safety.[10] The most common barriers to prescribing a biosimilar were indication extrapolation and a lack of data on tolerability. In Germany, clinicians reported preferring originator biologics, wanting more experience with them before moving to broader usage.[11, 12] They were not highly influenced by cost factors and only 60% were convinced of equivalent efficacy. In Belgium, barriers to uptake of biosimilars were reported as being lack of confidence in biosimilars, and uncertainty about interchangeability and substitutability and a hospital funding system that relied in part on profits from the hospital pharmacy.[13]

In the United Kingdom, 75% of gastroenterologists, rheumatologists, and diabetologists were aware of biosimilars on their local formulary and 77% believed they were important to save costs for the NHS.[14] Their main concerns were about safety and efficacy when switching patients to a biosimilar rather than starting biosimilars in biologic-naïve patients. Gastroenterologists had the most positive view of biosimilars followed by rheumatologists.

A New Zealand study of the attitudes of medical specialist to biosimilars in rheumatology, dermatology, gastroenterology, oncology, and hematology found that most had positive views.[15] While 54–74% were confident about safety, efficacy, manufacturing, and pharmacovigilance, they were less confident about indication extrapolation and switching between biologics and biosimilars. The situations where specialists would be reluctant to prescribe a biosimilar included: where there was lack of efficacy data, where there is evidence of adverse effects, where the patient is doing well on existing treatment, and where the patient has a complex medical history.

In terms of accessing information about biologic medicines, at least in Europe, clinicians use the regulatory "label" also known as the approved product information as a key information source and they have a preference for more information on the comparative data for biosimilars rather than less.[16] Professional guidelines and peer-reviewed publications were the two other major sources of information.

Interestingly, pharmacists, who in some health systems, are able to substitute a biosimilar medicine when a patient has been taking an originator biologic may also have concerns. In Poland, hospital pharmacists were concerned that the new drugs were not identical with the innovator, with respect to their immunogenicity and pharmacokinetic properties.

14.3.2 Patient Knowledge, Attitudes, and Beliefs

A large international survey of patients with inflammatory bowel disease, rheumatoid arthritis, psoriasis, or cancers and the general population found that overall knowledge about biologic medicines was low.[17] Awareness was significantly higher in people diagnosed with a condition, condition-specific advocacy groups, and caregiver groups (45–78%) compared with the general population (27%). Gaps in knowledge about biosimilars included safety, efficacy, and access to these medicines. Willingness to try a biosimilar was highly influenced by the identity of the manufacturer of the medicine.

Rheumatologists in the United States perceived patients to be accepting of a biosimilar without reluctance for 70% of cases where no previous biologic medicine had been used and 56% of cases where a switch from the innovator biologic was proposed without a clinical reason.[12] According to a self-administered patient questionnaire in the same study, a small majority of patients were indifferent to a change to a biosimilar (50–63%) and an additional 14–31% were somewhat or very happy to change. Patients receiving a biologic medicine were concerned that they did not know enough about the drug (36–41%) and side effects (26–32%). Patients receiving an innovator biologic were slightly more likely to be satisfied with their treatment than those on a biosimilar.

A US survey of adults with rheumatoid arthritis investigated patient preferences for different formulations.[18] It found that about half (53.1%) were open to both intravenous and subcutaneous injection and that 26.3% strongly preferred the subcutaneous route. In this study, rheumatologists tended to underestimate patients' acceptance of parenteral administration. In addition, 23% of patients taking a biologic medicine reported that they had initiated the conversation about moving to this therapy compared with 54% of cases where it was the rheumatologist.

In spite of patients being dissatisfied with traditional treatments for psoriasis, they also seem wary of biologic medicines. Patients in one study felt that biologics were burdensome because of anxiety related to the injection, inconvenience, and potential for side effects.[8] Cost and long-term health risks were also of concern. Of those patients who were taking biologic therapy only, 45% were satisfied and of all patients who had tried a biologic 45% had discontinued use. Results such as these should be understood in context, however, and indeed there is evidence that patients and dermatologists do not always agree on the most important symptoms to manage. Gonzalez et al.[19] noted marked differences in the perceived importance of both risks and benefits between patients and dermatologists. They also found that in more severe disease, commonly used measures of treatment success did not correlate well with improvements in patient well-being.

Several authors have pointed to indirect evidence for a nocebo effect with biosimilars.[20–22] The nocebo effect is a negative effect of a therapy that is induced by the patient's expectations and that is unrelated to the physiological action of the therapy. It is essentially the opposite of the placebo effect. Factors which have been reported to be strongly associated with the nocebo effect (across all therapeutics) include social observations, perceived dose, verbal suggestions such as warning about side effects, medication brand, and physiologically or cognitively based expectations.[20]

Evidence for the nocebo effect is generally indirect. For example, open-label studies have found disparities between subjective reports of adverse effects and physiologically measured effects and in addition, discontinuation rates to biosimilars compared with that of historical controls.[23–25] One study attempted to quantify the nocebo effect and found that among 125 patients with inflammatory bowel disease or rheumatoid arthritis that 12.8% experienced nocebo effects after switching to an infliximab biosimilar in spite of having agreed to the switch.[26]

Pouillon and colleagues[20] suggest the nocebo effect should be looked for by clinicians in a patient who does not respond to treatment with a biosimilar. They suggest positive framing and tailoring of information as well as a positive patient–physician relationship could improve patient acceptance of biosimilars and limit the risk of inappropriate negative biases.

14.3.3 Patient Characteristics, Industry, and Environmental Influences

A variety of external factors also have the potential to influence prescribing and use of medicines.

A Swedish study looked at whether a patient's geography affected the likelihood of receiving a biologic medicine.[27] While on average 9.7–11% of patients were switched to a biologic, the average patient in a low prescribing region was 2.5 times less likely to be offered a biologic medicine than a patient in the top prescribing region. The authors concluded that variation stemmed from regional differences in socioeconomic factors or differences in the composition of the healthcare tax base.

Sometimes, patient characteristics may be relevant in determining uptake and use of biologic medicines. For example, a review of prescribing for children with juvenile idiopathic arthritis (JIA) in the United Kingdom found that TNF inhibitors such as etanercept, adalimumab, and infliximab were first choice in most patients.[28] However, choice of a specific TNF inhibitor was influenced by presence of uveitis and patient preference for route of administration. A change in the formulation of adalimumab correlated with a shift in first-choice preferences. The subtype of JIA and family history were additional contributing factors that influenced prescribing.

Biologic medicine is big business. An IQVIA Institute report predicted that, in 2018, $19 billion of spending on innovator biologics would become exposed to competition from biosimilars for the first time, in one or more developed markets. This compared with $3 billion that became exposed in 2017, in addition to the $26 billion already facing competition. This level of competition is an incentive for promotional and patent protection tactics by companies to deter uptake of the competitor product.

Patent litigation is a primary means of slowing market entry.[29] In the United States, other tactics have included originator companies negotiating formulary exclusivity with payers in return for lower prices, rebates, or nonpayment of unsuccessful treatment; reinforcement of clinician scepticism regarding efficacy and safety of biosimilars; and misrepresentation of complex comparative analytics to concern and confuse patients.[29] Originator companies can create marketing messages that challenge the extent to which a different manufacturing process can yield a comparable product and thus whether a biosimilar is a suitable replacement for an innovator biologic.[30] Such messages have been noted in literature directed to relevant physicians and these have the potential to reinforce the underlying concerns and perceptions that slow uptake of biosimilars. The Working Group on Similar Biologic Medicinal Products of the European Medicines Agency noted that better understanding of the scientific principles of the biosimilar concept and access to unbiased information was important for clinicians making treatment choices.[31]

14.4 Policy to Support Appropriate Uptake of Biologic Medicines

Uptake of biologic medicines, including biosimilars, varies across Europe. A review of policies and initiatives in European countries to support uptake of biosimilars found that while biologics had been registered by the European Medical Agency, they were not marketed in all countries, usually because of a lack of availability of adequate funding sources.[32] National formularies and the reimbursement policies of insurers as third-party payers were also important influences on the uptake of biosimilars. Demand side measures such as physician education, prescribing quotas, and tendering were other measures used to increase uptake of biosimilars.

Substitution and interchangeability policies, definitions, and regulation also vary between jurisdictions, particularly between the United States and Europe, and this is likely to contribute to how quickly and extensively biosimilars are used. In Europe, decisions on interchangeability are made at the member state level, after regulatory approval by the European Medicines Agency. Therefore, decisions to switch are usually made by physicians or are policy driven at the local or national level. In the United States, the FDA requires at least one switching study before a product can be considered interchangeable with the innovator and subsequently, substitutable by a pharmacist.[33]

Some of the relevant policies[33] which influence if, when, and how biologics will be used as well as product selection include:

- Tendering. For example, in Norway, the responsible agency negotiates availability of a product to a limited list based on price. In 2014, biosimilar infliximab entered the tendering process resulting in an immediate price fall of 39% compared with the originator product and 69% in the following year.[34]
- Prescribing quotas. For example, in Germany, prescribing quotas have been implemented for erythropoietin, infliximab, and etanercept biosimilars. These quotas are complemented by information campaigns and local guidelines. Similar quota systems are used in Hungary, Italy, and Sweden.
- Gain-sharing agreements. For example, in the United Kingdom, an agreement negotiated with a local clinical commissioning group enabled a gastroenterology team at the hospital to directly reinvest some of the savings achieved from adoption of biosimilars.
- Substitution. For example, in France, legislation allows substitution on treatment initiation for a defined group of products.

In 2015, the Australian government committed $20 million over three years for an awareness and educational campaign intended to increase biosimilar uptake and reduce Pharmaceutical Benefits Scheme (PBS) expenditure. The initiative was intended to promote awareness of, and confidence in, biosimilar medicines in healthcare professionals and consumers. Despite these measures, biosimilar uptake in Australia has been limited. For example, the first biosimilar flagged as interchangeable in Australia, Inflectra® (infliximab), was listed on the PBS in December 2015, followed by Renflexis®, listed in August 2017. Prescriptions for the two biosimilars secured just 11.7% of the market in 2018 compared with the reference biologic (Source: PBS data, Australian Government 2019).

14.5 Formulary Management

Given that many decisions about substitution and interchangeability are made at the local level, hospitals, Pharmacy and Therapeutics Committees (PTCs), and third-party payers will be engaged in deciding how biologic medicines are used in their populations.

Formularies are a long-standing tool to assist these agencies manage quality and cost. It is to be expected, therefore, that formularies will be used as a means to guide prescriber selection of these expensive and complex therapies.

It is likely that formulary decision makers (FDMs) will need to invest more time and clinical resources into the assessment, approval, and ongoing support for appropriate use of these drugs.

At the outset, FDMs will need to fully understand the regulatory pathway for biologics and biosimilars in their jurisdiction. As previously mentioned, the FDA has different requirements to the EMA for biosimilars. In addition, responsibility for decisions on interchangeability varies across countries and sometime within countries. Clarity on the extent and type of assessment already undertaken by regulators and governments provides a sound starting point for the deliberations of FDMs.

Secondly, FDMs will need an appreciation of the basics of the manufacturing processes for biologic medicines. Small changes in the manufacturing process for a biologic medicine can have significant consequences for efficacy and safety and this caveat is relevant for both innovator biologics and biosimilars. Manufacturing of a biosimilar will always differ from that of the innovator product as information on the process is proprietary and access to identical cell lines is not available. Differences in the source of raw materials, temperature, pH, agitation, and contamination by compounds that leach from the equipment can lead to differences among innovator and biosimilar products.[35,36] Furthermore, over time a given biologic medicine may vary due to manufacturing changes and changes in the underlying process known as "drift." A stable production process and environment, therefore, will be of interest to FDMs. Similarly, the type of therapeutic protein is relevant to the immunogenicity of the product and proteins most similar to human proteins bring less chance of immune reactions.

Thirdly, FDMs should be aware of the clinical environment in which these medicines will be used. The level of knowledge about biologics may not be strong among prescribers and/or others in the healthcare team. An understanding of current practice and the therapies that biologic medicines will supplant is also useful.

The Council of Australian Therapeutic Advisory Groups[37] recommended nine guiding principles for the governance of biosimilars medicines in hospitals, including:

1) The governance of biologics/ biosimilars within the hospital system should be no different to that of any other medicine.

2) The selection of a biologic/biosimilar as first-line therapy in treatment-naïve patients should be subject to evidence of safety, efficacy, and cost-effectiveness.

3) Biologics/biosimilars should be prescribed by both the active ingredient name and the trade name.

4) A biologic maybe interchangeable with its biosimilar/s at dispensing only where it has been determined to be substitutable by a drug and therapeutics committee (DTC).

5) Patients should be fully informed when receiving treatment with a biologic/biosimilar.

6) Switching between a biologic and its biosimilars should be in accordance with a drug and therapeutics committee – approved treatment protocol that includes a monitoring plan.

7) The selection of a biologic/biosimilar as second-line therapy should be in accordance with a treatment pathway approved by the drug and therapeutics committee.

8) There should be a patient-centered pharmacovigilance framework within each hospital or health service to monitor and report outcomes and any adverse effects associated with biologic/biosimilar therapy.

9) The active ingredient and trade name of the biologic/biosimilar should be clearly communicated at all transitions of care.

These principles provide a useful framework for how biologic medicines are approved and governed in the hospital setting. Similarly, the Scottish national prescribing framework for biosimilar medicines is an excellent resource.[38]

14.5.1 Evaluation of Biologic Medicines by PTCs

Evaluation of biologic medicines should follow established PTC processes[36] and generally will include the same considerations although greater emphasis will apply to some factors such as immunogenicity. A strong assessment of efficacy, safety, and comparative effectiveness against relevant comparators should be at the heart of the evaluation. There will also be some additional data that may be relevant for biologic medicines (Table 14.1).

Biosimilar medicines may be registered for all or some of the approved indications of the innovator medicines as regulatory agencies differ in whether indication extrapolation for biosimilars is acceptable. PTCs must be clear about which indications they are evaluating in a biosimilar application.[36]

Table 14.1 Considerations for PTC members evaluating biosimilars for formulary inclusion.

Clinical considerations

- Indication(s)
- Evaluation of efficacy and safety using available data
- Immunogenicity

Product considerations

- Nomenclature
- Manufacturing and supply chain considerations
- Packaging, labeling, and storage

Institutional considerations

- Substitution and interchangeability
- Therapeutic interchange
- Transition of care
- Pharmacovigilance
- Cost
- Reimbursement
- Provider and patient education
- Information technology

Source: Adapted from Ventola.[36]

The extent of clinical studies available for biosimilars may also be limited. Reliance on *in vitro* and *in vivo* functional assays to show that the biosimilar is highly similar to the innovator biologic is usual for regulatory agencies.[36] The EMA, for example, talks about process being an essential marker of quality rather than outcomes from clinical studies. For given patient populations, the PTC may want to view clinical data and/or observational data collected in registries and large databases. However, an ability to both understand and communicate with clinicians about the analytical data will be a new consideration for many PTCs.

Immunogenicity is a critical factor to consider for all biologic medicines. Immune reactions are common leading to direct harm and/or loss of efficacy. Infection is the other common complication of biologic medicines. In determining how a biologic will be used, the PTC would consider the susceptibility of the patient population to these adverse effects as well as treatment-specific factors. For example, longer treatment periods are associated with more immune-mediated events and intramuscular and subcutaneous injections are more immunogenic than intravenous injections.[30]

PTCs will want to ensure that the manufacturer of a biologic medicine has processes in place to ensure consistent quality and uninterrupted supply. This will be particularly important if the PTC does not wish to expose patients to multiple switching between and among biologics and biosimilars, respectively. Global shortages in the supply of small molecule medicines are already increasingly common and given the added complexity of the manufacturing process for biologics, it is reasonable to seek assurances about reliability of supply. Security of the supply chain and precautions against counterfeiting are an additional consideration for most countries.

Practical matters such as the storage requirements, resources needed to safely administer the product, shelf-life, and adaptability of the existing electronic health record system to the naming convention for a biosimilar product may also be important in reaching a formulary decision.

14.5.2 Recommendations for Interchangeability at the Local Level

Therapeutic interchange is a strategy used by PTCs and hospitals to manage expenditure in pharmaceuticals. It is based on the premise that the drugs being interchanged are clinically equivalent, having similar efficacy and safety profile which should be differentiated from generic substitution, and entails determining which drug is the most appropriate to be prescribed according to the hospital's policies.[34]

Therefore, one of the most difficult and also most common decisions for PTCs are likely to revolve around interchangeability between an innovator biologic and a biosimilar, whether to only use biosimilars in biologic-naïve patients, and/or whether to stock only a biosimilar and switch all patients. Possible questions to ask,[30,36] include:

- Therapeutic and dose equivalence, efficacy, and safety when switching products.
- Cost advantages of one product over another.
- Potential for a clear interchange process and understanding by prescribers.
- Ability to opt out in specific circumstances.
- Ability to monitor and assess efficacy and safety.

Clinicians have expressed reservations about therapeutic interchange and switching and one of the most common concerns is that there could be an increased risk of immunogenicity arising from the switch. A review of 90 studies involving switching between the innovator biologic and a biosimilar included both randomized controlled trials and observational studies. This review did not find a loss of efficacy as a result of switching with only exceptional reports from registries where this had been reported.[39] The Norwegian and Danish authorities are promoting switching and collecting data systematically to support this

Table 14.2 Steps for implementation
of an interchangeability program for biosimilars.[34]

1) Review all published clinical evidence to assess the therapeutic equivalence of the drugs used in the interchange

2) Assess the indications approved by regulatory agencies

3) Analyze all efficacy and safety data for each indication and the various clinical situations

4) Consider the impact of the interchange in different patient populations

5) Establish clinical and economic outcomes, as well as, technical measures

6) Develop an education plan for prescribers and health professionals who deal with therapeutic interchange

7) Obtain approval for the therapeutic interchange from the institution

8) Implement a follow-up protocol for the therapeutic interchange impact on health outcomes

Source: Reproduced with permission of Springer Nature.

practice.[25,40] Italian data show routine switching between biologic medicines of the same class is frequent in routine care.[41] Conversely, there have been a small number of case reports of serum sickness and high drop-out rates with switching.[42–44]

Hence the evidence base to support switching between biologic medicines is growing and this will increasingly guide decisions on interchangeability. A review of 102 guidelines, position statements, and implementation processes for therapeutic interchange identified eight steps for obtaining suitable results from therapeutic interchange[34] (Table 14.2).

14.6 Sources of Information and Education

14.6.1 Objective Information

The availability of unbiased information about biologic medicines is critical for both health professionals and patients. This information should inform decision-making by being presented in the context of usual treatment pathways and for specific patient populations and conditions. Consideration of the place of any biologic medicine in therapy and whether it should be available as first-, second-, or third-line treatment would precede consideration of which biologic medicine may be more suitable – be that the biologic innovator or a biosimilar.

Guidelines for many conditions which include biologic medicines are now available from guideline organizations and peak bodies. As always, evidence-based guidelines that

use a structured process such as GRADE[45] should be preferred.[1]

Guidelines that recommend use of biologic medicines will need to consider some particular factors in more detail than would be necessary for small molecule medicines. These include risk of serious infection and reactivation or exacerbation of existing infection.

Serious infection rates can be increased with biologics such as the TNF inhibitors. For example, the British Society for Rheumatology Biologics puts the risk of serious infection with a TNF inhibitor at 4.2/100 patient-years of follow up (95% CI: 4.0, 4.4) compared with 3.2/100 patient-years of follow up (95% CI: 2.8, 3.6) for small molecule therapies.[46] Guidelines should include advice about commencing such medicines in patients already at high risk of infection and provide guidance on the appropriate steps should infection occur. Stopping biologic treatment may not be the best course of action, as with TNF inhibitor patients who switch to another class of biologic seem to have less recurrent infections than those who cease all biologic medicines.[46]

Patients with pre-existing hepatitis B, hepatitis C, HIV, tuberculosis, or with a history of malignancy may need specific considerations before commencing use of some biologic medicines. Pretreatment with pneumococcal and influenza vaccines is recommended for many biologic medicines. Some biologic medicines can be used in pregnancy and while breast feeding and guidelines should review the available evidence as is usual for small molecule drugs.[47] However, following the report of four deaths from disseminated BCG for tuberculosis infection after exposure to a TNF inhibitor *in utero*, some agencies advise that live attenuated vaccines are contraindicated in infancy up until six months following *in utero* exposure to biologics.[48]

14.6.2 Education and Training

Professional societies and peak bodies in medicine and pharmacy are making education about biologics a priority and online and face-to-face forums are regularly available

---

1 The Grading of Recommendations Assessment, Development and Evaluation (in short GRADE) working group began in 2000 as an informal collaboration of people with an interest in addressing the shortcomings of grading systems in health care. The working group has developed a common, sensible, and transparent approach to grading quality (or certainty) of evidence and strength of recommendations. The GRADE approach is now considered the standard in guideline development.

to upskill and bridge knowledge gaps for these health professionals. The American Society of Oncologists noted that "Continuous provider education is critical to inform, promote and use biosimilar products in a medically appropriate and cost-effective way to treat cancer. Also important is patient education about biosimilars provided by a knowledgeable health care professional. Public awareness and education and the use of standardized publically available materials from professional societies, government sources and patient advocacy groups will help ensure understanding."[49]

Given that regulatory review of biosimilars relies less on clinical data and more on structural, functional, and pharmacological data which health professionals may be less confident interpreting, assistance to understand, and assimilate the relevant facts should form part of education and training programs. Health professionals should be aware of the importance of pharmacovigilance and the specific monitoring requirements of biologic medicines. Education about biologic medicines should include information on:

- Local regulation and policy.
- Evidence for efficacy and safety of biologics and place in therapy.
- Can patients on one biologic medicine be switched to another, for example, a biosimilar.
- What different approaches to use of biologic medicines are required for different compounds and in different clinical conditions.
- Efficacy and safety concerns specific to biologic medicines.
- How to monitor ongoing therapy, including clinical outcomes, therapeutic drug monitoring, and use of clinical registries.
- How to prescribe biologic medicine and record product details, including use of brand names and batch numbers.
- Where to find information for patients and carers.
- How to switch from an innovator biologic medicine to a biosimilar, including when consent is relevant.

Patients may need specific education about biosimilars, especially if changing from an innovator biologic medicine, but they should be advised of the potential harms and benefits of being prescribed any biologic medicine. When patients are switching, it may be necessary to assess if the new formulation requires specific skills. For example, self-injection devices may vary. Where appropriate and balanced information is not readily available for patients, health professionals and PTCs may consider developing their own.[36]

14.6.3 Interventions to Influence Prescribing

Programs to influence prescribing, whether to support diffusion of innovation or to reinforce guidelines, exist for common conditions in both hospitals and primary care. Antimicrobial stewardship, academic detailing programs, audit and feedback, and plan-do-study-act cycles are familiar ways of attempting behavior change among clinicians.

In Australia, NPS MedicineWise has used suite of interventions to increase knowledge, shift attitudes, and change prescribing.[50] These programs bring together data from many sources to understand the context within which a condition is treated, including semi-structured interviews with clinicians, evaluations of previous programs, feedback from academic detailing staff, and relevant literature reviews and expert advisors. This background provides a basis for understanding the barriers and enablers for change and generally leads to deployment of multifaceted programs.

In its first program to discuss use of biologic medicines in rheumatoid arthritis, NPS MedicineWise partnered with the Australian Rheumatology Association (ARA) to ensure credibility and promote peer to peer influence. This program included a focus on appropriate use of methotrexate as first-line therapy, early diagnosis, and second- and third-line treatments. It was developed through a codesign process that engaged doctors, pharmacists, and patients. Placing the discussion about biologic innovators and biosimilars in the broader context to management was considered essential to support sound decisions. The program was popular among specialist physicians and evaluation results will be produced as data become available.[51]

14.7 Adherence and Persistence

Adherence to and persistence with medication for chronic conditions is a frequently reported issue for all pharmacological therapies and a number of factors have been found to be important. Indeed, in 2002, the World Health Organization contended that more than 50% of all medicines were prescribed, dispensed, or sold inappropriately and that 50% of patients fail to take them correctly.[52] Some common contributors to poor adherence include low socio-educational status, unaffordability of medicines, attitudes toward the condition being treated, poor relationship with the prescriber, and concerns about side effects and long-term effects of taking medicines. It is likely that many of these same factors are relevant for chronic treatments using biologic medicines and in addition, biologics may bring new challenges and opportunities.

A large retrospective cohort study of patients with psoriasis found that adherence to TNF inhibitors was better than for methotrexate and that adherence was better among those aged 55–64 years, men and people with health insurance.[53] However, high rates of discontinuation remain common when biologic medicines are used in patients with psoriasis with reports suggesting around 40% of patients will discontinue treatment although some agents perform better than others.[54] Factors such as insurance status, satisfaction with treatment, treatment efficacy, socioeconomic status, dosing frequency choice, concern about adverse effects, and positive personal interrelationships with physicians impact on adherence and persistence.

Some interventions that have been used in dermatology to improve adherence include use of an anchoring technique to make monthly injections appear more appealing than, for example, daily injections of insulin. Explaining the efficacy of biologics by comparing the lower risk from therapy with the risk of untreated disease is a means to address loss aversion. Regressive bias may respond to accurate explanations of the risks associated with biologic therapy and positive framing. Using the need for regular visits to administer therapy as an opportunity to build the therapeutic relationship should strengthen the personal relationship and promote adherence.[54]

The term "drug survival rates" has been coined to measure the combined impact of discontinuation of medicines from all causes: intolerance, lack of efficacy, nonadherence, and non-persistence. Using this measure, better results were seen in psoriasis treatment, all biologic medicines except infliximab showed better drug survival rates than traditional agents with the best being ustekinumab at 52.9 months.[55] Biologic medicines were most commonly discontinued because of lack of efficacy (6.8–28.3%) and less frequently for adverse effects (5.1–12.4%).

In rheumatoid arthritis, nonadherence affects the success of treatment, remission, and disease severity. Poor persistence has also been found to be associated with higher overall consumption of health resources and costs. An observational retrospective cohort analysis in Italy found better persistence with etanercept than infliximab or adalimumab and also that those on etanercept were less likely to shift to lower doses or switch therapies. The treatment costs for patients switching from initial treatment during the first year of follow-up were higher than for patients who did not switch.[56] For patients with rheumatoid arthritis receiving a subcutaneous preparation of a biologic medicine, the length of treatment and persistence were influenced by the type of biologic used but not adherence. Overall persistence at follow-up decreased with time from 81% at six months to 67% at one year and 65% at two years.[57]

As previously noted, the nocebo effect can be an important reason for discontinuing or switching from biosimilar medicines.[20–22] Some commentators have suggested that well-informed patients are more likely to accept biosimilars and persist for longer with treatment.[58] Providing patients with relevant information and good communication with the patient using a shared decision-making model may promote better acceptance and use of biosimilars.

14.8 Cost-effectiveness Considerations

Overall, biologic medicines are considerably more costly than small molecule medicines and biosimilars may be priced lower than the biologic innovator. As well as affordability considerations most payers will be interested in the value that can be expected in terms of better health outcomes and quality of life for the additional spend, that is, the cost-effectiveness of these medicines.

Early and regular monitoring of biologic medicines and ceasing medicines that are not effective or not well tolerated is an important means of reducing waste and enhancing the cost-effectiveness of treatment. Similarly, using less expensive treatments in patients with milder disease or specific sub-types of disease is important for ensuring that overall pharmacological treatments represent value for money.

As previously noted, switching patients to or starting all new patients on biosimilars is another means of attempting to reduce the overall cost of therapy. Reductions in acquisition costs when biosimilars enter the market range from around 10 to 40% depending on the product and the market.[59] Nonmedical switching of biologic innovators to a biosimilar is gaining acceptance and there is clear evidence that it does reduce costs.[60] In the United States, it has been forecasted that some USD44 billion in cost saving could be anticipated through greater use of biosimilars from 2014 to 2024.

Payers will also be interested in other aspects of the value chain when making decisions about biologic innovators and biosimilars. These will include confidence in product quality and supply chain reliability, anticounterfeit protection, convenience of the dose form and formulation, and the total resource utilization costs for supporting treatment.[59] As previously noted, higher overall health resource utilization is reported when treatment changes are required in the first year of therapy.

14.9 Monitoring and Real-World Experience

Measuring clinical outcomes is important for all medicines. For biologic medicines, it is important to consider monitoring clinical outcomes for individual patients and establishing population-based clinical registries.

Monitoring of effectiveness and for safety and tolerability is particularly important when treatment is commenced and when switching between treatments. Patient response should be measured using established parameters of disease activity and documented in the clinical record. Suspected adverse drug reactions should be reported to the national regulator and the institutional safety body using standard procedures. For biologic medicines, it is more important to specify the brand of medicine and the batch number where this is known.

In some jurisdictions, such as Scotland, therapeutic drug monitoring services are available as an additional means of monitoring selected biologic medicines and protocols have been developed for rheumatology and gastroenterology.[38]

National and international registries have been established for some biologic medicines as a way to systematically collect post-marketing data on biologic medicines, including biosimilars and the results of switching. Registries should enhance quality and capture patient clinical data and patient satisfaction to optimize clinical management although there are doubts about the long-term sustainability of registries.[38]

14.9.1 Therapeutic Drug Monitoring

The pharmacokinetics and pharmacodynamics of biologics are influenced by many factors: gender, weight, physiological function, age, disease type, presence of antidrug antibodies, albumin concentrations, and concomitant medications. Therapeutic drug monitoring of biologics is being increasingly used, taking account of the concentrations of drug and antidrug antibody, as means of individualizing dosage. For patients with inflammatory bowel disease, it has been postulated that therapeutic drug monitoring could identify patients who are eligible for dose-tapering, intensification of

treatment, cessation of treatment, and switching within- or out-of-class and switching to a biosimilar.[61] Indeed, a blinded review of 201 patients with inflammatory bowel disease found inclusion of therapeutic drug monitoring rather than clinical acumen alone would have led to a change in 29% of decisions and an additional 23 patients ceasing therapy.[62]

Therapeutic drug monitoring may be particularly useful for using biologics in children and young people. For example, proactive therapeutic drug monitoring for adolescents with Crohn's disease reported either dosage changes or a change in selection of agent for almost 25% of all occasions of monitoring.[63]

Current limitations for therapeutic drug monitoring include access to timely and accurate results in many healthcare settings, lack of standardization of testing techniques, and the expense of assays.[64] Also for some agents such as ustekinumab, no correlation has been shown between clinical efficacy and trough levels of the drug. Possible explanations could include relevance of trough levels, levels of competing interleukins, and study sample sizes.[65]

14.10 Conclusions

Biologic medicines offer new options for the treatments of many cancers and chronic immune conditions. These medicines also bring new challenges in terms of how they are used, their safety profiles, the impact of antidrug antibodies on efficacy, and their cost. Biosimilars are becoming increasingly available and this raises the level of therapeutic complexity again.

By applying clear quality use of medicines principles, it is possible to use biologic medicines in ways that achieve maximum benefit and minimizes harms. This requires a deep understanding of clinician and patient knowledge and attitudes within the specific healthcare setting. A range of external influences as well as the patient's condition and evidence of comparative effectiveness and safety will be relevant to the quality use of biologic medicines. Objective information, education, and training should be available to support shared decision-making and promote persistence with therapy.

Quality use of medicines promotes cost-effective use of the medicines while building the evidence base with real-world data to continue to provide patients with the best care. PTCs, insurers, and third-party payers will find these helpful in their formulary decisions and ongoing reviews of management pathways.

Ensuring the best use of biologic medicines will be a critical global health issue over the next decade.

References

1 Orbis Research (2019). *Global Biologics and Biosimilars Market 2019*. Dallas: Orbis Research.

2 IQVia Institute for Human Data Science Reports (2018). *2018 and Beyond: Outlook and Turning Points*. London: IQVia.

3 Editor (2018). Top 10 drugs 2017–18. *Australian Prescriber* 41: 194.

4 Commonwealth of Australia (2002). *The National Strategy for Quality Use of Medicines*, plain English edition.

5 Avorn, J. (2018). The psychology of clinical decision making – implications for medication use. *The New England Journal of Medicine* 378 (8): 689–691.

6 Rainbird, K., Sanson-Fisher, R.W., and Buchan, H. (2006). *Identifying Barriers to Evidence Uptake* (ed. National Institute of Clinical Studies). Melbourne: National Institute of Clinical Studies.

7 Abuabara, K., Wan, J., Troxel, A.B. et al. (2012). Variation in dermatologist beliefs about the safety and effectiveness of treatments for moderate to severe psoriasis. *Journal of the American Academy of Dermatology* 68 (2): 262–269.

8 Midura, M. and Garg, A. (2018). Patient and physician perspectives on traditional systemic and biologic therapies for psoriasis. In: *Biologic and Systemic Agents in Dermatology* (ed. P.S. Yamauchi), 61–64. Cham: Springer International Publishing.

9 Kalkan, A., Roback, K., Hallert, E., and Carlsson, P. (2014). Factors influencing rheumatologists' prescription of biological treatment in rheumatoid arthritis: an interview study. *Implementation Science: IS* 9: 153.

10 Beck, M., Michel, B., Rybarczyk-Vigouret, M.-C. et al. (2016). Rheumatologists' perceptions of biosimilar medicines prescription: findings from a French web-based survey. *BioDrugs* 30 (6): 585–592.

11 Sullivan, E., Piercy, J., Waller, J. et al. (2016). THU0599 key drivers in biosimilar prescription in inflammatory autoimmune diseases indications in Germany. *Annals of the Rheumatic Diseases* 75 (suppl. 2): 408.

12 Waller, J., Sullivan, E., Piercy, J. et al. (2017). Assessing physician and patient acceptance of infliximab biosimilars in rheumatoid arthritis, ankylosing spondyloarthritis and psoriatic arthritis across Germany (Original research) (Report). *Patient Preference and Adherence* 11: 519.

13 Dylst, P., Vulto, A., and Simoens, S. (2014). Barriers to the uptake of biosimilars and possible solutions: a Belgian case study. *PharmacoEconomics* 32 (7): 681–691.

14 Chapman, S.R., Fitzpatrick, R.W., and Aladul, M.I. (2017). Knowledge, attitude and practice of healthcare professionals towards infliximab and insulin glargine biosimilars: result of a UK web-based survey. *BMJ Open* 7 (6): e016730.

15 Hemmington, A., Dalbeth, N., Jarrett, P. et al. (2017). Medical specialists' attitudes to prescribing biosimilars. *Pharmacoepidemiology and Drug Safety* 26 (5): 570–577.

16 Hallersten, A., Fürst, W., and Mezzasalma, R. (2016). Physicians prefer greater detail in the biosimilar label (SmPC) – results of a survey across seven European countries. *Regulatory Toxicology and Pharmacology* 77: 275–281.

17 Jacobs, I., Singh, E., Sewell, L.K. et al. (2016). Patient attitudes and understanding about biosimilars: an international cross-sectional survey (Original research) (Survey). *Patient Preference and Adherence* 10: 937.

18 Bolge, S.C., Goren, A., Brown, D. et al. (2016). Openness to and preference for attributes of biologic therapy prior to initiation among patients with rheumatoid arthritis: patient and rheumatologist perspectives and implications for decision making. *Patient Preference and Adherence* 10: 1079–1090.

19 Gonzalez, J.M., Johnson, F.R., McAteer, H. et al. (2017). Comparing preferences for outcomes of psoriasis treatments among patients and dermatologists in the U.K.: results from a discrete-choice experiment. *The British Journal of Dermatology* 176 (3): 777–785.

20 Pouillon, L., Socha, M., Demore, B. et al. (2018). *The Nocebo Effect: A Clinical Challenge in the Era of Biosimilars*, 739–749. Taylor & Francis.

21 Rezk, M. and Pieper, B. (2017). Treatment outcomes with biosimilars: be aware of the nocebo effect. *Rheumatology and Therapy* 4 (2): 209–218.

22 Bakalos, G. and Zintzaras, E. (2019). Drug discontinuation in studies including a switch from an originator to a biosimilar monoclonal antibody: a systematic literature review. *Clinical Therapeutics* 41 (1): 155–173.e13.

23 Tweehuysen, L., Bemt, B.J.F., Ingen, I.L. et al. (2018). Subjective complaints as the main reason for biosimilar discontinuation after open-label transition from reference infliximab to biosimilar infliximab. *Arthritis & Rheumatology* 70 (1): 60–68.

24 Glintborg, B., Sørensen, I., Loft, A.G. et al. (2017). FRI0190 clinical outcomes from a nationwide non-medical switch from originator to biosimilar etanercept in patients with inflammatory arthritis after 5 months follow-up. Results from the danbio registry. *Annals of the Rheumatic Diseases* 76 (s2): 553.

25 Glintborg, B., Sørensen, I.J., Loft, A.G. et al. (2017). A nationwide non-medical switch from originator infliximab to biosimilar CT-P13 in 802 patients with inflammatory arthritis: 1-year clinical outcomes from the DANBIO registry. *Annals of the Rheumatic Diseases* 76 (8): 1426.

26 Boone, N., Liu, L., Romberg-Camps, M. et al. (2018). The nocebo effect challenges the non-medical infliximab switch in practice. *European Journal of Clinical Pharmacology* 74 (5): 655–661.

27 Calara, P.S., Althin, R., Carlsson, K.S., and Schmitt-Egenolf, M. (2017). Regional differences in the prescription of biologics for psoriasis in Sweden: a register-based study of 4168 patients. *BioDrugs* 31 (1): 75–82.

28 O'Hare, C., Kelly, I., Raimondo, V. et al. (2018). P37 prescribing of biologics for patients with juvenile idiopathic arthritis (JIA) in NHS Lothian. *Rheumatology* 57 (suppl. 8) key273.039, https://doi-org.ezproxy.library.uq.edu.au/10.1093/rheumatology/key273.039.

29 Shai, M.Z., Sarpatwari, A., and Kesselheim, A.S. (2019). Why are biosimilars not living up to their promise in the US? *AMA Journal of Ethics* 21 (8): E668–E678.

30 Lucio, D.S., Stevenson, G.J., and Hoffman, M.J. (2013). Biosimilars: implications for health-system pharmacists. *American Journal of Health-System Pharmacy* 70 (22): 2004–2017.

31 Weise, M., Bielsky, M.-C., De Smet, K. et al. (2012). Biosimilars: what clinicians should know. *Blood* 120 (26): 5111–5117.

32 Moorkens, E., Vulto, A.G., Huys, I. et al. (2017). Policies for biosimilar uptake in Europe: an overview. *PLoS One* 12 (12): e0190147. https://doi.org/10.1371/journal.pone.0190147.

33 O'Callaghan, J., Barry, S., Bermingham, M. et al. (2019). Regulation of biosimilar medicines and current perspectives on interchangeability and policy. *European Journal of Clinical Pharmacology* 75 (1): 1–11.

34 Adrover-Rigo, M., Fraga-Fuentes, M.-D., Puigventos-Latorre, F., and Martinez-Lopez, I. (2019). Systematic literature review of the methodology for developing pharmacotherapeutic interchange guidelines and their implementation in hospitals and ambulatory care settings. *European Journal of Clinical Pharmacology* 75 (2): 157–170.

35 Ventola, C.L. (2013). Biosimilars: part 2: potential concerns and challenges for p&t committees. *P&T: A Peer-Reviewed Journal for Formulary Management* 38 (6): 329.

36 Ventola, C.L. (2015). Evaluation of biosimilars for formulary inclusion: factors for consideration by P&T committees. *P&T: A Peer-Reviewed Journal for Formulary Management* 40 (10): 680–689.

37 Council of Australian Therapeutic Advisory Groups (2016). *Overseeing Biosimilar Use. Guiding Principles for the Governance of Biological and Biosimilar Medicines in Australian Hospitals*. CATAG.

38 Healthcare Improvement Scotland. Biosimilar Medicines (2018). *A National Prescribing Framework*. Edinburgh: NHS Scotland.

39 Cohen, H., Blauvelt, A., Rifkin, R. et al. (2018). Switching reference medicines to biosimilars: a systematic literature review of clinical outcomes. *Drugs* 78 (4): 463–478.

40 Jørgensen, K.K., Olsen, I.C., Goll, G.L. et al. (2017). Switching from originator infliximab to biosimilar CT-P13 compared with maintained treatment with originator infliximab (NOR-SWITCH): a 52-week, randomised, double-blind, non-inferiority trial. *The Lancet* 389 (10086): 2304–2316.

41 Ingrasciotta, Y., Sultana, J., Kirchmayer, U., and Trifirò, G. (2019). Challenges in post-marketing studies of biological drugs in the era of biosimilars: a report of the international society for pharmacoepidemiology 2019 mid-year meeting in Rome, Italy. *BioDrugs* 33 (4): 345–352.

42 Kang, Y.-S., Moon, H., Lee, S. et al. (2015). Clinical experience of the use of CT-P13, a biosimilar to infliximab in patients with inflammatory bowel disease: a case series. *Digestive Diseases and Sciences* 60 (4): 951–956.

43 Yazici, Y., Xie, L., Ogbomo, A. et al. (2017). A descriptive analysis of real-world treatment patterns in a Turkish rheumatology population that continued innovator infliximab (remicade) therapy or switched to biosimilar infliximab. *Annals of the Rheumatic Diseases* 76 (s2): 836.

44 Hanauer, S.B., Panes, J., Colombel, J.F. et al. (2010). Clinical trial: impact of prior infliximab therapy on the clinical response to certolizumab pegol maintenance therapy for Crohn's disease. *Alimentary Pharmacology & Therapeutics* 32 (3): 384–393.

45 Guyatt, G.H., Oxman, A.D., Kunz, R. et al. (2011). GRADE guidelines: 7. Rating the quality of evidence – inconsistency. *Journal of Clinical Epidemiology* 64 (12): 1294–1302.

46 Subesinghe, S., Rutherford, A.I., Byng-Maddick, R. et al. (2018). Biologic prescribing decisions following serious infection: results from the British society for rheumatology biologics register – rheumatoid arthritis. *Rheumatology* 57 (12): 2096–2100.

47 Flint, J., Panchal, S., Hurrell, A. et al. (2016). BSR and BHPR guideline on prescribing drugs in pregnancy and breastfeeding – part I: standard and biologic disease modifying anti-rheumatic drugs and corticosteroids. *Rheumatology* 55 (9): 1693–1697.

48 Hart, S. (2016). *Guideline for the Prescribing of Biologic Therapy in Adult Patients with Active and Progressive Psoriasis Arthritis with NHS Fife*. Fife: NHS: Fife Managed Service Drug and Therapeutics Committee.

49 Lyman, G.H., Balaban, E., Diaz, M. et al. (2018). American society of clinical oncology statement: biosimilars in oncology. *Journal of Clinical Oncology: Official Journal of the American Society of Clinical Oncology* 36 (12): 1260.

50 Weekes, L.M., Blogg, S., Jackson, S., and Hosking, K. (2018). NPS medicinewise: 20 years of change. *Journal of Pharmaceutical Policy and Practice* 11 (1): 19.

51 NPS MedicineWise (2019). *2018 Annual Evaluation Report*.

52 World Health Organization (2002). *WHO Policy Perspectives on Medicines – Promoting Rational Use of Medicines*. Geneva: WHO.

53 Dommasch, E.D., Lee, M.P., Joyce, C.J. et al. (2018). Drug utilization patterns and adherence in patients on systemic medications for the treatment of psoriasis: a retrospective, comparative cohort study (Report). *Journal of the American Academy of Dermatology* 79 (6): 1061.

54 Aleshaki, J.S., Cardwell, L.A., Muse, M.E., and Feldman, S.R. (2018). *Adherence and Resource Use Among Psoriasis Patients Treated with Biologics*, 609–617. Taylor & Francis.

55 Arnold, T., Schaarschmidt, M.-L., Herr, R. et al. (2016). Drug survival rates and reasons for drug discontinuation in psoriasis. *JDDG: Journal der Deutschen Dermatologischen Gesellschaft* 14 (11): 1089–1099.

56 Degli Esposti, L., Favalli, E.G., Sangiorgi, D. et al. (2016). Persistence, switch rates, drug consumption and costs of biological treatment of rheumatoid arthritis: an observational study in Italy. *ClinicoEconomics and Outcomes Research* 9: 9–17.

57 Alvarez-Madrazo, S., Kavanagh, K., Semple, Y. et al. (2017). Adherence and persistence of subcutaneous biologics delivered through homecare services to a cohort of patients with rheumatoid arthritis. *International Journal of Population Data Science* 1 (1) https://doi.org/10.23889/ijpds.v1i1.248.

58 Sigaux, J., Semerano, L., and Boissier, M.-C. (2018). Switch to a biosimilar: whatever the cost? *Joint, Bone, Spine* 85 (6): 651–654.

59 Smeeding, J., Malone, D.C., Ramchandani, M. et al. (2019). Biosimilars: considerations for payers. *P&T: A Peer-Reviewed Journal for Formulary Management* 44 (2): 54–63.

60 Liu, Y., Garg, V., Yang, M. et al. (2018). AB1276 economic impact of non-medical switching from originator biologics to biosimilars – a systematic literature review. *Annals of the Rheumatic Diseases* 77 (s2): 1731.

61 Gils, A. (2017). Combining therapeutic drug monitoring with biosimilars, a strategy to improve the efficacy of biologicals for treating inflammatory bowel diseases at an affordable cost. *Digestive Diseases (Basel, Switzerland)* 35 (1–2): 61.

62 Selinger, C., Lenti, M., Clark, T. et al. (2017). The impact of infliximab therapeutic drug monitoring on decisions made in a virtual biologics clinic for IBD. *Gut* 66 (suppl. 6): A148.

63 Rentsch, C.A., Sparrow, M., Ward, M.G. et al. (2018). Mo1854 – pharmacist-led proactive therapeutic drug monitoring with infliximab (proximo): utility of and cost saving associated with the use of a rapid assay for assessing drug level. *Gastroenterology* 154 (6): S-826.

64 Renton, W.D. and Ramanan, A.V. (2019). Biological therapeutic drug monitoring: a step towards precision medicine? *Archives of Disease in Childhood* 104 (3): 212.

65 Liau, M.M. and Oon, H.H. (2019). Therapeutic drug monitoring of biologics in psoriasis. *Biologics: Targets and Therapy* 13: 127–132.

15

Knowledge Areas and Competency Standards on Biologic Medicines for Pharmacists and Pharmacy Students

Iqbal Ramzan and Reza Kahlaee

Sydney Pharmacy School, Faculty of Medicine and Health, The University of Sydney, Sydney, New South Wales, Australia

---

KEY POINTS

- The current didactic material on biologic medicines in pharmacy degrees is not commensurate with their structural, functional, manufacturing, and clinical complexities compared to small molecule drugs (SMDs).
- Pharmacy degree programs should be critically appraised of prerequisite high school science and university-level foundation biomedical sciences.
- Pharmaceutical sciences relevant to innovator biologics, biosimilars, and biobetters and public health practice relevant to these complex and expensive medicines need to be included to promote QUM of biologics.

Abbreviations

Abbreviation	Full name
AACP	American Association of Colleges of Pharmacy
ACPE	Accreditation Council for Pharmacy Education
APC	Australian Pharmacy Council
BSA	Behavioral, Social, and Administrative
B Pharm	Bachelor of Pharmacy
BSc	Bachelor of Science
CAPE	Center for Advancement of Pharmacy Education
CCAPP	Canadian Council for Accreditation of Pharmacy Programs
EBHC	Evidence-Based Health Care
EPCF	European Pharmacy Competencies Framework
EU	European Union
GCC	Gulf Cooperation Council
GPA	Grade-Point Average
GPhC	General Pharmaceutical Council
IPAC	International Pharmacy Accreditation or Certification

Abbreviation	Full name
IPE	Inter-Professional Education
IPL	Inter-Professional Learning
M Pharm	Master of Pharmacy
NZPC	New Zealand Pharmacy Council
OSCEs	Objective Structured Clinical Examinations
PBL	Problem-Based Learning
PEG	Polyethylene Glycol
P/G	Postgraduate
Pharm D	Doctor of Pharmacy
PBA	Pharmacy Board of Australia
PhC	Pharmaceutical Chemist
PhG	Graduate in Pharmacy
QUM	Quality Use of Medicine
RMP	Risk Management Plan
SMD	Small Molecule Drug
TBE	Team-Based Education
UK	United Kingdom
US	United States

Biologics, Biosimilars, and Biobetters: An Introduction for Pharmacists, Physicians, and Other Health Practitioners, First Edition. Edited by Iqbal Ramzan.

15.1 Chapter Background

Evidence-based health care (EBHC) involves the active use of relevant and validated, up-to-date published research data to make decisions about patient care.[1,2] EBHC requires practitioners like pharmacists to understand and critically evaluate the primary literature. Pharmacists, as the only health practitioners who are experts in all aspects of medicines, are the best resource for evidence-based drug information and therefore must possess appropriate fundamental knowledge and comprehensive literature evaluation skills and be able to interpret research data to provide advice about medicines to doctors, other health practitioners, and patients.

As a result of rapid and massive changes in health care with the introduction of complex and expensive biologic medicines, the need for pharmacists to keep on top of the rapidly evolving information and data is even more pressing. The introduction of "highly similar" versions of innovator biologics, biosimilars, means pharmacists need to be knowledgeable about all aspects of biosimilars, from development to regulation, to pharmacy and public health practice, to clinical utility, and quality use of medicines. This becomes especially important as more biosimilars are approved and enter the market.

In addition, pharmacists must possess the skills and knowledge necessary to provide adequate health advice to diverse populations who have varying levels of health literacy about their ailments and drug therapy. Flexibility and understanding of cultural norms are also required to facilitate shared understanding between prescribers and patients in order to provide appropriate guidance for management and follow-up. The demographics of the pharmacy profession in some western countries, especially with the movement of refugees, probably with the notable exception of Australia, does not mirror the cultural, religious, or ethnic diversity of the general population. Coupling this disparity with the projected diversity of populations creates a challenge in communication and awareness between patients and their pharmacists.

This chapter therefore examines the knowledge areas and competency standards required by pharmacists and graduating pharmacy students on all aspects of biologic medicines.

15.2 Current Knowledge of Pharmacists and Pharmacy Students About Biologics

Several surveys exist of the current knowledge base of pharmacists about biologic medicines as well as about the adoption of biosimilars. A study that compared pharmacists' knowledge and views on biosimilars in Quebec and France concluded that the knowledge base was similar in Quebec and France. Most pharmacists understood the difference between SMD generics and biosimilars but the knowledge was less uniform with respect to biosimilarity and substitution.[3] The authors concluded that pharmacists should be knowledgeable about the peculiarities and critical issues of biosimilars as they will play a crucial role in their introduction into clinical practice. Another recent survey on perspectives of polish hospital pharmacists toward biosimilars in their general hospitals found that pharmacists expressed concerns about perceived differences between innovator biologics and biosimilars including their immunogenicity and pharmacist-led substitution.[4] The authors also suggested that greater communication is needed between pharmacists and physicians and that evidenced-based education is required to improve the use of biosimilars. Another report on professional pharmacy services among Australian community pharmacists in 2018 reported that only approximately a third of the respondents felt comfortable providing information to patients on biosimilars or comfortable about substitution with biosimilars.[5] There does not appear to be any published studies on such knowledge base of graduating pharmacy students.

15.3 Prescriber Knowledge of Biologic Medicines

A survey-based study[6] in US specialty physicians identified five significant knowledge gaps. These included defining biologics, biosimilars, and biosimilarity; understanding the approval process; and the use of "totality" of evidence to evaluate biosimilars. That safety and immunogenicity of biosimilars are similar to innovator biologics, the rationale for indication extrapolation and defining interchangeability and rules related to substitution were additional information gaps.[6]

There are also different perceptions of biosimilars across different physician groups in Europe; rheumatologists were more aware of biosimilars, but gastroenterologists were more confident in their safety and efficacy.[7] Interestingly, there was a relationship between duration of rheumatology practice and likelihood of use of biosimilars.

A recent systematic review involving 20 studies of US and European providers examined factors affecting health-care provider knowledge and acceptance of biosimilars. This study[8] concluded that health-care providers still approach biosimilars with caution, due to limited biosimilar knowledge, low prescribing comfort, and safety and efficacy concerns as the main deterrents to biosimilar use. The authors also suggested that targeted prescriber education was urgently required.

15.4 Pharmacist's Role in Biologic Medicine Education

Pharmacy is a knowledge[9] and science-based[10] profession and pharmacists are acknowledged experts on all aspects of drugs from their preparation, chemical properties, analytical tools to ensure their purity and potency, their pharmacology/pharmacodynamics, their disposition/pharmacokinetics, and clinical application.[11] As more and more biosimilars are approved, pharmacists, especially those working in hospitals, regulatory and industry environments have a crucial role in all aspects of biosimilars including design, development, procurement, appropriate and safe clinical use, policy development, regulation, education and patient counseling, and post-marketing surveillance and pharmacovigilance.[11,12] For these reasons, it has been claimed that biosimilars are a pharmacist's opportunity[13] and patients need pharmacists' guidance on biosimilars.[14] In addition, pharmacists have a key role in the education of different speciality prescribers since as already discussed, clinicians in the United States and Europe are cautious about biosimilar use and do not predominantly support biosimilar use in patients already receiving innovator biologics.

15.5 General Background on Biologic Medicines

As already discussed in Chapter 1, biologic medicines are complex molecules; generally, proteins or mixtures of proteins compared to small molecule drugs (SMDs) and their generic equivalents are exact copies of innovator SMDs.[15] Conventional SMDs are generally either synthesized chemically or isolated from natural sources compared to biologics which are produced through a (or many) biological process(es), often involving biotechnology methods and invariably including numerous steps which have the potential to all inherently display variability.[16] There are diverse categories of biologics including antibodies, vaccines, and more recently gene and cell therapies, which raise many scientific and logistical challenges for clinicians as well patients.[17]

Biosimilars follow-on from innovator biologics via a similarity exercise while biobetters are active improvements on innovator biologics and biosimilars to enhance a particular property like improved pharmacokinetic behavior or reduced immunogenicity. The chemical, functional, therapeutic, pharmacokinetic, and manufacturing complexity of biologic medicines compared to SMDs (and their generic equivalents) translates to a more extensive data and literature base of knowledge and competencies required for biologic medicines compared to conventional drugs. In addition, these drugs are significantly more expensive and have higher efficacy and safety concerns than their predecessors. So, biologics are exponentially complex, have more significant manufacturing and formulation challenges, are manyfold more expensive, and are the fastest-growing sector of the drug market.[18] Purely on these simple metrics, pharmacists, who are the most competent health practitioners on medicines, need to know a lot more than they currently do on biologic medicines either innovator biologics, biosimilars, or biobetters. This chapter focuses on the appropriate knowledge areas and competencies that are crucial for current and future pharmacists and the current pharmacy students, about innovator biologics, biosimilars, and biobetters.

15.6 History of Pharmacy Education

Pharmacy education in the west has undergone many and varied stepwise changes over the last 100–200 years, the principal evolutions being the shift from a non-degree "training" to a university degree and a change from a compounding-activity-related training program to one focusing on pharmacy practice and clinical services.

In Europe, today the predominant pharmacy qualification required for registration and practice as a pharmacist is either a Bachelor's, Master's degrees, or a doctorate, typically 5–6 years in duration. This is very different from the early days of pharmacy education in Europe. In the United Kingdom, for example, in the 1800s, the pharmacy training was voluntary with the Pharmacy Act of 1841 allowing the Pharmaceutical Society of Great Britain to grant the title of Pharmaceutical Chemist to those individuals passing its exams. It was not until 1904 that the first university pharmacy qualification (BSc in Pharmacy) was offered (by the University of Manchester). Now the predominant registrable qualification in the United Kingdom is a four-year Master of Pharmacy (M Pharm) degree program.

In the United States, there is a somewhat similar evolution in pharmacy education. In the 1800s, the pharmacy qualification involved either a two-year program (Graduate in Pharmacy, PhG) or a three-year degree program (Pharmaceutical Chemist, PhC), offered at the Philadelphia College of Pharmacy. The latter then evolved into a four-or five-year Bachelor of Science in Pharmacy (BSc, Pharmacy) degree with the five-year program serving as a precursor to the Doctor of Pharmacy degree. California, unlike other US states, opted for a six-year Doctor of Pharmacy degree program in the 1950–1960s. Now, all states in the United States have opted for a Pharm D as the entry-level qualification for the pharmacy profession. Canada has followed this trend in the last 10–20 years with a universal Pharm D in all its provinces.

Pharmacy education in India traditionally has been industry and product-oriented with graduates preferentially seeking industry positions. To practice as a pharmacist in India, one needs at least a 2.25-year diploma in pharmacy; these diploma pharmacists are the mainstay of pharmacy practice in India. A three-year Bachelor of Pharmacy qualification is also available and a practice-based Doctor of Pharmacy (Pharm D) degree program was started in some private institutions in 2008.[19] More recently in some selected universities like JSS College of Pharmacy, a Pharm D akin to the US Pharm D has been introduced. In India, the Pharmacy Council of India has agreed to the use of the prefix "Dr" for Doctor of Pharmacy degrees from recognized universities.

In Australia, Pharmacy education started as an apprenticeship system involving some formal didactical and practical training in addition to workplace experience. For example, at the University of Sydney, this resulted in a Materia Medica qualification as part of the discipline of pharmacology, a forerunner to the Department of Pharmacy. This led to the introduction of a three-year B Pharm degree in the early 1960s followed by its evolution into a four-year bachelor's program in 1997. The principal pharmacy degree now remains this four-year program, but some universities also offer a graduate-entry master's degree (Master of Pharmacy, M Pharm), delivered over six semesters covering two years.[20]

In summary, historically, as well as currently, there have been dramatic increases in the number of pharmacy education institutions worldwide in both developed and developing countries. There is also an increase in diversity of pharmacy programs, which differ in nature, scope, curriculum focus, degrees offered, and length of study by region, country, and university.

15.7 Accreditation of Pharmacy Degrees

The purpose of accreditation is to assure the quality of the education and training provided to pharmacy students and to promote continuous improvement in the quality of pharmacy degree programs. The primary intended goal of pharmacy accreditation is to safeguard the public by ensuring the program graduates achieve the required competence to practice as pharmacists and contribute to enhanced health outcomes for consumers. The focus of the accreditation process is, therefore, on both the pharmacy program and the education provider.

Most pharmacy degree programs around the world are accredited either via a government regulatory framework or a professional body framework. For example, in the United States, the Accreditation Council for Pharmacy Education (ACPE) accredits and pre-accredits schools offering Pharm D degrees and providers of continuing pharmacy education. ACPE is made up of professionals from the American Council on Education, the American Association of Colleges of Pharmacy (AACP), the American Pharmacists Association, and the National Association of Boards of Pharmacy. Similarly, in Canada, the Canadian Council for Accreditation of Pharmacy Programs, CCAPP, is responsible for the accreditation of the professional degree programs in pharmacy at the universities in Canada and of pharmacy technician programs in the college-level system.

In the United Kingdom (England, Wales, and Scotland), pharmacy schools are accredited to offer the M Pharm program by the General Pharmaceutical Council (GPhC).[21] It also sets standards and regulates pharmacy technicians. In Europe, the accreditation is against the European Pharmacy Competencies Framework (EPCF),[22] developed in the EU-sponsored Phar-QA (Quality Assurance in European Pharmacy) and Phar-IN (Competences for Industrial Pharmacy Practice in Biotechnology) projects. EPCF covers accreditation for all pharmacists (community, hospital, and industrial).

The Australian Pharmacy Council (APC) on behalf of Australian Health Practitioner Regulation Agency (AHPRA) via a contract with the Pharmacy Board of Australia (PBA) and The New Zealand Pharmacy Council (NZPC) accredits all pharmacy programs in Australia and New Zealand, respectively. APC has recently revised these competency standards for pharmacy degrees which have been in place since 2014. These standards, "Accreditation Standards for Pharmacy Programs in Australia and New Zealand 2020"[23] came into effect in January 2020. A unique feature is that Social Accountability underpins the standards. APC has also provided Performance Outcomes Framework[24] to accompany its standards as well as two resources for education providers (Implementation and Transition Process Guidance Document and Self-assessment Tool and Action Plan – Template).

APC has also published a review of the international literature on pharmacy and related health professional accreditation.[25]

In other regions, like the Gulf Cooperation Council (GCC) countries, International Pharmacy Accreditation or Certification (IPAC) has become increasingly popular in the absence of uniform mandatory national bodies to accredit pharmacy degrees.[26]

Accreditation generally involves the alignment of degree programs to accreditation competencies which are compiled after consultation with the profession including key pharmacy organizations and individual education providers. A competency may be defined as a general statement describing the desired knowledge, skills, and behaviors of a graduate of a program or a course of study.[27, 28] Each standard is a statement of expectation written in active form. Competencies generally define the skills and knowledge that enable a practitioner to perform his or her professional role. A learning outcome in the context of accreditation, in contrast, is a specific statement that describes specifically what a student will be able to do in an assessable manner. There may be more than one measurable learning outcome for a given competency.[28]

Competencies and learning outcomes are thus two related but different educational terms within the context of accreditation; they do not have the same meaning as discussed above. When the term competence is used, it should be defined to ensure clarity of meaning in a particular context, and the required competencies should be expressed in terms of achievement of specific learning outcomes by students or graduates.[29]

Competencies are generally grouped into several domains; for example, in the Australian context, the groupings include Governance, Structure and Administration, Resource Allocation and Management, Curriculum, Program Students & Quality and Risk Management. In addition, the standards are accompanied by performance (learning) outcomes framework to assist education providers.

Accreditation agencies also provide guidance on the indicative curriculum for pharmacy degree programs (for example, as was the case by the Royal Pharmaceutical Society of Great Britain in 2002) or pharmacy learning domains as in Australia by the APC

which were based on the indicative curriculum in the United Kingdom since there is a more significant move by pharmacy education providers for an integrated curriculum. These APC learning domains include: The health-care consumer; Medicines - drug action; Medicines - the drug substance; Medicines - the medicinal product; Health-care systems and the roles of professionals; and the broader context. It is essential to understand the inter-relationship between indicative curriculum (or knowledge areas) and learning domains and the codependency of competencies and learning domains and their mutual impact. It is also essential to understand that the curriculum is indicative rather than prescribed to avoid stifling curriculum innovation.

Accreditation, of course, is ultimately about patient safety but it is not easy to elucidate the relationship between either generic or specific competencies and patient outcomes although generic competencies have been shown to impact on workplace performance.[30] It is also essential to think about the relationship between competencies and assessment.[31]

15.8 Content of Pharmacy Degree Curricula

There have always been discussions within the pharmacy profession on the content of pharmacy registrable qualifications and the competencies required for pharmacists. For example, there have been robust debates on the balance of fundamental biomedical/pharmaceutical sciences compared to clinical pharmacy content. In most countries, this balance has shifted toward clinical pharmacy in the last several decades.

Interestingly, there have been many similar past curriculum discussions leading to adoption/addition of seemingly less important or critical or less evidence-based topics into pharmacy degree curriculum like aromatherapy science, homeopathy, cosmeceutical compounding, primary and clinical science aspects of vitamins and minerals or other herbal and dietary supplements, complementary and alternative medicines, wellness programs, and pharmacy history. Of course, some recent additions to pharmacy curriculum have been arguably more logical, including areas of societal need like palliative or end-of-life care and cultural competency; newer areas of innovation in health care such as health informatics, pharmacogenomics and personalized medicine; drug information programs; and specialized areas of pharmacy practice like veterinary pharmacy, for example.

Table 15.1 Core knowledge areas for pharmacists and pharmacy students.

Biomedical/pharmaceutical sciences	Enabling pharmacy and public health practice
Fundamental biochemical and biophysical properties of proteins and large molecules including vaccines and antibodies	Efficacious, safe, and cost-effective therapeutic use of biologic medicines and the differences between biologics and SMDs
Suite of complex biophysical/functional tests	Regulatory frameworks for approval of biosimilars; their naming conventions
Formulation and mode(s) of administration	QUM and behavior change concepts for enhanced adoption of biosimilars
PK/PD of proteins/large molecules (their distinct elimination pathways and modes[s] of action)	Pharmacoeconomics of biologics and understanding that biosimilars are not priced like SMD generics
Cell biology and basic immunology and immunogenicity from proteins and antibodies	Pharmacoepidemiology and post-marketing pharmacovigilance and safety plans

Pharmacy education leaders continuously review pharmacy degree curricula content and knowledge areas to meet contemporary pharmacy practice and evolving professional roles for pharmacists. New treatment modalities impact on a pharmacist's role, their contribution to health systems, and ability to enhance patient well-being and outcomes. Similarly, pharmacy degree accreditation agencies also have an interest in whether graduating pharmacists have the appropriate knowledge areas on newer therapies like biologic medicines including biosimilars and biobetters.

The marked structural, functional, and clinical complexities of innovator biologics, biosimilars, and biobetters, together with their exponential growth in modern therapeutics as well as their expense make it mandatory that pharmacists fully understand the underlying science and clinical issues for these agents. This includes biochemical/biophysical properties, unique disposition characteristics, and immunologic responses. The enabling pharmacy practice and public health concepts, which are unique to the therapeutic utility of protein drugs, pharmacovigilance plans, and economic considerations are also essential and should always underpin the fundamental and prerequisite sciences. Such science should also be infused into the curriculum and integrated with public health concepts. In addition, naming conventions of biosimilars and behavior-modifying concepts to enhance wider adoption of biosimilars should also be included to ensure QUM of innovator biologics and biosimilars. The recommended core knowledge areas as they relate to all biologic medicines are summarized in Table 15.1. These areas have been chosen for recommendation in the pharmacy degree curricula on the following premise:

1) Pharmacy is a knowledge-[9] and science-based[10] profession with a fundamental clinical component.

2) It follows from 1 above that Pharmacy's curriculum needs to be evidence-based and data-driven.

3) These suggested knowledge areas are critical to maintaining the pre-eminent standing of pharmacists as the medicines experts in all medicines within the health sector, including new and emerging therapies.

4) The pharmacy curriculum needs to prepare pharmacy graduates for not only traditional or current roles but more importantly for emerging and more challenging professional roles, opportunities, and responsibilities.[32]

5) Graduates should have the opportunity to gain detailed knowledge, understanding, and skills to enable them to achieve the practitioner outcomes for practice.

Four broad curriculum areas are identified: prerequisite or fundamental underpinning sciences; biomedical and pharmaceutical sciences, clinical sciences, public health, and pharmacy practice. These core knowledge areas will now be discussed in greater detail.

15.8.1 High School Basic Science Prerequisites

There have been many discussions over many decades on the enabling sciences that are required for entry into pharmacy degree programs. In the past, there have been far more stringent requirements with respect to high school science-based subjects for entry into pharmacy programs. These have included requirements for mathematics (different levels including standard or advanced mathematics), physics, chemistry, general biology, to mention a few. In the last decade, in most countries and universities, offering pharmacy degree programs, such requirements, as part of admissions requirements, have been relaxed. Especially in the western world, there are few if any such requirements remaining in a move to provide greater flexibility in entry requirements and less restrictive practices generally for students entering science-based health professional university courses. In fact, in the last decade, it has been theoretically possible to be admitted to a health professional degree including pharmacy without any high school science. However, this has generally come at a cost and many universities now offer short intensive additional courses. These typically occur during summer or winter schools outside of standard university semesters/terms to bring incoming entrants to a level of science literacy that would assist them with normal progression with their first year of university foundation science courses.

With the introduction of more complex biologic medicines and emerging biotherapies including cell and/or tissues personalized therapies, it might be timely for the pharmacy education sector, the pharmacy degree accrediting agencies, the pharmacist registering authorities, and the pharmacy professional more generally to have a robust, transparent, and objective discussion about what high school science subjects might be mandatory for entry into a pharmacy degree. These might include science subjects like human biology, chemistry, and certainly mathematics. It can also be argued that conversations with doctors, patients, and other health practitioners require sound communication skills and that some social science-based high school subjects might also assist entry-level pharmacy students in their training as pharmacists.

15.8.2 University-Level Basic Sciences

Most North American (US and Canada) pharmacy programs admit incoming students based mainly on completion of pre-pharmacy university courses. The pre-pharmacy grade-point average (GPA) is often used as an entry criterion but GPA, of course, is a composite measure of overall pre-pharmacy performance. It is interesting to note that a recent publication that examined the relationship between prerequisite grades and types of academic performance in pharmacy degrees found a consistent relationship between biology-based prerequisites and academic performance in pharmacy degrees.[33] Interestingly, this prerequisite was better at predicting academic performance in the pharmaceutical sciences as opposed to performance in clinical practice or behavioral, social, and administrative (BSA) aspects of pharmacy. Based on their study results, the authors suggested a higher weighting of biology prerequisites in admissions for GPA calculations.

In countries like Australia, generally, the first year of an undergraduate pharmacy degree (four-year Bachelor of Pharmacy degree) involves university-level science courses like mathematics, chemistry, and biology (generally human biology) for all incoming students regardless of whether these students have previously taken science-based courses at high school. Even for the postgraduate (P/G) master's-level pharmacy degree (Master of Pharmacy, M Pharm), which is considered equivalent to the bachelor's program for registration as a pharmacist, many science-based prerequisites must be met or completed before commencing pharmacy-specific components of the degree.

15.8.3 Enabling Biomedical Sciences

Some two decades ago, it was shown that pharmacy students' knowledge and fluency in chemistry and the biological sciences were significant determinants of success in both professional pharmacy curriculum and contemporary pharmacy practice.[34] This led to calls to the ACPE and the AACP for standardizing basic science prerequisites across all preprofessional programs in the United States. Biochemistry and pathophysiology instruction was also seen to be inadequate in the US pharmacy students.[35]

The pipeline of new medicines to enter the therapeutics arena are far more complex than the current biologics and SMDs that pharmacists that trained in the last three or so decades have had to deal with in their professional careers. This will continue to be the case in the coming decades as game-changing cell, and tissue-based personalized treatments make their entry into therapeutics.

Biologic medicines are certainly far more complex structurally and functionally as well as in manufacturing attributes and the analytical methods to characterize their biosimilars are far more complex. Biochemistry is undoubtedly required and including content on protein and large macromolecule concepts are critical. Such courses need to be explicitly tailored to pharmacy students as opposed to general biochemistry for science students majoring in biochemistry as part of pure science education. The biophysical and functional tests needed to characterize proteins as might be required for approval of biosimilars also need to be included in pharmacy degree programs. These analytical topics are not always popular with pharmacy students, however. Since most of these courses are service taught to pharmacy students by biochemistry departments, such material needs to be targeted and relevant to pharmacy students. This requires a collaborative effort on behalf of pharmacy schools and general science departments. Reliance on teaching income for such service teaching schemes should not be the primary driver of the course content, which appears to be the case in many universities. What is required and necessary for pharmacy professional programs in terms of the enabling biomedical sciences, whether taught within a pharmacy school or by an external service department, should drive the curriculum content.

In the human biology area, emphasis needs to be placed on protein immunology and immunogenicity concepts as these are the principal drivers of the adverse effects from protein biologic drugs that are currently in the market and those that are likely to enter the market. An understanding of the science behind the use of cells, bacteria, yeast, or other living systems which are the workhorse manufacturing tools to produce biologic medicines is also essential. Molecular biology techniques that are used currently and those that are emerging are also critical concepts for pharmacy students.

15.8.4 Enabling Pharmaceutical Sciences

Pharmaceutical sciences relevant to biologic medicines required for pharmacy students may be broadly classified into two groups. The first involves concepts of formulation science for macromolecules like proteins and vaccines, considering that minor changes in manufacture including drift, divergence, and evolution[36] involving process efficiencies in manufacturing science[37] all impact the consistency of not only innovator biologics but also biosimilars and biobetters. This may also include both upstream processing issues (choice of master and working cell banks; growth of the cells and expression of the protein and process control of the physiologic environment in the bioreactor where the cells are grown) and downstream issues that involve all bioprocess steps that occur after cessation of cell culture beginning with harvesting of the culture from the bioreactor. These involve harvesting and purifying the protein, concentration, and other formulation steps including fill and finish procedures.[37,38] Pharmacists and pharmacy students also need to appreciate formulation differences between SMDs (and their generics) and protein or protein-like drugs including stability and post-translational modifications following manufacture and/or during storage and transport especially in climate-challenged countries. An understanding of concepts related to delivery devices for biologic medicines should include their rheologic (flow) properties because of the viscosity of protein solutions and the methods of administration and thus physical/mechanical propen-

sity for instability. An appreciation of the naming and labeling conventions for innovator biologics and biosimilars is also required as this has significant implications for efficacy and safety monitoring as part of post-marketing pharmacovigilance (PVP) and risk management plans (RMPs) with respect to traceability of the exact biologic product used. Pharmacists also need to understand the need for innovation in delivery modes. A move to delivering such medicines orally, i.e. not only by injections or infusions as is universally the case now is required.[39] Patients, of course, prefer oral dosing for these drugs.[40]

The second aspect of pharmaceutical sciences of great importance to pharmacists is the different, complex, and novel ways that biologic medicines are cleared or removed from the body (catabolized by endogenous processes and/or renally excreted depending on their molecular size) after parenteral administration.[41] Biologics have very different elimination pathways compared to SMDs which may also be subject to endogenous feedback mechanisms. In addition, understanding how during the discovery and development process for biobetters, specific strategies like increasing their molecular size by linking to PEG molecules, PEGylation[42] are employed to prolong their exposure in the body by reducing their clearances or increasing their pharmacokinetic half-lives is critical. The accompanying pharmacodynamic aspects, including time-course of biologic drug effects, drug-antibody formation, and unwanted immunogenicity effects, are also essential. It must be noted that graduating pharmacy students are not expected to retain (or remember) all pharmaceutical science knowledge after graduation; variable levels of pharmaceutical science knowledge are retained.[43]

In summary, the pharmaceutical sciences content in pharmacy degree curricula needs to drastically shift to protein drugs and newer emerging biotherapies from SMDs. This needs to happen quickly to keep pace with the exponential developments in emerging biologic therapies.

15.8.5 Enabling Pharmacy and Public Health Practice

All pharmacy practice concepts related to biologic medicines need to be seen in the overall context of clinical sciences and public health and not just from the lens of pharmacists, pharmacy students, or the pharmacy profession. The unique mechanisms of action and role of biologics in pharmacotherapies need to be understood and nuanced in the context of the shifting treatment paradigm from SMDs to biologics. The positive and negative feedback mechanisms that exist naturally in the body and their perturbation by biologic medicines needs to be carefully understood so they can be translated to the pharmacists' current and expanding professional roles.

Pharmacists also need to understand how naming conventions of biosimilars may affect pharmacovigilance or RMPs and these need to be communicated at the right level to doctors and patients. The various ways of monitoring safety and efficacy and gathering/identifying safety signals for these from adverse drug effect reporting systems need to be understood and should be an emerging priority area for pharmacists. Efficacy and safety recording and reporting methods and their advantages and pitfalls also need to be understood by pharmacists.

Pharmacists also need to be very familiar with current and rapidly evolving and changing regulatory frameworks for innovator biologics, biosimilars, and biobetters in major regulatory jurisdictions. How procurement, distribution, payment, and subsidy frameworks accompany regulatory guidelines and policies are also important. Some understanding of the ethics of the affordability of these costly medicines and who can afford these is also relevant. Guiding principles for the "rationing" of these high-cost medicines is also relevant.

Clinical issues related to nocebo effects (compared with placebo effects more commonly seen with SMDs) as well as indication extrapolation including prescriber and patient views on these are essential as well. In addition, design concepts related to clinical studies implemented to evaluate the effect of single or multiple switches in therapy from one biologic to another in the same class or different class in patients already receiving biologics or in biologic-naive patients are essential concepts for pharmacists to grasp and understand fully.

The pharmacoeconomics of biologic medicines compared with SDM alternatives and innovator products compared with biosimilars are important concepts for pharmacists to understand and communicate to doctors, patients, patient advocacy groups, and their carer's.

Quality use of medicine (QUM) concepts and how these fit into national and international medicines policies analogous to SMDs need to be fully appreciated, and consideration should be given to how these need to be refined to fit the introduction of all biologic medicines within national contexts. Lessons learnt from the introduction of SMD generics (100% adoption has not been achieved in most countries despite massive concerted efforts by many including governments and other third-party insurers) need to be fully appreciated by pharmacists.[44,45]

Understanding the facilitators (availability of information to patients and doctors, interchange guidelines and incentives to use generics for doctors, pharmacists, and patients) and barriers (negative perceptions about generics and lack of coherent policies about generics) also need to be fully appreciated. Understanding behavior-change concepts including the motivations of doctors, patients, and other stakeholders are needed to facilitate greater adoption of biosimilars as has been the case with SMD generics. Pharmacists, therefore, need to work collaboratively with prescribers, public health experts, psychologists, and other behavior-change experts to drive this agenda and educate patients and their prescribers for the greater adoption of biosimilars.

15.9 Recommended Knowledge Areas on Biologics in the Published Literature

There appears to be only one publication that has specifically addressed biosimilars education in a pharmacy degree program. This paper[46] provides a specific framework for integrating biosimilars knowledge areas into the requirements for the Doctor of Pharmacy program, which is the entry-level pharmacy registrable qualification in the United States. While the context of this review is somewhat narrow in that it only applies directly to the US degree, nonetheless, it is an excellent publication on this topic as it links different pharmaceutical sciences and pharmacy practice biosimilar concepts (like proteins as therapeutic molecules, immunogenicity of biologics, and manufacturing differences from SMDs, to mention just a few here), to ACPE standards and Center for the Advancement of Pharmacy Education (CAPE) outcome domains. In addition, and even more importantly, sample learning objectives, associated examples of learning activities, and more critically examples of appropriate assessment types are also provided. Where some of these activities and tasks fit into the overall curriculum (which year of the curriculum) is also discussed.

These authors advocate for the integration of biosimilar concepts into the four main US domains of pharmacy education: biomedical sciences, pharmaceutical sciences, clinical sciences, and social/administrative/behavioral sciences. They propose four "key educational parameters" for biosimilars that could be taught throughout the Pharm D program. These include:

- What is a biologic product? This beginner-level concept is recommended in the curriculum within the first year.
- How are biosimilars developed and regulated? This intermediate-level material should be offered in the first two years of the four-year doctoral program.
- US state pharmacy practice laws. This intermediate-level material would be included in other courses about medication.
- Pharmacy practice management issues. This advanced-level concept would be taught in context with other clinical therapeutics courses as a way of tying together the previously learned biosimilars information.[46]

Any pharmacy education leader wanting to initiate a discussion within their school about biosimilars knowledge areas and curriculum content would benefit significantly from this review as a way of background material to drive their curriculum reform/renewal considering the exponential growth of biologic medicines into pharmacotherapy over SMDs and their generics.

15.10 Modes of Learning and Teaching Delivery and Assessments

Apart from identifying the relevant curriculum areas and competencies on biologic medicines, it is also essential for pharmacy educators to identify the most appropriate delivery methods for such material and the relevant assessment to accompany these methods to demonstrate competency of the students with such material. Education has now evolved from the conventional lecture classroom to a more student-centered learning and delivery environment involving various modes like flipped classrooms, virtual learning environments, blended learning involving mix of online, face-to-face teaching, virtual interactive sessions, self-directed web-based learning modules, problem-based learning (PBL), inter-professional learning (IPL) or education (IPE), and team-based education (TBE), where student teams practice their critical thinking skills to solve problems likely to be faced as pharmacists. How to get full participation from all members of the team, is of course, a challenge with TBE.

Another consideration for pharmacy educators is whether the curriculum material on biologic medicines needs to be core material or be provided as electives? Some curriculum material, whether well-established areas like complementary and alternative medicine[47] or relatively new areas like pharmacoeconomics[48] may be delivered satisfactorily as electives. However, for such fundamental areas like the science of biologic medicines and the relevant scientific and clinical aspects of biosimilars, the material must be core material embedded in the entire curriculum.

International exchange programs may also assist in that within-country multicultural issues may also be mitigated by lessons from the challenges of teaching and learning in a transcultural setting as well as language barriers for students and academics and differences in belief systems about disease and medicines.[48] Such programs may also provide the platform for schools in different countries to share their expertise in curricular development and research initiatives and enhance academic staff and student's multicultural experiences of knowledge about all biologic medicines including biosimilars.

Whether there should be additional P/G modules on biologics or biosimilars or a separate higher-level credentialing of biologic medicines knowledge is also to be considered by education leaders and/or the profession more generally.

Educators need to consider the role of innovation in education delivery of such programs on biosimilars in transforming to competency-based curricula.

With respect to assessment, use of an appropriate mix of assessment tools including formative and summative assessments, OSCEs, and in-class quizzes need to be used in tandem with the various delivery modes. Resources are available in the literature for designing competency-based assessments.[31,49]

In both delivery of the essential material and in assessment of competencies, rapid-cycle innovation and improvement (quality improvement that identifies, implements, and measures changes made to improve a process/system) should be used to guide curriculum development and the outcomes framework.[50–52]

15.11 Urgent Need for Curriculum Reform

The evidence thus far suggests that current levels of knowledge on innovator biologic medicines and biosimilars are not uniform across the pharmacy profession despite the anticipated growth in their clinical use. Also, there does not appear to be any information on the perceptions of pharmacy students or pharmacists on future importance and impact of biologics knowledge on their professional roles and whether they feel prepared for their ever-increasing role in such therapy.

The majority of pharmacists and graduating pharmacy students would agree that pharmacists have a pivotal role to ensure that biologic medicines are used safely and effectively and that they should provide evidence-based, objective information to doctors, patients, and patient carers' and patient advocacy groups to facilitate an informed choice about innovator biologics and biosimilars. It has been suggested that pharmacy educators should lead the way in helping pharmacy students understand the complexities of innovator biologics and biosimilars.[53]

Competency-based pharmacy education is now firmly in place internationally.[54] The main driver is to prepare pharmacists for their evolving professional roles in improving health care and patient safety. A structural framework for such curriculum reform to include all aspects of biologics, particularly biosimilars, is summarized in Figure 15.1. It includes the roles and responsibilities of educators, the accrediting agencies, and the profession. Key guiding principles are also provided. This should form the basis of discussions on how pharmacy education needs to change to reflect the increasing role of biologic medicines in the pharmacotherapy of many diseases.

15.12 Concluding Remarks

Considering the relatively low current levels of biosimilars knowledge among community pharmacists and recognizing the increased use of innovator biologic

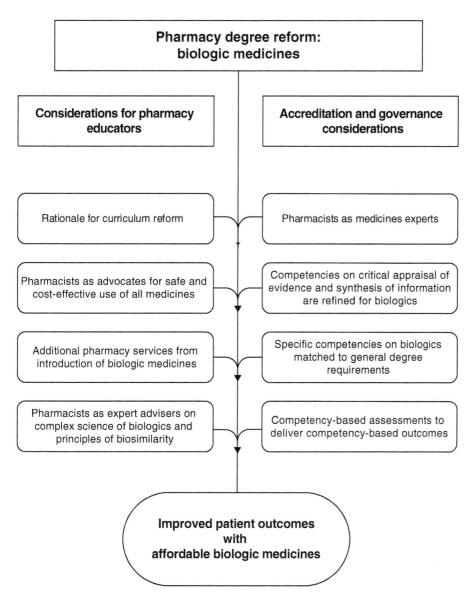

Figure 15.1 Pharmacy curriculum reform framework on biologics.

medicines and biosimilars, their content in pharmacy degrees needs urgent review. Accrediting agencies, the pharmacy profession, pharmacy education leaders, and pharmacy academic staff and pharmacy students as stakeholders all should urgently call for more biologics and especially biosimilar education in pharmacy degree curricula for the benefit of patients receiving innovator biologics and biosimilars.

References

1 Green, S. (2005). Systematic reviews and meta-analysis. *Singapore Medical Journal* 46: 270–273.

2 Gopalakrishnan, S. and Ganeshkumar, P. (2013). Systematic reviews and meta-analysis: understanding the best evidence in primary healthcare. *Journal of Family Medicine and Primary Care* 2 (1): 9–14.

3 Ade, A., Bourdon, O., and Bussieres, J.F. (2017). A survey of pharmacists' knowledge and views of biosimilars in Quebec and France. *Annales Pharmaceutiques Françaises* 75 (4): 267–275.

4 Pawlowska, I., Pawlowski, L., Krzyzaniak, N., and Kocic, I. (2019). Perspectives of hospital pharmacists towards biosimilar medicines: a survey of polish pharmacy practice in general hospitals. *BioDrugs* 33: 183–191.

5 UTS Pharmacy (2019). The 2018 pharmacy barometer: professional services in the community pharmacy on the rise. https://www.uts.edu.au/about/graduate-school-health/pharmacy/news/professional-services-community-pharmacy-rise (accessed 20 February 2020).

6 Cohen, H., Beydoun, D., Chien, D., and Lessor, T. (2016). Awareness, knowledge, and perceptions of biosimilars among speciality physicians. *Advances in Therapy* 33: 2160–2172.

7 Shah-Manek, B., Baskett, A., Baynton, E. et al. (2018). Perceptions of biosimilars across physician specialities in Europe. *Value in Health* 21: S265.

8 Leonard, E., Wascovich, M., Oskouei, S. et al. (2019). Factors affecting health care provider knowledge and acceptance of biosimilar medicines: a systematic review. *Journal of Managed Care and Speciality Pharmacy (JMCP)* 25 (1): 102–112.

9 Waterfield, J. (2010). Is pharmacy a knowledge-based profession? *American Journal of Pharmaceutical Education* 74 (3): Article 50.

10 Skau, K. (2007). Pharmacy is a science-based profession (viewpoints). *American Journal of Pharmaceutical Education* 71 (1): Article 11.

11 Jarrett, S. and Dingermann, T. (2015). Biosimilars are here: a hospital phamacist's guide to educating health care professionals on biosimilars. *Hospital Pharmacy* 50 (10): 884–893.

12 Kunter, I., Balogun, H., and Sahin, G. (2018). The role of the pharmacist from development to pharmacovigilance of biosimilars. *Marmara Pharmaceutical Journal (now Journal of Research in Pharmacy)* 22 (4): 469–473.

13 McShea, M., Borns, M., and Pollom, D. (2016). Biosimilars and follow-on biologics: a pharmacist opportunity. Speciality Pharmacy Times Web site. https://www.pharmacytimes.com/publications/issue/2016/november2016/biosimilars-and-followon-biologics-a-pharmacist-opportunity (accessed 16 June 2020).

14 Balick, R. (2016). As biosimilars reach the shelves, patients need pharmacists' guidance. *Pharmacy Today* 22 (1): 21–23.

15 GaBI Online (2012). Small molecule versus biological drugs. http://www.gabionline.net/Biosimilars/Research/Small-molecule-versus-biological-drugs (accessed 20 February 2020).

16 Kabir, E., Moreino, S., and Sharif Siam, M. K. (2019). The breakthrough of biosimilars: a twist in the narrative of biological therapy. *Biomolecules* 9: 410.

17 Morrow, T. and Felcone, L. (2004). Defining the difference: what makes biologics unique. *Biotechnology Healthcare* (September): 24–26, 28–29.

18 Declerck, P., Danesi, R., Petersel, D., and Jacobs, I. (2017). The language of biosimilars: clarification, definitions, and regulatory aspects. *Drugs* 77 (6): 671–677.

19 Basak, S. and Sathyanarayana, D. (2010). Pharmacy education in India. *American Journal of Pharmaceutical Education* 74 (4): Article 68.

20 Marriott, J., Nation, R., Roller, L. et al. (2008). Pharmacy practice in the context of australian practice. *American Journal of Pharmaceutical Education* 72 (6): Article 131.

21 General Pharmaceutical Council (2016). About us: who are we: the GPhC Council. www.pharmacyregulation.org (accessed 20 February 2020).

22 European Expertise Centre for Pharmacy Education and Training (2016). Competency framework. https://eec-pet.eu/pharmacy-education/competency-framework. (accessed 20 February 2020).

23 Australian Pharmacy Council (2020). Accreditation standards for pharmacy programs in Australia and New Zealand 2020. https://www.pharmacycouncil.org.au/standards (accessed 20 February 2020).

24 Australian Pharmacy Council (2020). Accreditation standards 2020 for pharmacy programs in Australia and New Zealand – performance outcomes framework. https://www.pharmacycouncil.org.au/standards/standards-review/performance-outcomes-draft.pdf (accessed 20 February 2020).

25 Australian Pharmacy Council (2018). Health profession accreditation practices – international literature review. https://www.pharmacycouncil.org.au/standards/standards-review (accessed 22 February 2020).

26 Alkhateeb, F., Arkle, S., McDonough, S., and Latif, D. (2018). Review of national and international accreditation of pharmacy programs in the Gulf Cooperation Council Countries. *American Journal of Pharmaceutical Education* 82 (10): Article 5980.

27 Hartel, R. and Foegeding, E. (2004). Learning: objectives, competencies, or outcomes. *Journal of Food Science Education* 3: 69–70.

28 Gosselin, D. (2013). Competencies and learning outcomes. https://serc.carleton.edu/integrate/programs/workforceprep/competencies_and_LO.html (accessed 20 February 2020).

29 Kennedy, D., Hyland, Á., and Ryan, N. (2009). Learning outcomes and competencies. *Using Learning Outcomes: Best of the Bologna Handbook* 33: 59–76.

30 Moy, J. (1999). The impact of generic competencies on workplace performance: review of research. https://www.ncver.edu.au/research-and-statistics/publications/all-publications/the-impact-of-generic-competencies-on-workplace-performance-review-of-research (accessed 20 February 2020).

31 Hudson, E. (2018). How to design a competency-based assessment. https://medium.com/@ejhudson/how-to-design-a-competency-based-assessment-39f312235bde (accessed 20 February 2020).

32 Brazeau, G., Meyer, S., Belsey, M. et al. (2009). AACP curricular change summit supplement: preparing pharmacy graduates for traditional and emerging career opportunities. *American Journal of Pharmaceutical Education* 73 (8): Article 157.

33 Cor, M. and Brocks, D. (2018). Examining the relationship between prerequisite grades and types of academic performance in pharmacy school. *Currents in Pharmacy Teaching & Learning* 10: 695–700.

34 Brown, B., Skau, K., and Wall, A. (2009). Learning across the curriculum: connecting the pharmaceutical sciences to practice in the first professional year. *American Journal of Pharmaceutical Education* 73: Article 36.

35 Cruthirds, D., Cretton-Scott, E., Wilborn, T. et al. (2011). Biochemistry and pathophysiology instruction in U.S. colleges and schools of pharmacy: faculty and course characteristics and faculty perceptions of student preparedness. *Currents in Pharmacy Teaching & Learning* 3: 137–147.

36 Ramanan, S. and Grampp, G. (2014). Drift, evolution, and divergence in biologics and biosimilars manufacturing. *BioDrugs* 28 (4): 363–372.

37 Niazi, S. (2017). Reinventing commercial biomanufacturing. *European Pharmaceutical Review* https://www.europeanpharmaceuticalreview.com/article/77395/reinventing-commercial-biomanufacturing (accessed 16 June 2020).

38 Vulto, A. and Jaquez, O. (2017). The process defines the product: what really matters in biosimilar design and production? *Rheumatology* 56: iv14–iv29.

39 Mantaj, J. and Vllasaliu, D. (2020). Recent advances in the oral delivery of biologics. *The Pharmaceutical Journal* 15 (10): 759–770.

40 Eek, D., Krohe, M., Mazar, I. et al. (2016). Patient-reported preferences for oral versus intravenous adminstration for the treatmnet of cancer: a review of the literature. *Patient Preference and Adherence* 10: 1609–1621.

41 Ovacik, M. and Lin, K. (2018). Tutorial on monoclonal antibody pharmacokinetics and its considerations in early development. *Clinical and Translational Science* 11: 540–552.

42 Santos, J., Torres-Obreque, K., Meneguetti, G. et al. (2018). Protein PEGylation for the design of biobetters: from reaction to purification processes. *Brazilian Journal of Pharmaceutical Sciences* 54 (spe): e01009. https://doi.org/10.1590/s2175-97902018000001009.

43 Unni, E., Mandal, M., Radhakrishnan, R. et al. (2016). Knowledge retention of basic pharmaceutical sciences in a PharmD program. *Currents in Pharmacy Teaching & Learning* 8: 827–832.

44 Hassali, M., Alrasheedy, A., Mclachlan, A. et al. (2014). The experiences of implementing generic medicine policy in eight countries: a review and recomendations for a successful promotion of generic medicine use. *Saudi Pharmacetical Journal* 22: 491–503.

45 Moe-Byrne, T., Chambers, D., Harden, M., and McDaid, C. (2014). Behaviour change interventions to promote prescribing of generic drugs: a rapid evidence synthesis and systematic review. *BMJ Open* 4: e004623.

46 Li, E., Liu, J., and Ramchandani, M. (2017). A framework for integrating biosimilars into the didactic core requirements of a doctor of pharmacy curriculum. *American Journal of Pharmaceutical Education* 81 (3): Article 57.

47 Scaletta, A., Ghelani, N., and Sunny, S. (2017). Complementary and alternative medicine education in U. S. schools and colleges of pharmacy. *Currents in Pharmacy Teaching & Learning* 9: 521–527.

48 Kamal, K., Sakamaki, A., Bricker, J., and Kawasaki, K. (2010). An elective course in pharmacoeconomics for pharmacy students in Japan – a transcultural experience. *Currents in Pharmacy Teaching & Learning* 2: 218–227.

49 Idrissi, M.K., Hnida, M., and Bennani, S. (2017). Competency-based assessment: from conceptual model to operational tool. In: *Innovative Practices for Higher Education Assessment and Measurement*, 57–78. Hershey, PA, USA: IGI Global.

50 Christensen Institute (2016). Rapid-cycle innovation should be the holy grail of education. https://www.christenseninstitute.org/blog/rapid-cycle-innovation-holy-grail-education (accessed 20 February 2020).

51 Serdyukov, P. (2017). Innovation in education: what works, what doesn't, and what to do about it? *Journal of Research in Innovative Teaching & Learning* 10 (1): 4–33.

52 The Office of the National Coordinator for Health Information Technology (2019). How do I use a rapid-cycle improvement strategy? HealthIT.gov. https://www.healthit.gov/faq/how-do-i-use-rapid-cycle-improvement-strategy (accessed 20 February 2020).

53 Malcom, D.R. and Al-Ghananeem, A.M. (2015). Biotechnology in practice: call for incorporation of biosimilars into pharmacy education. *Currents in Pharmacy Teaching & Learning* 7 (2): 256–258.

54 Koster, A., Schalekamp, T., and Meijerman, I. (2017). Implementation of competency-based pharmacy education (CBPE). *Pharmacy* 5: Article 10.

16

A Checklist for Pharmacists on Biologics and Biosimilars: Tips to Enhance Patient-Centered Discussions

Sanja Mirkov[1,2] and Johan Rosman[3]

[1] School of Pharmacy, The University of Auckland, Auckland, New Zealand
[2] Ramsay Pharmacy Services, Melbourne, Victoria, Australia
[3] Medical School, Curtin University, Perth, Western Australia, Australia

---

KEY POINTS

- Innovator biologics, biosimilars, and biobetters offer great benefits in many diseases that are otherwise not fully managed with small molecule drugs. Their use, however, is not without clinical risks and exorbitant costs to health systems.
- Improvements in biosimilar uptake may be achieved by behavior change strategies that increase staff and patient capability, opportunities for desired prescribing behaviors, and motivate clinicians and patients.
- A checklist is provided to support clinician and patient decision-making with respect to therapy selection, prescribing, monitoring to reduce risks and improve patient outcomes.

Abbreviations

Abbreviation	Full name
ADAs	Anti-Drug Antibodies
ADR	Adverse Drug Reaction
AE	Adverse Event
AIDET	Acknowledge, Introduce, Duration, Explain, Thank you
AusPAR	Australian Public Assessment Report
BPMH	Best Possible Medication History
CRS	Cytokine Release Syndrome
CTCAE	Common Terminology Criteria for Adverse Events
EMA	European Medicines Agency
ESA	Erythropoiesis Stimulating Agents
FDA	U.S. Food and Drug Administration
ISBAR	Identify, Situation, Background, Assessment, Recommendation

Abbreviation	Full name
mAbs	monoclonal Antibodies
MeDRA	Medical Dictionary for Regulatory Activities
MUE	Medicines Use Evaluation
NIOSH	National Institute for Occupational Safety and Health
PBAC	Pharmaceutical Benefits Advisory Committee
PBS	Pharmaceutical Benefits Scheme
PEG	Polyethylene Glycol
QUM	Quality Use of Medicines
SMDs	Small Molecule Drugs
SmPC	Summary of Product Characteristics
TGA	Therapeutic Goods Administration
TNF	Tumor Necrosis Factor
WHO	World Health Organization

16.1 Context of This Checklist for Pharmacists and Health Professionals

The preceding chapters of this book have comprehensively provided the very complex science as well as the pharmacy practice and public health issues surrounding innovator biologics and their biosimilars where they exist and are marketed. Ultimately, all these complex issues surrounding these very expensive but very potent and enormously useful drugs must be interpreted by pharmacists to facilitate quality use of medicines (QUMs) of all biologic medicines for greater patient outcomes and enhanced quality of life.

Checklists are a handy tool that assist busy health practitioners in many daily professional interactions and communication. With the aim of assisting in structured and respectful discussions with physicians and patients, and to reduce critical omissions, errors or misunderstandings, a comprehensive checklist is developed in this chapter for use in discussions about innovator biologics and biosimilars when a physician first prescribes a biologic for a new patient or when a patient is switched to a biosimilar or a follow-on biologic or biobetter.

Among other pertinent issues, we have used the following topics to develop the checklist: preclinical and clinical rationale for the prescribing of biologics; any prior small molecule treatment; who should take a biologic and for which clinical condition; questions to ask the physician and the patient; the questions likely to be posed to the pharmacist; the duration, frequency, and mode of administration; and the storage and transport issues especially for patients in rural/remote areas. The Do's, Don'ts, and the real risks (clinically relevant adverse events [AE]) for biologic drugs (and biosimilars) are also discussed and guidance provided.

This checklist also incorporates the steps/items that require a discussion and the likely queries from physicians/patients; this should also allow checking-off items as the discussion proceeds and evolves. Most importantly, this checklist should facilitate shared collaborative discussions between pharmacists and physicians, pharmacists and patients, and pharmacists, physicians, and patients.

16.2 Summary of Biologics, Biosimilars, and Biobetters

Biologics (*synonyms: biologic therapies, biological medicinal products, biotechnological products, biologic therapies, originator biologics, reference biologics, innova-*tor biologics) are used for a wide range of chronic inflammatory and immune diseases and for treatment and supportive care of cancer.[1,2] Biologic therapies include recombinant human hormones, e.g. erythropoietin, insulin, somatostatin; cytokines, e.g. filgrastim; monoclonal antibodies (the mAbs), e.g. infliximab, rituximab, trastuzumab; fusion proteins, e.g. etanercept; and antibody plus drug combinations or antibody conjugates, e.g. trastuzumab plus emtansine (Kadcyla®).

Biologics are protein-based molecules with complex structures with unique three-dimensional shape, manufactured by genetic engineering from living cells.[3] Depending on their molecular size, biologics can be simple or small, e.g. insulin (5808 Da) or erythropoietin (30 400 Da); and complex or large, e.g. monoclonal antibody (150 000 Da), compared to small molecule drugs (SMDs) like aspirin at 180 Da.[4,5]

Yeast and filamentous fungi glycosylated proteins differ from those produced in mammalian cells.[6] Transfected bacteria, e.g. *Escherichia coli* or yeast, are useful as hosts for the production of nonglycosylated recombinant proteins such as growth hormone, insulin, calcitonin, and albumin. On the other hand, mammalian cells are required for the production of sophisticated glycosylated complex proteins such as erythropoietin, Factor VII, and tissue plasminogen activator (tPa). The choice of the host cell influences the glycosylation pattern of glycoproteins and thus glycoform properties of proteins differ between Chinese hamster ovary (CHO), Syrian baby hamster kidney (BHK), and human cells.[7,8]

mAbs or immunoglobulins are genetically engineered protein molecules and can be murine (produced from mouse cells), chimeric (part mouse, part human), humanized (contain a bare minimum of non-human amino acid sequences), and completely human.[1] mAbs are the Y-shaped protein consisting of two heavy and two light polypeptide chains with a molecular size of approximately 150 000. mAbs have Fc and Fab regions that bind to different molecular targets resulting in different biological activities.[9] For example, antigen neutralization requires only binding through the Fab region, whereas antibody-dependent cell-mediated cytotoxicity (ADCC) requires binding to antigen through both the Fab and Fc regions.

Biosimilars (*synonyms: similar biologic medicinal product [EU], similar biotherapeutic product [WHO], subsequent entry products [Canada], follow-on products [FDA], biogenerics, generic biopharmaceuticals, and comparable biologicals*) are biological products that are designed to be highly similar to a reference biologic product. Due to the complex structure of a large molecule innovator biologic, identical copies cannot be made; therefore, biosimilars, unlike small-molecule generics are highly similar but not identical to their reference biologic.

The development of biosimilars requires confirmation of the structure and function, pharmacology, pharmacokinetics in most sensitive populations, pharmacodynamic studies with dose–response equivalence, and clinical confirmation in Phase 1 and Phase 3 clinical studies. In summary, the highly similar nature of biosimilars must be demonstrated in clinical trials for approval by the European Medicines Agency (EMA) and the U.S. Food and Drug Administration (FDA) and it is based on "totality of evidence" with respect to comparability and similarity.

Biobetters (*second-generation biologics*) are enhanced versions of an original biologic.[10] Therefore, biobetters have the same epitope as marketed antibodies, but have been engineered to further improve their safety and/or adverse profile or improving their pharmacokinetic profile such as a longer half-life.[1,7,11] Modified glycosylation and pegylation are used to obtain products with longer half-life compared to innovator biologics, decreasing the frequency of administration.[7,10] Glycosylation increases stability by inhibiting aggregation, degradation, or denaturation of the protein. Pegylation is covalent binding of polyethylene glycol (PEG) moiety to a protein. PEG, either linear or branched chain, is a non-immunogenic, hydrophilic, biodegradable molecule that has been used to increase molecular mass and hydrodynamic radius of biologics, in order to decrease their elimination via glomerular filtration in the kidney. An example of a biobetter produced by modified glycosylation is Aranesp® (darbepoetin-α). Examples of pegylated biobetters are Neulasta® (PEGylated version of filgrastim) and Mircera® (a PEGylated form of epoetin-β).[6,7] Due to improved pharmacokinetic profiles, these products require less frequent dosing.

Fusion proteins are most commonly obtained by fusion of the biologic protein or peptide to human serum albumin (HSA), fusion to the constant fragment (Fc) domain of a human immunoglobulin (Ig) G, or fusion to nonstructured polypeptides such as XTEN (recombinant PEG or rPEG).[7] Etanercept is an example of human TNF receptor p75 Fc fusion protein produced by recombinant DNA technology using CHO cells.

Furthermore, humanization of murine or chimeric mAbs has led to improved immunogenicity profile.[10] For example, Obinutuzumab (Gazyva®), a fully humanized mAb is a biobetter to chimeric mAb Rituximab while Golimumab (Simponi®) is a human mAb.

16.3 Storage and Handling of Biologics

Special consideration needs to be given to the storage, handling, preparation, administration, and disposal of biotechnology medications.[12] Innovator biologics, biosimilars, and biobetters are protein molecules that are unstable in the presence of proteases, sensitive to temperature extremes, and their solutions support bacterial growth. The products often do not contain a preservative, due to the risk of protein degradation.[12] For these reasons, they need to be stored and shipped chilled in the form of freeze-dried powders and require refrigeration. The cold chain must be maintained during storage and distribution. For small volumes and short transportation times (less than three hours), insulated containers should be used. For longer transportation, gel or ice packs should be added to the insulated containers; however, they must not be in direct contact with the products.[13]

Aseptic technique must be used for reconstitution, e.g. isolator cabinet, closed system drug transfer devices, or cytotoxic drug safety cabinet for monoclonal antibody–drug conjugates.[14] Solutions have to be prepared by slowly adding the diluent and mixing gently and vigorous shaking of the products, both reconstituted solutions, bags or bottles must be avoided. Furthermore, the products may require special syringes for reconstitution (e.g. silicone-free syringe provided with abatacept [Orencia®]), intravenous containers and administration sets as proteins may adhere to polyvinyl chloride. Following reconstitution, mAbs must be used immediately or alternatively stored refrigerated and usually used within 24 hours or discarded.[2]

The information available on the occupational exposure and toxicity of mAbs is limited.[15] Therefore, pregnant personnel and personnel with compromised immune function should avoid handling of mAbs.[14] Biotechnology products may contain viral vectors or genetically modified material, may be teratogenic, mutagenic, carcinogenic, or immunogenic and therefore may need to be handled and disposed of as hazardous materials, by specially trained staff.[12,14] mAbs that are conjugated to cytotoxic agents must be handled as cytotoxic products, e.g. trastuzumab emtansine (Kadcyla®), brentuximab vedotin (Adcetris®). Teratogenicity of trastuzumab has been reported in the post-marketing human case reports; bevacizumab, ipilimumab, pertuzumab, and brentuximab vedotin were teratogenic in animal studies; brentuximab vedotin is mutagenic based on animal studies; and chimeric mAbs rituximab and cetuximab are highly immunogenic.[14]

mAbs are large molecules and therefore, the risk of exposure via intact skin and oral route are low; however, it increases with skin conditions such as dermatitis and other skin damage.[15] There is a moderate risk of exposure via inhalation and mucosa during the preparation of doses for administration where staff may be exposed to powdered aerosolized liquid particles.[14] Thus, reconstitution and compounding of mAbs require protective eyewear, respiratory mask, and gloves should be worn alongside effective hand hygiene to protect against aerosolization, contamination, and infection; however, a protective gown is not essential. The disposal of unused products and waste should be the same as for clinical waste, not like cytotoxic waste disposal.

In contrast, mAbs® that are conjugated to a cytotoxic agent, e.g. trastuzumab emtansine (Kadcyla®), brentuximab vedotin (Adcetris®), pertuzumab, fusion proteins, or radioisotopes, are hazardous products and the guidelines for safe preparation, administration, and disposal of cytotoxics or radiopharmaceuticals have to be followed.[16,17] Full personal protective equipment must be worn including gowns, double gloves, protective eyewear, and a respiratory mask. In addition, when handling radiopharmaceuticals, protective shields and gowns, disposable gowns and gloves, and tools and shields for maximizing the distance from the source must be used.[18]

The U.S. National Institute for Occupational Safety and Health (NIOSH)[16] has published a list of antineoplastic and other hazardous drugs in healthcare settings. In addition, the eviQ website[19] lists hazardous cytotoxic antineoplastics that require full personal protective equipment. Furthermore, any agent lacking sufficient information such as investigational products used in clinical trials are considered as high risk until additional information becomes available.[14] Although, the NIOSH guidelines list pertuzumab with a black box warning in Australia due to the potential to cause birth defects (Pregnancy Category D), pertuzumab has not been classified as a hazardous drug following the assessment by the TGA.[19] For detailed information on individual product reconstitution, storage, and handling requirements, refer to the manufacturer's information.

The therapeutic class, mode of action, target pathway, preparations, cold chain requirements, and disease area for selected biologics, biobetters, and available biosimilars are presented in Table 16.1.

16.4 Regulatory Requirements

The regulatory requirements for biosimilars is already comprehensively covered earlier in the book (see Chapter 10). In the context of this chapter, Table 16.2 provides a summary of regulatory differences between biosimilars and generics of SMDs.

For biobetters, the regulation follow the same pathway as for the new chemical entity requiring Phase I–IV clinical trials including Phase III clinical trials for each indication (non-inferiority trials).[10] There is no specific regulatory framework for biobetters, the data required for authorization are negotiated with the regulatory agencies on a case-to-case basis.

16.5 Naming and Labeling of Biologic Therapies

The predetermined naming scheme for differentiating biosimilars from reference products is important for ensuring traceability for monitoring safety and effectiveness of biological products.[23] However, currently, there is no consensus worldwide on naming conventions for biosimilars.[3,29] Some groups have advocated for a nonproprietary name distinguishable from the reference product to avoid difficulties tracing AE. The major disadvantage of this system is that it could assume that the biosimilar might have a different mechanism of action and efficacy profile compared to the reference product.[3,30]

In Europe, the labeling of a biosimilar is the same as for the reference biologic and a summary of product characteristics (SmPC) is a copy of the SmPC of the innovator biologic with the exception of the pharmaceutical particulars.[30,31] The World Health Organization (WHO) indicated that the use of identical nonproprietary names may lead to inadvertent switching between the reference biologic and biosimilars.[3] Currently, the same International Nonproprietary Names (INN) are assigned to the biosimilar that has the identical amino acid sequence (nonglycosylated biologics) and a unique prefix or suffix is assigned to the root INN of the reference biologics for glycosylated biologics and biosimilars (e.g. a Greek letter suffix to indicate different glycosylation patterns, e.g. epoetin alfa).[3] In addition, the WHO guidelines also recommend prescribing information for biosimilars to clearly state the omitted indications and the reasons for exclusion of indications.[29,32] In Australia, the Therapeutic Goods Administration (TGA) supports the WHO program on INN, and biosimilars with products labeled using the Australian biological name (ABN) without a specific biosimilar identifier suffix.[33]

In 2014, the WHO Pharmaceutical Substances Expert Group on INN suggested a two-part naming system for biosimilars – the first part is the INN and the second part is a four-letter biological qualifier (BQ)[34] which is randomly assigned. The adoption of the BQ scheme is

Table 16.1 Summary information on selected biologic therapies.[1,2,5,7,20-25]

Agent	Type of molecule	Mode of action	Biologic brand name/year approved	Biosimilar brand name/year approved	Preparation	Mode of administration	Storage conditions/ stability	Disease area
Rheumatology/gastroenterology/dermatology								
Abatacept	Fusion protein produced by recombinant DNA technology in Chinese hamster ovary cells. Human cytotoxic T lymphocyte antigen 4 (CTLA4) is linked to the modified FC domain of human IgG1	T-cell co-stimulation blocker, binds to CD80 and CD86 and prevents second signal	Orencia® 2005	Not available. The patent expired in 2017 in Europe and 2019 in the United States. M834 biosimilar under development	IV: 250 mg vial lyophilized powder: single use vial for IV infusion with a silicone-free disposable syringe. Dilute to 100 mL of 0.9% sodium chloride, infuse over 30 minutes using low-protein binding filter (pore size 0.2–1.2 μm). SC: 125 mg/mL as a 1 mL prefilled glass syringe or a 1 mL single-dose disposable ClickJect® prefilled autoinjector	SC weekly or IV monthly	2–8 °C, refrigerate – do not freeze. Protect from light. Do not shake. After dilution of lyophilized powder for injection, stable for 24 hours at 2–8 °C.	Rheumatoid arthritis Psoriatic arthritis Juvenile idiopathic arthritis
Adalimumab[a]	Recombinant human immunoglobulin IgG1	Anti-TNFα	Humira® 2002	Amgevita® 2017	Prefilled syringe, pen, vial 20 mg/0.2 mL 20 mg/0.4 mL 80 mg/0.8 mL	SC fortnightly	2–8 °C, refrigerate – do not freeze. Protect from light. Do not shake. A single syringe or pen can be stored at room temperature for 14 days	Rheumatoid arthritis Psoriatic arthritis Ankylosing spondylitis Non-radiographic axial spondylitis Crohn's disease Ulcerative colitis Psoriasis Uveitis Hidradenitis suppurativa
Certolizumab pegol	Recombinant humanized antibody Fab fragment expressed in *E. Coli* bacterial expression system, subsequently purified and conjugated to glycol	Anti-TNFα	Cimzia® 2009	Not available. Patent expires in 2021 in Europe and 2024 in US. Biosimilar PF688 under development	Pre-filled syringe 200 mg/1 mL	SC 2– 4 weekly	2–8 °C, refrigerate – do not freeze. Protect from light. Do not shake. If required, may be stored at room temperature up to a maximum of 25 °C for a single period of 10 days	Rheumatoid arthritis Psoriatic arthritis Ankylosing spondylitis Non-radiographic axial spondylitis Plaque psoriasis
Etanercept[a]	Fusion protein. Human TNF receptor p75 Fc fusion protein produced by recombinant DNA technology in Chinese hamster ovary (CHO) mammalian expression system	Anti-TNFα	Enbrel® 1998	Benepali® 2016 Brenzyx® 2016 Erelzi® 2017	25 and 50 mg Prefilled syringe. 50 mg solution for injection in autoinjector. 25 mg vial powder for injection with prefilled syringe with 1 mL water for injection.	SC weekly	2–8 °C, refrigerate – do not freeze. Protect from light. Do not shake. After reconstitution, stable at room temperature up to 25 °C for up to 14 days	Adult rheumatoid arthritis Active polyarticular juvenile idiopathic arthritis Adult psoriatic arthritis Adult active ankylosing spondylitis Adult plaque psoriasis

(Continued)

Table 16.1 (Continued)

Agent	Type of molecule	Mode of action	Biologic brand name/year approved	Biosimilar brand name/year approved	Preparation	Mode of administration	Storage conditions/ stability	Disease area
Golimumab	Human monoclonal antibody IgG1	Anti-TNFα	Simponi® 2009	Not available. Patent expires in 2024. Biosimilar ONS-3035 under development	Prefilled syringe 100 mg/1 mL, 50 mg/0.5 mL	SC 4 weekly	2–8 °C, refrigerate – do not freeze. Protect from light. Do not shake	Rheumatoid arthritis Psoriatic arthritis Ankylosing spondylitis Axial spondylitis Non-radiographic axial spondylitis Ulcerative colitis
Infliximab[a]	Chimeric human-murine IgG1 monoclonal antibody	Anti-TNFα	Remicade® 1998	Inflectra® 2013 Remisma® 2013 Renflexis® 2016	Vial 100 mg powder for injection Reconstitute with SWFI, then dilute with 0.9% sodium chloride	IV every 8 weeks	2–8 °C, refrigerate – do not freeze. Protect from light. Do not shake. Diluted infusion solution stable for 24 hours at 2–8 °C, unless reconstitution and dilution takes place under aseptic conditions	Adult active rheumatoid arthritis, active ankylosing spondylitis, adult and pediatric patients with active Crohn's disease, adult fistulizing Crohn's disease, active ulcerative colitis, psoriatic arthritis, adult plaque psoriasis
Rituximab[a]	Chimeric human-murine IgG1 monoclonal antibody against CD20 antigen produced by mammalian CHO cell suspension culture containing the antibiotic gentamicin	B cell blockade (CD20)	MabThera® 1997 MabThera SC® 2016	Riximyo® 2017 Truxima® 2018	Vial IV: 100 mg in 10 mL, 500 mg in 50 mL. No reconstitution. Dilution with 0.9% sodium chloride or 5% dextrose. Vial SC: 1400 mg (NHL only) Vial SC: 1600 mg (CLL only)	IV 6–12 monthly	2–8 °C, solutions for infusion are stable for 24 hours at 2–8 °C and up to 8 hours at room temperature	Rheumatoid arthritis (RA) Granulomatosis with polyangiitis (Wegener's GPA) Microscopic polyangiitis (MPA)
Tocilizumab	Recombinant humanized antibody of the IgG1 which binds to the interleukin 6	IL-6 receptor blocker	Actemra® / RoActemra® 2009	Patent expired in 2015 in the United States and 2017 in Europe. Biosimilar BOW070 under development	Via 80 mg/4 mL 200 mg/4 mL 400 mg/20 mL. Prefilled syringes, pens 162 mg/0.9 mL	SC weekly or IV every 4 weeks	2–8 °C, refrigerate – do not freeze. Protect from light. Do not shake. Prepared infusion is stable for 24 hours at 30 °C	Rheumatoid arthritis (IV, SC), Giant cell arteritis (SC) Polyarticular juvenile idiopathic arthritis (IV, SC), Systemic juvenile idiopathic arthritis (IV) Cytokine release syndrome (IV)

Oncology/Hematology

Drug	Description	Mechanism	Brand (year)	Patent	Formulation	Administration	Storage	Indications
Bevacizumab	Recombinant humanized monoclonal IgG1 antibody	Anti-vascular endothelial growth factor (VEGF) inhibitor	Avastin® 2004	Not available. Patent expire(d)(s) in July 2019 in the United States and in January 2022 in Europe	100 mg/4 mL single-dose vial 400 mg/16 mL single-dose vial No reconstitution. Dilution with 0.9% sodium chloride. Do not freeze or shake	IV every 2 or 3 weeks	2–8 °C, refrigerate – do not freeze. Protect from light. Do not shake. Solutions for infusion stable for up to 24 hours at 2–8 °C	Metastatic colorectal cancer HER2-negative metastatic breast cancer Locally advanced, metastatic or recurrent non-small cell lung cancer Advanced or metastatic renal cell cancer Relapsed high-grade glioma Epithelial ovarian, fallopian tube, or primary peritoneal cancer Cervical cancer
Cetuximab	Chimeric monoclonal antibody of the IgG1, produced in mammalian cell culture by mouse myeloma cells Sp2/0	Anti-epidermal growth receptor (EGFR) inhibitor	Erbitux® 2008	Not available. Patent expired in Europe and in the United States	Supplied in ready-to-use vials	IV weekly	2–8 °C, solutions for infusion are stable for 24 hours at 2–8 °C and up to 8 hours at room temperature	Colorectal cancer, squamous cell carcinoma of the head and neck
Rituximab[a]	Chimeric human-murine IgG1 monoclonal antibody against CD20 antigen produced by mammalian CHO cell suspension culture containing the antibiotic gentamicin	CD20 inhibitor	MabThera® 1997 MabThera SC® 2016	Riximyo® 2017 Truxima® 2018	Vial IV: 100 mg in 10 mL 500 mg in 50 mL No reconstitution. Dilution with 0.9% sodium chloride or 5% dextrose. Vial SC: 1400 mg (NHL only). Vial SC: 1600 mg (CLL only)	IV infusion weekly on day one of each chemotherapy cycle	2–8 °C, refrigerate - do not freeze. Protect from light. Do not shake. Solutions for infusion are stable for 24 hours at 2–8 °C and up to 8 hours at room temperature	Non-Hodgkin's lymphoma (NHL) Chronic lymphocytic leukemia (CLL)
Trastuzumab[a]	Recombinant humanized IgG1 monoclonal antibody produced by mammalian CHO cells	Anti-HER2	Herceptin® 1998 Herceptin SC® 2013	Herzuma® 2018 Ogivri® 2018 Ontruzant® 2019 Kanjinti® 2019 Trazimera® 2018	Powder for IV infusion 60 mg and 150 mg. Reconstitute with SWFI. Add required volume into 250 mL of sodium chloride 0.9%. Solution for SC injection 600 mg/5mL.	IV infusion, SC	2–8 °C, infusion stable for 24 hours at 30 °C	Early-stage HER2-overexpressing breast cancer HER2-over expressing metastatic breast cancer
Trastuzumab emtansine	Antibody–drug conjugate. Recombinant humanized IgG1 monoclonal antibody is linked by MCC linker to the small molecule DM1	HER2 inhibitor and cytotoxic DM1 microtubule inhibitor	Kadcyla® 2013	Not available	Cytotoxic - Vial: 100 and 160 mg in the cytotoxic sterile cabinet reconstitute with SWFI. Add required volume into 250 mL of sodium chloride 0.9% or sodium chloride 0.45%.	IV infusion every 3 weeks	2–8 °C, refrigerate - do not freeze. Protect from light. Do not shake	Early-stage HER2-overexpressing breast cancer HER2-over expressing metastatic breast cancer

(Continued)

Table 16.1 (Continued)

Agent	Type of molecule	Mode of action	Biologic brand name/year approved	Biosimilar brand name/year approved	Preparation	Mode of administration	Storage conditions/ stability	Disease area
Oncology supportive care								
Filgrastim[a]	Recombinant human G-CSF	G-CSF	Neupogen® 1998	Nivestim® 2010 Tevagrastim® 2008 Zarzio® 2009 Grastofil® 2013	Vials. Prefilled syringes 300 µg/0.5 mL 480 µg/1.6 mL	Daily SC or IV infusion	2–8 °C, refrigerate – do not freeze. Protect from light. Do not shake. A single exposure temperatures from −20 to 30 °C up to 3 days. Solutions withdrawn under aseptic conditions may be stored in polypropylene syringes at 2–8 °C for 24 hours	To decrease the incidence of infection, as manifested by febrile neutropenia, in patients with non-myeloid malignancies receiving myelo suppressive anticancer drugs
Pegfilgrastim[a]	Biobetter. Pegylated form of recombinant human G-CSF	Long-acting pegylated G-CSF	Neulasta® 2002 Neulastim® 2002	Fulphila® 2016 Pelmeg® 2017 Ziextenzo® 2017 Pelgraz® 2018 Ristempa® 2017	Prefilled glass syringe 6 mg/0.6 mL	SC with each cycle of chemotherapy, 24 hours following cytotoxic chemotherapy	2–8 °C, refrigerate – do not freeze. Protect from light. Do not shake. Stable at room temperature up to 30 °C for a maximum single period of up to 72 hours	Reduction in duration of neutropenia and incidence of febrile neutropenia in cytotoxic chemotherapy for malignancy
Nephrology								
Basiliximab	Chimeric mAb (IgG1k) directed against interleukin-2 receptor alpha chain (CD25 antigen) on the surface of T-lymphocytes	Anti-IL2 receptor	Simulect® 1998	Patent expired in 2013 in Europe. Biosimilar STI-003 under development	Freeze-dried powder 20 mg 1 ampoule of 5 mL WFI	IV bolus or infusion over 20–30 minutes	2–8 °C, refrigerate – do not freeze. Protect from light. Do not shake. Reconstituted solution stable for up to 24 hours at 2–8 °C	Immuno suppressive agent Prophylaxis of kidney transplant rejection
Eculizumab	Humanized glycosylated hybrid IgG2-IgG4 kappa immunoglobulin	mAb against the C5 complement protein, terminal complement inhibitor	Soliris® 2007	Patent expired (d)(s) in May 2020 in Europe and March 2021 in the United States.Biosimilars being developed ABP959 and BOW080	30 mL vial, 10 mg/mL	IV infusion	2–8 °C, refrigerate – do not freeze. Protect from light. Do not shake. Diluted solution is stable for 24 hours	Atypical hemolytic uremic syndrome (aHUS), Paroxysmal nocturnal hemoglobinuria (PNH) Membranoproliferative glomerulonephritis (off-label use)

Generic name	Description	Category	Brand® Year	Patent	Presentation	Dosing	Storage	Indications
Epoetin alfa[a]	Glycoprotein, rHuEPO, produced by recombinant DNA technology by inserting the gene into the CHO cells	Hormone, ESA	Eprex® 1988 Epogen® 1989 Procrit® 1989	Retacrit® 2018	Prefilled syringes 1000 IU/0.5 mL 2000 IU/0.5 mL 3000 IU/0.3 mL 4000 IU/0.4 mL 5000 IU/0.5 mL 6000 IU/0.6 mL 8000 IU/0.8 mL 10 000 IU/1.0 mL 40 000 IU/1.0 mL	IV 2–3 times per week	2–8 °C, refrigerate – do not freeze. Protect from light. Do not shake. If required, it may be stored at room temperature up to a maximum of 25 °C for a single period of 7 days	Anemia associated with chronic kidney disease. Anemia associated with non-myeloid malignancies due to chemotherapy. Adult patients with mild to moderate anemia scheduled for surgery with an expected moderate blood loss. To augment autologous blood collection in anemic patients prior to major surgery
Epoetin beta	Glycoprotein rHuEPO produced by recombinant DNA technology in CHO cells	Hormone, ESA	Recormon® 1990 Neo-Recormon® 1997		Prefilled syringes 2000 IU/0.3 mL 3000 IU/0.3 mL 4000 IU/0.3 mL 5000 IU/0.3 mL 10 000 IU/0.6 mL	IV, SC 2–3 times per week	2–8 °C, refrigerate – do not freeze. Protect from light. Do not shake. If required, it may be stored at room temperature up to a maximum of 25 °C for a single period of 3 days	Anemia associated with chronic kidney disease. Anemia associated with non-myeloid malignancies due to chemotherapy. Prevention of anemia in premature infants
Darbepoetin alfa	Biobetter. Glycoprotein with two additional oligosaccharide chains, produced by recombinant DNA technology in CHO cells and modified glycosylation	ESA	Aranesp® 2001	Not available. Patent expired in Europe in 2016, will expire in May 2024 in the United States	Injection syringe within a pen injector ranges from 10 µg/0.4 mL to 500 µg/1.0 mL. Injection syringe with automatic needle guard ranges from 15 µg/0.38 mL to 500 µg/1.0 mL. Injection syringe ranges from 10 µg/0.4 mL to 500 µg/1.0 mL	SC Fortnightly	2–8 °C, refrigerate – do not freeze. Protect from light. Do not shake. Stable at temperatures from −20 °C to +30 °C for 2 days	Anemia associated with chronic renal disease. Anemia associated with non-myeloid malignancies due to chemotherapy
Methoxy polyethylene glycol epoetin beta	Biobetter. Pegylated epoetin beta, produced by recombinant DNA technology in CHO cells and pegylation	ESA	Mircera® 2007	Patent expires in 2025 in the United States	Prefilled syringe ranges from 30 µg/0.3 mL to 360 µg/0.6 mL	IV, SC monthly	2–8 °C, refrigerate – do not freeze. Protect from light. Do not shake. If required, it may be stored at room temperature up to a maximum of 25 °C for up to 1 month	Anemia associated with chronic kidney disease

[a] biosimilar available.

CHO, Chinese hamster ovary; ESA, erythropoiesis stimulating agent; G-CSF, granulocyte colony stimulating factor; IL2, interleukin 2; IV, intravenous; HER2, human epidermal growth factor receptor 2; mAb, monoclonal antibody; SC, subcutaneous; rHuEPO, recombinant human erythropoietin; SWFI, sterile water for injection; TNF, tumor necrosis factor.

Table 16.2 Regulatory differences – comparison of biosimilars and generics of SMDs, an example from the United States.[3, 20, 26–28]

	Biosimilars	Generics
Regulatory requirements	Demonstrated structural and functional similarity to the reference product "Totality of the evidence" concept Clinical trials are necessary	Demonstrated structural identity to the reference product Clinical trials are not required for approval
Regulatory Pathways under the Food Drug and Cosmetic Act (FDCA) and license applications	Biologic Competition and Innovation Act 2010 (BPCI) 1) Biologics License Application (BLA) 2) Biosimilar License Application by demonstrating the absence of clinically meaningful difference in clinical trials FDA cannot accept an application for a biosimilar until 4 years after the licensure of the biologic and cannot approve a biosimilar until 12 years after the biologic was approved (8 years in the EU) Biosimilar applicant must share its application with the originator reference product sponsor	Drug Price Competition and Patent Term Restoration Act 1984 (Hatch-Waxman Act) 1) New Drug Application (NDA) 2) Abbreviated New Drug Application (ANDA) Pathway for the approval of generics by demonstrating bioequivalence, clinical trials are not required
Active ingredient, dosage form, route of administration, strength, conditions of use, and labeling	Highly similar, but may not be the same as originator reference product Cannot automatically claim all indications of the reference product Clinical studies required to demonstrate safety, purity, and potency for one or more appropriate conditions of use for which the reference product is licensed May or may not be therapeutically equivalent	Same as originator reference product Generic has the same ingredient, dosage form, route of administration, strength and conditions of use, and labeling as the originator reference product Therapeutically equivalent
Interchangeability	May or may not be interchangeable with the reference products Requires evidence from multiple switching studies Only interchangeables are substitutable	Interchangeable with the reference product, assuming similar purity and bioequivalence has been demonstrated
Automatic substitution	May or may not necessarily be automatically substituted with the reference product	Allowed
Nomenclature	International naming system for biosimilars varies, universal regulation for biosimilar naming are under development	Have the same International non-proprietary name as the reference product
Development cost	US$100–250 million	US$1–4 million
Development time	Approximately 5–9 years	Approximately 2 years
Price discounts	10–35%	70–80%

Table 16.3 The principles of naming monoclonal antibodies.[35]

Prefix	Substem A Antibody target		Substem B Antibody source		Suffix
Variable	c(i) l(i) s(o) tu(m)	cardiovascular immunomodulating bone tumor	o u xi zu	mouse human chimeric humanized	mab

Examples: beva**ci**zumab, deno**s**umab, pembro**li**zumab, ri**tu**ximab, tras**tu**zumab.

voluntary and is intended to assist in the identification of biological substance for prescribing, dispensing, pharmacovigilance, and globally transferring prescriptions.[34]

In 2008, the INN Working Group reviewed and streamlined the nomenclature for mAbs.[35] The INN for mAbs are composed of a prefix, a substem A, a substem B, and a suffix. A prefix is variable, a substem A refers to antibody target (e.g. molecule, cell, organ, tumor, or system), a substem B refers to the species on which the immunoglobulin sequence of the mAb is based (e.g. mouse, chimeric, humanized, human), and a suffix is "mab." The general principles for nomenclature for mAbs are summarized in Table 16.3.

16.6 Funding of Biosimilars: Australian Approach

In Australia, the evaluation of biosimilars is conducted by the TGA by using quality, safety, and efficacy data to determine whether the biosimilar is equally safe and effective as the reference biologic.[33,36] The TGA has adopted a number of European guidelines and the International Conference of Harmonisation (ICH) guideline on the assessment of comparability.

Following the approval by the TGA, the Pharmaceutical Benefits Advisory Committee (PBAC) evaluates the biosimilar for addition to the Pharmaceutical Benefits Scheme (PBS). The evaluation of biosimilars is conducted on a case-by-case basis on whether to permit substitution by doctors or pharmacists. The PBAC recommends whether the biosimilar and its reference medicine should be substitutable, i.e. dispensing one brand of the medicine instead of another equivalent and interchangeable brand of the same medicine at the pharmacy level without needing to refer to the prescriber.[33]

During the dispensing process, pharmacists should ensure continuous product traceability in clinical practice.[37] The INN and brand name and batch number of the biosimilar should be recorded in the electronic dispensing system to allow for traceability. The batch numbers are not included in barcodes and therefore the batch number needs to be recorded manually.

In 2015, the PBAC made a world-first recommendation to allow clinicians and pharmacists to give patients the option of substituting an expensive innovator biologic medicine with a less costly replacement biosimilar if available.[38] Clinical assessment by the prescriber evaluates appropriateness for substitution for each individual patient. Brand substitution is a choice for the patient and the prescriber.[39] The prescriber may tick the "brand substitution not permitted" box when writing a prescription, if they do not think brand substitution is appropriate.[33] Pharmacists follow substitution principles with respect to preference of the prescriber and patient. An example of biosimilar infliximab approved by the TGA and subsidized on the PBS[40] is presented in Table 16.4. For "a" – flagged biosimilars Inflectra® and Renflexis® – brand substitution may be undertaken by pharmacists at the point of dispensing without anticipating differences in clinical effect.[41]

Likewise, rituximab biosimilar (Ryximyo®) was approved by the TGA[42] and PBAC[43] in 2018. Rituximab is a chimeric monoclonal antibody that binds to the CD20 surface marker on B cells leading to the death of B cells. In Australia, it is funded for treatment of the following hematological and autoimmune disorders: non-Hodgkin's lymphoma (NHL), chronic lymphocytic leukemia (CLL), rheumatoid arthritis (RA), granulamatosis with polyangiitis (Wegener's GPA), and microscopic polyangiitis (MPA). Rituximab Riximyo®, a biosimilar medicine, has been assessed by the TGA to be highly similar to MabThera®; it provides the same health outcome and it is safe and effective as MabThera®. The

Table 16.4 An example of infliximab listing on the Pharmaceutical Benefits Scheme (PBS).

Active ingredient	Reference brand (sponsor)	Biosimilar brand (sponsor)	Biosimilar medicine subsidized by PBS	Therapeutic class / therapeutic area
Infliximab	Remicade®(Janssen-Cilag) "a" flagged *Also registered under the brand name Jaximab®*	Inflectra® (Pfizer) "a" flagged *Also registered under the brand names Remsima®, Emisima®, Flixceli®*	Yes	Immunosuppressant Monoclonal antibody/ arthritis, inflammatory bowel disease, psoriasis
		Renflexis® (Samsung Bioepis AU) 'a' flagged	Yes	

PBAC recommends it being used for the same indications as innovator biologic rituximab MabThera® under the S100 Efficient Funding of Chemotherapy and Highly Specialized Drugs Program. Therefore, Riximyo® and MabThera® should be considered equivalent for the purposes of substitution at the pharmacy level ("a"-flagged) with respect to the PBS.[40,41]

16.7 Pharmacovigilance

Pharmacovigilance legislation mandates that all biological medicines, including biosimilar medicines, are subject to additional post-authorization monitoring for safety as comparative clinical trials are not long enough to detect rare AE.[44] Postmarketing pharmacovigilance surveillance programs have been established with the goal to promptly identify and evaluate safety signals, such as unanticipated adverse effects or a lack of efficacy, so the risks can be appropriately managed.[23]

For every adverse drug reaction (ADR), the medication name, including brand name, and a batch number have to be reported to ensure that ADR is linked to the correct product. For this reason, the protocols have to be established to ensure the traceability of biologics and biosimilars in clinical practice. In Australia, ADRs are reported to the Therapeutics Goods Administration (TGA) at https://aems.tga.gov.au/privacy. A report should describe the type of reaction, indication, the INN, the ABN, brand name (trade name), the AUST R number, the batch number and expiry date, the dosage form and presentation, manufacturer's name, and country of origin.[45]

Adverse reactions to prescription drugs and biological products from controlled trials, post-authorization safety studies, and spontaneous reporting are described in the SmPC.[46] The summary of the safety profile contains the most serious and/or most frequently occurring adverse reactions, the timing, and frequency of ADR. In addition, a tabulated list of adverse drug reactions (ADRs) according to the Medical Dictionary for Regulatory Activities (MeDRA) system organ classification (SOC) and within each organ class, frequency category in order of decreasing seriousness: Very common ($\geq 1/10$); common ($\geq 1/100$ to $< 1/10$); uncommon ($\geq 1/1000$ to $< 1/100$); rare ($\geq 1/10\,000$ to $< 1/1000$); very rare ($< 1/10\,000$). Furthermore, the SmPC also contains the description of selected serious and/ or frequently occurring adverse reactions alongside with the measures to be taken to avoid them and action to be taken if a specific adverse reaction occurs.

In addition, for clinical trials in oncology and HIV, The National Cancer Institute's (NCI) Common Terminology Criteria for Adverse Events (CTCAE) descriptive terminology for grading all AE is commonly used.[47] The CTCAE grading (severity) scale lists the clinical description of AE in increased severity: Grade 1 are Mild AEs that are asymptomatic or mild, necessitating clinical or diagnostic observations only and interventions are not required. Grade 2 AEs are moderate, requiring local or noninvasive intervention as those AEs are limiting age-appropriate instrumental activities of daily living. Grade 3 AEs are severe or medically significant, but not immediately life-threatening, resulting in hospitalization or prolongation of hospitalization, they are disabling, and limit self-care activities of daily living. Grade 4 AEs are life-threatening, requiring urgent intervention. Grade 5 AEs cause death related to AE.[47]

ADRs represent a subset of AE that can be directly attributed to a drug and its physiologic properties.[48] Type A ADRs are dose-dependent and predictable, caused by augmentation of known pharmacologic effect(s). Immunomodulatory biological therapies are associated with the following Type A ADR: serious infections such as tuberculosis (e.g. rituximab), fungal infections, meningitis (e.g. eculizumab), malignancies (anti-tumor necrosis factor-alpha agents), progressive multifocal leukoencephalopathy (natalizumab, rituximab), compromised wound healing or arterial thromboembolic events (angiogenesis inhibitors, e.g. bevacizumab), dermatological toxicities (epidermal growth factor receptor inhibitors, e.g. cetuximab, panitumumab), and B-cell lymphocyte depletion from anti-CD20 antibodies (rituximab).[37]

In contrast, Type B ADRs are not related to pharmacologic action of the drug, they are uncommon, unpredictable, and independent of the dose, e.g. immunologic and idiosyncratic reactions.[48] Unlike small molecules, that must form haptens in order to induce an allergic reaction, biologics do not need to be biotransformed into hapten–protein complexes to induce an allergic reaction. Therefore, an immune response is a major concern with biologic therapy and may include anaphylaxis, organ-specific immunopathy, autoimmune reactions, and systemic hypersensitivity. Clinical manifestations of immunogenicity range from the transient appearance of anti-drug antibodies (ADAs) without any clinical significance to severe life-threatening conditions and may include: a lack of efficacy, pharmacokinetic alteration, development of antibodies that neutralize the product, hypersensitivity, infusion reactions, and development of antibodies toward an endogenous protein resulting in a serious adverse event.[37,49]

Cytokine release syndrome (CRS),[37] or "cytokine storm" or "sterile sepsis,"[4,47] is severe or life-threatening reaction that occurs due to immune activation and release of inflammatory cytokines. CRS has been reported with infliximab, rituximab, alemtuzumab, and with chimeric antigen receptor T-cell therapy.[37]

Infusion reactions are related to cytokine release and can be also caused by the idiosyncratic reaction.[50] The symptoms include fever, chills, rigors, nausea, vomiting, urticaria, itch, headache, bronchospasm, angioedema, rhinitis, and hypotension. In clinical practice, it is hard to distinguish between IgE-mediated reactions (e.g. anaphylaxis, CRS) and non-IgE mediated reactions (e.g. severe cardiac and pulmonary events). Infusion-related reactions have been associated with mAbs, most commonly with rituximab, and trastuzumab.[50] Rechallenge following infusion reaction largely depends on the underlying etiology of an infusion reaction.[50] For example, infusion reactions with rituximab are mostly cytokine-mediated, and therefore managed by resuming the infusion at 50% of the infusion rate once the symptoms completely resolve. However, with cetuximab, rechallenge is not recommended as severe infusion reactions appear to be IgE-mediated.

In addition, the product quality complaints indicate a potential quality issue with the product or packaging. The examples of product quality complaints include instances when the product performs differently to what is expected, appears physically different (e.g. different color, odor, foreign material), packaging or labeling defect (e.g. leaking, broken package, suspected tempering), suspected contamination (e.g. product causes infection), device malfunction (e.g. failure to auto-inject), or is a counterfeit product.

Furthermore, incidents such as medication errors and near misses, overdose, off-label use, drug interaction, suspected transmission of infection, accidental or occupational exposure, and suspected abuse or misuse should be reported via healthcare organization's incident management database. Special care needs to be taken to prevent medication errors for a look-alike, sound-alike medications, e.g. trastuzumab (Herceptin®) and trastuzumab emtansine (Kadcyla®) and different formulations for IV vs. SC use, e.g. rituximab, trastuzumab.

16.8 Medicines Use Evaluation

Expanded pharmacovigilance strategies have been suggested, e.g. collection of the data via an electronic platform to monitor clinical response to biosimilar introduction and related clinical outcomes to manage the issues associated with their use.[51] Medicines use evaluation (MUE) is a performance improvement method that focuses on evaluating and improving medication-use processes with the goal of optimal patient outcomes.[52] As biologic therapies are used relatively rarely, a multicenter MUE of biosimilars using interdisciplinary and intersectoral collaborative approach in clinical research, training, and funding may be beneficial.[53] Gathering information on biosimilar prescribing practices before and after the switching from originator biologic is recommended for evaluation of the impact of the switching program to determine the degree of its success. In addition to the disease-specific patient outcomes and treatment-related harm, the following data may be collected[54]: spend on the originator biologic drug, spend on biosimilar drugs, overall drug cost savings, specialty where biologic is used, indications for which biosimilar is used, number of patients treated using originator biologic, the number of new patients on the biosimilar, number of patients switched to the biosimilar medicine part-way through current treatment, for the approved indication alongside with the reasons why patients may not be receiving the biosimilar.

16.9 Reducing Clinical Risks from Innovator Biologics, Biosimilars, and Biobetters

Biologic therapies are associated with the following clinical risks:

- Suboptimal regulatory assessment[30] due to a lack of the sound licensing process in some countries ("pharmerging markets": China, Algeria, Brazil, Argentina, India, Egypt, Colombia, Indonesia, Mexico, Turkey, Saudi Arabia, Pakistan Thailand, and Venezuela; or biosimilar markets with little or no presence, e.g. China, Russia[55]), misleading nomenclature and pronunciation[56] and clinicians' knowledge gaps.[3,28,51]
- Immune response, toxicity or loss of efficacy due to increased immunogenicity[49,57] caused by repetitive switching between biologic and biosimilar or substitution at the pharmacy level, or inadvertent switching due to a lack of traceability.[37]
- Compromised patient safety due to anaphylaxis, hypersensitivity reactions and infusion reactions,[50] CRS, or other severe adverse reactions.[37]
- Compromised patient safety caused by erroneous prescribing of biologic therapy in patients with a history

of hypersensitivity reaction to biologics, acute infection or other contraindication or receiving therapy with known drug–drug interactions with biologic therapy.

- Preventable severe infection due to a lack of screening and vaccination.
- Prescribing, dispensing, and administration errors due to look-alike, sound-alike names of biologics/biosimilar or incorrect route of administration.
- Modified or loss of effect due to "nocebo" effect.[58]
- Compromised safety of healthcare staff due to inadvertent occupational exposure.

Recent research[20,51,59] on health professionals' attitudes toward innovator biologics and biosimilars indicate that there is low awareness of the processes for approval and funding for these treatments. Significant trust has been placed on the prescriber's decision-making regarding treatment. In addition to specialist physician initiating therapy, primary care physicians are increasingly involved in prescribing and community pharmacists in dispensing of innovator biologics and biosimilars. The following concerns have been raised among prescribers, pharmacists, and consumers.[28,51]

Prescribers' primary concerns were related to knowledge/awareness gaps in information about therapies and outcomes, restrictions with respect to funding or formulary access, or access to biologic therapies for off-label use, indication extrapolation, and immunogenicity. Additional concerns included the risks associated with traceability due to a lack of nonproprietary naming convention, substitution without prescribers' knowledge, and issues pertinent to communication between prescriber and pharmacist and prescribers and patients.

Further to the concerns related to access to biosimilars and communication issues, pharmacists' concerns were related to extended pharmacovigilance requirements such as issues related to traceability and recording of batch numbers, affordability of therapy, including the cost of the drug, the cost of handling multiple forms of high-cost drugs and the cost of storage in a pharmacy, and the cost if the medicine is wasted. Additional concerns were related to procurement, e.g. due to the geographical delay in supply, maintaining cold chain, and the need for training and certification for sterile compounding.

Patients/consumers concerns were related to access due to funding restriction, overall cost and benefit, and side effect profile of biologic medicines. Understandably, patients who were stable on therapy were reluctant to make any changes as achieving adequate disease control or remission can be challenging. Patients/consumers with low health literacy or language difficulties were not able to understand the benefits and risks of biologic therapies, indicating the need for multicultural and multilanguage educational resources.

16.10 Checklist on Biologics

The opportunities for improving the quality use of biosimilars lie in the implementation of a multimodal strategy focusing on interventions and policies to change behavior.[60] The improvements can be achieved by focusing on increasing staff and patients' capability, increasing the opportunities for the desired prescribing behaviors to take place, and by motivating clinicians and patients.

Capability. Increasing staff and patient capability can be achieved by developing the resources for education, training, and by involving patients in resource development, establishing the process that promote open communication to identify the barriers to patient understanding, and addressing the needs of vulnerable populations, ultimately improving patient experience and outcomes. Since interchangeability is largely viewed as a cost-saving measure, it may be associated with negative treatment expectations that could potentially reduce a medicine's effectiveness or increase side effects. This is known as the "nocebo effect"[58] and has been associated with negative viewpoints held by health professionals. Therefore, healthcare professionals must be fully informed and confident about the use of biosimilars to be able to effectively address patients' queries and concerns.[61] Biosimilar Awareness Initiative is a Government-funded initiative to promote uptake of biosimilars in Australia. The information on biosimilar product availability alongside the fact sheets for health professional and consumers are available at https://www1.health.gov.au/internet/main/publishing.nsf/Content/biosimilar-awareness-initiative

Opportunity. Improving the workflow for the desired actions to take place can support clinicians to make evidence-based, safe, and cost-effective choices in prescribing and dispensing. Hence, policies, guidelines, procedures, and checklists can be developed to support clinicians' decision-making. Checklists are cognitive aids that condense a large quantity of knowledge in a concise fashion.[62] They are both mnemonic tools that ensure important steps in patient care are not omitted and evaluative tools that allow the cognitive workload to be redistributed enabling processing and calm consideration of the evidence.

Motivation. Increasing staff and patient motivation can be achieved by change leadership,[63] utilizing social marketing, persuasion, and role modeling to engage with a sense of shared purpose, empathy, and solidarity. A socioethical approach to behavior changes that draws on social, rhetorical, practical, and ideologic understanding has been suggested.[64,65] Instead of using controls or carrots and sticks to motivate people, hospitals should use systems that rely on engagement and a sense of common purpose to build cooperative systems by encouraging communication, ensuring authentic framing, fostering empathy and solidarity, guaranteeing fairness and morality, using rewards that appeal to intrinsic motivations, relying on reputation, and ensuring flexibility.[66] Transparent communication and reporting of patient-centered performance indicators by providing peer comparisons to emphasize positives and celebrate success can have a positive effect on motivation and engagement and can be also used for education, training, and clinical practice improvement.

A checklist for pharmacists on biologic therapies is developed to enhance patient-centered discussions when a physician first prescribes a biologic for a new patient or when a patient is switched to a biosimilar. The aim of this checklist is to assist in structured and collegial discussions with doctors, pharmacists, and patients; to reduce omissions and errors; and to ensure safe and effective use of these therapies. While the checklist perhaps is most relevant to the Australian healthcare landscape, it contains many common elements that are relevant internationally. The checklist may be modified without difficulty based on individual national policies and guidelines.

The checklist has six sections; an introduction section guides the clinicians through the evaluation of the primary literature for biosimilars, followed by the five risk management sections focusing on preventing AE, enabling effective communication between physicians and pharmacists, safe pharmacy dispensing processes, and effective communication for engaging with patients/caregivers. The final section, on QUM, is an overarching section blending the concepts of the QUM[67] and clinical governance[68] with behavioral change,[60] teaming,[69] and change leadership[63] (Table 16.5).

16.11 Practice Points: Anticipated Questions from Health Professionals

16.11.1 How Safe Are Biosimilars?

Biosimilars have demonstrated similar quality, safety, and efficacy to the original reference biologic in clinical trials and have been used for more than 10 years. As both biologics and biosimilars are proteins or protein antibodies, they are associated with high immunogenicity and their immunogenicity profiles are found to be comparable.[84–87]

16.11.2 What Are the Potential Advantages to Using Biosimilars?

Biosimilar medicines are reducing the cost of subsidized biologic therapies. Due to their abbreviated approval pathway and reduced mandatory testing requirements, biosimilars may cost 20–40% less than their reference products.[3] This introduces price competition and reduces the cost of subsidizing biological therapies. This provides better access and supply, greater choice for treatment, and better patient health outcomes. This would free-up healthcare resources allowing investment in new innovative treatments.[88] In addition, access to different biosimilars or brands of biosimilars reduces the risk of drug shortages.

16.11.3 Can Patients Be Switched from Reference Biologics to Biosimilars?

Switching between originator biologic and biosimilar raises the theoretical concern that repeated switching may increase immunogenicity.[29] To date, the switch-related immunogenicity was found to be associated with alteration of a manufacturing process of the biologic resulting in an suboptimal version of the biosimilar or due to altered storage conditions[89] and there is no clinical evidence suggesting that a switch itself would directly cause immunogenicity. In addition, the incidence of AE and antibody titers in patients who have received biosimilar throughout the study and the patients who were switched from innovator biologic to biosimilar were found to be comparable during the one-year follow-up period.[85,87,90]

As available clinical evidence is limited, switching patients currently treated with innovator biologic to a biosimilar is recommended in stable patients after six months of treatment with innovator biologic for economic reasons only.[91] Patients already receiving the innovator biologic should have a discussion with their doctor about switching from innovator biologic to a biosimilar. Therefore, the appropriateness of switching should be evaluated on a case-to-case basis and it is the ultimate decision of the physician with the consent of the patient. However, the pharmacist, as the medicines expert, can provide unbiased objective advice based on published data for the appropriate decision. Of course,

Table 16.5 Checklist for pharmacists for safe and effective use of biologic therapies. [12,23,45,49,51,70–72]

Section/item	Item no	Description	Setting
Evaluation of the primary literature for biosimilars [37,45,49,73–76]			
Comparability. *Biosimilar is neither superior nor inferior to the reference biologic – both increased or reduced immunogenicity relative to the reference biologic would question biosimilarity*	Ia	Similarity study design 1) Parallel blinded study (head-to-head) in treatment-naïve patients 2) Subsequently, a subset of patients should be evaluated in a crossover study with the reference biologic to the proposed biosimilar to evaluate the risk of hypersensitivity, immunogenicity, and other adverse drug reactions (ADRs)	Desktop
	Ib	Assays Sensitivity and specificity – capable of detecting the anti-drug antibodies (ADAs). The incidence, nature, and ADA titers are measured and assessed	
Patient population	IIa	Clinically relevant	
	IIb	The most sensitive patient population used, i.e. the population at the highest risk of developing adverse immune reactions, clinically meaningful differences between biosimilar and biologic are most likely to be detected. Any extrapolated indications are for uses and patient populations with lower risk	
Power/sample size	IIIa	Sufficiently powered to detect potential differences between biosimilar and innovator biologic	
	IIIb	Appropriate statistical methodology	
	IIIc	95% confidence intervals and *p*-value <0.05 is used for statistical significance	
Dose	IV	Dose and route are consistent with the reference biologic	
End-points	V	Relevant to the disease state, sensitive enough to detect clinically relevant differences in efficacy and safety	
Efficacy	VI	Efficacy measures within the pre-specified acceptable margin of similarity	
Safety	VIIa	Comparable ADA positivity	
	VIIb	Comparable incidence and types of treatment-emergent adverse effects (TEAEs) between biosimilar and reference biologic	
	VIIc	Comparable incidence of augmentation of the known pharmacologic actions, e.g. serious infections (e.g. tuberculosis) and malignancies	
	VIId	Known and potential safety concerns are highlighted in the application dossier	
	VIIe	Risk management plan (RMP), pharmacovigilance guidelines, and periodic safety update reports are available	
Study duration	VIII	Appropriate to detect clinical effects	
	IX	It depends on the duration of the treatment course, disappearance of the product from the circulation, and the time for emergence of humoral immune response – at least 4 weeks. Chronic administration – 1 year of data, 6 months might be justified based on the immunogenicity profile of the reference biologic	
Risk management: clinical pharmacy – preventing adverse events			
Patient unique identifiers	1	Full name, date of birth, address, national health index number; insurance number	Bedside
Patient demographics	2	Age, gender, body weight, body surface area (BMI if potentially obese), pregnancy, breastfeeding	

Prescription check	3a	Has innovator biologic or biosimilar been prescribed?
		Check the INN, brand name, dose, route, frequency, duration and quantity, cycle number of treatment
	3b	Confirm the indication
	3c	Check the prescription is legible, i.e. date, doctor's name, signature, and contact phone number
Medical history	4	Check patient's medical history for contraindications and precautions.
		E.g. severe infection, blood disorders, heart failure, avoid anti-TNF-α agents in patients with tuberculosis, multiple sclerosis, heart failure, and cancer within 5 years
Laboratory results	5	Check the relevant laboratory data and any companion test results
Medication history	6a	Check hypersensitivities and adverse drug reactions (ADRs)
	6b	Obtain the best possible medication history (BPMH)
	6c	First time dispensed or subsequent dispensing?
		Is this the first prescription for treatment-naïve patients or a subsequent prescription?
		Initial prescription: biosimilar should be prescribed and dispensed if available. Prescribing biosimilar for treatment-naïve patients may be encouraged by the Government
Drug interactions	7	Check for potential drug–drug interactions.
		E.g. avoid live vaccines, concomitant administration of anti-TNF-α with mAbs due to neutropenia and severe infection, rituximab and clozapine due to increased risk of neutropenia, trastuzumab and anthracyclines (daunomycin, doxorubicin), increased risk of heart failure, trastuzumab emtansine – increased exposure with azole antifungals, clarithromycin, protease inhibitors (e.g. indinavir, nelfinavir, ritonavir, saquinavir), filgrastim and cytotoxic agents, increased risk of myelosuppression (filgrastim should not be used 24 hours before and until 24 hours after cytotoxic chemotherapy)[77]
Adverse drug reactions	8	What are the potential serious adverse effects?
		Type A: Serious infections (e.g. tuberculosis, fungal infections) and malignancies (with anti-TNF-α agents), progressive multifocal leukoencephalopathy (natalizumab, rituximab), compromised wound healing or arterial thromboembolic events (angiogenesis inhibitors, e.g. bevacizumab), dermatological toxicities (epidermal growth factor receptor inhibitors, e.g. cetuximab, panitumumab), and B-cell lymphocyte depletion from anti-CD20 antibodies (rituximab).[37]
		Type B: Immunogenicity – hypersensitivity, infusion reactions, and development of antibodies toward an endogenous protein resulting in a serious adverse event, cytokine release syndrome (CRS), lack of efficacy, pharmacokinetic alteration, development of antibodies that neutralize the product.
		• Murine mAbs the most immunogenic > chimeric > humanized > completely human mAbs the least immunogenic.
		• Less immunogenic: short-term treatment, intravenous route of administration, continuous administration.
		• More immunogenic: long-term treatment, inhalation, intradermal, subcutaneous and intramuscular route of administration, intermittent treatment, re-exposure and long treatment-free intervals
Premedications	9	Ensure premedication therapy is prescribed if required, e.g. for rituximab: antipyretic, antihistamine, glucocorticoids

(Continued)

Table 16.5 (Continued)

Section/item	Item no	Description	Setting
Preventative measures	10a	Screening for Hepatitis B and C, tuberculosis[21]	
	10b	Vaccination: Specialist advice should be sought. Live vaccines are contraindicated. Immunization is contraindicated with immune checkpoint inhibitors. Immunization should be delivered prior to elective immunosuppression. Inactive influenza vaccine, Hepatitis B vaccine, Pneumococcal vaccine (every 5 years). The meningococcal vaccine should be given at least 2 weeks prior to eculizumab	Clinical
Risk management: collaboration between physicians and pharmacists			
Preventing adverse events	11	Follow-up on any issues using the ISBAR communication framework[78]. Identify – Situation – Background – Assessment – Recommendation	
Interchangeability (switching and substitution)	12a	Specific agent. Has a reference biologic or a biosimilar been prescribed?	
	12b	Approved indication or indication extrapolation? Check the Pharmaceutical Benefits Advisory Committee (PBAC) website for approved indications	
	12c	Therapeutically naïve patient vs. patient already receiving biologic?	
	12d	Acute vs. chronic therapy	
	12e	Similarity/equivalence. Check the Australian Public Assessment Report (AusPAR) at the Therapeutic Goods Administration (TGA) website	
	12f	Efficacy and safety. Check the AusPAR	
	12g	Published studies for additional information. Follow the checklist for evaluation of the primary literature for biosimilars, items I–VIII	
	12h	Cost advantage. What is the cost of reference biologic and biosimilar? Check at the Pharmaceutical Benefits Scheme (PBS) website	
	12i	Switch from reference biologic to biosimilar (evidence-based in stable patients >6 months) or Switch from biosimilar to reference biologic or Switch from one biosimilar to another biosimilar	
	12j	Does a biologic have "a-flagged" biosimilar? Check at the PBS website. E.g. Pharmaceutical Benefits Advisory Committee (PBAC) recommends that the etanercept brands Enbrel® and Brenzys® could be marked as equivalent in the Schedule of Pharmaceutical Benefits ("a" flagged) for the purposes of substitution by the pharmacist at the point of dispensing for all the circumstances (restrictions) for which both brands are listed[40]	

12k	Specialty monitoring surrounding switching or substitution
	E.g. Nephrology: hemoglobin levels are required when (i) switching from a reference biologic to a biosimilar erythropoiesis stimulating agent (ESA); (ii) switching from one biosimilar ESA molecule to another; (iii) when switching from a different brand of the same molecule.[4]
12l	Verify prescriber's intent for "Brand substitution not permitted" box if relevant.
12m	If necessary, provide prescriber information on biosimilars.
	Australian Government Department of Health. The basics for healthcare professionals https://www1.health.gov.au/internet/main/publishing.nsf/content/biosimilar-awareness-initiative/$File/Biosimilar-medicines-the-basics-for-healthcare-professionals-Brochure.pdf
Outcome of pharmacist–physician discussion	
13	Document the outcome of discussion

Risk management: pharmacists' process for high-risk medications

Procurement	
14a	How long will it take to obtain?
14b	Plan for subsequent doses
14c	What to do if product is not available or out of stock
Formulation considerations	
15a	Is product being compounded or ordered from an external supplier?
	Check if the product needs to be stored, handled, administered, and disposed of as a hazardous material by trained staff, e.g. the eviQ Hazardous drugs table[19] or the National Institute for Occupational Safety and Health (NIOSH) list of antineoplastic and other hazardous drugs.[16]
	Consider occupational health and safety risks, operational and clinical factors. Follow the Australian consensus guidelines14 and COSA position statement15 for the safe handling of mAbs.
	Aseptic technique: isolator cabinet or closed system drug transfer devices or cytotoxic safety cabinet plus:
	1) Monoclonal antibodies (mAbs) – wear protective eyewear, respiratory mask and gloves.
	2) Monoclonal antibody conjugates with cytotoxics drugs (e.g. Adcetris®, Kadcyla®), fusion proteins or radiopharmaceuticals, cytotoxics and hazardous products.
	• Cytotoxics and hazardous products: full personal protective equipment (PPE): gown, double gloved, protective eyewear, respiratory mask.
	• Radiopharmaceuticals – protective shields and gowns, disposable gowns and gloves, and tools and shields for maximizing the distance from the source
15b	Diluent and stability data: refer to the product information
15c	Delivery devices, appropriate route of administration intravenous (IV) vs. subcutaneous (SC)
15d	Consumables, e.g. polyvinyl chloride-free IV containers and administration sets
15e	Storage, handling, and disposal,[12,13,17] e.g. an insulated container, if transport longer than 3 hours add gel or ice packs; disposed as clinical waste, cytotoxic waste for monoclonal antibody drug conjugates

Table 16.5 (Continued)

Section/item	Item no	Description	Setting
Dispensing	16a	Check the three patient unique identifiers	
	16b	Check the drug name, dose, form, and route of administration	
		Alert: Look-alike, sound-alike drugs: trastuzumab and trastuzumab emtansine (Kadcyla®), brentuximab and brentuximab vedotin (Adcetris®).	
		Alert: SC vs. IV formulations (e.g. rituximab, trastuzumab)	
	16c	Check product INN, brand name, strength, form, route of administration, expiry date, and quantity	
	16d	Prepare medication label and updated medication list outlining INN, brand name, and the appropriate patient-centered description of biosimilar regimen and monitoring	
	16e	Traceability	
		Record the brand name and batch number to ensure continuous product traceability	
	16f	Attach appropriate ancillary labels, e.g. Refrigerate – do not freeze, Do not shake, Cytotoxic label for monoclonal antibody conjugates	
Quality control	17	Establish local protocols (standard operating procedures, SOPs) to avoid inadvertent drug interchange, or switching administration route (intravenous versus subcutaneous)	
Subsequent dispensing	18a	Where is the prescription?	
	18b	Confirm generic, brand name, and route of administration	
	18c	If box is not ticked will I substitute?	
		Discuss with the patient – 24d	
	18d	Has patient been on any new medication that could now interact with reference biologic/biosimilar?	
	18e	Therapeutic drug monitoring, e.g. infliximab[79,80]	
	14–16	Follow steps 14–16 of this checklist	
Report any adverse drug reactions to the TGA https://aems.tga.gov.au/privacy	19a	Description of ADR	
		– Frequency category: Very common (≥1/10); common (≥1/100 to <1/10); uncommon (≥1/1000 to <1/100); rare (≥1/10000 to <1/1000); very rare (<1/10000).[46]	
		– Oncology: Common Terminology Criteria for Adverse Events (CTCAE v5.0)[47]	
	19b	Indication	
	19c	International nonproprietary name (INN) = Australian biologic number (ABN)	
	19d	Brand name (Trade name)	
	19e	The AUST R number	
	19f	The batch number	
	19g	The expiry date	
	19h	The dosage form and presentation	
	19i	Manufacturer's name	
	19j	Country of origin	

		Bedside
	Report product quality incidents	20	Report product quality incidents to the manufacturer
	Report incidents and near-misses	21	Report incidents and near-misses via the healthcare organization's incident management system
	Follow up	22	Monitor for effectiveness and safety, e.g. clinical response, laboratory tests, therapeutic drug monitoring, ADR, adherence. Schedule a follow up

Risk management: engagement between pharmacist and patient/caregiver[81,82]

Tailor advice to the patient's/caregiver level of health literacy; use show and tell and teach-back techniques

Acknowledge	23	Greet the patient/caregiver and introduce yourself – make them feel comfortable
Identify	24	Identify the patient – check the three unique patient identifiers
Duration	25	Give indication of duration of counseling session
Explain	26a	Verify with patient what was ordered and what is expected
	26b	Confirm indication
	26c	INN and brand name of the medication
	26d	Discussion between patient and pharmacist regarding brand selection (substitution)
	26e	The dose, form, route, and frequency of administration
	26f	Duration of therapy
	26g	Storage and transport issues. Refrigerate – do not freeze; insulated container; Do not shake; Cytotoxic, e.g. monoclonal antibody conjugates
	26h	Ask patient if they have had any issues/adverse drug reactions previously
	26i	Explain potential adverse effects and action plan
	26j	Monitoring requirements
	26k	Check adherence, advice not to stop biosimilar/reference biologic except on the advice of a health professional
	26l	Advice not to share biosimilar/reference biologic with anyone
	26m	Provide patient education brochure for biosimilar https://www1.health.gov.au/internet/main/publishing.nsf/content/biosimilar-awareness-initiative/$File/Biosimilar-medicines-the-basics-for-consumers-and-carers-Bochure.pdf
		Provide a consumer medicines information https://www.ebs.tga.gov.au/ebs/picmi/picmirepository.nsf/PICMI?OpenForm&t=PI&k=0&r=https://www.ebs.tga.gov.au
	26n	Enroll in a patient support program
	26o	Encourage the patient to take the photo of his/her reference biologic/biosimilar medicine, displaying brand name and batch number
	26p	Encourage the patient to carry a medication list when interacting with healthcare professionals
	26q	Confirm patient/caregiver understanding – teach-back technique
	26r	Clarify any concerns or patient worries or anxiety about any issues
	26s	Encourage patient to attend the same dispensing pharmacist
	26t	Dispose syringes in sharps container.
		Advise on returning unwanted medications to the pharmacy for disposal

(Continued)

Table 16.5 (Continued)

Section/item	Item no	Description	Setting
Thank you	27	Thank the patient/caregiver and offer further assistance – give your contact details	
Quality use of medicines (61, 63, 67–69)			**Organization**
Clinical governance framework to support implementation	A1	Develop system-based strategies.	
		E.g. develop policies, guidelines, and checklists to support clinicians to make clinically and cost-effective choices in prescribing of biological medicines, and if necessary, consider formulary restrictions and automation via electronic medication management	
	A2	Develop educational strategies	
		E.g. medicines bulletin, presentations, one-on-one sessions (academic detailing), social media, posters	
	A3	Develop communication plan.	
		Establish communication network.	
		Communication plan in place to alert providers to new and better value innovator biologic and biosimilar.[24]	
		Prepare advocacy strategy for providers, e.g. use of better value biosimilar enables funding for new therapies	
	A4	Set achievable targets and monitor performance.E.g. for specialized services: 90% of new patients being on the best value biological medicine within three months of product launch and 80% of existing patients with 12 months[71]	
Engagement and change leadership	B1	Identify and engage key opinion leaders and early adopters	
	B2	Provide staff education and training	
	B3	Provide access to the resources, staff and patient information, checklists	
	B4	Implement the communication plan	
	B5	Provide information on performance; for multiple sites, provide peer comparisons	
	B6	Social activities, e.g. organize campaigns, celebrate successes	
Monitoring for effectiveness	C1	Clinical outcomes – medicines use evaluation (MUE) for innovator biologics and biosimilars	
	C2	Patient experience survey	
	C3	Patient-reported outcome measures (PROMs)[83]	
	C4	Staff experience survey	
	C5	Economic evaluation	
Monitoring for safety	D1	Traceability audit	
	D2	Pharmacovigilance report	
	D3	Analyze near-misses and incidents, provide feedback, and disseminate the lessons learned	

switching at the pharmacy (pharmacist) level is sometimes enshrined in national policies.

16.11.4 In a Case of Nonresponse to a Biosimilar, Can the Patient Be Switched Back to the Reference Biologic?

A lack of response to treatment with the biosimilar may be due to poor adherence, poorly responsive disease, development of neutralizing antibodies, or due to the nocebo effect. The reason for nonresponse should be determined and if neutralizing antibodies are detected, the patient should not be switched to the reference product, due to cross-reactivity. The recommended strategies are dose modification, the addition of concomitant therapy, discontinuation of the biosimilar, and switching to an alternative treatment.[29] Multiple switch clinical studies would be useful. The pharmacist (and doctor) has a key role in minimizing the nocebo effect by positive framing of the conversation about biosimilars.

16.12 Practice Points: Anticipated Questions from Patients

16.12.1 What Is a Biologic?

Biologics are highly complex large molecules produced by biotechnology using living cells. They are used to treat serious chronic diseases such as cancer, diabetes, severe psoriasis, arthritis, intestinal diseases, multiple sclerosis, and kidney disease. The manufacturing process for biologic therapies is naturally variable and no two batches of biologics are expected to be the same.

16.12.2 What Is a Biosimilar?

Biosimilar is a biologic product that is "highly similar" to the existing reference biologic. Biosimilar is produced using a similar manufacturing process once the patent for the reference biologic has expired. A biosimilar has to demonstrate the same quality, efficacy, and safety as the reference biologic to be approved by the regulatory bodies

following rigorous testing. The first biosimilar was approved in 2006 in Europe.

16.12.3 Why Are Biosimilars Valuable in Therapy?

The benefits of using biosimilars are related to their lower cost while not compromising quality. Biosimilars are therefore more affordable and they reduce the cost of subsidizing biologic therapies. As a result, more patients can be treated with these therapies and it also allows the Government or other third-party payers to fund other treatments. In addition, this reduces the risk of drug shortages as variety of products will be available.

16.12.4 Who Chooses Whether the Biosimilar or Reference Biologic Is Used?

Talk to your doctor and/or pharmacist about choosing biosimilar medicines. You should discuss with your doctor which medicine is right for you. Your doctor may decide which biologic and/or which brand to prescribe or leave this to the pharmacist to offer you the choice of different products. A doctor can also indicate on the prescription that only a particular biologic (innovator or biosimilar) brand can be dispensed.

16.12.5 Will the Treatment Outcomes Be the Same Using a Biosimilar?

Yes, treatment outcomes are the same. Biosimilars have been assessed to be as safe and effective as their reference biologic. They have been used for more than 10 years and have demonstrated the same treatment outcomes as their reference biologics. In addition, scientific studies demonstrate no difference in treatment outcomes following switching from an innovator biologic to a biosimilar.

16.12.6 Are the Side Effects the Same for a Biosimilar and a Reference Biologic?

Yes, adverse effects of biosimilars are similar to those from reference biologics.

References

1 Revers, L. and Furczon, E. (2010). An introduction to biologics and biosimilars. Part I: biologics: what are they and where do they come from? *Canadian Pharmacists Journal/Revue des Pharmaciens du Canada* 143 (3): 134–139.

2 Revers, L. and Furczon, E. (2010). An introduction to biologics and biosimilars. Part II: subsequent entry biologics: biosame or biodifferent? *Canadian Pharmacists Journal/Revue des Pharmaciens du Canada* 143 (4): 184–191.

3 Camacho, L.H., Frost, C.P., Abella, E. et al. (2014). Biosimilars 101: considerations for U.S. oncologists in clinical practice. *Cancer Medicine* 3 (4): 889–899.

4 National Kidney Foundation (2014). Biosimilar drugs frequently asked questions. https://www.kidney.org/sites/default/files/02-10-6762_HBE_Biosimilars_Booklet_v2.pdf.

5 Mellstedt, H. (2013). Clinical considerations for biosimilar antibodies. *EJC Supplements* 11 (3): 1–11.

6 Jelkmann, W. (2007). Recombinant EPO production – points the nephrologist should know. *Nephrology, Dialysis, Transplantation* 22 (10): 2749–2753.

7 Strohl, W.R. (2015). Fusion proteins for half-life extension of biologics as a strategy to make biobetters. *BioDrugs: Clinical Immunotherapeutics, Biopharmaceuticals and Gene Therapy* 29 (4): 215–239.

8 Tariman, J.D. (2018). Biosimilars: exploring the history, science, and progress. *Clinical Journal of Oncology Nursing* 22 (5): 5–12.

9 Feagan, B.G., Choquette, D., Ghosh, S. et al. (2014). The challenge of indication extrapolation for infliximab biosimilars. *Biologicals* 42 (4): 177–183.

10 Kumar, S., Chawla, S., and Dutta, S. (2018). Biobetters: betting on the future [Internet]. *Journal of Rational Pharmacotherapeutics and Research* 4 (2): 13–21. http://isrpt.co.in/j18pdfs/issue2/Review-Article-Biobetters-Betting-on-the-Future.pdf (accessed 21 November 2019).

11 Beck, A. (2011). Biosimilar, biobetter and next generation therapeutic antibodies. *MAbs* 3 (2): 107–110.

12 Johnson, P.E. (2008). Implications of biosimilars for the future. *American Journal of Health-System Pharmacy* 65 (14 Suppl 6): S16–S22.

13 Todd, S. (2008). Refrigerated medicinal products: what pharmacists need to know. *The Pharmaceutical Journal* 281: 449.

14 Alexander, M., King, J., Bajel, A. et al. (2014). Australian consensus guidelines for the safe handling of monoclonal antibodies for cancer treatment by healthcare personnel. *Internal Medicine Journal* 44 (10): 1018–1026.

15 Siderov, J. (2013). Position statement: safe handling of monoclonal antibodies in healthcare settings, pp. 1–5. https://www.cosa.org.au/media/173517/cosa-cpg-handling-mabs-position-statement_-november-2013_final.pdf (accessed 19 November 2019).

16 Connor, T.H., MacKenzie, B.A., DeBord, D.G. et al. (2016). *NIOSH List of Antineoplastic and Other Hazardous Drugs in Healthcare Settings, 2016.* Cincinnati, OH: US Department of Health and Human Services, Centers for Disease Control and Prevention, National Institute for Occupational Safety and Health, DHHS (NIOSH) Publication Number 2016-161 (Supersedes 2014-138) [Internet] https://www.cdc.gov/niosh/docs/2016-161/pdfs/2016-161.pdf?id=10.26616/NIOSHPUB2016161.

17 Society of Hospital Pharmacist of Australia (SHPA) (2015). SHPA standards of practice for the safe handling of cytotoxic drugs in pharmacy departments. *Journal of Pharmacy Practice and Research* 35 (1): 44–52.

18 European Association of Nuclear Medicine (EANM) (2008). *The Radiopharmacy a Technologist's Guide.* European Association of Nuclear Medicine https://www.eanm.org/content-eanm/uploads/2016/11/tech_radiopharmacy.pdf.

19 Cancer Institute NSW (2011). Hazardous drug table. ID:909 V4 (updated 20 February 2019). https://www.eviq.org.au/clinical-resources/administration-of-antineoplastic-drugs/909-hazardous-drugs-table.

20 Zalcberg, J. (2018). Biosimilars are coming: ready or not. *Internal Medicine Journal* 48 (9): 1027–1034.

21 Jones, G., Nash, P., and Hall, S. (2017). Advances in rheumatoid arthritis. *The Medical Journal of Australia* 206 (5): 221–224.

22 Wish, J.B. (2019). Biosimilars – emerging role in nephrology. *Clinical Journal of the American Society of Nephrology* 14 (9): 1391–1398.

23 Jarrett, S. and Dingermann, T. (2015). Biosimilars are here: a hospital pharmacist's guide to educating health care professionals on biosimilars. *Hospital Pharmacy* 50 (10): 884–893.

24 Derbyshire, M. and Shina, S. (2017). Patent expiry dates for biologicals: 2018 update. *Generics and Biosimilar Initiative Jouranl (GaBI Journal)* [Internet] 6 (1): 27–30. http://gabi-journal.net/patent-expiry-dates-for-biologicals-2018-update.html.

25 Dörner, T., Isaacs, J., Gonçalves, J. et al. (2017). Biosimilars already approved and in development. *Considerations in Medicine.* 1 (1): 7–12.

26 Rak Tkaczuk, K. and Jacobs, I. (2014). Biosimilars in oncology: from development to clinical practice. *Seminars in Oncology* 41: S3–S12.

27 Lucio, S.D., Stevenson, J.G., and Hoffman, J.M. (2013). Biosimilars: Implications for health-system pharmacists. *American Journal of Health-System Pharmacy: AJHP: Official Journal of the American Society of Health-System Pharmacists* 70 (22): 2004–2017.

28 Weise, M., Bielsky, M.-C., De Smet, K. et al. (2012). Biosimilars: what clinicians should know. *Blood* 120 (26): 5111–5117.

29 Feldman, S.R. (2015). Inflammatory diseases: Integrating biosimilars into clinical practice. *Seminars in Arthritis and Rheumatism* 44 (6 Suppl): S16–S21.

30 World Health Organisation (2014). Guidance on scientific principles for regulatory risk assessment of biotherapeutic products. https://www.who.int/biologicals/WHO_Risk_Assessment_for_BTP_2nd_PC_10_Dec_2014.pdf (accessed 15 July 2019).

31 Jensen, A.J. (2017). Biosimilar product labels in Europe: what information should they contain? *Generics and Biosimilars Initiative Journal (GaBI Journal)* [Internet] 6 (1): 38–40. http://gabi-journal.net/biosimilar-product-labels-in-europe-what-information-should-they-contain.html (accessed 20 December 2019).

32 Socinski, M.A., Curigliano, G., Jacobs, I. et al. (2015). Clinical considerations for the development of biosimilars in oncology. *MAbs* 7 (2): 286–293.

33 Therapeutic Goods Administration (2018). *Biosimilar Medicines Regulation*. Canberra, ACT: Australian Government Department of Health https://www.tga.gov.au/publication/biosimilar-medicines-regulation.

34 World Health Organisation (2015). Biological quantifier an INN proposal. https://www.who.int/medicines/services/inn/WHO_INN_BQ_proposal_2015.pdf?ua=1 (accessed 17 November 2019).

35 World Health Organization (2009). General policies for monoclonal antibodies. https://www.who.int/medicines/services/inn/Generalpoliciesformonoclonalantibodies2009.pdf.

36 Council of Australian Therapeutic Advisory Groups (2016). Overseeing biosimilar use. Guiding principles for the governance of biological and biosimilar medicines in Australian hospitals. http://www.catag.org.au/wp-content/uploads/2012/08/OKA10798-CATAG-Overseeing-biosimilar-use-Version-2-final.pdf.

37 Ingrasciotta, Y., Cutroneo, P.M., Marciano, I. et al. (2018). Safety of biologics, including biosimilars: perspectives on current status and future direction. *Drug Safety* 41 (11): 1013–1022.

38 Ley, H. (2015). PBAC world-first biosimilar drug decision. The Australian Government [Internet]. https://www.google.com.au/webhp?sourceid=chrome-instant&ion=1&espv=2&ie=UTF-8#q=PBAC+world+first+biosimilar+drug+decision.

39 Pharmaceutical Society of Australia (2015). Biosimilar medicines: position statement. https://chf.org.au/sites/default/files/position-statement-on-biosimilars_august-2015.pdf (accessed 11 November 2019).

40 The Department of Health (2019). Which biosimilar medicines are available in Australia?. https://www.health.gov.au/internet/main/publishing.nsf/Content/biosimilar-which-medicines-are-available-in-australia (accessed 15 July 2019).

41 NPS Medicinewise – RADAR (2013). Brand equivalence – "a" flagging explained. https://www.nps.org.au/radar/articles/brandequivalence-a-flagging-explained (accessed 10 November 2019).

42 Therapeutic Goods Administration. AusPAR: Rituximab. Australian Public Assessment Report. https://www.tga.gov.au/auspar/auspar-rituximab-62018 (accessed 15 November 2019).

43 Pharmaceutical Benefits Advisory Committee (PBAC) (2018). Rituximab Public Summary Document (PSD) March 2018 PBAC Meeting. http://www.pbs.gov.au/info/industry/listing/elements/pbac-meetings/psd/2018-03/Rituximab-infusion-psd-march-2018 (accessed 15 November 2019).

44 European Medicines Agency (EMA) (2017). Guideline on Immunogenicity assessment of therapeutic proteins. https://www.ema.europa.eu/en/documents/scientific-guideline/guideline-immunogenicity-assessment-therapeutic-proteins-revision-1_en.pdf (accessed 10 November 2019).

45 Mirkov, S. (2018). Biosimilars: how similar is a biosimilar – a practical guide for pharmacists. *Journal of Pharmacy Practice and Research* 48 (5): 442–449.

46 European Commission (2009). A guideline on Summary of Product Characteristics (SmPC). https://ec.europa.eu/health//sites/health/files/files/eudralex/vol-2/c/smpc_guideline_rev2_en.pdf (accessed 11 November 2019).

47 U.S. Department of Health and Human Services, Food Drug Administration (2017). Common Terminology Criteria for Adverse Events (CTCAE) v5.0. https://ctep.cancer.gov/protocolDevelopment/electronic_applications/docs/CTCAE_v5_Quick_Reference_5x7.pdf (accessed 12 November 2019).

48 Schatz, S. and Weber, R. (2015). Adverse drug reactions. In: *PSAP 2015 Book 2 CNS/Pharmacy Practice: Pharmacotherapy Self-Assessment Program* (eds. J. Murphy and W.-L. Lee). American College of Clinical Pharmacy.

49 Mirkov, S. and Hill, R. (2016). Immunogenicity of biosimilars. *Drugs & Therapy Perspectives* 32 (12): 532–538.

50 Chung, C.H. (2008). Managing premedications and the risk for reactions to infusional monoclonal antibody therapy. *The Oncologist* 13: 725–732.

51 AusBiotech (2016). Biologic and biosimilar medicines 2020. Report 2016. https://www.ausbiotech.org/documents/item/48.

52 American Society of Health-System Pharmacists (ASHP) (1996). ASHP guidelines on medication-use

evaluation. *American Journal of Health-System Pharmacy* 53: 1953–1955.

53 NSW Therapeutic Advisory Group (NSW TAG) (2017). Revitalising hospital evaluation of medicines use – the case of DUE. http://nswtag.gradestage.com/wp-content/uploads/2017/07/due-survey-results-jan-2017.pdf (accessed 19 November 2019).

54 Midlands and Lancashire Commissioning Support Unit (2018). Toolkit for implementing best-value adalimumab. https://www.sps.nhs.uk/wp-content/uploads/2018/07/AGA_7962_Biosimilar-Adalimumab-Toolkit-Interactive-PDF_Final2.pdf (accessed 16 June 2020).

55 Kabir, E.R., Moreino, S.S., and Sharif Siam, M.K. (2019). The breakthrough of biosimilars: a twist in the narrative of biological therapy. *Biomolecules* 9 (9): 410.

56 Frank, D.S. (2018). I'm talking to you-mab-how to pronounce the new, unpronounceable pharmaceuticals. *JAMA Internal Medicine* 178 (3): 319–320.

57 Reinisch, W. and Smolen, J. (2015). Biosimilar safety factors in clinical practice. *Seminars in Arthritis and Rheumatism* 44 (6 Suppl): S9–S15.

58 Odinet, J.S., Day, C.E., Cruz, J.L., and Heindel, G.A. (2018). The biosimilar nocebo effect? A systematic review of double-blinded versus open-label studies. *Journal of Managed Care & Specialty Pharmacy* 24 (10): 952–959.

59 Murby, S. and Reilly, M. (2017). A survey of Australian prescribers' views on the naming and substitution of biologicals. *Generics and Biosimilars Initiative Journal (GaBI Journal)* [Internet] 6 (3): 107–112. http://gabionline.net/Biosimilars/Research/Australian-prescribers-views-on-biologicals-naming-and-substitution (accessed 19 November 2019).

60 Michie, S., van Stralen, M.M., and West, R. (2011). The behaviour change wheel: a new method for characterising and designing behaviour change interventions. *Implementation Science* 6 (1): 42.

61 Australian Government Department of Health Pharmaceutical Benefits Scheme. Biosimilar awareness initiative. https://www1.health.gov.au/internet/main/publishing.nsf/Content/biosimilar-awareness-initiative?Open=&utm_source=health.gov.au&utm_medium=redirect&utm_campaign=digital_transformation&utm_content=biosimilars (accessed 19 November 2019).

62 Hales, B., Terblanche, M., Fowler, R., and Sibbald, W. (2007). Development of medical checklists for improved quality of patient care. *International Journal for Quality in Health Care* 20 (1): 22–30.

63 Kotter, J.P. (2012). Accelerate! *Harvard Business Review* 90: 35–58.

64 Mah, M.W., Deshpande, S., and Rothschild, M.L. (2006). Social marketing: a behavior change technology for infection control. *American Journal of Infection Control* 34 (7): 452–457.

65 Mah, M.W. and Meyers, G. (2006). Toward a socioethical approach to behavior change. *American Journal of Infection Control* 34 (2): 73–79.

66 Benkler, Y. (2011). The unselfish gene. *Harvard Business Review* (July-August): 77–85.

67 The Department of Health (2019). Quaity use of medicines. https://www1.health.gov.au/internet/main/publishing.nsf/Content/nmp-quality.htm (accessed 15 December 2019).

68 Australian Commission on Safety and Quality in Health Care. The NSQHS standards: medication safety standard. https://www.safetyandquality.gov.au/standards/nsqhs-standards/medication-safety-standard (accessed 19 November 2019).

69 Mirkov, S. (2018). Teamwork for innovation in pharmacy practice: from traditional to flexible teams. *Drugs & Therapy Perspectives* 34 (6): 274–280.

70 Preston, C.L. (2019). *Stockley's Drug Interactions: A Source Book of Interactions, Their Mechanisms, Clinical Importance and Management*, 12e. London: Pharmaceutical Press.

71 NHS England (2017). Commissioning framework for biological medicines (including biosimilar medicines). https://www.england.nhs.uk/wp-content/uploads/2017/09/biosimilar-medicines-commissioning-framework.pdf (accessed 14 November 2019).

72 Boucher, A., Dhanjal, S., Kong, J., and Ho, C. (2018). Medication incidents associated with patient harm in community pharmacy: a multiincident analysis. *Pharm Connect* [Internet] https://pharmacyconnection.ca/ismp-multi-incident-analysis-winter-2018 (accessed 16 December 2019).

73 Mikhail, A. and Farouk, M. (2013). Epoetin biosimilars in Europe: five years on. *Advances in Therapy* 30 (1): 28–40.

74 Zuniga, L. and Calvo, B. (2010). Biosimilars: pharmacovigilance and risk management. *Pharmacoepidemiology and Drug Safety* 19 (7): 661–669.

75 Alten, R. and Cronstein, B.N. (2015). Clinical trial development for biosimilars. *Seminars in Arthritis and Rheumatism* 44 (6 Suppl): S2–S8.

76 Haynes, R. (2006). Forming research questions. In: *Clinical Epidemiology: How to do Clinical Practice Research*, 3e (eds. R. Haynes, D. Sacket, G. Guyatt and

P. Tugwell), 3–14. Philadelphia, PA: Lippincott Williams & Wilkins.

77 Baxter, K. and Preston, C.L. *Stockley's Drug Interactions* [online]. London: Pharmaceutical Press nzf.org.nz (accessed 13 November 2019).

78 Haig, K.M., Sutton, S., and Whittington, J. (2006). SBAR: a shared model for improving communication between clinicians. *Journal of Quality and Patient Safety* 32 (3): 167–175.

79 Moore, C., Corbett, G., and Moss, A.C. (2016). Systematic review and meta-analysis: serum infliximab levels during maintenance therapy and outcomes in inflammatory bowel disease. *Journal of Crohn's & Colitis* 10 (5): 619–625.

80 Vande Casteele, N., Ferrante, M., Van Assche, G. et al. (2015). Trough concentrations of infliximab guide dosing for patients with inflammatory bowel disease. *Gastroenterology* 148 (7): 1320–1329.e3.

81 Thériault, G., Bell, N.R., Grad, R. et al. (2019). Teaching shared decision making. An essential competency. *Canadian Family Physician* 65 (7): 514–516.

82 Studer Group. AIDET patient communication. https://www.studergroup.com/aidet (accessed 17 November 2019).

83 Kingsley, C. and Patel, S. (2017). Patient-reported outcome measures and patient-reported experience measures. *BJA Education* 17 (4): 137–144.

84 Yoo, D.H., Hrycaj, P., Miranda, P. et al. (2013). A randomised, double-blind, parallel-group study to demonstrate equivalence in efficacy and safety of CT-P13 compared with innovator infliximab when coadministered with methotrexate in patients with active rheumatoid arthritis: the PLANETRA study. *Annals of the Rheumatic Diseases* 72 (10): 1613–1620.

85 Yoo, D.H., Prodanovic, N., Jaworski, J. et al. (2017). Efficacy and safety of CT-P13 (biosimilar infliximab) in patients with rheumatoid arthritis: comparison between switching from reference infliximab to CT-P13 and continuing CT-P13 in the PLANETRA extension study. *Annals of the Rheumatic Diseases* 76 (2): 355–363.

86 Park, W., Hrycaj, P., Jeka, S. et al. (2013). A randomised, double-blind, multicentre, parallel-group, prospective study comparing the pharmacokinetics, safety, and efficacy of CT-P13 and innovator infliximab in patients with ankylosing spondylitis: the PLANETAS study. *Annals of the Rheumatic Diseases* 72 (10): 1605–1612.

87 Park, W., Yoo, D.H., Miranda, P. et al. (2017). Efficacy and safety of switching from reference infliximab to CT-P13 compared with maintenance of CT-P13 in ankylosing spondylitis: 102-week data from the PLANETAS extension study. *Annals of the Rheumatic Diseases* 76 (2): 346–354.

88 O'Callaghan, J., Barry, S.P., Bermingham, M. et al. (2019). Regulation of biosimilar medicines and current perspectives on interchangeability and policy. *European Journal of Clinical Pharmacology* 75 (1): 1–11.

89 Kurki, P. (2015). Biosimilars for prescribers. *Generics and Biosimilars Initiative Journal (GaBI Journal)* [Internet] 4 (1): 33–35. http://gabi-journal.net/biosimilars-for-prescribers.html (accessed 19 November 2019).

90 Jorgensen, K.K., Olsen, I.C., Goll, G.L. et al. (2017). Switching from originator infliximab to biosimilar CT-P13 compared with maintained treatment with originator infliximab (NOR-SWITCH): a 52-week, randomised, double-blind, non-inferiority trial. *Lancet* 389 (10086): 2304–2316.

91 Canadian Agency for Drugs and Technologies in Health (2015). Switching from innovator to biosimilar (subsequent entry) infliximab: a review of the clinical effectiveness, cost-effectiveness, and guidelines. Canadian Agency for Drugs and Technologies in Health Rapid Response Report 2015. https://www.cadth.ca/media/pdf/htis/mar-2015/RC0635%20Infliximab%20Switching%20Final.pdf (accessed 19 November 2019).

Index

Biologics, Biosimilars, and Biobetters: An Introduction for Pharmacists, Physicians, and Other Health Practitioners,
First Edition. Edited by Iqbal Ramzan.
© 2021 John Wiley & Sons, Inc. Published 2021 by John Wiley & Sons, Inc.